# Pig Genomics and Genetics

# Pig Genomics and Genetics

Editors

**Katarzyna Piórkowska**
**Katarzyna Ropka-Molik**

MDPI • Basel • Beijing • Wuhan • Barcelona • Belgrade • Manchester • Tokyo • Cluj • Tianjin

*Editors*

Katarzyna Piórkowska
Animal Molecular Biology
National Research Institute of Animal Production
Cracow
Poland

Katarzyna Ropka-Molik
Animal Molecular Biology
National Research Institute of Animal Production
Cracow
Poland

*Editorial Office*
MDPI
St. Alban-Anlage 66
4052 Basel, Switzerland

This is a reprint of articles from the Special Issue published online in the open access journal *Genes* (ISSN 2073-4425) (available at: www.mdpi.com/journal/genes/special_issues/Pig_Genomics_Genetics).

For citation purposes, cite each article independently as indicated on the article page online and as indicated below:

LastName, A.A.; LastName, B.B.; LastName, C.C. Article Title. *Journal Name* **Year**, *Volume Number*, Page Range.

**ISBN 978-3-0365-5358-0 (Hbk)**
**ISBN 978-3-0365-5357-3 (PDF)**

# Contents

MDPI

*Editorial*

# Pig Genomics and Genetics

**Katarzyna Piórkowska * and Katarzyna Ropka-Molik**

National Research Institute of Animal Production, Animal Molecular Biology, 31-047 Cracow, Poland; katarzyna.ropka@iz.edu.pl

* Correspondence: katarzyna.piorkowska@iz.edu.pl

The pig (Sus scrofa) is the most popular large farm animal in the world. They are frequently used as animal models for human medical research due to high biological similarity to humans, such as body proportions [1,2], metabolic process [3,4], adipose tissue distribution and adipocyte size [5]. In addition, both species reveal also a high genetic analogy: the human genome is composed of 3.5 billion bp, the pig genome of 3.0 billion bp; 21,630 protein-coding genes were identified in pigs, while in humans, this is 20,310 [6,7].

On the other hand, molecular biology methods assist agricultural progress, for example, in pig production and breeding. In addition, since the reference genome sequence of the domestic pig was assembled in 2012, the identification processes of crucial phenotypic traits and search of genetic markers for selection have been significantly refined, including the newest wide-range high-throughput techniques. The use of these new genomic tools has the advantage of generating information about multiple genes and gene products in parallel, which makes it possible to identify pathways and gene interactions [8,9]. This approach provides insight into the epistatic effects of genes that could improve understanding of the genetic component of pig phenotype. At first, DNA microarray that is broadly used to date, supports livestock production by predicting the potential genetic breeding value of farm animals [10]. Microarray approach also serves as a research tool in pig breeding, as well. For example, Lee et al. [11] used it to prove that the porcine immune system was affected by different breeding environments, suggesting the importance of controlling microbes in the animal room for qualified research. Another kind of microarray is used to identification of gene expression. In the Sun et al. [12] study, the authors applied cDNA microarray to identify differentially expressed genes (DEGs) between two Chinese pig breeds, pinpointing the association between BAX and BMPR1B genes with litter size. Such methodology allows to highlight potential genetic markers which can used in the pig industry. However, this method, due to numerous limitations in data analysis [13] and the possibility to identify only profiles predefined transcripts/genes through hybridization [14], is eagerly replenished by 'omic' approaches. Omic methods integrate structural and functional genomics and relate them with phenotypic data for farm animals, including pigs [8]. They offer the comprehensive detection of the whole transcriptome, genome, proteome, etc. [15]. Next-generation sequencing (NGS) methods using high throughput platforms identify genetic and transcriptomic components by sequencing long hundred-nucleotide reads and then mapping them to the reference genome [15]. Using this approach, Piórkowska et al. [16] pinpointed a new gene cluster involved in porcine meat quality determination via regulating cell proliferation and differentiation and calcium-binding. In turn, more advanced tool PacBio sequencing platform providing ultra-long sequencing reads, allow in the more precise manner identifying gene mutations, new transcripts and gene candidates throughout the whole genome, transcriptome, or epigenome and estimating quantitative traits important for breeding as well as the genetic backgrounds of inherited diseases. However, PacBio is a very expensive method and for now it is applied mainly to improve genome reference, also included pig genome [17].

In this Special Issue, we will present the state of the art in the field of pig genetics and genomics, including the identification of gene candidates linked to important pig

**Citation:** Piórkowska, K.; Ropka-Molik, K. Pig Genomics and Genetics. *Genes* **2021**, *12*, 1692. https://doi.org/10.3390/genes12111692

Received: 15 October 2021
Accepted: 21 October 2021
Published: 25 October 2021

traits and to nutritional modifications, with the aim of collecting the most recent advances. Manuscripts focusing on high-throughput methodologies, such as RNA sequencing, ATAC-seq, MACE-seq, chip-seq and RRBS and covering other fields of pig genetics are included.

**Conflicts of Interest:** None of the authors has a financial or other relationship with other people or organizations that may inappropriately influence this work.

## References


1. Groth, C.G. The potential advantages of transplanting organs from pig to man: A transplant surgeon's view. *Indian J. Urol.* **2007**, *23*, 305–309. [CrossRef]
2. Lelovas, P.P.; Kostomitsopoulos, N.G.; Xanthos, T.T. A comparative anatomic and physiologic overview of the porcine heart. *J. Am. Assoc. Lab. Anim. Sci.* **2014**, *53*, 432–438. [PubMed]
3. Mota-Rojas, D.; Orozco-Gregorio, H.; Villanueva-Garcia, D.; Bonilla-Jaime, H.; Suarez-Bonilla, X.; Hernandez-Gonzalez, R.; Roldan-Santiago, P.; Trujillo-Ortega, M.E. Foetal and neonatal energy metabolism in pigs and humans: A review. *Vet. Med.* **2011**, *56*, 215–225. [CrossRef]
4. Nielsen, K.L.; Hartvigsen, M.L.; Hedemann, M.S.; Lræke, H.N.; Hermansen, K.; Bach Knudsen, K.E. Similar metabolic responses in pigs and humans to breads with different contents and compositions of dietary fibers: A metabolomics study. *Am. J. Clin. Nutr.* **2014**, *99*, 941–949. [CrossRef]
5. Stachowiak, M.; Szczerbal, I.; Switonski, M. Genetics of Adiposity in Large Animal Models for Human Obesity—Studies on Pigs and Dogs. In *Progress in Molecular Biology and Translational Science*; Academic Press: Cambridge, MA, USA, 2016; Volume 140, pp. 233–270. ISBN 9780128046159.
6. Lunney, J.K. Advances in swine biomedical model genomics. *Int. J. Biol. Sci.* **2007**, *3*, 179–184. [CrossRef]
7. Ensembl, GCA_000003025.6, Sscrofa11.1. Available online: https://www.ensembl.org/Sus_scrofa/Info/Index (accessed on 15 October 2021).
8. Davoli, R.; Braglia, S. Molecular approaches in pig breeding to improve meat quality. *Brief. Funct. Genom. Proteom.* **2007**, *6*, 313–321. [CrossRef] [PubMed]
9. Tuggle, C.K.; Dekkers, J.C.M.; Reecy, J.M. Integration of structural and functional genomics. *Anim. Genet.* **2006**, *37*, 1–6. [CrossRef] [PubMed]
10. Vandenplas, J.; Windig, J.J.; Calus, M.P.L. Prediction of the reliability of genomic breeding values for crossbred performance. *Genet. Sel. Evol.* **2017**, *49*, 43. [CrossRef] [PubMed]
11. Lee, J.Y.; Kim, S.E.; Lee, H.T.; Hwang, J.H. Comparative analysis of immune related genes between domestic pig and germ-free minipig. *Lab. Anim. Res.* **2020**, *36*, 44. [CrossRef]
12. Sun, X.; Mei, S.; Tao, H.; Wang, G.; Su, L.; Jiang, S.; Deng, C.; Xiong, Y.; Li, F. Microarray profiling for differential gene expression in PMSG-hCG stimulated preovulatory ovarian follicles of Chinese Taihu and Large White sows. *BMC Genom.* **2011**, *12*, 111. [CrossRef] [PubMed]
13. Lee, M.T.; Whitmore, G.A. Power and sample size for DNA microarray studies. *Stat. Med.* **2002**, *21*, 3543–3570. [CrossRef] [PubMed]
14. Rao, M.S.; Van Vleet, T.R.; Ciurlionis, R.; Buck, W.R.; Mittelstadt, S.W.; Blomme, E.A.G.; Liguori, M.J. Comparison of RNA-Seq and Microarray Gene Expression Platforms for the Toxicogenomic Evaluation of Liver from Short-Term Rat Toxicity Studies. *Front. Genet.* **2019**, *9*, 636. [CrossRef] [PubMed]
15. Suravajhala, P.; Kogelman, L.J.A.; Kadarmideen, H.N. Multi-omic data integration and analysis using systems genomics approaches: Methods and applications in animal production, health and welfare. *Genet. Sel. Evol.* **2016**, *48*, 38. [CrossRef] [PubMed]
16. Piórkowska, K.; Żukowski, K.; Ropka-Molik, K.; Tyra, M.; Gurgul, A. A comprehensive transcriptome analysis of skeletal muscles in two Polish pig breeds differing in fat and meat quality traits. *Genet. Mol. Biol.* **2018**, *41*, 125–136. [CrossRef] [PubMed]
17. Warr, A.; Affara, N.; Aken, B.; Beiki, H.; Bickhart, D.M.; Billis, K.; Chow, W.; Eory, L.; Finlayson, H.A.; Flicek, P.; et al. An improved pig reference genome sequence to enable pig genetics and genomics research. *Gigascience* **2020**, *9*, giaa051. [CrossRef] [PubMed]

*Article*

# Phenotypic and Genomic Analysis of Cystic Hygroma in Pigs

**Anna Letko [1], Alexandria Marie Schauer [2], Martijn F. L. Derks [3,4], Llorenç Grau-Roma [2], Cord Drögemüller [1] and Alexander Grahofer [5,*]**

[1] Institute of Genetics, Vetsuisse Faculty, University of Bern, 3012 Bern, Switzerland; anna.letko@vetsuisse.unibe.ch (A.L.); cord.droegemueller@vetsuisse.unibe.ch (C.D.)
[2] Institute of Animal Pathology, Vetsuisse Faculty, University of Bern, 3012 Bern, Switzerland; alexandria.schauer@vetsuisse.unibe.ch (A.M.S.); llorenc.grauroma@vetsuisse.unibe.ch (L.G.-R.)
[3] Topigs Norsvin Research Center, 6640 AA Beuningen, The Netherlands; martijn.derks@topigsnorsvin.com
[4] Animal Breeding and Genomics Group, Wageningen University, 6700 Wageningen, The Netherlands
[5] Clinic for Swine, Vetsuisse Faculty, University of Bern, 3012 Bern, Switzerland
* Correspondence: alexander.grahofer@vetsuisse.unibe.ch

**Abstract:** Cystic hygroma is a malformation of the lymphatic and vascular system and is recognized as a benign congenital tumor that affects humans and animals in the perinatal period. This congenital disorder is rarely described in animals, and until today, cystic hygroma in pigs has not been described in the literature. In a purebred Piètrain litter with twelve live-born piglets, cystic hygroma was noticed on the rump of two male pigs within the first week of life. In addition, a third case of a crossbred weaner (Large White × Landrace) was detected during a herd examination. To rule out common differential diagnoses, e.g., abscess or hematoma, further clinical and pathological investigations were conducted. During clinical examination, a painless and soft mass, which was compressible, was detected on the rump of all affected animals. The ultra-sonographic examination revealed a fluid-filled and cavernous subcutaneous structure. In addi-tion, a puncture of the cyst was conducted, revealing a serosanguinous fluid with negative bacte-riological culture. In all cases, a necropsy was performed, showing that the animals had fluid-filled cysts lined by well-differentiated lymphatic endothelium. Based on the clinicopathological examination, cystic hygroma was diagnosed. Furthermore, SNP array genotyping and whole-genome sequencing was performed and provided no evidence for a chromosomal disorder. In the Piètrain family, several genome regions were homozygous in both affected piglets. None-theless, a dominant acting de novo germline variant could not be ruled out, and therefore differ-ent filtering strategies were used to find pathogenic variants. The herein presented lists of pri-vate variants after filtering against hundreds of control genomes provide no plausible candidate and no shared variants among the two sequenced cases. Therefore, further studies are needed to evaluate possible genetic etiology. In general, systematic surveillance is needed to identify ge-netic defects as early as possible and to avoid the occurrence of losses in the pig population.

**Keywords:** *Sus scrofa*; precision medicine; lymphatic system; whole-genome sequencing; SNP array genotyping

**Citation:** Letko, A.; Schauer, A.M.; Derks, M.F.L.; Grau-Roma, L.; Drögemüller, C.; Grahofer, A. Phenotypic and Genomic Analysis of Cystic Hygroma in Pigs. *Genes* **2021**, *12*, 207. https://doi.org/10.3390/genes12020207

Academic Editors: Katarzyna Piórkowska and Katarzyna Ropka-Molik

Received: 22 December 2020
Accepted: 28 January 2021
Published: 31 January 2021



## 1. Introduction

Cystic hygroma, often also referred to as 'cystic lymphangioma', is one of the most commonly presenting lymphangioma in human medicine. It is a well-known congenital malformation of the lymphatic system characterized as single or multiloculated fluid-filled cavities due to a lack of communication between the lymphatic and venous systems [1–4]. Cystic hygroma occurs with an incidence of ~1:1000–6000 births and 1:750 miscarriages in humans [5,6]. Even though cases of cystic hygroma are rarely described in animals [7–11], an estimation of the prevalence of this malformation in animals is still missing. Cystic hygromas can manifest anywhere in the body but are often found in the neck, clavicle, and axillary regions in humans [1–4,12]. In approximately half of the reported cases, cystic hygromas are present directly after birth, whereas the other cases occur within the age of two years [13]. Until today, the exact etiology of cystic hygroma in humans and animals has been unclear, but an association with chromosomal aberrations and genetic syndromes, such as the Noonan syndrome (OMIM PS163950), has been described [1–3,12]. Abnormal karyotype was found in 29% to 60% of the cases [14], whereas congenital disorders with normal karyotype ranged from 25% to 53% [15]. However, submicroscopic chromosomal abnormalities that are missed by conventional karyotyping are also described in cystic hygroma [2]. In addition, cases of familial cystic hygroma with normal karyotype have been described and suggest that both recessively as well as dominantly inherited genetic variants are involved in the phenotype [2,16–19].

Until now, no information regarding the occurrence, the etiology, and the genetic background of cystic hygroma in the pig population has been available. The clinical phenotype of cystic hygroma in pigs resembles the human condition as well as reports of similarly affected individuals of other domestic animal species. To the authors' knowledge, this is the first report of the hygroma cyst in pigs. Therefore, this report describes the phenotypic findings of hygroma cysts in three pigs of different breeds and the subsequent preliminary genomic analysis, including SNP genotyping and whole-genome sequencing, to evaluate a possible inherited cause.

## 2. Materials and Methods

### 2.1. Ethics Statement

All animal experiments were performed according to the local regulations. The study was approved by the Cantonal Committee for Animal Experiments (Canton of Bern; permit 109/18) at the University of Bern.

### 2.2. Animals and DNA Samples

Three male cases with cystic hygroma on the rump were used in this study. Two cases were littermates from a purebred Piètrain litter. The third case, a crossbred weaner (Large White × Landrace), was observed during a herd examination in a fattening herd. Blood samples were obtained from all cases as well as from the Piètrain sow, boar, and all 20 healthy siblings from two independent litters after repeated mating for further genetic investigations. Genomic DNA was isolated from EDTA blood samples using the Maxwell RSC Whole Blood DNA Kit (Promega AG, Dübendorf, Switzerland).

### 2.3. Clinical and Further Examination

A total of three male cases with cystic hygroma on the rump were examined in this study. Two cases (case 1 and 2) were littermates from a purebred Piètrain litter born at the Clinic for Swine in Bern. The pregnant Piètrain sow, artificially inseminated from a Piètrain boar of a boar study, was bought from a nucleus farm in Switzerland and farrowed at the Clinic for Swine. The sow was raised under conventional conditions. At the beginning of gestation, the sow was kept in a group house system with straw as a bedding material according to legal requirements and received a conventional feed diet and water ad libitum. At the Clinic for Swine, the animal was housed in a single pen with contact to other pigs. The pen was interspersed with straw and sawdust. In addition to the conventional feed, the sow also received hay. Within the first week after farrowing, cystic hygroma occurred in two (male) out of eleven (3 female, 8 male) piglets. A rebreeding of the sire and dam under experimental conditions was conducted to investigate a possible genetic effect. The litter size of the second litter was twelve (7 female, 5 male), but no more affected piglets were observed during the lactation period of four weeks. Due to leg weakness, no further breeding with the sow was possible, and therefore, no further rebreeding was conducted.

The third case (case 3) was observed during a herd examination in a fattening herd. The farmer reported that this weaner (Large White × Landrace, male castrated) already arrived with the cystic hygroma to his farm from a piglet producer. No information about the mother and father could be obtained. However, this animal was referred to the Clinic for Swine for further investigation. A clinical examination was conducted in all affected animals. Further investigations, including ultrasonography and cytology of the fluid in the cyst, were conducted to clarify the phenotype of the cystic hygroma and rule out differential diagnoses.

### 2.4. Postmortem Examination, Histology, and Bacteriology

A full postmortem was performed immediately after euthanasia on the two affected Piètrain littermates (cases 1 and 2), and one crossbred weaner pig (case 3). Tissue samples from the skin, subcutaneous lesions, skeletal muscle, superficial inguinal lymph nodes, and internal organs including lung, liver, and kidney from all pigs and from a mass observed within the radius in case 1 were fixed in 10% buffered formalin, pH 7.2, overnight. All tissues were routinely processed for histopathology and stained with hematoxylin and eosin (H&E). Immunohistochemical examination was performed using rabbit polyclonal antibodies against LYVE1 (Abcam, Cambridge, UK), at the dilution of 1:3000, and counterstained with hematoxylin. Inguinal lymph nodes were included as an internal positive control.

Sterile samples for bacteriological investigations of the fluid present within the cystic masses were collected before sectioning the tissues in all three cases. The fluid samples were cultured aerobically on blood agar with and without ammonia and on MacConkey agar for 48 h. As part of the national surveillance system, blood samples were tested for the presence of antibodies against African swine fever virus (ASFV), Classical swine fever virus (CSFV), and Porcine reproductive and respiratory syndrome virus (PRRSV) by ELISA.

### 2.5. Genetic Analyses

From both of the purebred Piètrain litters, 24 animals (dam, sire, 2 affected piglets, and 20 apparently normal littermates) were genotyped using the Illumina PorcineSNP60 BeadChip (Illumina, San Diego, CA, USA) containing 50,915 SNPs. Basic quality control filtering steps of the SNP array genotyping data and parentage confirmation were carried out using PLINK v1.9 [20]. Markers with call rates <90% were excluded, and all individuals had call rates >90%. The dataset was additionally pruned for low minor allele frequency (0.05) and failure to meet Hardy–Weinberg equilibrium (0.0001), resulting in 31,054 markers. The dataset was also scanned for Mendelian errors using the –mendel option of PLINK to reveal any deviations from expected values based on per-individual, per-family, and per-SNP error rates.

Merlin software [21] was used to test for cosegregation of any chromosomal regions and the cystic hygroma phenotype in one complete family representing the first litter (sire, dam, two affected, and eight unaffected offspring) by performing parametric linkage analysis under a fully penetrant, recessive model of inheritance. Assuming identity-by-descent (IBD), the autozygosity mapping approach in PLINK v1.9 [20] was used to discover homozygous intervals with alleles shared by both affected piglets (using –homozyg-match 0.95 for allelic matching between both cases). Additionally, individual homozygous intervals were determined in the 22 control animals for comparison. All plots were constructed in R environment v3.6.0 [22].

### 2.6. Whole-Genome Sequencing

Whole-genome sequence (WGS) data were obtained from five pigs including four animals of the purebred Piètrain litter (case 1, its dam, sire, and one normal littermate), as well as the unrelated affected crossbred pig (case 3), after preparation of PCR-free fragment libraries with approximately 400 bp inserts that were sequenced for paired-end reads of $2 \times 150$ bp length. All five animals were sequenced on the Illumina NovaSeq 6000 System at an average coverage of $21\times$ (Supplementary Table S1). The obtained reads were mapped to the pig reference genome assembly Sscrofa11.1 using the Burrows-Wheeler Aligner v0.7.15 [23] with default settings. Picard v2.9 [24] was used to sort the mapped reads by the sequence coordinates and to label the read duplicates. Genome Analysis Toolkit v3.8

(GATK) [25] was used to perform local realignment and to produce a cleaned BAM file. The single nucleotide variants (SNVs) and small indels were identified using genotypeGVCFs of GATK, and the prediction of their functional effects was performed with SnpEff v4.3 [26] using the NCBI Annotation Release 106. The generated files were merged with 10 other unrelated pig genomes (Supplementary Table S1), which were generated previously during the course of different studies and are publicly available, into a final variant call format (VCF) file, including all individual variants and their functional annotations. The VCF file was then used for determining private variants of the affected animals.

As an independent validation, the obtained variants were additionally searched in a cohort of 756 commercial pig genomes from four breeds (Duroc, Large White, Landrace, Piètrain). This cohort of sequenced animals was analyzed according to Derks et al., 2019 [27]. In short, sequence reads were mapped using Burrows–Wheeler aligner against the Sscrofa11.1 reference genome, and SAMtools [23] was used to sort, merge, and index BAM files. We performed variant calling using FreeBayes with the setting: –min-base-quality 10 –min-alternate-fraction 0.2 –haplotype-length 0 –min-alternate-count 2 [28]. We discarded variants with a Phred quality <20. The resulting sequence variants were functionally annotated using the Ensembl Variant Effect Predictor pipeline (v99) [29].

The Integrative Genomics Viewer (IGV) [30] was used for visual inspection and screening for structural variants in the regions of interest in the WGS of the affected pigs. Additionally, coverage for the five animals was calculated using the function bedcov of the program Samtools [23] by a sliding window approach with the window size of 300 kb moving for half the window size and including reads with mapping quality greater than 15. The average coverage over every window was plotted for each pig with the qqman package [31] in R environment v3.6.0 [22].

### 2.7. Availability of Data and Material

All positions refer to the pig reference genome assembly Sscrofa11.1 and NCBI Annotation Release 106. The herein generated WGS data are freely available at the European Nucleotide Archive (ENA) under study accession number PRJEB29465, and individual sample accession numbers are available in Supplementary Table S1. The genotypes of the commercial pig genomes used as the validation cohort within the candidate regions are available on reasonable request.

## 3. Results

### 3.1. Clinical and Further Examination

A clinical examination of all three cases revealed that the animals were alert and in good body condition. The vital signs were within the reference values, and no locomotion or neurological disorders were determined. In all three animals, a painless, soft, and compressible mass on the rump was detected (Figure 1, Supplementary Video S1, and Supplementary Video S2); and an ultrasonographic examination revealed a cystic or multicystic lesion with internal septations (Figure 2). In addition, after sterile preparation, an aspiration of the fluid was conducted in all three animals and a cytological investigation was performed. Results from the cytological examination are listed in Table 1.

Based on the findings of the clinical examination, the diagnosis of cystic hygroma was made in all three affected pigs. In addition, a clinical examination of all controls revealed that the animals were alert and in good body condition. The vital parameters were all within the reference range, and examination of the locomotion and neurological system revealed no pathological findings. Based on the results of the clinical examination, all controls were healthy.

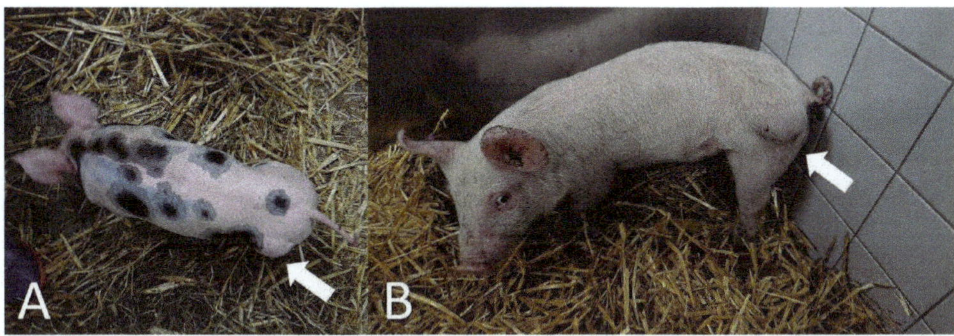


**Figure 1.** Phenotypic appearance of cystic hygroma in pigs. A soft and compressible mass on the rump of two cases is visible ((**A**) = case 1; (**B**) = case 2).

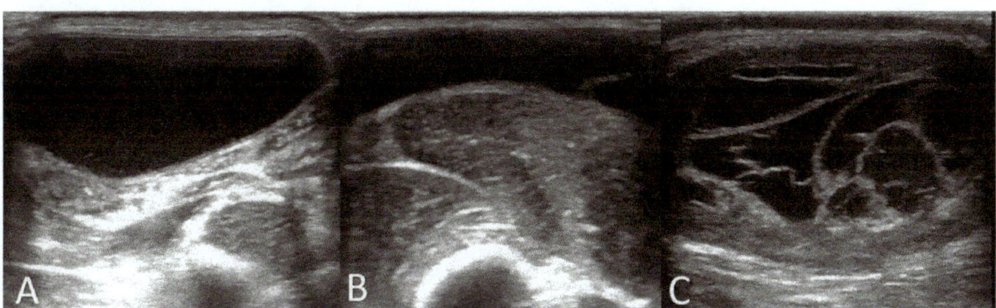


**Figure 2.** Ultrasonographic imaging of cystic hygroma in pigs. Fluid-filled cysts ((**A**) = case 1; (**B**) = case 2; (**C**) = case 3) with cavernous structure (**C**) could be detected.

**Table 1.** Cytological examination.

| Parameters | Case 1 | Case 2 | Case 3 |
|---|---|---|---|
| **Colour** | Light red | Red | Red |
| **Transparency** | Cloudy | Cloudy | Cloudy |
| **Protein (g/l)** | 16 | 26 | 44 |
| **Specific weight** | 1.016 | 1.022 | 1.032 |
| **Cell count (10 × $10^9$/l)** | 4.50 | 2.70 | 13.03 |
| **Conclusion** | Hygroma or seroma | Hygroma or seroma and blood | Hygroma or seroma and mild inflammation |

### 3.2. *Postmortem Examination, Histology, and Bacteriology*

The gross and histological appearance of the masses was similar in all three pigs. The masses were grossly palpable, fluctuant, well-demarcated, and located subcutaneously at the level of the rump. The masses were 9 × 7 × 3 cm, 10 × 8 × 4 cm, and 17 × 7 × 2 cm in cases 1 to 3, respectively. In the cut sections, the masses were surrounded by a thick, fibrous capsule with a width of 0.3–0.5 cm and contained multiple cystic cavities filled with serosanguineous fluid and a few aggregates of coagulated protein (Figure 3).

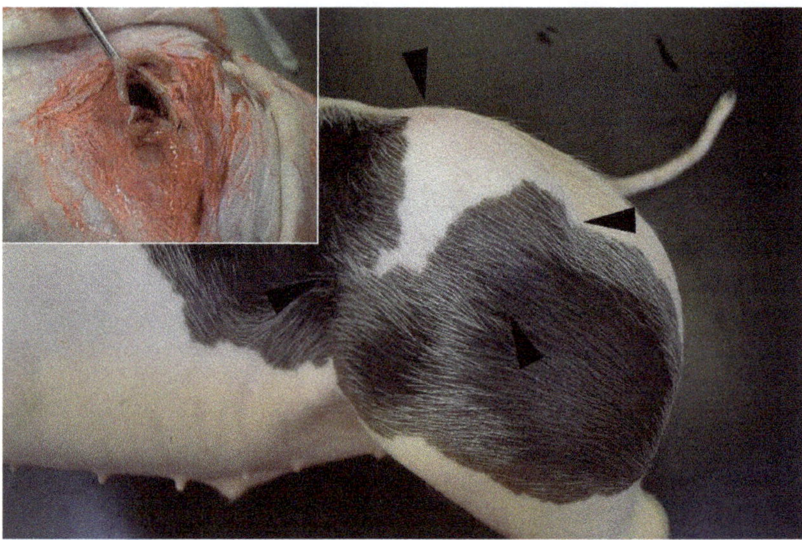

**Figure 3.** Porcine cystic hygroma (case 1). Note the well-demarcated, subcutaneous mass located at the level of the left rump (arrowheads). Inset: The mass is surrounded by a thick, fibrous capsule and contains a central cystic cavity filled with serosanguineous fluid.

Histologically, the masses were limited to the subcutis, and the cystic cavities were surrounded by a thick capsule comprised of mature fibrous connective tissue. The innermost layer of the capsule was lined by a single layer of cells that were mostly flat but multifocally plump to cuboidal. Cells contained a moderate amount of pale, eosinophilic cytoplasm and one central, round to ovoid nucleus. The central cavity was filled with pale, extracellular eosinophilic material, a low amount of fibrin, and a few scattered neutrophils and lymphocytes. Immunohistochemistry for LYVE-1 demonstrated diffuse and strongly positive staining within the cells lining the cystic masses, which was consistent with lymphatic endothelium (Figure 4). The endothelial cells lining the afferent and efferent lymphatic vessels of the lymph node as well as the subcapsular and peritrabecular sinuses showed similar positive staining, indicating the success of the positive control.

No other relevant lesions were observed in the three affected pigs. No bacterial growth was obtained in the bacteriological investigations of any of the collected samples. The serology for ASFV, CSFV, and PRRSV was negative.

### 3.3. Genetic Analyses

A total of 24 pigs from two Piètrain litters were genotyped using the SNP array. The same parentage of both litters was confirmed by the IBD estimates for all pairs of individuals. No increased number of Mendelian errors between the parents and the two affected offspring (case 1 and case 2) was observed, indicating no larger structural variants in the genome of the affected piglets.

Based on the pedigree structure, initially, an autosomal recessive inheritance was suspected. Parametric linkage analysis for a recessive trait in the first litter resulted in ten genome regions with a positive logarithm of the odds (LOD) score (Supplementary Table S2). Moreover, ten runs of homozygosity shared by the two Piètrain cases were detected on nine autosomes (chr 1, 3, 5, 6, 11, 12, 14, 17, and 18). Only one homozygous segment overlapped with a linked interval on chromosome 12, forming a ~4.6 Mb (chr 12: 178,275–4,826,688) region of interest (Supplementary Table S2). However, from 9% to 100% of the 22 control pigs were also homozygous over the ten detected homozygous regions (Supplementary Figure S1, Supplementary Table S2).

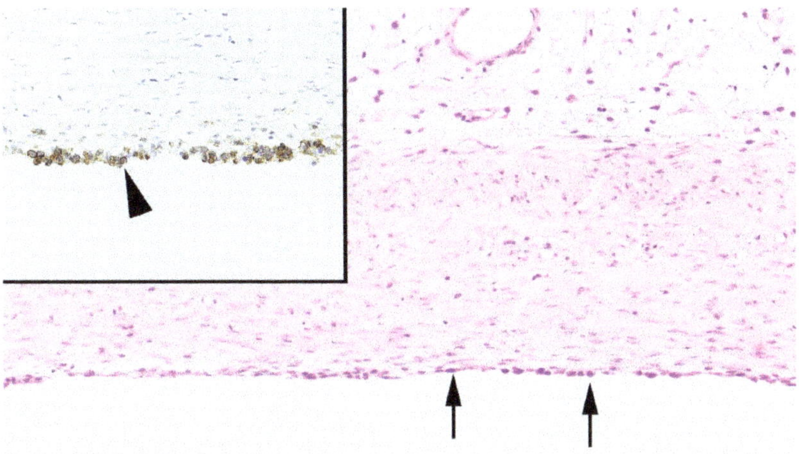

**Figure 4.** Histological phenotype of porcine cystic hygroma (case 1). Detail of the subcutaneous cystic mass surrounded by a thick capsule of mature connective tissue and lined by a single layer of well-differentiated cells with a squamous to cuboidal morphology (arrows). Inset: Immunohistochemistry for LYVE-1. Most of the epithelial cells lining the cystic cavity are diffusely and strongly positively stained with hematoxylin and eosin (arrowhead).

As both affected piglets of the Piètrain litter were males, an X-linked recessive inheritance could also be assumed. SNP genotyping showed that both cases received three IBD segments on chromosome X (Supplementary Table S2). Subsequently, we noticed that these haplotypes also occurred in several unaffected littermates as in these three segments, the unaffected male siblings received the identical maternal X chromosomes (Supplementary Table S3).

### 3.4. Whole-Genome Sequencing

The genomes of one affected Piètrain piglet (case 1), both parents, one healthy littermate, and the unrelated case 3 were sequenced. All five cleaned BAM files were inspected for coverage differences, and no evidence for chromosomal imbalance was detected (Supplementary Figure S2).

Subsequently, we focused genome-wide on variants in annotated genes and loci, assuming a protein-changing variant was causing the observed phenotype. For the sequenced Piètrain family consisting of four animals, we applied three different strategies to filter for disease-associated variants assuming various modes of inheritance, including autosomal recessive (AR), X-linked recessive (XR), and trio filtering for possible autosomal dominant (AD) de novo variants (Table 2). For the crossbred case 3, we searched for private heterozygous and homozygous variants that were absent in all other available genomes (Table 2).

Filtering for SNVs and small indels was performed against 10 unrelated control pigs (Supplementary Table S1) and revealed 104 coding variants detected in the purebred Piètrain case 1 based on different filtering strategies (Table 2, Supplementary Table S4). After a second round of filtering using the independent validation cohort of 756 commercial pig genomes, 22 coding variants of which 8 were predicted to be protein-changing remained. These were present only in the genome of the affected piglet from the Piètrain litter and absent in all controls (Table 2). In light of the outcome of the previously performed genetic analyses, only a single private protein-changing variant in *LOC110255918* mapped to the IBD genome region on chromosome 12 (Supplementary Table S4).

**Table 2.** Results of different filtering strategies.

| Strategy [1] | Purebred Piètrain | | | Crossbred | 10 Unrelated Controls | No. of Variants with Different Predicted Impact | | | after Filtering against Validation Cohort | | |
|---|---|---|---|---|---|---|---|---|---|---|---|
| | Case 1 | Dam/Sire of Case 1 | Full-sib Of Case 1 | Case 3 | | High | Moderate | Low | High | Moderate | Low |
| Piètrain-specific simple AR | 1/1 | 0/1 | 0/1 or 0/0 | 0/0 | 0/0 | 1 | 33 | 37 | 0 | 2 | 5 |
| Piètrain-specific compound heterozygous AR | 0/1 | 0/1 or 0/0 | 0/1 or 0/0 | 0/0 | 0/0 | 0 | 4 | 9 | 0 | 1 | 1 |
| Piètrain-specific XR | 1/- | 0/1/0/- | 0/- | 0/0 | 0/0 | 0 | 2 | 5 | 0 | 2 | 5 |
| Piètrain-specific AD (de novo) | 0/1 | 0/0 | 0/0 | 0/0 | 0/0 | 0 | 6 | 7 | 0 | 3 | 3 |
| Case 3-specific AR/AD | 0/0 | 0/0 | 0/0 | 0/1 or 1/1 | 0/0 | 50 | 872 | 1663 | 12 | 189 | 299 |

[1] AR denotes autosomal recessive, XR denotes X-linked recessive, and AD denotes autosomal dominant modes of inheritance.

In the WGS data of case 3, we found 2585 variants; after the second round of filtering, this list was reduced to 500 private SNVs and small indels, including 2 nonsense, 6 frameshift, 184 missense, 216 synonymous, and 86 intronic variants, as well as 3 inframe insertions, 2 inframe deletions, and 1 bidirectional gene fusion (Table 2, Supplementary Table S5).

No private variants shared between the two sequenced affected pigs were found. Two genes located on chromosome X were shared in case 1 and case 3, harboring independent private coding variants. The two different variants found in the chloride voltage-gated channel 4 (*CLCN4*) gene located in a homozygous region (Supplementary Table S4) were synonymous in both sequenced cases. In addition, the two different variants in the myotubularin related protein 1 (*MTMR1*) gene located in a linked region (Supplementary Table S4) were silent in case 1 and only predicted to alter the protein sequence in case 3.

## 4. Discussion

This is the first case report of cystic hygroma in pigs of different breeds in combination with further downstream analyses, including a comprehensive but still preliminary genomic analysis based on short-read whole-genome sequencing. Hence, a novel congenital disorder in pigs was described. In comparison with similar cases described in humans showing cystic hygroma in the neck region [1–4,12], in all three herein described porcine cases, the cystic hygroma was located at the level of the hind limb. Interestingly, no breathing problems, which are described in infants, could be observed in the pigs because the cystic hygroma was only located on the hind limb. However, all the pigs showed a higher risk of injuries on the rump, and therefore, this congenital disorder has implications for animal welfare. Moreover, the cystic hygroma on the limb tremendously influenced the meat quality, especially the dry ham production [32]. If the incidence of cystic hygroma in the pig population increased due to the high use of carrier boars, it would have a significant impact on the pig industry economy. Therefore, the results of this preliminary study are of major importance.

Due to the high use of semen from boar studs for artificial insemination, there is an increased risk of distributing a hereditary disease in the pig population [33–35]. Therefore, a thorough diagnostic and rapid analysis of this genetic disorder was conducted to avoid spreading of the disease in the pig population. The outcome of genetic analysis depends on the number of cases with a confirmed phenotype and the causative genetic variant. Hence, the diagnosis of the phenotype is of major importance, although systematic surveillance in the pig production is mostly lacking. In this study, all affected animals were confirmed by clinical examination and further pathological investigations to receive a valid phenotype for the subsequent genomic analysis. The clinical examination revealed all typical signs for hygroma cysts, including a painless, compressible, and soft mass on the rump [1,2]. To confirm this diagnosis, further investigations, including ultrasonographic examination, aspiration of the mass, and pathological examination were conducted. Findings showed cystic or multicystic lesions with internal septations and cloudy, reddish fluid with a low content of proteins, which was in line with the diagnosis of cystic hygroma [1,2]. Although half of the cases in human medicine are described at birth [13], all three herein described porcine cases developed the malformation within several days after birth, indicating a congenital condition.

As a familial occurrence of cystic hygroma is hypothesized in human medicine [16–18], we hypothesized a possible genetic origin for the observed cases in pigs. As the mode of inheritance was unclear, we evaluated different possible scenarios, such as monogenic recessive, X-linked, and dominant inheritance. As all three cases were male, a sex-linked inheritance seemed to be likely, and therefore, a second litter of the Piètrain sow with the same boar was produced under experimental conditions but revealed no further affected piglets. As all piglets of the second litter were apparently showing no signs of cystic hygroma until weaning, it might be speculated that due to the low number of male offspring, no further cases occurred. However, a resolution of the malformation on the lymphatic tissue has been proven by ultrasonographic examination during pregnancy [17]. This could not be ruled out in this study because the examination of all piglets in utero with ultrasonography is not possible in sows.

The outcome of the genetic and genomic analyses were inconclusive. There was at least a single genome region linked to the phenotype and showing shared homozygosity within the Pietrain family, and a single private protein-changing variant was found in that region on chromosome 12. Additionally, the condition may not be fully penetrant as the possibility that two normal littermates were identically homozygous for that genome region could not be ruled out. Using MutPred2 [36], the silico prediction of possible deleterious consequences for this missense variant in LOC110255918 revealed a score of 0.046, indicating no pathogenic effect. Nonetheless, the impact of this missense mutation in the uncharacterized locus remains unclear.

Furthermore, we opted for a possible sex-linked inheritance approach, which showed three shared IBD regions on chromosome X. However, the unaffected male littermates also shared these haplotypes. Only a single normal littermate shared the identical haplotypes in all three regions of the X chromosome, indicating that this offspring received the identical nonrecombinant copy of the maternal chromosome. Nonetheless, several further normal piglets received recombinant versions of the maternal X chromosome in the three distant chromosomal segments, resulting in several normal littermates sharing the respective haplotypes seen in both affected piglets. Therefore, the X-recessive mode of inheritance seems to be less likely. Furthermore, we could not rule out a possible paternal mosaicism of the X chromosome as an explanation. Filtering in the three regions revealed a variant in the CLCN4 gene that is known to be associated with the X-linked dominant inherited Raynaud-Claes syndrome, a very rare neurodevelopmental disease in humans (OMIM 300114). In addition to the fact that this synonymous variant does not alter the encoded protein, we postulate that the variant also found in the dam of the cystic hygroma-affected Pietrain piglet is most likely not causative. Indeed, sequencing the genome of the second affected littermate would have been beneficial in the identification of a possible shared causal variant. Recently it was evaluated that sequencing of each additional family member helped to narrow down the number of variants by 50%–75% [37]

In addition, the sequence analysis performed for the independent crossbred pig revealed no plausible candidate variant for the observed cystic hygroma phenotype that was highly similar to the two Pietrain cases. The private missense variant in MTMR1 located on the X chromosome seems to be less likely causative as the closely related MTM1 gene is known to be associated with forms of X-linked inherited myopathies (OMIM 300415). Furthermore, the long list of remaining private variants after filtering against hundreds of control genomes clearly illustrates that without sequencing of close relatives, ideally parents, as done for the Pietrain case, it is nearly impossible to limit the number of relevant variants if no obvious candidate gene is known. Furthermore, the chosen short-read genome sequencing method is known to be limited for the detection of structural variants. Therefore, long-read sequencing might be interesting in future cases to identify these kinds of genetic variations. Finally, the generated genome data revealed no indication of the presence of submicroscopic chromosomal abnormalities that were described in human cystic hygroma.

## 5. Conclusions

For the first time, this report provides a comprehensive description of cystic hygroma in pigs, including a preliminary genomic evaluation of possible inherited causes. It could be assumed that the genetic origin is heterogeneous as no shared variants across the two whole-genome sequenced affected pigs are found. Further targeted matings might help to elucidate the mode of inheritance. Systematic surveillance is needed to identify congenital defects as early as possible and to avoid the occurrence of further losses in the pig population.

**Supplementary Materials:** The following are available online at https://www.mdpi.com/2073-4425/12/2/207/s1, Figure S1: Runs of homozygosity across the autosomes of 24 analyzed Piètrain pigs, including two cases (red lines) and 22 controls (blue lines). Figure S2: Average coverage over 300 kb sliding windows as determined by whole-genome sequencing (WGS) of the analyzed Piètrain family and the unrelated case 3; note that the red lines indicate an individual's average coverage over the whole genome. Table S1: Sample designations and breed information of the whole-genome sequenced pigs. Table S2: Genomic regions of interest identified by homozygosity and linkage analyses in the Piètrain family. Table S3: Haplotype analysis for three segments on chromosome X. Table S4: Genomic variants detected in the WGS of the Piètrain case 1 based on different filtering strategies. Table S5: Genomic variants detected in the WGS of the unrelated case 3. Video S1: Video illustrating the clinical phenotype of case 2. Video S2: Video illustrating the clinical phenotype of case 3.

**Author Contributions:** Conceptualization, A.G. and C.D.; methodology, A.G., A.L. and L.G.-R.; formal analysis, A.G. and A.L.; investigation, A.L., A.M.S. and L.G.-R.; data curation, A.L.; software, A.L.; validation, M.F.L.D.; writing—original draft preparation, A.G., A.L., A.M.S., L.G.-R. and C.D.; writing—review and editing, A.G., A.L., M.F.L.D., A.M.S., L.G.-R. and C.D.; visualization, A.G., A.L., A.M.S. and L.G.-R.; supervision, A.G. and C.D.; funding acquisition, C.D. All authors have read and agreed to the published version of the manuscript.

**Funding:** Authors have no external funding to report.

**Institutional Review Board Statement:** Not applicable.

**Informed Consent Statement:** Not applicable.

**Data Availability Statement:** All data generated or analyzed during this study are available from the corresponding author on reasonable request.

**Acknowledgments:** The Next Generation Sequencing Platform and the Interfaculty Bioinformatics Unit of the University of Bern are acknowledged for performing the WGS and providing high-performance computational infrastructure.

**Conflicts of Interest:** The authors declare that they have no competing interests.

## References

1. Mirza, B.; Ijaz, L.; Saleem, M.; Sharif, M.; Sheikh, A. Cystic hygroma: An overview. *J. Cutan. Aesthet. Surg.* **2010**, *3*, 139. [CrossRef]
2. Noia, G.; Maltese, P.E.; Zampino, G.; D'Errico, M.; Cammalleri, V.; Convertini, P.; Marceddu, G.; Mueller, M.; Guerri, G.; Bertelli, M. Cystic hygroma: A preliminary genetic study and a short review from the literature. *Lymphat. Res. Biol.* **2019**, *17*, 30–39. [CrossRef]
3. Chen, Y.N.; Chen, C.P.; Lin, C.J.; Chen, S.W. Prenatal Ultrasound Evaluation and Outcome of Pregnancy with Fetal Cystic Hygromas and Lymphangiomas. *J. Med. Ultrasound* **2017**, *25*, 12–15. [CrossRef]
4. Levy, A.T.; Berghella, V.; Al-Kouatly, H.B. Outcome of 45, X fetuses with cystic hygroma: A systematic review. *Am. J. Med. Genet. Part A* **2020**, *185*, 26–32. [CrossRef]
5. Fisher, R.; Partington, A.; Dykes, E. Cystic hygroma: Comparison between prenatal and postnatal diagnosis. *J. Pediatr. Surg.* **1996**, *31*, 473–476. [CrossRef]
6. Chervenak, F.A.; Isaacson, G.; Blakemore, K.J.; Breg, W.R.; Hobbins, J.C.; Berkowitz, R.L.; Marge, T.; Mayden, K.; Mahoney, M.J. Fetal Cystic Hygroma—Cause and Natural History. *N. Engl. J. Med.* **1983**, *309*, 822–825. [CrossRef] [PubMed]
7. Williams, J.H. Lymphangiosarcoma of dogs: A review. *J. S. Afr. Vet. Assoc.* **2005**, *76*, 127–131. [CrossRef] [PubMed]
8. Thornton, H. Hygromas in cattle. *Cent. Afr. J. Med.* **1970**, *16*, 66–67. [PubMed]
9. Mauldin, E.A.; Peters-Kennedy, J. Integumentary System. In *Jubb, Kennedy and Palmer's Pathology of Domestic Animals*, 6th ed.; Grant Maxie, M., Ed.; Elsevier: St. Louis, MO, USA, 2016.
10. Savage, V.L.; Cudmore, L.A.; Russell, C.M.; Railton, D.I.; Begg, A.P.; Collins, N.M.; Adkins, A.R. Intra-abdominal cystic lymphangiomatosis in a Thoroughbred foal. *Equine Vet. Educ.* **2018**, *30*, 403–408. [CrossRef]
11. Gehlen, H.; Wohlsein, P. Cutaneous lymphangioma in a young Standardbred mare. *Equine Vet. J.* **2000**, *32*, 86–88. [CrossRef] [PubMed]
12. Chen, C.P.; Liu, F.F.; Jan, S.W.; Lee, C.C.; Town, D.D.; Lan, C.C. Cytogenetic evaluation of cystic hygroma associated with hydrops fetalis, oligohydramnios or intrauterine fetal death: The roles of amniocentesis, postmortem chorionic villus sampling and cystic hygroma paracentesis. *Acta Obstet. Gynecol. Scand.* **1996**, *75*, 454–458. [CrossRef] [PubMed]
13. Kumar, N.; Kohli, M.; Pandey, S.; Tulsi, S.P.S. Cystic hygroma. *Natl. J. Maxillofac. Surg.* **2001**, *1*, 528. [CrossRef] [PubMed]
14. Shulman, L.P.; Emerson, D.S.; Felke, R.E.; Phillips, P.O.; Simpson, L.J.; Elias, S. High frequency of cytogenetic abnormalities in fetuses with cystic hygroma diagnosed in the first trimester. *Obstet. Gynecol.* **1992**, *80*, 80–82. [PubMed]

15. van Zalen-Sprock, R.M.; van Vugt, J.M.G.; van Geijn, H.P. First-trimester diagnosis of cystic hygroma—Course and outcome. *Am. J. Obstet. Gynecol.* **1992**, *167*, 94–98. [CrossRef]
16. Baxi, L.; Brown, S.; Desai, K.; Thaker, H. Recurrent cystic hygroma with hydrops. *Fetal Diagn. Ther.* **2009**, *25*, 127–129. [CrossRef] [PubMed]
17. Rotmensch, S.; Celentano, C.; Sadan, O.; Liberati, M.; Lev, D.; Glezerman, M. Familial occurrence of isolated nonseptated nuchal cystic hygromata in midtrimester of pregnancy. *Prenat. Diagn.* **2004**, *24*, 260–264. [CrossRef]
18. Teague, K.E.; Eggleston, M.K.; Muffley, P.E.; Gherman, R.B. Recurrent fetal cystic hygroma with normal chromosomes: Case report and review of the literature. *J. Matern. Fetal Med.* **2000**, *9*, 366–369.
19. Maltese, P.E.; Rakhmanov, Y.; Zulian, A.; Notarangelo, A.; Bertelli, M. Genetic testing for cystic hygroma. *Eur. Biotechnol. J.* **2018**, *2*, 22–25. [CrossRef]
20. Chang, C.C.; Chow, C.C.; Tellier, L.C.A.M.; Vattikuti, S.; Purcell, S.M.; Lee, J.J. Second-generation PLINK: Rising to the challenge of larger and richer datasets. *Gigascience* **2015**, *4*, 7. [CrossRef]
21. Abecasis, G.R.; Cherny, S.S.; Cookson, W.O.; Cardon, L.R. Merlin—Rapid analysis of dense genetic maps using sparse gene flow trees. *Nat. Genet.* **2002**, *30*, 97–101. [CrossRef]
22. R Development Core Team. *R: A Language and Environment for Statistical Computing*; R Foundation for Statistical Computing: Vienna, Austria, 2019.
23. Li, H. A statistical framework for SNP calling, mutation discovery, association mapping and population genetical parameter estimation from sequencing data. *Bioinformatics* **2011**, *27*, 2987–2993. [CrossRef] [PubMed]
24. Broad Institute. "Picard Toolkit." GitHub Repository. 2019. Available online: http://broadinstitute.github.io/picard/ (accessed on 1 October 2020).
25. McKenna, A.; Hanna, M.; Banks, E.; Sivachenko, A.; Cibulskis, K.; Kernytsky, A.; Garimella, K.; Altshuler, D.; Gabriel, S.; Daly, M.; et al. The Genome Analysis Toolkit: A MapReduce framework for analyzing next-generation DNA sequencing data. *Genome Res.* **2010**, *20*, 1297–1303. [CrossRef] [PubMed]
26. Cingolani, P.; Platts, A.; Wang, L.L.; Coon, M.; Nguyen, T.; Wang, L.; Land, S.J.; Lu, X.; Ruden, D.M. A program for annotating and predicting the effects of single nucleotide polymorphisms, SnpEff: SNPs in the genome of Drosophila melanogaster strain w1118, iso-2; iso-3. *Fly (Austin)* **2012**, *6*, 80–92. [CrossRef] [PubMed]
27. Derks, M.F.L.; Gjuvsland, A.B.; Bosse, M.; Lopes, M.S.; Van Son, M.; Harlizius, B.; Tan, B.F.; Hamland, H.; Grindflek, E.; Groenen, M.A.M.; et al. Loss of function mutations in essential genes cause embryonic lethality in pigs. *PLoS Genet.* **2019**, *15*, 1008055. [CrossRef]
28. Garrison, E.; Marth, G. Haplotype-based variant detection from short-read sequencing. *arXiv* **2012**, arXiv:1207.3907.
29. McLaren, W.; Gil, L.; Hunt, S.E.; Riat, H.S.; Ritchie, G.R.S.; Thormann, A.; Flicek, P.; Cunningham, F. The Ensembl Variant Effect Predictor. *Genome Biol.* **2016**, *17*, 122. [CrossRef]
30. Thorvaldsdóttir, H.; Robinson, J.T.; Mesirov, J.P. Integrative Genomics Viewer (IGV): High-performance genomics data visualization and exploration. *Brief. Bioinform.* **2013**, *14*, 178–192. [CrossRef]
31. Turner, S. qqman: An R package for visualizing GWAS results using Q-Q and manhattan plots. *J. Open Source Softw.* **2018**, *3*, 731. [CrossRef]
32. Čandek-Potokar, M.; Škrlep, M. Factors in pig production that impact the quality of dry-cured ham: A review. *Animal* **2012**, *6*, 327–338. [CrossRef]
33. Knox, R.V. Artificial insemination in pigs today. *Theriogenology* **2016**, *85*, 83–93. [CrossRef]
34. Grahofer, A.; Letko, A.; Häfliger, I.M.; Jagannathan, V.; Ducos, A.; Richard, O.; Peter, V.; Nathues, H.; Drögemüller, C. Chromosomal imbalance in pigs showing a syndromic form of cleft palate. *BMC Genom.* **2019**, *20*, 349. [CrossRef] [PubMed]
35. Grahofer, A.; Nathues, H.; Gurtner, C. Multicystic degeneration of the Cowper's gland in a Large White boar. *Reprod. Domest. Anim.* **2016**, *51*, 1044–1048. [CrossRef] [PubMed]
36. Pejaver, V.; Urresti, J.; Lugo-Martinez, J.; Pagel, K.A.; Lin, G.N.; Nam, H.-J.; Mort, M.; Cooper, D.N.; Sebat, J.; Iakoucheva, L.M.; et al. MutPred2: Inferring the molecular and phenotypic impact of amino acid variants. *Nat. Commun.* **2020**, *11*, 5918. [CrossRef] [PubMed]
37. Alfares, A.; Alsubaie, L.; Aloraini, T.; Alaskar, A.; Althagafi, A.; Alahmad, A.; Rashid, M.; Alswaid, A.; Alothaim, A.; Eyaid, W.; et al. What is the right sequencing approach? Solo VS extended family analysis in consanguineous populations. *BMC Med. Genom.* **2020**, *13*, 103. [CrossRef] [PubMed]

 genes 

*Article*

# Genes and SNPs Involved with Scrotal and Umbilical Hernia in Pigs

**Ariene Fernanda Grando Rodrigues [1], Adriana Mércia Guaratini Ibelli [2,3], Jane de Oliveira Peixoto [2,3], Maurício Egídio Cantão [2], Haniel Cedraz de Oliveira [4], Igor Ricardo Savoldi [1], Mayla Regina Souza [1,5], Marcos Antônio Zanella Mores [2], Luis Orlando Duitama Carreño [6] and Mônica Corrêa Ledur [1,2,*]**

1 Programa de Pós-Graduação em Zootecnia, Departamento de Zootecnia, Centro de Educação Superior do Oeste (CEO), Universidade do Estado de Santa Catarina, UDESC, 89815-630 Chapecó, Brazil; ariene.grando@gmail.com (A.F.G.R.); igorsaavoldii@gmail.com (I.R.S.); mayla.zootecnista@gmail.com (M.R.S.)
2 Embrapa Suínos e Aves, Distrito de Tamanduá, 89715-899 Concórdia, Brazil; adriana.ibelli@embrapa.br (A.M.G.I.); jane.peixoto@embrapa.br (J.d.O.P.); mauricio.cantao@embrapa.br (M.E.C.); marcos.mores@embrapa.br (M.A.Z.M.)
3 Programa de Pós-Graduação em Ciências Veterinárias, Departamento de Ciências Veterinárias, Universidade Estadual do Centro-Oeste, 85015-430 Guarapuava, Brazil
4 Animal Science Department, Universidade Federal de Viçosa, 36570-900 Viçosa, Brazil; hanielcedraz@gmail.com
5 Programa de Pós-Graduação em Zootecnia, Departamento de Zootecnia, Universidade Federal do Rio Grande do Sul, UFRGS, 91540-000 Porto Alegre, Brazil
6 BRF SA, 82305-100 Curitiba, Brazil; luis.carreno@brf-br.com
* Correspondence: monica.ledur@embrapa.br or monica.ledur@udesc.br; Tel.: +55-49-3441-0411

**Citation:** Rodrigues, A.F.G.; Ibelli, A.M.G.; Peixoto, J.d.O.; Cantão, M.E.; Oliveira, H.C.d.; Savoldi, I.R.; Souza, M.R.; Mores, M.A.Z.; Carreño, L.O.D.; Ledur, M.C. Genes and SNPs Involved with Scrotal and Umbilical Hernia in Pigs. *Genes* **2021**, *12*, 166. https://doi.org/10.3390/genes12020166

Academic Editor: Katarzyna Piórkowska

Received: 18 December 2020
Accepted: 22 January 2021
Published: 27 January 2021



**Abstract:** Hernia is one of the most common defects in pigs. The most prevalent are the scrotal (SH), inguinal (IH) and umbilical (UH) hernias. We compared the inguinal ring transcriptome of normal and SH-affected pigs with the umbilical ring transcriptome of normal and UH-affected pigs to discover genes and pathways involved with the development of both types of hernia. A total of 13,307 transcripts was expressed in the inguinal and 13,302 in the umbilical ring tissues with 94.91% of them present in both tissues. From those, 35 genes were differentially expressed in both groups, participating in 108 biological processes. A total of 67 polymorphisms was identified in the inguinal ring and 76 in the umbilical ring tissue, of which 11 and 14 were novel, respectively. A single nucleotide polymorphism (SNP) with deleterious function was identified in the integrin α M (*ITGAM*) gene. The microtubule associated protein 1 light chain 3 γ (*MAP1LC3C*), vitrin (*VIT*), aggrecan (*ACAN*), alkaline ceramidase 2 (*ACER2*), potassium calcium-activated channel subfamily M α 1 (*KCNMA1*) and synaptopodin 2 (*SYNPO2*) genes are highlighted as candidates to trigger both types of hernia. We generated the first comparative study of the pig umbilical and inguinal ring transcriptomes, contributing to the understanding of the genetic mechanism involved with these two types of hernia in pigs and probably in other mammals.

**Keywords:** gene expression; congenital defects; RNA-Seq; transcriptomics; swine

## 1. Introduction

Pig production is one of the most important livestock activities in the world and its evolution and expansion are mainly due to the development of technologies that combine genetics, management, nutrition and well-being [1], which increase productivity and bring the final product closer to what the consumer idealizes. Meat quality and feed efficiency are traits that have been prioritized by genetic breeding programs [2–4]. However, in recent years, studies have also been carried out to improve our knowledge on diseases that persist in production, which bring losses to the entire chain [5–9]. Among these, scrotal (SH)/inguinal (IH) and umbilical hernias (UH) are birth defects often found in pigs [10], causing pain and discomfort to the animals and, consequently, economic losses related to reduced performance [11,12] and increased risk of death [13].

Scrotal hernia is mainly characterized by the displacement of intestinal loops to the scrotal sac, resulting from an abnormality in the inguinal ring [11]. Failure to obliterate the vaginal process [14], impairment of the innervations that act on the site [15] or the failed involution of the internal inguinal ring [16] are processes associated to the manifestation of this defect. Moreover, SH can arise as a result of low resistance of the inguinal region [17] caused by disturbances in the metabolism and hydrolysis of the extracellular matrix components, such as collagen and muscle fiber structures [18], compromising the repair of connective tissue [19]. The incidence of SH in pigs is influenced by genetics and environmental factors. Sevillano et al. [20] have shown that the incidence of SH/IH in Landrace and Large White pigs was 0.34% and 0.42%, respectively.

Umbilical hernia is characterized by the passage of abdominal contents (mainly intestine) to the hernial sac, in the umbilical region [12]. The discomfort and pain can be aggravated when secondary factors are associated with the defect, and the pig welfare is compromised [12]. UH can be related to genetic and non-genetic factors such as navel infections, lesions at the site, obesity and incorrect umbilical cord cutting [21,22]. This defect has been associated with a failure in the process of umbilical ring closure [23,24]. The incidence of UH ranges from 0.55% [25] to 1.5% [23] and it varies according to the management, breed line and production lot [26]. Usually, the UH is not observed at birth, and this defect appears when the pigs are already in the growing period [27]. This demonstrates the difficulty that breeding companies and pig farmers have to eliminate such defect from their herds.

The heritability for SH/IH and UH were estimated in 0.31 [20] and 0.25 [28], respectively, which shows moderate genetic influence in the appearance of these defects. The knowledge of the genetic mechanisms associated with the formation of these anomalies is important to understand their underlying causes, regardless of the type of hernia studied. In pigs, quantitative trait loci (QTL) were detected for the occurrence of SH in pig chromosomes (SSC) 2, 4, 8 (locus SW 933), 13 and 16 [29]. In addition, suggestive QTL for IH and SH were found in seven chromosomes (SSC1, 2, 5, 6, 15, 17 and X) when comparing healthy and herniated pigs [11]. Moreover, candidate genes involved with the manifestation of SH, such as Insulin-like receptor 3 (*INSL3*), Müllerian inhibiting substance (*MIS*) and Calcitonin gene-related peptide (*CGRP*) were also identified [11]. Twenty-two single nucleotide polymorphisms (SNPs) located on chromosomes 1, 2, 4, 10 and 13 in Landrace pigs, and 10 SNPs on SSC 3, 5, 7, 8 and 13 of Large White pigs, were associated with the incidence of SH/IH in these populations [20]. To date, there is a record of 116 QTL/associations related to the appearance of SH/IH in pigs in the Animal QTL database [30].

Regarding UH, there are 55 QTL/associations related to its manifestation in the Pig QTL database (accessed in 17th December 2020) [30]. A SNP in the *CAPN9* gene (Calpain 9) on SSC14 significantly associated with UH has already been detected [31]. In commercial pigs, four candidate genes were identified in QTL regions associated with UH, namely *TBX15* (T-Box 15) and *WARS2* (Tryptophanyl TRNA Synthetase 2, Mitochondrial) in chromosome 4, and *LIPI* (Lipase I) and *RBM11* (RNA Binding Motif Protein 11) in chromosome 13 [32]. A QTL on chromosome 14 of the Landrace breed was highly correlated with UH, where the *LIF* (Leukemia inhibitory factor) and *OSM* (Oncostatin M) genes were identified as candidates for this defect [25].

Another approach that has recently been used to investigate the genetic mechanisms involved with these defects is the whole transcriptome study. Functional candidate genes were prospected for scrotal [6] and umbilical [7] hernias based on differentially expressed (DE) genes between affected and unaffected pigs in the inguinal and umbilical ring tissues, respectively, using RNA sequencing (RNA-Seq). Information about gene expression in those tissues is scarce and there are still many gaps to be elucidated. Interestingly, these previous studies have shown some genes and pathways that seem to be shared between SH and UH, many of them involved with muscle. Therefore, new and systematic analyses on tissues characterization and on common processes involved with both types of hernia are needed.

Although several genetic studies have been performed, no comparison between large-scale gene expression profile of pigs affected with SH from those affected with UH was reported to date. Since SH and UH are related to some extent to muscle dysfunction, our hypothesis is that some muscle-related genes could be involved in the development of both types of hernia. Therefore, the objective of this study was to investigate the common molecular mechanisms and genes involved with these two types of hernia by comparing the SH and UH transcriptomes, as well as to identify and characterize polymorphisms in those transcriptomes.

## 2. Materials and Methods

This study was performed with the approval of the Embrapa Swine and Poultry National Research Center Ethical Committee of Animal Use (CEUA) under the protocol number 011/2014. A diagram summarizing the experiment and the analyses performed can be seen in the Supplementary Materials: Figure S1.

### *2.1. Animals and Sample Collection*

Eighteen pigs were selected from a Landrace female line from the same nucleus farm, located in Santa Catarina State, SC, Brazil. From those, five were females affected (case) with UH, five healthy (control) females for UH, four males with SH and four control males for SH, as described by Souza et al. [7] and Romano et al. [6]. Control animals were healthy pigs, without any type of hernia, came from hernia-free litter and were from the same management group of their respective cases. The animals were selected at approximately 60 days of age for SH and 90 days of age for UH. At the Embrapa Swine and Poultry National Research Center, located in Concórdia, SC, Brazil, the pigs were euthanized by electrocution stunning for ten seconds, followed by bleeding, in accordance with the practices recommended by the Ethics Committee. The necropsy was performed for general evaluation and to confirm the both types of hernia. In the pathological analysis, the two groups of animals were confirmed: hernia-affected or without hernias (Figure 1). Tissue samples from the inguinal and umbilical rings were collected for investigating the scrotal and umbilical hernias, respectively (Figure 1). Samples were immediately placed in liquid nitrogen and stored in ultra-freezer (−80 °C) for RNA extraction. Samples from those tissues were also collected and placed in 4% paraformaldehyde for histopathological analysis.

### *2.2. Histopathological Analysis of the Inguinal and Umbilical Ring Tissues*

The samples previously collected were routinely processed for histopathology as described by Souza et al. [7]. Briefly, sample tissues were dehydrated in increasing concentrations of ethanol, diaphanized with xylol and embedded in paraffin. Tissue sections were cut with an automatic microtome, stained using the hematoxylin and eosin (HE) and analyzed by optical microscopy.

### *2.3. RNA Extraction, RNA-Seq Libraries Preparation and Sequencing*

Total RNA extraction was initiated using the Trizol reagent (Invitrogen, Carlsbac, CA, USA) according to the manufacturer's instructions. A 100 mg of the tissues were macerated in liquid nitrogen with a mortar and added to a polypropylene tube with 1 mL of the Trizol reagent, mixed using the vortex, and then incubated for five minutes at room temperature (RT, 25 °C). Then, 200 µL of chloroform were added, shaked vigorously and incubated at RT for another five minutes, followed by centrifugation. Approximately 600 µL of the aqueous phase were transferred to a new tube and 600 µL of 70% ethanol were added and homogenized by inversion. This volume was then added to a mini RNeasy silica column (Qiagen, Düsseldorf, Germany) following the manufacturers' protocol. The quantity and quality of total RNA were assessed by quantification in a Biodrop spectrophotometer (Biodrop, Cambridge, UK), in a 1% agarose gel and in the Bioanalyzer Agilent 2100 (Agilent

Technologies, Santa Clara, CA, USA). Samples with an RNA integrity number (RIN) > 8.0 were used to prepare the RNA-Seq libraries.

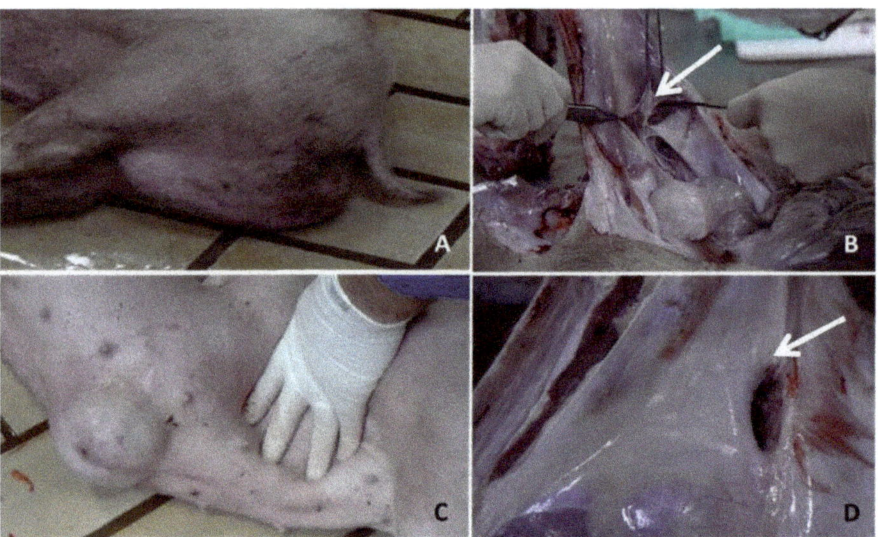


**Figure 1.** Pathological analysis. Legend: (**A**) swine affected with scrotal hernia. (**B**) Region affected with scrotal hernia (the white arrow shows the inguinal ring). (**C**) Swine affected with umbilical hernia. (**D**) Region affected with umbilical hernia (the white arrow shows the umbilical ring).

The RNA of four normal and four SH-inguinal ring and five normal and five UH-umbilical ring tissues were submitted to library preparation with the TruSeq mRNA Stranded Sample Preparation kit (Illumina, Inc., San Diego, CA, USA). For this, 2 μg of total RNA was purified using magnetic microspheres for poly-A selection. The libraries were quantified by qPCR and evaluated in Bioanalyzer (Agilent, Santa Clara, CA, USA), prior to sequencing. Libraries passing the quality control were sent for RNA sequencing in HiSeq 2500 equipment (Illumina, San Diego, CA, USA) at the Functional Genomics Center of ESALQ/USP, in a paired-end protocol (2 × 100 bp). All samples of each hernia and their respective controls were placed in the same lane. Additional information can be found in Romano et al. [6] and Souza et al. [7]. The transcriptome sequences for the scrotal and umbilical hernias used in this study were previously deposited in the SRA database with BioProject numbers PRJNA350530 and PRJNA445856, respectively.

### 2.4. *Quality Control and Differentially Expressed Genes*

The quality control analysis and mapping were performed using the Bioinformatics Analysis for Quality Control and Mapping (BAQCOM) pipeline available in the Github repository [33]. BAQCOM pipeline uses Trimmomatic tool [34] to remove short reads (<70 bp), reads with low quality (QPhred < 24) and adapter sequences. The sequences were mapped against the swine reference genome (Sus scrofa, assembly 11.1) available on the Ensembl [35] version 95 using the STAR (version 2.7.0) program [36]. To verify the consistency of the expression pattern between the sample groups, principal component analysis (PCA) plots were performed in RStudio [37] (version 1.1.463) from R language [38] (version 3.5.3). The EdgeR package [39] from R was used for the differentially expressed analysis. Differentially expressed (DE) genes in the analyzed tissues (case and control for each hernia) were selected based on the level of false discovery rate (FDR < 0.05) after the Benjamini–Hochberg (BH) method for multiple correction tests [40].

### 2.5. Transcriptomes Characterization of Scrotal and Umbilical Hernia

Initially, SH and UH-related transcriptomes were characterized as the total number of transcripts, number of protein-encoding genes, miRNAs, lncRNAs and non-characterized genes in the swine genome using the Biomart tool available in Ensembl 95 [35]. An annotation of non-characterized genes in the porcine genome was performed using DAVID 6.8 database [41]. The comparison between both transcriptomes was carried out to identify genes expressed in both types of hernia, SH and UH.

### 2.6. Comparison of Differentially Expressed Genes and In Silico Functional Analysis

The classification of DE genes was performed according to the database available in Ensembl 95 and further enrichment in DAVID 6.8 database [41]. A comparison of DE genes between both conditions was performed to verify if the genes were involved in the manifestation of the two types of hernia, and whether there was agreement or not between gene expression profiles in both conditions. In order to verify if DE transcripts named as uncharacterized proteins in Ensembl 95 were similar to genes known in other genomes, the sequences were aligned against the UniProt database [42] using BLASTp tool from the NCBI [43,44]. A gene interaction network was built with the DE genes common to both types of hernia using the STRING database [45,46]. The gene ontology (GO) was evaluated in the Panther [47] and DAVID 6.8 [41] databases, followed by clusterization in the REVIGO tool [48]. Furthermore, it was verified if the DE genes were located in QTL regions previously reported in the Pig QTL database [30].

### 2.7. Identification of Polymorphisms

For the polymorphisms discovery between the transcriptome of animals affected with each type of hernia, the Genome Analysis Tool Kit (GATK) program (version 3.8) [49] was used with the Picard (version 2.5) toolkit [50]. The search for SNPs and insertions or deletions (InDels) was carried out following the filtering parameters and sequence quality suggested by the best practices protocol [51,52]. The polymorphisms annotation for the two hernias studied was performed using the variant effect predictor (VEP) tool (version 3.8) [53] with standard parameters available in the Ensembl 95 database and using the KEGG Pathway database [54]. Therefore, this annotation allowed the discovery of new polymorphisms, as well as to verify their location and possible function in the genome. Using the Biomart data mining tool [35], miRNAs were observed, and a manual comparison was made to identify common miRNAs between SH and UH. Subsequently, a search in the miRBase database (version 22) [55,56] was performed to obtain individual information for each miRNA.

## 3. Results

### 3.1. Histopathological Analysis of the Inguinal and Umbilical Ring Tissues

In the microscopic evaluation, the group affected with SH showed a larger number of connective tissue fibers compared to the control group (Figure 2A,B). In the UH-affected group, the connective tissue was denser than in the control group (Figure 2C,D) as shown by Souza et al. [7].

### 3.2. Sequencing and Mapping

The RNA sequencing of all samples ($n$ = 18) generated approximately 465 million paired-end reads and, after the quality control analyses, 13.84% of these were removed, resulting in approximately 400 million reads (86.16%) (Additional file 1: Table S1). The PCA plots show the separation between the affected and control samples from the two evaluated types of hernia (Supplementary Materials 2: Figure S2A–C). Around 93.50% of reads were mapped against the swine reference genome (Sus scrofa 11.1) (Ensembl 95), with individual samples ranging from 87.73% to 96.05%, distributed between the groups of healthy and affected by SH or UH. From those, 78.77% of all reads were mapped in genes. From the 25,880 annotated genes in the swine reference genome (Sus scrofa 11.1, Ensembl

95) [35], 13,307 (51.42%) genes were expressed in the inguinal ring and 13,302 (51.40%) in the umbilical ring.

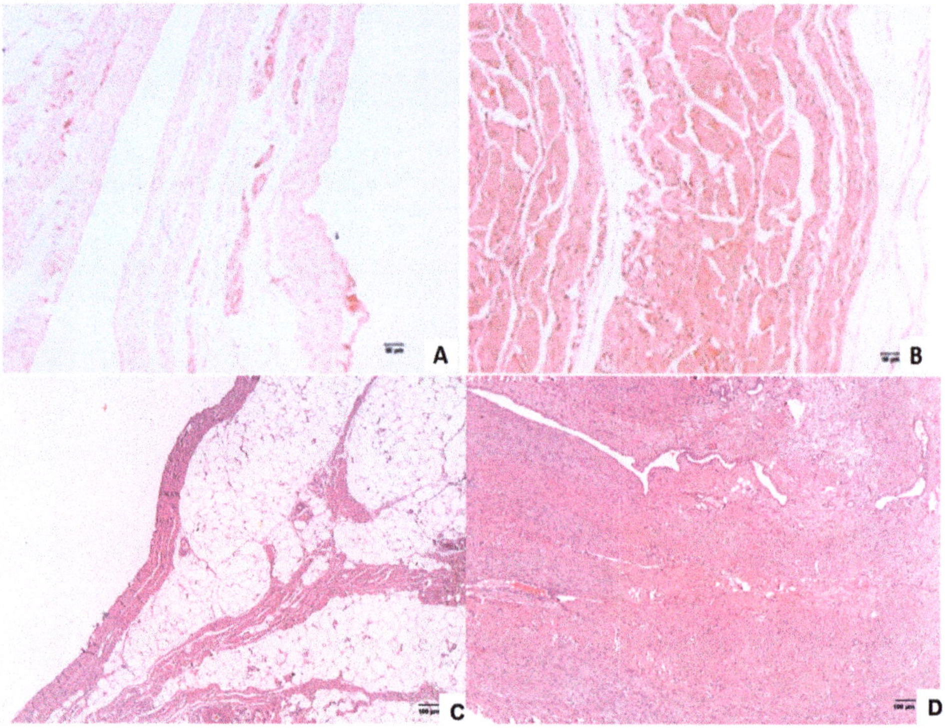

**Figure 2.** Histopathological slide stained with hematoxylin and eosin (HE). Legend: (**A**) sample from the scrotal hernia (SH)-control group and (**B**) sample from the SH-affected group. A larger number of connective fibers is observed in the sample of the SH-affected group than in the sample from the SH-control group. (**C**) Sample from the umbilical hernia (UH)-control group and (**D**) sample from the UH-affected group. Connective tissue interspersed with adipose tissue is observed in the sample of the UH-control group, while in the sample from the UH-affected group, only proliferated connective tissue is observed.

### 3.3. Characterization of the Scrotal and Umbilical Hernia Transcriptomes

The transcripts from the inguinal and the umbilical ring tissues were classified according to the Ensembl 95 database [35] (Table 1). After comparing the two transcriptomes (SH and UH), 94.91% of the genes were identified in both groups (Figure 3). The Venn diagram also presents the number of transcripts expressed exclusively in each type of tissue.

### 3.4. Differentially Expressed Genes

In the pig inguinal ring transcriptome, 627 genes were differentially expressed (FDR $< 0.05$) between the control and the SH-affected group. Out of those, 435 genes (69.38%) were downregulated and 192 (30.62%) were upregulated in the SH-affected pigs compared to the normal animals. Regarding the genes expressed in the umbilical ring, 199 were DE between normal and UH-affected pigs. From those, 129 were downregulated (64.82%) and 70 (35.18%) upregulated in the UH-affected pigs when compared to the normal ones. In the samples from the SH group, 98.09% of the DE genes were characterized as protein coding genes, 0.64% as lncRNA, 0.32% as pseudogenes, 0.32% as C immunoglobulins, 0.32% as miscRNA, 0.16% as encoding immunoglobulins V and 0.16% as ribozyme. In the

UH transcriptome, 92.46% were protein coding genes, 3.52% immunoglobulin C coding genes, 1.51% pseudogenes, 1.01% miscRNAs, 1.01% mitochondrial ribosomal RNA and 0.50% lncRNA.

**Table 1.** Characterization of the transcripts identified in the inguinal and umbilical ring samples.

| Annotated Transcripts | SH | | UH | |
|---|---|---|---|---|
| LncRNA | 68 | 0.53% | 77 | 0.60% |
| MiRNA | 5 | 0.04% | 4 | 0.03% |
| Mt rRNA | 2 | 0.02% | 2 | 0.02% |
| Mt tRNA | 1 | 0.01% | 1 | 0.01% |
| Processed pseudogene | 1 | 0.01% | 0 | 0% |
| Protein-coding | 12,601 | 98.55% | 12,598 | 98.50% |
| Pseudogene | 90 | 0.70% | 91 | 0.71% |
| Ribozime | 1 | 0.01% | 1 | 0.01% |
| ScaRNA | 1 | 0.01% | 1 | 0.01% |
| SnoRNA | 13 | 0.10% | 14 | 0.11% |
| SnRNA | 1 | 0.01% | 0 | 0% |
| Y RNA | 2 | 0.02% | 1 | 0.01% |
| Total annotated transcripts | 12,786 | | 12,790 | |

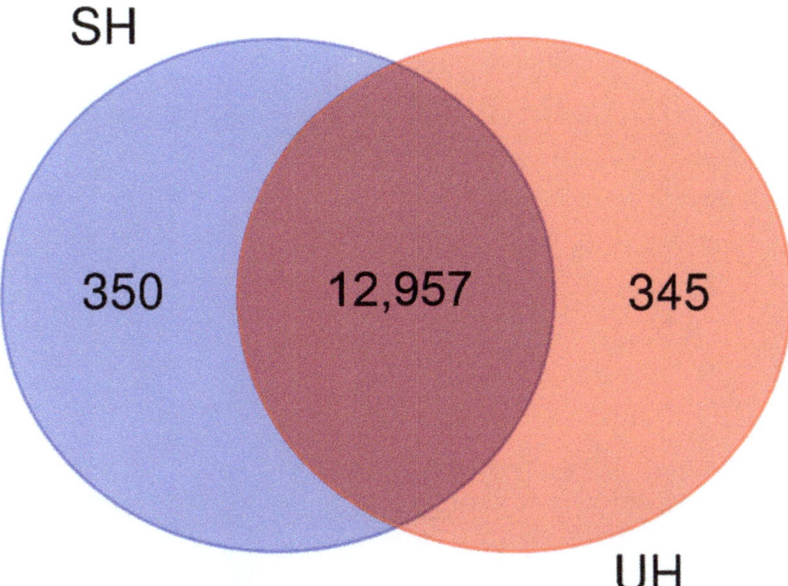


**Figure 3.** Distribution of transcripts identified in the pig inguinal and umbilical ring tissue samples. Legend: SH—scrotal hernia group; UH—umbilical hernia group. For the SH, the inguinal ring tissue was evaluated, and for the UH, the umbilical ring tissue was analyzed.

### *3.5. Differentially Expressed Genes Common to Both SH and UH Transcriptomes*

Comparing the DE genes identified in the SH and UH groups, 35 DE genes were present in both transcriptomes (Table 2).

**Table 2.** Differentially expressed genes identified in both scrotal (SH) and umbilical hernia (UH) groups.

| ENSEMBL ID | Gene Symbol | Chromosome | Gene Name | SH-logFC | SH-FDR | UH-logFC | UH-FDR |
|---|---|---|---|---|---|---|---|
| ENSSSCG00000001832 | *ACAN* | 7 | Aggrecan | 2.913 | 0.001 | 2.788 | 0.040 |
| ENSSSCG00000034213 | *ACER2* | 1 | Alkaline ceramidase 2 | −3.066 | 0.001 | −2.373 | 0.004 |
| ENSSSCG00000036223 | *ACKR1* | 4 | Atypical chemokine receptor 1 (Duffy blood group) | −1.119 | 0.030 | −1.023 | 0.034 |
| ENSSSCG00000010370 | *ANXA8* | 14 | Annexin A8 | 2.026 | 0.000 | 1.744 | 0.004 |
| ENSSSCG00000032709 | *ARL4A* | 9 | ADP ribosylation factor like GTPase 4A | −1.308 | 0.001 | −1.199 | 0.031 |
| ENSSSCG00000033350 | *BCHE* | 13 | Butyrylcholinesterase | 1.011 | 0.028 | 1.557 | 0.009 |
| ENSSSCG00000028567 | *BTNL9* | 2 | Butyrophilin like 9 | −1.356 | 0.010 | −2.268 | 0.016 |
| ENSSSCG00000002662 | *C16orf74* | 6 | Chromosome 16 open reading frame 74 | 1.818 | 0.025 | 1.661 | 0.008 |
| ENSSSCG00000006736 | *CD2* | 4 | CD2 molecule | 2.275 | 0.028 | −1.990 | 0.002 |
| ENSSSCG00000009138 | *CFI* | 8 | Complement factor I | 2.254 | 0.000 | 1.904 | 0.029 |
| ENSSSCG00000011524 | *CHL1* | 13 | Cell adhesion molecule L1 like | −1.776 | 0.004 | −1.321 | 0.009 |
| ENSSSCG00000021588 | *DAPK2* | 1 | Death associated protein kinase 2 | −1.347 | 0.026 | −1.473 | 0.013 |
| ENSSSCG00000012126 | *GPM6B* | X | Glycoprotein M6B | −1.047 | 0.009 | −1.244 | 0.022 |
| ENSSSCG00000002847 | *GPT2* | 6 | Glutamic–pyruvic transaminase 2 | −3.752 | 0.002 | 1.571 | 0.029 |
| ENSSSCG00000036438 | *GPX3* | 16 | Glutathione peroxidase 3 | −2.184 | 0.000 | −2.762 | 0.006 |
| ENSSSCG00000017010 | *INSYN2B* | 16 | Inhibitory synaptic factor family member 2B | 3.436 | 0.000 | 1.667 | 0.020 |
| ENSSSCG00000002245 | *KATNBL1* | 7 | Katanin regulatory subunit B1 like 1 | 0.877 | 0.040 | 1.147 | 0.002 |
| ENSSSCG00000010325 | *KCNMA1* | 14 | Potassium calcium-activated channel subfamily M $\alpha$ 1 | 0.982 | 0.042 | 1.652 | 0.001 |
| ENSSSCG00000034838 | *MAP1LC3C* | 10 | Microtubule associated protein 1 light chain 3 $\gamma$ | 6.715 | 0.000 | 3.819 | 0.002 |
| ENSSSCG00000004191 | *MOXD1* | 1 | Monooxygenase DBH like 1 | −2.980 | 0.010 | 2.570 | 0.033 |
| ENSSSCG00000011133 | *PFKFB3* | 10 | 6-phosphofructo-2-kinase/fructose-2,6-biphosphatase 3 | −1.135 | 0.008 | −1.450 | 0.010 |
| ENSSSCG00000007528 | *PHACTR3* | 17 | Phosphatase and actin regulator 3 | 1.727 | 0.012 | 1.986 | 0.028 |
| ENSSSCG00000007470 | *RIPOR3* | 17 | RIPOR family member 3 | −1.119 | 0.049 | −1.377 | 0.042 |
| ENSSSCG00000009111 | *SYNPO2* | 8 | Synaptopodin 2 | −1.293 | 0.006 | −1.194 | 0.045 |
| ENSSSCG00000014834 | *UCP3* | 9 | Uncoupling protein 3 | −2.693 | 0.024 | −3.262 | 0.016 |
| ENSSSCG00000008501 | *VIT* | 3 | Vitrin | −2.317 | 0.004 | −3.709 | 0.029 |
| ENSSSCG00000015766 | *WDR17* | 15 | WD repeat domain 17 | −1.761 | 0.000 | −1.773 | 0.019 |
| ENSSSCG00000013714 | | 2 | | 2.087 | 0.048 | 1.813 | 0.019 |
| ENSSSC00000037142 | | AEMK02000452.1 | Cysteine-rich protein 1 | 1.222 | 0.000 | −4.322 | 0.010 |
| ENSSSCG00000031037 | | 14 | | 3.813 | 0.000 | −4.021 | 0.029 |
| ENSSSCG00000032582 | | 14 | | 3.873 | 0.000 | −3.822 | 0.012 |
| ENSSSCG00000036224 | | 3 | | 3.369 | 0.001 | −3.010 | 0.045 |
| ENSSSCG00000036983 | | AEMK02000452.1 | | 4.568 | 0.001 | −5.951 | 0.001 |
| ENSSSCG00000037009 | | AEMK02000452.1 | | 3.679 | 0.024 | −3.785 | 0.031 |
| ENSSSCG00000039111 | | AEMK02000452.1 | | 4.596 | 0.000 | 1.180 | 0.041 |

From the 35 DE genes found in both tissues (inguinal and umbilical ring), 34 were protein coding and one was an immunoglobulin C coding gene. Moreover, eight transcripts (22.86%) were uncharacterized proteins (Table 3), of which six were similar to the amino acid sequences of the pig immunoglobulin and other was similar to another predicted protein in pigs (Table 3). When the relative expression of the 35 common DE genes from each group that represents a type of hernia was compared based on the log2 fold-change (logFC), 26 of these genes had a similar expression profile in the two types of hernia (Figure 4A), and nine had opposite expression profiles considering both types of hernia (Figure 4B).

**Table 3.** Differentially expressed genes in the inguinal and umbilical ring annotated as uncharacterized protein.

| Gene ID | Description | e-Value | Query Cover (%) | Identity (%) | Accession (RefSeq) |
|---|---|---|---|---|---|
| ENSSSCG00000013714 | Mucin-16 [*Sus scrofa*] | $8 \times 10^{-157}$ | 100 | 93.1 | XP_020940777.1 |
| ENSSSCG00000036224 | Ig kappa chain V-C region (PLC18) [*Sus scrofa domesticus*] | $5 \times 10^{-89}$ | 82 | 85 | PT0219 |
| ENSSSCG00000031037 | Immunoglobulin lambda-like polypeptide 5 precursor [*Sus scrofa*] | $3 \times 10^{-71}$ | 99 | 99.09 | NP_001230248.1 |
| ENSSSCG00000032582 | Immunoglobulin lambda-like polypeptide 5 precursor [*Sus scrofa*] | $7 \times 10^{-68}$ | 94 | 92.04 | NP_001230248.1 |
| ENSSSCG00000036983 | IgG heavy chain precursor [*Sus scrofa*] | 0.0 | 74 | 79.78 | BAM75547.1 |
| ENSSSCG00000037009 | IgG heavy chain precursor [*Sus scrofa*] | 0.0 | 100 | 100 | BAM75542.1 |
| ENSSSCG00000039111 | IgG heavy chain constant region [*Sus scrofa*] | $1 \times 10^{-74}$ | 100 | 100 | BAM66306.1 |
| ENSSSCG00000037142 | Cysteine-rich protein 1 [*Camelus dromedarius*] | $3 \times 10^{-41}$ | 36 | 94.37 | KAB1277051.1 |

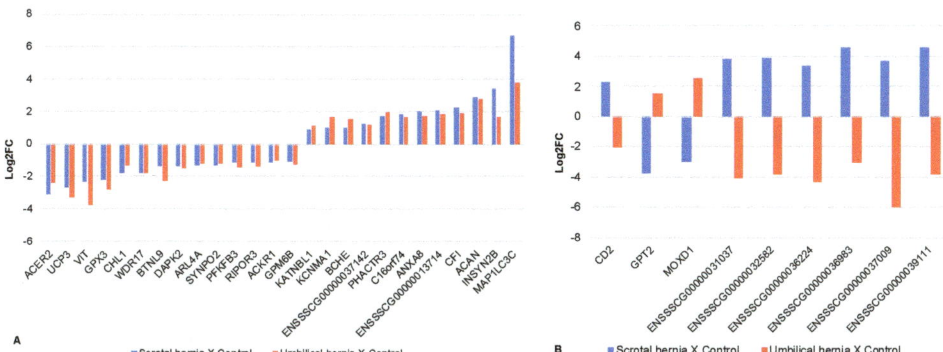


**Figure 4.** Common differentially expressed genes for scrotal and umbilical hernias and their respective control groups. Legend: (**A**) Genes with similar expression profile and (**B**) with opposite expression profile in the two types of hernia based on the Log2 Fold Change (log2FC).

From the 35 genes DE in both types of hernia, a network with 27 of them was built and the *MAP1LC3C* and *MUC16* genes grouped the two largest clusters of the network (Figure 5). One cluster was related to macroautophagy including the *MAP1LC3C*, *ATG3*, *ATG5* and *ATG12* genes (Figure 5) and the other cluster was composed by the mucin gene family (*MUC4*, *MUC6*, *MUC16* and *MUC20*) (Figure 5), which plays an important role protecting against environmental stress. A third group was related to the complement and coagulation cascade composed of genes *C3*, *CFH* and *CFI* (Figure 5).

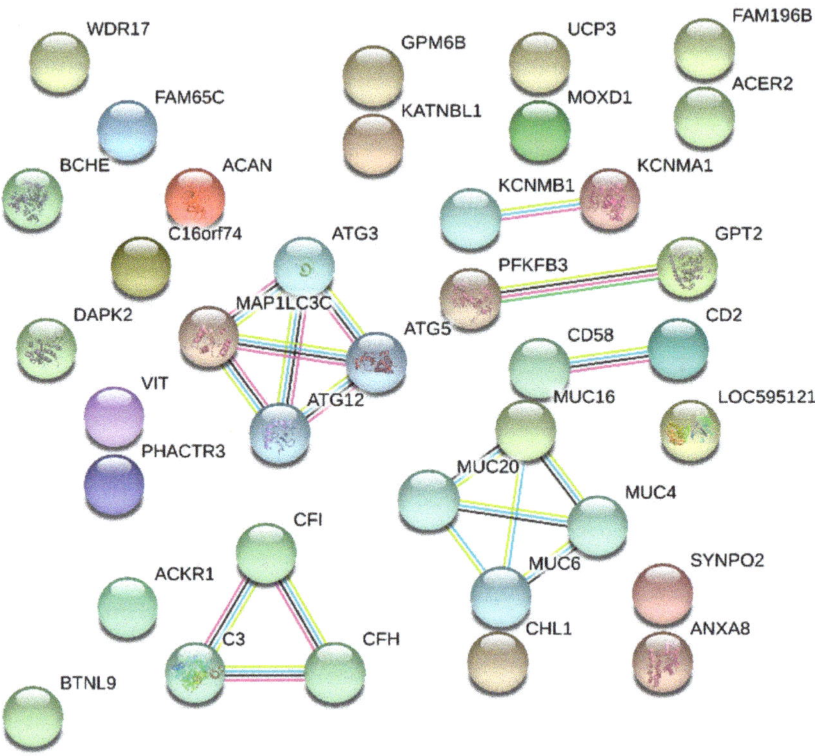


**Figure 5.** Gene interaction network with differentially expressed genes common to both scrotal and umbilical hernias. Legend: gene network built with 27 of the 35 differentially expressed genes common to both types of hernia obtained with the STRING database using information from *Sus scrofa* proteins.

Four metabolic pathways were enriched with the 35 genes DE in both types of hernia using the PANTHER database [47]: Huntington's disease (P00029) (*ARL4A*); muscarinic receptor signaling pathway 1 and 3 of acetylcholine (P00042) (*BCHE*); acetylcholine muscarinic receptor 2 and 4 signaling pathway (P00043) (*BCHE*) and acetylcholine receptor nicotinic signaling pathway (P00044) (*BCHE*). The enrichment of this set of 35 DE genes using the DAVID 6.8 database [41] indicated that those genes participate in 108 biological processes (BP) (Additional file 1: Table S2). The *KCNMA1* gene (potassium calcium-activated channel subfamily M $\alpha$ 1) was the most enriched in BP, appearing in 18 of them (Supplementary Materials 1: Table S2). These BP were clustered in nine macro biological processes (superclusters) using the REVIGO tool [48] (Table 4).

The 26 DE genes with similar expression profile enriched 99 BP (Supplementary Materials 1: Table S3). Considering the molecular function, the set of 35 genes was present in 57 different molecular functions mainly comprising binding, catalytic activity, molecular function regulator, structural molecule activity and transport activity. Using the Pig QTL database [30], two DE genes in both groups of hernias studied here were located in QTL regions already identified as being associated to SH hernia in pigs: the *ACAN* and *BCHE* genes were mapped, respectively, in the QTLs 55892 (SSC7) and 8794 (SSC13).

**Table 4.** Macro biological processes (superclusters) enriched with the 35 differentially expressed genes common to both types of hernia.

| Superclusters | Genes | | |
|---|---|---|---|
| | Upregulated in Both Groups | Downregulated in Both Groups | Opposite Expression Profile |
| Cell adhesion (GO:0022610) | | *CHL1* | *CD2* |
| Biological regulation (GO:0065007) | *ANXA8* | *ACKR1, SYNPO2* | *MOXD1*, ENSSSCG00000036224, ENSSSCG00000039111 |
| Cellular process (GO:0009987) | *PHACTR3, ANXA8, MAP1LC3C* | *VIT, SYNPO2, BTNL9* | *GPT2, MOXD1, CD2*, ENSSSCG00000039111, ENSSSCG00000036224 |
| Development (GO:0032502) | *PHACTR3* | *VIT* | |
| Immune system process (GO:0002376) | | *ACER2, BTNL9* | *CD2*, ENSSSCG00000032582, ENSSSCG00000039111, ENSSSCG00000031037, ENSSSCG00000036224 |
| Location (GO:0051179) | *ANXA8, MAP1LC3C* | *ARL4A, UCP3* | ENSSSCG00000036224, ENSSSCG00000039111 |
| Metabolic process (GO:0008152) | | *UCP3, PFKFB3* | *GPT2, MOXD1*, ENSSSCG00000036224, ENSSSCG00000039111 |
| Multicellular organismal process (GO:0032501) | *ACAN, PHACTR3* | *ACKR1, CHL1, GPM6B, VIT* | |
| Response to stimulus (GO:0050896) | | *ACKR1, GPX3, UCP3* | ENSSSCG00000037009, ENSSSCG00000036983, ENSSSCG00000036224, ENSSSCG00000039111 |

### 3.6. Identification of Polymorphisms

Using the GATK program with all sequences obtained from the RNA-Seq analyses, 67 polymorphisms were identified in the inguinal ring tissue between SH-affected and unaffected samples (Supplementary Materials 1: Table S4) and 76 in the umbilical ring tissue between UH-affected and unaffected samples (Supplementary Materials 1: Table S5). Comparing the transcriptomes of pigs affected with each type of hernia, the polymorphisms were then classified (Table 5). From the 67 polymorphisms related to scrotal hernia, 56 (83.58%) have already been described in VEP tool and 11 (16.42%) are considered new (Supplementary Materials 1: Table S4). Of the 76 polymorphisms referring to umbilical hernia, 62 (81.58%) have been previously described in VEP tool and 14 (18.42%) are new (Supplementary Materials 1: Table S5).

Considering the whole transcriptome of the two tissues, the variants detected for SH and UH were classified according to the functional region indicating their possible consequences in gene regulation (Table 6). Most of the SNPs in the SH group (37.74%) were classified as synonymous variants (Additional file 1: Table S4), and in the UH group, most were of the UTR3′ type (44.78%) (Supplementary Materials 1: Table S5). In the SH group, two observed variants had calculated SIFT (sorting intolerant from tolerant) score classified as tolerant (SIFT score > 0.05) (Table 7). One of them has already been described in the dbSNP database [35] and the other was classified as new. From the variants belonging to the UH group, six had the calculated tolerance prediction score (SIFT) detected, one of them being deleterious (SIFT $\leq$ 0.05) and five tolerant (SIFT > 0.05) (Table 7), all of which were already present in the dbSNP database [35]. These six variants belong to six genes, two of which were enriched for metabolic pathways in the KEGG Pathway database [54] (Table 8). The frameshift type variants were located in two genes (*NCOA7* and *SEC62*), of which one was enriched for a metabolic pathway in the same database [54] (Table 8). The SNPs of the SH group were observed in 17 different genes, which enriched nine BP (Table 9) in the DAVID 6.8 database [41]. The SNPs found in the UH group were mapped in 24 genes, which enriched six biological processes (Table 10).

**Table 5.** Classification of polymorphisms found in samples from the inguinal and umbilical ring tissues from normal, and scrotal and umbilical hernia-affected pigs, respectively.

| Polymorphism Type | Scrotal Hernia | | Umbilical Hernia | |
|---|---|---|---|---|
| | N° | (%) | N° | (%) |
| Insertion | 10 | 14.93 | 6 | 7.90 |
| Deletion | 4 | 5.97 | 3 | 5.26 |
| SNP | 53 | 79.10 | 67 | 86.84 |
| Total | 67 | 100 | 76 | 100 |

**Table 6.** Variants annotated in different functional classes in samples from inguinal and umbilical ring tissues.

| Variant Type | Scrotal Hernia (%) | Umbilical Hernia (%) |
|---|---|---|
| Intronic | 23.88 | 5.26 |
| Synonym | 29.85 | 23.68 |
| Missense | 2.99 | 7.89 |
| Splicing | 1.49 | - |
| UTR5′ | 5.97 | 6.58 |
| UTR3′ | 34.33 | 47.37 |
| Downstream | 1.49 | 6.58 |
| Frameshift | - | 2.63 |

**Table 7.** Missense variants observed in groups with sorting intolerant from tolerant (SIFT) score calculated in the dbSNP database (Ensembl).

| Group | Variant | Location | Impact | Gene | SIFT |
|---|---|---|---|---|---|
| Scrotal hernia | rs325370594 | 16:20418972-20418972 | Moderate | *RAI14* | Tolerant (1) |
| | - | 7:64303141-64303141 | Moderate | *RALGAPA1* | Tolerant (0.63) |
| Umbilical hernia | rs325089032 | 6:81571496-81571496 | Moderate | *ELOA* | Tolerant (0.1) |
| | rs327289001 | 3:17254444-17254444 | Moderate | *ITGAM* | Deleterious (0.01) |
| | rs789266896 | 3:17628688-17628688 | Moderate | *RNF40* | Tolerant (0.6) |
| | rs330957838 | 3:17468302-17468302 | Moderate | *SETD1A* | Tolerant low confidence (0.34) |
| | rs337670844 | 3:17399477-17399477 | Moderate | *ZNF646* | Tolerant (0.08) |
| | rs323115420 | 3:16964045-16964045 | Moderate | *ZNF713* | Tolerant (0.65) |

**Table 8.** Genes with elevated impact variants enriched in metabolic pathways with the KEGG Pathway Database.

| Variant | Gene | Pathway (ssc [1]) |
|---|---|---|
| New (Frameshift) | *SEC62* | Protein exports (ssc03060); Protein processing in the endoplasmic reticulum (ssc04141). |
| rs327289001 (Missense) | *ITGAM* | Rap1 signaling path (ssc04015); Phagosome (ssc04145); Cell adhesion molecules (CAMs) (ssc04514); Hematopoietic cell line (ssc04640); Transendothelial migration of leukocytes (ssc04670); Regulation of the actin cytoskeleton (ssc04810); Whooping cough (ssc05133); Legionellosis (ssc05134); Leishmaniasis (ssc05140); Amebiasis (ssc05146); Infection by Staphylococcus aureus (ssc05150); Tuberculosis (ssc05152); Incorrect regulation of transcription in cancer (ssc05202) |
| rs330957838 (Missense) | *SETD1A* | Lysine degradation (ssc00310) |

[1] Metabolic pathway identifying code described for Sus scrofa by the KEGG Pathway.

**Table 9.** Biological processes enriched with genes harboring SNPs in the scrotal hernia group.

| David Term | Biological Process | Enriched Genes |
|---|---|---|
| GO:0010604 | Positive regulation of the metabolic process of macromolecules | *TBX3*, ***MYRF*, *MYLIP*, *PARP3*** |
| GO:0055088 | Lipid homeostasis | *ACACA*, ***MYYLIP*** |
| GO:0009893 | Positive regulation of the metabolic process | *TBX3*, ***MYRF*, *MYLIP*, *PARP3*** |
| GO:0043170 | Metabolic process of the macromolecule | *TBX3*, ***MYRF*, *MYLIP*, *PARP3***, *ACACA*, ***DDB2*** |
| GO:0065008 | Regulation of biological quality | *TBX3*, *ACACA*, ***MYLIP*, *PARP3*** |
| GO:0045935 | Positive regulation of the compound metabolic process containing nucleobase | *TBX3*, ***MYRF*, *PARP3*** |
| GO:0051173 | Positive regulation of the metabolic process of nitrogen compounds | *TBX3*, ***MYRF*, *PARP3*** |
| GO:0042592 | Homeostatic process | *ACACA*, ***MYLIP*, *PARP3*** |
| GO:0033554 | Cellular stress response | *TBX3*, ***DDB2*, *PARP3*** |
| GO:0060249 | Anatomical structure homeostasis | *ACACA*, ***PARP3*** |

The genes in bold were upregulated in the scrotal hernia-affected group.

In the SH group, the genes corresponding to exonic regions, in which the variants were observed, were not enriched by the KEGG Pathway database [54]. Considering the type of impact caused by the variants, the results were distributed as shown in Figure 6A,B for the scrotal and umbilical hernias, respectively. These figures show that more than 60% of the variants represent variations of modifying impact for both types of hernia.

Among the transcripts present in the analyzed samples, five miRNAs were identified in the SH transcriptome and four in the UH transcriptome (Table 11). From these, three were expressed in both types of hernia. No DE miRNAs belonging to the evaluated samples were identified.

**Table 10.** Biological processes enriched with genes harboring single nucleotide polymorphisms (SNPs) in the umbilical hernia group.

| GO Term | Biological Process | Enriched Genes |
|---|---|---|
| GO:0010468 | Regulation of gene expression | *RPRD1A, RNF40, ELOA,* ***VIM****,* *ZNF629,* ***ZNF713*** |
| GO:0010977 | Negative regulation of the development of neuron projection | ***EPHB2, VIM*** |
| GO:0060255 | Regulation of the metabolic process of macromolecules | ***ITGB2****, RPRD1A, RNF40, ELOA,* ***VIM****, ZNF629,* ***ZNF713*** |
| GO:0031345 | Negative regulation of cellular projection organization | ***EPHB2*** *e* ***VIM*** |
| GO:0019222 | Regulation of the metabolic process | ***ITGB2****, RPRD1A, RNF40, ELOA,* ***VIM****, ZNF629,* ***ZNF713*** |
| GO:0045665 | Negative regulation of neuron differentiation | ***EPHB2, VIM*** |

The genes in bold were upregulated in the umbilical hernia-affected group.

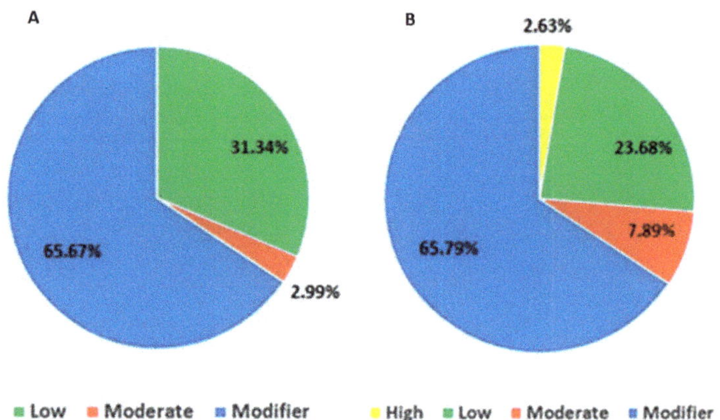


**Figure 6.** Impact caused by variants and its frequency. Legend: (**A**) samples from the scrotal hernia group. (**B**) Samples from the umbilical hernia group.

**Table 11.** miRNAs identified in the transcriptomes of the pig inguinal and umbilical ring tissues.

| ENSEMBL ID | Name/Symbol | miRBase | Location | Group |
|---|---|---|---|---|
| ENSSSCG00000018513 | ssc-mir-145 (MIR145) | MI0002417 | 2: 150.580.126-150.580.211 | SH |
| ENSSSCG00000018758 | ssc-mir-214 (MIR214) | MI0002441 | 9: 114.527.990-114.528.101 | SH/UH |
| ENSSSCG00000019065 | ssc-mir-186 (MIR186) | MI0002456 | 6: 141.943.328-141.943.409 | SH |
| ENSSSCG00000034634 | ssc-mir-6782 | MI0031620 | AEMK02000489.1: 40.305-40.379 | UH |
| ENSSSCG00000035742 | - | - | 12: 14.578.144-14.578.210 | SH/UH |
| ENSSSCG00000037094 | ssc-mir-9810 | MI0031577 | 4: 83.070.363-83.070.457 | SH/UH |

SH stands for scrotal hernia and UH for umbilical hernia.

## 4. Discussion

Some studies investigating genes involved with the occurrence of hernias have been performed using candidate genes and GWAS approaches [11,25,29,31,32,57–60]. More recently, functional candidate genes were prospected for scrotal [6] and umbilical [7] hernias by our group using RNA-Seq approach. Nevertheless, since the molecular mechanisms involved with these anomalies are not yet completely understood, a comparison between the transcriptome of umbilical and scrotal hernias was performed here, allowing the identification of several common genes differentially expressed in both conditions. Moreover, several SNPs involved with these conditions were also identified and characterized.

### *4.1. Transcriptome Characterization*

Gene expression studies obtained from samples of the inguinal and the umbilical ring tissue are quite recent. Lorenzetti et al. [5] and Romano et al. [6] performed gene expression analyses from the pig inguinal ring and Souza et al. [7] performed analyses with umbilical ring samples to investigate scrotal and umbilical hernias, respectively. Information about gene expression in those tissues is scarce and there are still many gaps to be elucidated in this field. Those authors found DE genes and pathways in each type of hernia. Since some genes and biological processes seem to be shared between SH and UH, new analyses were performed with focus on tissues characterization and on common processes involved with both types of hernia.

From the transcripts characterization of the two tissues (Table 1), a great similarity between the groups of both types of hernia was observed when comparing the number of each class of transcripts, implying that the appearance of both hernias may be related to the same set of genes or family of genes. This large number of transcripts that are expressed in both groups can also be seen in the Venn diagram (Figure 3). With the exception of processed pseudogenes and snRNA, which were not identified in the UH group, the percentage of each type of transcript was similar. Thus, the expression profile of the genes in the inguinal ring tissue was very similar to the profile found in the umbilical ring, being compatible with the histopathological composition of these two tissues (Figure 2).

### *4.2. Common Differentially Expressed Genes in Scrotal and Umbilical Hernias*

From the DE genes observed in each type of hernia, 35 were common to both groups. Among these, nine genes (*CD2, GPT2, MOXD1*, ENSSSCG00000031037, ENSSSCG00000032582, ENSSSCG00000036224, ENSSSCG00000036983, ENSSSCG00000037009 and ENSSSCG00000039111) had different expression profiles when comparing both types of hernia. This behavior may have occurred due to the expression be tissue specific (inguinal ring and umbilical ring) for those genes. Other reasons could be the differences in sex and age between the groups evaluated for the two types of hernia. The other 26 DE genes have shown similar expression in both types of hernia, of which 14 genes were downregulated and 12 were upregulated in pigs affected by both types of hernia.

From the gene interaction network (Figure 5), three DE genes were enriched in both types of hernia. The *MAP1LC3C* (microtubule associated protein 1 light chain 3 $\gamma$) interacted in the group of the macroautophagy BP (GO: 0016236) [61]. Macroautophagy is the main path involved in inducing general renewal of cytoplasmic constituents in eukaryotic cells and is essential for cell survival, development, differentiation and homeostasis [62–66]. The Gene Ontology (GO) annotations related to this gene include the assembly and maturation of the autophagosome (Supplementary Materials 1: Table S4). Marcelino et al. [67] indicated the *MAP1LC3C* gene as a candidate for the formation of UH in pigs since this gene was upregulated in the affected compared to the normal pigs. In our research, the *MAP1LC3C* gene exhibited the same behavior, being upregulated in affected animals of both types of hernia when compared to the control groups (Figure 4A). Moreover, this expression profile can be one of the causes of the hernia onset, since the high expression of this gene can cause excessive autophagy and interfere with normal tissue development [68].

The *CFI* gene (Complement Factor I) was grouped in the cluster of the coagulation cascade metabolic pathway and complement system (Figure 5). The coagulation cascade is a sequence of interconnected reactions in order to clot the local blood when a blood vessel injury occurs [69]. The complement system is a proteolytic cascade in blood plasma and a mediator of innate immunity [70]. The GO annotations related to this gene include endocytosis (content absorption through membrane invagination process) and proteolysis (protein degradation process) (Supplementary Materials 1: Table S2). The CFI gene, like the *MAP1LC3C*, was upregulated in animals with hernia when compared to the control group (Figure 4A). The *CFI* encodes the trypsin-like protein serine-protease [42], which plays an essential role in regulating the immune response, controlling all the complement pathways [71]. The participation of the *CFI* in these pathways and processes, taken together with its expression profile, suggests that this gene could be involved with the consequence of these disorders.

The *MUC16* (Mucin-16) gene encodes a protein of the mucin family, which are O-glycosylated proteins found in the apical surfaces of the epithelium and play an important role in the formation of a protective mucous barrier [72]. This gene was enriched in the gene network (Figure 5) as a participant in processing O-glycan BP (GO: 16266) [61]. This process is related to the gradual addition of carbohydrate residues or carbohydrate derivatives to form the O-glycan structure [73]. The *MUC16* gene was also enriched as an integral cell membrane component BP (GO: 16021) [61]. According to Blalock et al. [74], the MUC16 build a protective barrier to the epithelial cell surface, where binding proteins are associated with its tail, linking it to the actin cytoskeleton. This gene was upregulated in the affected group of both types of hernia compared to their respective control groups (Figure 4A), thus configuring a defense system that might have arisen as a consequence of the hernias formation.

### 4.3. Enriched Biological Processes

When the 99 BP enriched by the 26 DE genes with an equivalent profile in both types of hernia (Supplementary Materials 1: Table S3) were evaluated, the BP of cell adhesion, apoptosis, organization of the actin cytoskeleton and organization of collagen fibrils can be highlighted, because they are generally linked to the formation of hernias [5–7]. The enriched genes for cell adhesion, *VIT* (Vitrin), *ACER2* (Alkaline Ceramidase 2) and *CHL1* (Cell Adhesion Molecule L1 Like) were downregulated in the affected animals compared to the control groups and *ACAN* (Aggrecan) was upregulated in the affected animals. The cell adhesion BP allows the interaction among cells, and between cells and the extracellular matrix [75]. This BP has already been related to tissue maintenance and cell differentiation [76,77]. The *CHL1* gene was enriched with the process of homophilic cell adhesion via plasma membrane adhesion molecules. *ACER2* participates in the specific BP of negative regulation of cell adhesion mediated by integrin and negative regulation of cell matrix adhesion. The *VIT* gene enriched the process of positive regulation of cell substrate adhesion. Thus, the reduced expression of these genes that actively participate in cell adhesion interferes with the integrity of tissues, which can be determinant for the appearance of both types of hernia.

The *KCNMA1* gene (Potassium Calcium—Activated Channel Subfamily M α 1), which was upregulated in animals affected with hernia, and *ACER2* (downregulated) were enriched in the apoptosis BP. This process is related to the regulation of programmed cell death, which is extremely important for the maintenance of the development of living beings [78]. The overexpression of the *KCNMA1* gene can compromise the tissue as a result of an accumulation of immature cells in the region, which can influence the appearance of hernias, especially when associated with unfavorable environmental factors. *ACER2* was enriched with the specific process of activating cysteine-type endopeptidase activity, involved in the apoptotic process, and the *KCNMA1* was enriched for positive regulation of the apoptotic process. This last gene was also enriched for the relaxation process of the vascular smooth muscle that is related to the negative regulation of the contraction of this muscle. The relaxation is mediated by a decrease in the phosphorylation state of the myosin light chain [79]. As the expression of this gene was higher in herniated than in normal pigs, the *KCNMA1* can be pointed out as a candidate gene for the formation of umbilical and scrotal hernia, since the lack of local muscle contraction facilitates the passage of the abdominal content through the rings.

Biological processes that regulate the activities of collagen and its structures have been indicated in the enrichment of the *ACAN* and *VIT* genes. The first gene was related to the condensation of mesenchymal cells that differentiate into chondrocytes, organization of collagen fibrils and the development of chondrocytes [61]. The *VIT* gene, on the other hand, was related to the morphogenesis of chondrocytes in the cartilage of the growth plate, in which the structures of a chondrocyte are generated and organized [80]. The *ACAN* gene was upregulated in animals affected with hernia, which is in accordance with the histopathological analyses that evidenced a larger amount of collagen compared to normal pigs. Moreover, *ACAN* upregulation in animals affected with hernia can generate an exaggerated collagen production, which has already been related to hernia previously [81].

Regarding the organization of the cytoskeleton, especially those processes related to actin, two genes were enriched, *SYNPO2* (Synaptopodin 2) and *ENSSSCG00000037142*. *SYNPO2* has been enriched specifically for the process of positive regulation of the actin filament bundles set. The organization of the actin cytoskeleton is carried out at the cellular level and results in the assembly, disposition of the constituent parts or disassembly of the structures, including filaments and their associated proteins [61]. In our study, this gene was downregulated in animals with hernia. This negative regulation can be a predisposing factor to hernia, since the non-assembly and organization of the structures that constitute the tissue can make it less resistant [18].

### 4.4. DE Genes Located in QTL Regions for Hernias and Polymorphisms Characterization

Several studies have been carried out to identify QTL regions related to umbilical and scrotal hernia [11,20,29,81,82]. Among the DE genes in the two types of hernia, *ACAN* and *BCHE* (Butyrylcholinesterase) are highlight since they have already been located in QTL regions associated to scrotal/inguinal hernia [20,29]. Even with scientific reports relating these two genes only with QTL regions for scrotal hernia, in our study, the expression profile of these two genes was equivalent in both types of hernia, being upregulated in the affected animals. Souza et al. [7] have recently indicated *ACAN* as a strong candidate gene for triggering umbilical hernias in pigs.

Our results have shown that variations in the transcripts may be related to the manifestation of the different types of hernia. In both groups, most of the polymorphisms detected were SNPs, followed by insertions and deletions (Supplementary Materials 1: Tables S4 and S5). In the SH group, a new SNP was identified on chromosome 13 (13: 34083960-34083960), which is located within a QTL region (QTL ID 55898) associated with scrotal hernia [11,20,29]. This SNP was mapped in the *PARP3* gene (member of the poly ADP-ribose 3 polymerase family), which acts in the repair pathways by base excision, apoptosis and necroptosis, participating in biological processes of DNA repair [4]. Moreover, Piórkowska et al. [83] carried out research with Polish Landrace and Pulawska pigs and pointed out the participation of the PARP3 gene in the regulation of the actin cytoskeleton BP. The muscle tissues belonging to the regions where the hernias occur are classified as skeletal striatum, which are formed by myofibrils composed by actin and myosin. As mentioned by Bendavid [18], disturbances in the structures of muscle fibers cause low resistance in the inguinal region, which can lead to scrotal hernia.

The 53 SNPs observed in the SH group were located in 17 genes (Additional file 1: Table S4), which have been enriched in nine biological processes (Table 9). Most of these BP were related to homeostasis, which are processes that maintain the stability of the structure of the analyzed tissue (GO:0042592) [61]. The *ACACA* gene, enriched in these BP, participates in processes that maintain the stability of anatomical structures of the site [43]. *ACACA* was downregulated in the SH-affected animals (Table 9) indicating the development of an unstable structure of the inguinal ring, which can influence the development of hernia. In humans, this gene participates in the fatty acid synthesis BP [43], which reinforces the histopathological findings that showed greater amount of adipose tissue in normal than in SH-affected pigs (Figure 2A,B). From the SNPs found in the SH group, all variants were tolerated (Table 7). According to the SIFT score, two had a moderate impact classification, so they can alter the effectiveness of the encoded protein. This means that the function of the proteins resulting from these sites has not been altered, since the SIFT score is a tool that predicts whether the variant affects the function of the protein or not [35]. These SNPs were located in two genes, *RAI14* and *RALGAPA1*; the first has already been annotated and the second has no identification in the VEP tool. No high impact polymorphisms were identified in the SH group.

The 67 SNPs found in the UH group were mapped in 24 genes (Supplementary Materials 1: Table S5). These genes were enriched in six biological processes (Table 10) [41], all of which were related to some type of regulation, mainly metabolic. The *EPHB2* and *VIM* genes were enriched in BP that interrupts the processes of cellular projections formation (GO: 0031345) [61]. These two genes were upregulated in animals affected with UH when compared to the control group (Table 10). The *VIM* gene encodes an intermediate filament protein that is part of the cytoskeleton [43]. Lazarides [84] reported that high amount of this filament is observed in the early stages of myogenesis in humans, and is hardly identified in adult muscles. Thus, the levels of this protein indicate functionality feature. The upregulation of *VIM* in the umbilical ring tissue of the UH-affected animals suggests that this gene may be involved with a consequence of UH since Miller et al. [12] reported that the appearance of hernia can be a consequence of a muscular defect.

Polymorphisms that had a high impact rating in the UH group (Supplementary Materials 1: Table S5) were identified in two genes (*NCOA7* and *SEC62*). These variants still do not have identification in the tool used, but they were classified as insertions of the Frameshift type. Therefore, they can cause an interruption in the translation reading frame, because the number of inserted nucleotides is not multiple of three [35]. The *NCOA7* (Nuclear receptor coactivator 7) is involved in the biological process of RNA polymerase II transcription and negative regulation of the cellular response to oxidative stress. The *SEC62* (Preprotein translocation factor) is related to the regulation of post-translational protein transport to the membrane BP and was mapped in a QTL region for stillborn pigs [85]. The detection of these polymorphisms is important because they can alter not only processes related to hernias, but all important processes for biological maintenance, possibly resulting in transcription failures or disruption in the transport of translated proteins by lack of regulation.

The SNPs classified as having moderate impact for the UH group were found in six genes (Supplementary Materials 1: Table S5), with the SNP rs327289001 being highlighted due to its deleterious SIFT score. This SNP is located in the *ITGAM* gene that participates in the biological process of ectodermal cell differentiation [35]. This process is related to the specialization of previously non-specialized cells, which acquire structure and functioning of ectodermic cells. This differentiation integrates the processes involved in the commitment of a cell to its specific purpose (GO: 0010668) [61]. In the embryonic gastrulation phase, the formation of germ layers (ectoderm, mesoderm and endoderm) occurs, which will give rise to specific tissues and organs [86]. The ectoderm is the external layer of a developing embryo and gives rise to epidermis, hair, nails, cutaneous and mammary glands, tooth enamel, inner ear, lens, and the anterior part of the pituitary gland, besides others related to the neural tube and neural crest [86]. A SNP with deleterious SIFT score indicates that the function of the protein can be altered due to the polymorphism, which in this case can result in non-differentiated cells, compromising the formation of resistant tissues, which, when associated with environmental factors such as obesity, can lead to hernia. SNPs located in QTL regions associated with UH were not found in the current study.

The SNP rs339972872 from the SH group and the SNPs rs324236192 and rs340781986 from the UH group were located in the same gene (*ACACA*) (Supplementary Materials 1: Tables S4 and S5). These are synonym SNPs and were classified as low impact. According to Stachowiak et al. [87], the *ACACA* gene is involved with performance traits in pigs.

From the expressed miRNAs, three were identified in the groups of both hernias. One of them, ssc-mir-214, plays an important role in the regulation of ovarian function and in the induction of granular ovarian cells to induce follicular development [88]. The ssc-mir-145, which was identified only in samples from the SH group, is involved in the development of adipose tissue [89].

We conducted the first comparative study of the pig inguinal and umbilical ring tissue transcriptomes. The results demonstrated similarities related to the expression profile of the whole transcriptome and DE genes in both types of hernia. The *ACAN* gene, which had already been associated to the appearance of scrotal hernia, showed similar behavior in the data obtained from the umbilical hernia group. Moreover, the *MAP1LC3C*, *VIT*, *ACER2*, *KCNMA1* and *SYNPO2* genes were highlighted as candidates for the formation of the two types of hernias evaluated in our studied for presenting equivalent expression in both hernias and for being involved in biological processes such as cell adhesion, cytoskeleton organization, collagen production, muscle relaxation and autophagy. Furthermore, the differential expression of some of those genes, such as *MAP1LC3C*, *VIT*, *ACER2* and *ACAN*, has already been confirmed using qPCR [6,7]. However, further studies are needed to identify the expression profile of these same genes in younger animals to improve our interpretation of the gene regulation mechanisms triggering the formation of hernias. The knowledge of the genetic factors that control the manifestation of both scrotal and umbilical hernia brings possibilities to the pig production chain to develop actions to reduce the appearance of these defects in their herds, aiming to reduce economical losses and favoring the animal welfare.

## 5. Conclusions

The expression profile of the inguinal and umbilical ring transcriptomes showed great similarity. Thirty-five differentially expressed genes between normal and affected samples were common to both types of hernia. The *MAP1LC3C*, *ACAN*, *VIT*, *ACER2*, *KCNMA1* and *SYNPO2* genes are indicated as strong candidates for the appearance of both defects. A total of 11 and 14 new SNPs were identified in the samples related to the scrotal hernia and umbilical hernia, respectively. Moreover, a SNP with predicted deleterious function was identified in the *ITGAM* gene, which might be related to the appearance of umbilical hernia in pigs. Finally, the expression profile of these genes possibly interferes with the normal development of the tissues, causing weakness and decreasing the resistance of the site, which can lead to the formation of both types of hernia in pigs.

**Supplementary Materials:** The following are available online at https://www.mdpi.com/2073-4425/12/2/166/s1, Table S1: average of reads sequenced, removed in the quality control analysis and mapped in each group of samples, Table S2: biological processes of the 35 differentially expressed genes between normal and affected pigs common to both scrotal and umbilical hernias using DAVID database. Table S3: enrichment for biological process of the 26 DE genes with equivalent expression profile between both types of hernias using DAVID database. Table S4: polymorphisms identified in samples of the pig inguinal ring. Table S5: polymorphisms identified in samples of the pig umbilical ring. Figure S1: diagram summarizing the experiment and analyses performed. Figure S2: principal component analysis (PCA) plot S2A showing the separation of control (c) and affected (a) samples used to generate the transcriptome of the inguinal ring for scrotal hernia (SH), S2B showing the separation of control (c) and affected (a) samples used to generate the transcriptome of the umbilical ring for umbilical hernia (UH), and S2C with all samples together showing the separation of samples from both SH and UH transcriptomes.

**Author Contributions:** Conceptualization, A.F.G.R., A.M.G.I., J.d.O.P., M.E.C. and M.C.L.; methodology, A.M.G.I., J.d.O.P. and M.C.L.; software, M.E.C. and A.M.G.I.; validation, M.E.C., A.M.G.I., J.d.O.P. and M.C.L.; formal analysis, A.F.G.R., A.M.G.I., J.d.O.P., M.E.C., L.O.D.C., I.R.S., M.R.S., M.A.Z.M., H.C.d.O. and M.C.L.; investigation, A.F.G.R., A.M.G.I., J.d.O.P., M.E.C. and M.C.L.; resources, M.C.L.; data curation, A.F.G.R., A.M.G.I., J.d.O.P., M.E.C. and M.C.L.; writing—original draft preparation, A.F.G.R., A.M.G.I., J.d.O.P., L.O.D.C. and M.A.Z.M.; writing—review and editing, A.F.G.R., A.M.G.I., J.d.O.P., I.R.S., M.R.S., M.E.C., M.A.Z.M., H.C.d.O., L.O.D.C. and M.C.L.; visualization, A.F.G.R., A.M.G.I., J.d.O.P. and M.C.L.; supervision, M.C.L.; project administration, M.C.L.; A.M.G.I., funding acquisition, M.C.L. All authors have read and agreed to the published version of the manuscript.

**Funding:** This research was funded by National Council of Scientific and Technological Development (CNPq), from the Brazilian Government, grant number #476146/2013.

**Institutional Review Board Statement:** This study was performed with the approval of the Embrapa Swine and Poultry National Research Center Ethical Committee of Animal Use (CEUA) under the protocol number 011/2014 in 07/11/2014.

**Informed Consent Statement:** Not applicable.

**Data Availability Statement:** The datasets used or analyzed during the current study are available from the corresponding author on reasonable request. The transcriptome sequences for the scrotal and umbilical hernias are available in the SRA database with BioProject numbers PRJNA350530 and PRJNA445856, respectively. The SNP information is available in the EVA database with Project number PRJEB42670, Analyses number ERZ1737910.

**Acknowledgments:** The authors are grateful to Alexandre L. Tessmann for technical assistance. I.R. Savoldi was sponsored by a PROMOP/Udesc scholarship and M.R. Souza by a CAPES/FAPESC scholarship. MCL is recipient of a productivity fellowship from the National Council for Scientific and Technological Development (CNPq).

**Conflicts of Interest:** The authors declare no conflict of interest.

## Abbreviations

| | |
|---|---|
| BP | Biological process |
| DE | Differentially expressed |
| GATK | Genome analysis Tool Kit |
| GO | Gene ontology |
| HE | Hematoxylin and eosin |
| IH | Inguinal hernia |
| MDS | Multidimensional scaling |
| QTL | Quantitative trait loci |
| SH | Scrotal hernia |
| SIFT | Sorting intolerant from tolerant |
| SNP | Single nucleotide polymorphism |
| UH | Umbilical hernia |
| VEP | Variant effect predictor |

## References

1. Heck, A. Fatores que influenciam o desenvolvimento dos leitões na recria e terminação. *Acta Sci. Vet.* **2009**, *37* (Suppl. 1), 211–218.
2. Duarte, D.A.S.; Fortes, M.R.S.; Duarte, M.D.S.; Guimarães, S.E.F.; Verardo, L.L.; Veroneze, R.; Ribeiro, A.M.F.; Lopes, P.S.; De Resende, M.D.V.; E Silva, F.F. Genome-wide association studies, meta-analyses and derived gene network for meat quality and carcass traits in pigs. *Anim. Prod. Sci.* **2018**, *58*, 1100. [CrossRef]
3. González-Prendes, R.; Quintanilla, R.; Cánovas, A.; Manunza, A.; Cardoso, T.F.; Jordana, J.; Noguera, J.L.; Pena, R.N.; Amills, M. Joint QTL mapping and gene expression analysis identify positional candidate genes influencing pork quality traits. *Sci. Rep.* **2017**, *7*, 39830. [CrossRef] [PubMed]
4. Ding, R.; Quan, J.; Yang, M.; Wang, X.; Zheng, E.; Yang, H.; Fu, D.; Yang, Y.; Yang, L.; Li, Z.; et al. Genome-wide association analysis reveals genetic loci and candidate genes for feeding behavior and eating efficiency in Duroc boars. *PLoS ONE* **2017**, *12*, e0183244. [CrossRef]
5. Lorenzetti, W.R. Análise da Expressão Gênica em Suínos Normais e Afetados com Hérnia Escrotal. Available online: http://sistemabu.udesc.br/pergamumweb/vinculos/000047/00004770.pdf(2018) (accessed on 29 January 2020).

6. Romano, G.D.S.; Ibelli, A.M.G.; Lorenzetti, W.R.; Weber, T.; Peixoto, J.D.O.; Cantão, M.E.; Mores, M.A.Z.; Morés, N.; Pedrosa, V.B.; Cesar, A.S.M.; et al. Inguinal Ring RNA Sequencing Reveals Downregulation of Muscular Genes Related to Scrotal Hernia in Pigs. *Genes* **2020**, *11*, 117. [CrossRef] [PubMed]
7. Souza, M.R.; Ibelli, A.M.G.; Savoldi, I.R.; Cantão, M.E.; Peixoto, J.D.O.; Mores, M.A.Z.; Lopes, J.S.; Coutinho, L.L.; Ledur, M.C. Transcriptome analysis identifies genes involved with the development of umbilical hernias in pigs. *PLoS ONE* **2020**, *15*, e0232542. [CrossRef] [PubMed]
8. Nietfeld, F.; Höltig, D.; Willems, H.; Valentin-Weigand, P.; Wurmser, C.; Waldmann, K.-H.; Fries, R.; Reiner, G. Candidate genes and gene markers for the resistance to porcine pleuropneumonia. *Mamm. Genome* **2020**, *31*, 54–67. [CrossRef]
9. Yang, T.; Zhang, F.; Zhai, L.; He, W.; Tan, Z.; Sun, Y.; Wang, Y.; Liu, L.; Ning, C.; Zhou, W.; et al. Transcriptome of Porcine PBMCs over Two Generations Reveals Key Genes and Pathways Associated with Variable Antibody Responses post PRRSV Vaccination. *Sci. Rep.* **2018**, *8*, 1–12. [CrossRef] [PubMed]
10. Mattsson, P. Prevalence of Congenital Defects in Swedish Hampshire, Landrace and Yorkshire Pig Breeds and Opinions on Their Prevalence in Swedish Commercial Herds. 2011. Available online: https://stud.epsilon.slu.se/2390/1/mattsson_p_110330.pdf (accessed on 29 January 2020).
11. Grindflek, E.; Moe, M.; Taubert, H.; Simianer, H.; Lien, S.; Moen, T. Genome-wide linkage analysis of inguinal hernia in pigs using affected sib pairs. *BMC Genet.* **2006**, *7*, 25. [CrossRef]
12. Miller, P.A.; Mezwa, D.G.; Feczko, P.J.; Jafri, Z.H.; Madrazo, B.L. Imaging of abdominal hernias. *RadioGraphics* **1995**. [CrossRef]
13. Straw, B.; Bates, R.; May, G. Anatomical abnormalities in a group of finishing pigs: Prevalence and pig performance. *J. Swine Health Prod.* **2009**, *17*, 28–31.
14. Clarnette, T.D.; Lam, S.K.; Hutson, J.M. Ventriculo-peritoneal shunts in children reveal the natural history of closure of the processus vaginalis. *J. Pediatr. Surg.* **1998**, *33*, 413–416. [CrossRef]
15. Amato, G.; Agrusa, A.; Romano, G.; Salamone, G.; Gulotta, G.; Silvestri, F.; Bussani, R. Muscle degeneration in inguinal hernia specimens. *Hernia* **2011**, *16*, 327–331. [CrossRef]
16. Clarnette, T.D.; Hutson, J.M. Is the ascending testis actually? stationary?? Normal elongation of the spermatic cord is prevented by a fibrous remnant of the processus vaginalis. *Pediatr. Surg. Int.* **1997**, *12*, 155–157. [CrossRef] [PubMed]
17. Beuermann, C.; Beck, J.; Schmelz, U.; Dunkelberg, H.; Schütz, E.; Brenig, B.; Knorr, C. Tissue Calcium Content in Piglets with Inguinal or Scrotal Hernias or Cryptorchidism. *J. Comp. Pathol.* **2009**, *140*, 182–186. [CrossRef] [PubMed]
18. Bendavid, R. The Unified Theory of hernia formation. *Hernia* **2004**, *8*, 171–176. [CrossRef]
19. Franz, M.G. The Biology of Hernia Formation. *Surg. Clin. N. Am.* **2008**, *88*, 1–15. [CrossRef]
20. Sevillano, C.A.; Lopes, M.S.; Harlizius, B.; Hanenberg, E.H.A.T.; Knol, E.F.; Bastiaansen, J.W.M. Genome-wide association study using deregressed breeding values for cryptorchidism and scrotal/inguinal hernia in two pig lines. *Genet. Sel. Evol.* **2015**, *47*, 18. [CrossRef]
21. Petersen, H.H.; Nielsen, E.O.; Hassing, A.-G.; Ersboll, A.K.; Nielsen, J.P. Prevalence of clinical signs of disease in Danish finisher pigs. *Veter Rec.* **2008**, *162*, 377–382. [CrossRef]
22. Warren, T.R.; Atkeson, F.W. Inheritance of Hernia. *J. Hered.* **1931**, *22*, 345–352. [CrossRef]
23. Searcy-Bernal, R.; Gardner, I.A.; Hird, D.W. Effects of and factors associated with umbilical hernias in a swine herd. *J. Am. Vet. Med. Assoc.* **1994**, *204*, 1660–1664. [PubMed]
24. Young, G.B.; Angus, K. A note on the genetics of umbilical hernia. *Veter Rec.* **1972**, *90*, 245–247. [CrossRef] [PubMed]
25. Grindflek, E.; Hansen, M.H.S.; Lien, S.; Van Son, M. Genome-wide association study reveals a QTL and strong candidate genes for umbilical hernia in pigs on SSC14. *BMC Genom.* **2018**, *19*, 1–9. [CrossRef] [PubMed]
26. Sobestiansky, J.; Carvalho, L.F.O.S. Malformações. In *Doenças dos Suínos*; Sobestiansky, J., Barcellos, D.J., Eds.; Cânone Editorial: Goiânia, Brazil, 2007; pp. 527–538.
27. Sutradhar, B.C.; Hossain, M.F.; Das, B.C.; Kim, G.; Hossain, M.A. Comparison between open and closed methods of herniorrhaphy in calves affected with umbilical hernia. *J. Vet. Sci.* **2009**, *10*, 343–347. [CrossRef]
28. Thaller, G.; Dempfle, L.; Hoeschele, I. Maximum likelihood analysis of rare binary traits under different modes of inheritance. *Genetics* **1996**, *143*, 1819–1829. [CrossRef] [PubMed]
29. Ding, N.S.; Mao, H.R.; Guo, Y.M.; Ren, J.; Xiao, S.J.; Wu, G.Z.; Shen, H.Q.; Wu, L.H.; Ruan, G.F.; Brenig, B.; et al. A genome-wide scan reveals candidate susceptibility loci for pig hernias in an intercross between White Duroc and Erhualian. *J. Anim. Sci.* **2009**, *87*, 2469–2474. [CrossRef] [PubMed]
30. Pig QTL Database. Available online: https://www.animalgenome.org/cgi-bin/QTLdb/SS/index (accessed on 4 February 2020).
31. Li, X.; Xu, P.; Zhang, C.; Sun, C.; Han, X.; Li, M.; Qiao, R. Genome-wide association study identifies variants in the CAPN9 gene associated with umbilical hernia in pigs. *Anim. Genet.* **2019**, *50*, 162–165. [CrossRef] [PubMed]
32. Fernandes, L.T.; Ono, R.K.; Ibelli, A.M.G.; Lagos, E.B.; Morés, M.A.Z.; Cantão, M.E.; Lorenzetti, W.R.; Peixoto, J.D.O.; Pedrosa, V.B.; Ledur, M.C. Novel putative candidate genes associated with umbilical hernia in pigs. In *11th World Congress on Genetics Applied to Livestock Production*; WCGALP, Massey University: Auckland, New Zealand, 2018; Available online: https://www.alice.cnptia.embrapa.br/alice/bitstream/doc/1093239/1/final8696.pdf (accessed on 29 January 2020).
33. Github Repository. Available online: https://github.com/hanielcedraz/BAQCOM (accessed on 6 November 2019).
34. Bolger, A.M.; Lohse, M.; Usadel, B. Trimmomatic: A flexible trimmer for Illumina sequence data. *Bioinformatics* **2014**, *30*, 2114–2120. [CrossRef]

35. Ensembl. Available online: http://www.ensembl.org/index.html (accessed on 4 February 2020).
36. Dobin, A.; Davis, C.A.; Schlesinger, F.; Drenkow, J.; Zaleski, C.; Jha, S.; Batut, P.; Chaisson, M.; Gingeras, T.R. STAR: Ultrafast universal RNA-Seq aligner. *Bioinformatics* **2013**, *29*, 15–21. [CrossRef]
37. Team, RStudio. *RStudio: Integrated Development for R*; RStudio, Inc.: Boston, MA, USA; Available online: http://www.rstudio.com/ (accessed on 16 March 2020).
38. R Core Team. R: A Language and Environment for Statistical Computing. R Foundation for Statistical Computing, Vienna, Austria. Available online: https://www.R-project.org/ (accessed on 25 April 2020).
39. Robinson, M.D.; Mccarthy, D.J.; Smyth, G.K. edgeR: A Bioconductor package for differential expression analysis of digital gene expression data. *Bioinformatics* **2010**, *26*, 139–140. [CrossRef]
40. Benjamini, Y.; Hochberg, Y. Controlling the false discovery rate: A practical and powerful approach to multiple testing. *J. Royal Stat. Soc. B* **1995**, *57*, 289–300. [CrossRef]
41. DAVID Bioinformatics Resources 6.8. Available online: https://david.ncifcrf.gov/tools.jsp (accessed on 4 February 2020).
42. UniProt. Available online: https://www.uniprot.org/ (accessed on 4 February 2020).
43. NCBI. Available online: https://www.ncbi.nlm.nih.gov/ (accessed on 4 February 2020).
44. NCBI (BLAST Glossary). Available online: https://www.ncbi.nlm.nih.gov/books/NBK62051/ (accessed on 23 March 2020).
45. String. Available online: https://string-db.org/cgi/input.pl?sessionId=3nHd3fJD9mAv&input_page_active_form=multiple_identifiers (accessed on 4 February 2020).
46. Szklarczyk, D.; Gable, A.L.; Lyon, D.; Junge, A.; Wyder, S.; Huerta-Cepas, J.; Simonovic, M.; Doncheva, N.T.; Morris, J.H.; Bork, P.; et al. STRING v11: Protein-protein association networks with increased coverage, supporting functional discovery in genome-wide experimental datasets. *Nucleic Acids Res.* **2018**, *47*, D607–D613. [CrossRef] [PubMed]
47. Panther. Available online: https://www.pantherdb.org/ (accessed on 4 September 2019).
48. REVIGO. Available online: http://revigo.irb.hr/ (accessed on 4 September 2019).
49. McKenna, A.; Hanna, M.; Banks, E.; Sivachenko, A.; Cibulskis, K.; Kernytsky, A.; Garimella, K.; Altshuler, D.; Gabriel, S.B.; Daly, M.J.; et al. The genome analysis toolkit: A MapReduce framework for analyzing next-generation DNA sequencing data. *Genome Res.* **2010**, *20*, 1297–1303. [CrossRef]
50. Wysoker, A.; Tibbetts, K.; Fennel, T. Picard Tools Version 2.5. 2013. Available online: http://broadinstitute.github.io/picard/ (accessed on 6 January 2020).
51. Depristo, M.A.; Banks, E.; Poplin, R.; Garimella, K.V.; Maguire, J.R.; Hartl, C.; Philippakis, A.A.; Del Angel, G.; Rivas, M.A.; Hanna, M.; et al. A framework for variation discovery and genotyping using next-generation DNA sequencing data. *Nat. Genet.* **2011**, *43*, 491–498. [CrossRef]
52. Van Der Auwera, G.A.; Carneiro, M.O.; Hartl, C.; Poplin, R.; Del Angel, G.; Levy-Moonshine, A.; Jordan, T.; Shakir, K.; Roazen, D.; Thibault, J.; et al. From FastQ Data to High-Confidence Variant Calls: The Genome Analysis Toolkit Best Practices Pipeline. *Curr. Protoc. Bioinform.* **2013**, *43*, 1–33. [CrossRef]
53. McLaren, W.; Gil, L.; Hunt, S.E.; Riat, H.S.; Ritchie, G.R.S.; Thormann, A.; Flicek, P.; Cunningham, F. The Ensembl Variant Effect Predictor. *Genome Biol.* **2016**, *17*, 1–14. [CrossRef] [PubMed]
54. KEGG Pathway Database. Available online: https://www.genome.jp/kegg/pathway.html (accessed on 4 February 2020).
55. miRbase. Available online: http://www.mirbase.org/ (accessed on 4 February 2020).
56. Griffiths-Jones, S.; Grocock, R.J.; Van Dongen, S.; Bateman, A.; Enright, A.J. miRBase: microRNA sequences, targets and gene nomenclature. *Nucleic Acids Res.* **2006**, *34*, D140–D144. [CrossRef] [PubMed]
57. Lago, L.V.; Da Silva, A.N.; Zanella, E.L.; Marques, M.G.; Peixoto, J.D.O.; Da Silva, M.V.G.; Ledur, M.C.; Zanella, R. Identification of Genetic Regions Associated with Scrotal Hernias in a Commercial Swine Herd. *Vet. Sci.* **2018**, *5*, 15. [CrossRef] [PubMed]
58. Joaquim, L.B.; Chud, T.C.S.; Marchesi, J.A.P.; Savegnago, R.P.; Buzanskas, M.E.; Zanella, R.; Cantão, M.E.; Peixoto, J.O.; Ledur, M.C.; Irgang, R.; et al. Genomic structure of a crossbred Landrace pig population. *PLoS ONE* **2019**, *14*, e0212266. [CrossRef]
59. Liao, X.J.; Li, L.; Zhang, Z.Y.; Long, Y.; Yang, B.; Ruan, G.R.; Su, Y.; Ai, H.S.; Zhang, W.C.; Deng, W.Y.; et al. Susceptibility loci for umbilical hernia in swine detected by genome-wide association. *Russ. J. Genet.* **2015**, *51*, 1000–1006. [CrossRef]
60. Long, Y.; Su, Y.; Ai, H.; Zhang, Z.; Yang, B.; Ruan, G.; Xiao, S.; Liao, X.; Ren, J.; Huang, L.; et al. A genome-wide association study of copy number variations with umbilical hernia in swine. *Anim. Genet.* **2016**, *47*, 298–305. [CrossRef]
61. QuickGO. Available online: https://www.ebi.ac.uk/QuickGO/ (accessed on 4 February 2020).
62. Levine, B.; Mizushima, N.; Virgin, H.W. Autophagy in immunity and inflammation. *Nat. Cell Biol.* **2011**, *469*, 323–335. [CrossRef] [PubMed]
63. Levine, B.; Klionsky, D.J. Development by Self-Digestion. *Dev. Cell* **2004**, *6*, 463–477. [CrossRef]
64. Kroemer, G. Autophagy: A druggable process that is deregulated in aging and human disease. *J. Clin. Investig.* **2015**, *125*, 1–4. [CrossRef] [PubMed]
65. Zhang, H.; Baehrecke, E.H. Eaten alive: Novel insights into autophagy from multicellular model systems. *Trends Cell Biol.* **2015**, *25*, 376–387. [CrossRef] [PubMed]
66. Yang, Z.; Klionsky, D.J. Eaten alive: A history of macroautophagy. *Nat. Cell Biol.* **2010**, *12*, 814–822. [CrossRef]
67. Marcelino, D.E.P.; Souza, M.R.; Auler, M.E.; Savoldi, I.R.; Ibelli, A.M.G.; Peixoto, J.D.O. Expressão dos genes MAP1LC3C e EPYC em suínos normais e afetados com hérnia umbilical. In *Anais da 13thJornada de Iniciação Científica*; Embrapa Suínos e Aves: Concórdia, UNC, 2019; pp. 34–35.

68. Van Limbergent, J.; Stevens, C.; Nimmo, E.R.; Wilson, D.C.; Satsangi, J. Autophagy: From basic science to clinical application. *Mucosal Immunol.* **2009**, *2*, 315–330. [CrossRef]
69. Carlos, M.M.L.; Freitas, P.D.F.S. Estudo da cascata de coagulação sanguínea e seus valores de referência. *Acta Vet. Bras.* **2007**. [CrossRef]
70. Iturry-Yamamoto, G.; Portinho, C. Sistema complemento: Ativação, regulação e deficiências congênitas e adquiridas. *Revista Associação Médica Brasileira* **2001**, *47*, 41–51. [CrossRef]
71. Xue, X.; Wu, J.; Ricklin, D.; Forneris, F.; Di Crescenzio, P.; Schmidt, C.Q.; Granneman, J.; Sharp, T.H.; Lambris, J.D.; Gros, P. Regulator-dependent mechanisms of C3b processing by factor I allow differentiation of immune responses. *Nat. Struct. Mol. Biol.* **2017**, *24*, 643–651. [CrossRef]
72. Winterfeld, G.A.; Khodair, A.I.; Schmidt, R.R. O-Glycosyl Amino Acids by 2-Nitrogalactal Concatenation – Synthesis of a Mucin-Type O-Glycan. *Eur. J. Org. Chem.* **2003**, *2003*, 1009–1021. [CrossRef]
73. Brockhausen, I. Pathways of O-glycan biosynthesis in cancer cells. *Biochim. Biophys. Acta (BBA) Gen. Subj.* **1999**, *1473*, 67–95. [CrossRef]
74. Blalock, T.D.; Spurr-Michaud, S.J.; Tisdale, A.S.; Heimer, S.R.; Gilmore, M.S.; Ramesh, V.; Gipson, I.K. Functions of MUC16 in Corneal Epithelial Cells. *Investig. Opthalmol. Vis. Sci.* **2007**, *48*, 4509–4518. [CrossRef] [PubMed]
75. Alberts, B.; Johnson, A.; Lewis, J.; Morgan, D.; Raff, M.; Roberts, K.; Walter, P. *Molecular Biology of the Cell*, 6th ed.; Garland Science: New York, NY, USA, 2014.
76. Huang, S.; Ingber, D.E. The structural and mechanical complexity of cell-growth control. *Nat. Cell Biol.* **1999**, *1*, E131–E138. [CrossRef] [PubMed]
77. Khalili, A.A.; Ahmad, M.R. A review of cell adhesion studies for biomedical and biological applications. *Int. J. Mol. Sci.* **2015**, *16*, 18149–18184. [CrossRef] [PubMed]
78. Grivicich, I.; Regner, A.; Rocha, A.B. Morte celular por apoptose. *Rev. Bras. Cancerol.* **2007**, *53*, 335–343.
79. Puetz, S.; Lubomirov, L.T.; Pfitzer, G. Regulation of smooth muscle contraction by small GTPases. *Physiology* **2009**, *24*, 342–356. [CrossRef]
80. Sato, T.; Konomi, K.; Yamasaki, S.; Aratani, S.; Tsuchimochi, K.; Yokouchi, M.; Masuko-Hongo, K.; Yagishita, N.; Nakamura, H.; Komiya, S.; et al. Comparative analysis of gene expression profiles in intact and damaged regions of human osteoarthritic cartilage. *Arthritis Rheum.* **2006**, *54*, 808–817. [CrossRef]
81. Zhao, X.; Du, Z.-Q.; Vukasinovic, N.; Rodriguez, F.; Clutter, A.C.; Rothschild, M.F. Association of HOXA10, ZFPM2, and MMP2 genes with scrotal hernias evaluated via biological candidate gene analyses in pigs. *Am. J. Veter Res.* **2009**, *70*, 1006–1012. [CrossRef]
82. Du, Z.-Q.; Zhao, X.; Vukasinovic, N.; Rodríguez, F.; Clutter, A.C.; Rothschild, M.F. Association and Haplotype Analyses of Positional Candidate Genes in Five Genomic Regions Linked to Scrotal Hernia in Commercial Pig Lines. *PLoS ONE* **2009**, *4*, e4837. [CrossRef]
83. Piórkowska, K.; Żukowski, K.; Ropka-Molik, K.; Tyra, M. Detection of genetic variants between different Polish Landrace and Puławska pigs by means of RNA-seq analysis. *Anim. Genet.* **2018**, *49*, 215–225. [CrossRef] [PubMed]
84. Lazarides, E. Intermediate Filaments: A Chemically Heterogeneous, Developmentally Regulated Class of Proteins. *Annu. Rev. Biochem.* **1982**, *51*, 219–250. [CrossRef] [PubMed]
85. Onteru, S.K.; Fan, B.; Du, Z.-Q.; Garrick, D.; Stalder, K.J.; Rothschild, M.F. A whole-genome association study for pig reproductive traits. *Anim. Genet.* **2011**, *43*, 18–26. [CrossRef] [PubMed]
86. Moore, K.L.; Persaud, T.V.N.; Torchia, M.G. *Embriologia Básica*, 9th ed.; Elsevier: Rio de Janeiro, Brazil, 2016; p. 463. Available online: https://ia801004.us.archive.org/30/items/Embriologia_bsica_9._ed._-_www.meulivro.biz/Embriologia_bsica_9._ed._-_www.meulivro.biz.pdf (accessed on 16 December 2019).
87. Stachowiak, M.; Nowacka-Woszuk, J.; Szydlowski, M.; Switonski, M. The ACACA and SREBF1 genes are promising markers for pig carcass and performance traits, but not for fatty acid content in the longissimus dorsi muscle and adipose tissue. *Meat Sci.* **2013**, *95*, 64–71. [CrossRef] [PubMed]
88. Tian, M.; Zhang, X.; Ye, P.; Tao, Q.; Zhang, L.; Ding, Y.; Chu, M.; Zhang, X.; Yin, Z. MicroRNA-21andmicroRNA-214play important role in reproduction regulation during porcine estrous. *Anim. Sci. J.* **2018**, *89*, 1398–1405. [CrossRef]
89. Guo, Y.; Chen, Y.; Zhang, Y.; Zhang, Y.; Chen, L.; Mo, D. Up-regulated miR-145 Expression Inhibits Porcine Preadipocytes Differentiation by Targeting IRS1. *Int. J. Biol. Sci.* **2012**, *8*, 1408–1417. [CrossRef]

*Article*

# Weighted Single-Step GWAS Identified Candidate Genes Associated with Growth Traits in a Duroc Pig Population

**Donglin Ruan [1,2,†], Zhanwei Zhuang [1,2,†], Rongrong Ding [1,2], Yibin Qiu [1,2], Shenping Zhou [1,2], Jie Wu [1,2], Cineng Xu [1,2], Linjun Hong [1,2], Sixiu Huang [1,2], Enqin Zheng [1,2], Gengyuan Cai [1], Zhenfang Wu [1,2,*] and Jie Yang [1,2,*]**

[1] National Engineering Research Center for Breeding Swine Industry, College of Animal Science, South China Agricultural University, Guangzhou 510642, China; ruandl@stu.scau.edu.cn (D.R.); zwzhuang@outlook.com (Z.Z.); drr_scau@foxmail.com (R.D.); 13422157044qyb@gmail.com (Y.Q.); shenpingzhou1109@163.com (S.Z.); wujiezi163@163.com (J.W.); cnxu@stu.scau.edu.cn (C.X.); linjun.hong@scau.edu.cn (L.H.); sxhuang815@scau.edu.cn (S.H.); eqzheng@scau.edu.cn (E.Z.); cgy0415@163.com (G.C.)

[2] Lingnan Guangdong Laboratory of Modern Agriculture, Guangzhou 510642, China

* Correspondence: wzf@scau.edu.cn (Z.W.); jieyang@scau.edu.cn (J.Y.)

† These authors contributed equally to this work.

**Abstract:** Growth traits are important economic traits of pigs that are controlled by several major genes and multiple minor genes. To better understand the genetic architecture of growth traits, we performed a weighted single-step genome-wide association study (wssGWAS) to identify genomic regions and candidate genes that are associated with days to 100 kg (AGE), average daily gain (ADG), backfat thickness (BF) and lean meat percentage (LMP) in a Duroc pig population. In this study, 3945 individuals with phenotypic and genealogical information, of which 2084 pigs were genotyped with a 50 K single-nucleotide polymorphism (SNP) array, were used for association analyses. We found that the most significant regions explained 2.56–3.07% of genetic variance for four traits, and the detected significant regions (>1%) explained 17.07%, 18.59%, 23.87% and 21.94% for four traits. Finally, 21 genes that have been reported to be associated with metabolism, bone growth, and fat deposition were treated as candidate genes for growth traits in pigs. Moreover, gene ontology (GO) and Kyoto Encyclopedia of Genes and Genomes (KEGG) enrichment analyses implied that the identified genes took part in bone formation, the immune system, and digestion. In conclusion, such full use of phenotypic, genotypic, and genealogical information will accelerate the genetic improvement of growth traits in pigs.

**Keywords:** Duroc pigs; growth traits; weighted single-step GWAS; SNP

**Citation:** Ruan, D.; Zhuang, Z.; Ding, R.; Qiu, Y.; Zhou, S.; Wu, J.; Xu, C.; Hong, L.; Huang, S.; Zheng, E.; et al. Weighted Single-Step GWAS Identified Candidate Genes Associated with Growth Traits in a Duroc Pig Population. *Genes* **2021**, *12*, 117. https://doi.org/10.3390/genes12010117

Received: 23 October 2020
Accepted: 11 January 2021
Published: 19 January 2021



## 1. Introduction

Pork is the primary source of protein for humans, with global pork consumption exceeding 110 metric kilotons per year [1]. Growth traits are economically important traits in porcine breeding programs, as accelerating the genetic process of growth-related traits can increase the supply of pork. At present, the age to 100 kg, average daily gain, backfat thickness, and lean meat percentage for a specific stage are vital indicators to measure the growth rate and carcass fat content of pigs due to their significant impact on production efficiency [2]. Furthermore, both genetic and non-genetic effects can affect growth traits, including pig breed, feeding behavior, and nutrition level. However, the above four traits have moderate heritability [3], suggesting that they could be improved by the genetic method.

Since the first genome-wide association study (GWAS) for age-related macular degeneration was published in 2005, GWAS has been widely used to identify quantitative trait loci (QTL) and to map candidate genes for complex traits in humans [4] and domestic animals [5]. Until now, 2036 QTL for growth traits have been reported in the pig QTL database (https://www.animalgenome.org/cgi-bin/QTLdb/SS/summary, release 27 August 2020). These findings have provided a certain number of molecular markers to porcine breeding for growth traits—for instance, Jiang et al. [6] performed a GWAS in a total of 2025 American and British Yorkshire pigs using PorcineSNP80 bead chip and detected five significant SNPs for days to 100 kg and the other five significant SNPs for 10th rib backfat thickness. Qiao et al. [7] found 14 QTL significantly associated with growth-related traits for White Duroc × Erhualian F2 and Sutai (Chinese Taihu × Western Duroc) populations. Although many studies have contributed to complex quantitative traits by GWAS, the genetic mechanisms of growth traits in pigs remain unclear. Additionally, some single marker GWAS analyses result in a weak power for QTLs detection and low accuracy for mapping. Moreover, most studies on GWAS for growth traits used the limited population size of genotyped animals and neglected the pedigree relationship. To overcome the limitation of the traditional GWAS approach, the weighted single-step GWAS (wssGWAS) proposed by Wang et al. [8] is preferable for livestock breeding, for which phenotypic and genealogical information is available for the vast majority of individuals and the small size of individuals genotyped.

The GWAS under the single-step genomic best linear unbiased prediction (ssGBLUP) framework is called ssGWAS, which intermixes genotypes, pedigree, and phenotypes data in a single analysis without creating pseudo-phenotypes [9]. However, when some traits are affected by significant QTL in practice, it is improper to account for all SNPs to explain the same proportion of genetic variance in ssGBLUP [10]. In that case, the wssGWAS can be adopted, which weighs SNPs according to their effects that were calculated genomic estimated breeding values (GEBVs) via ssGBLUP. The wssGWAS method has been successfully applied to detect supplementary QTLs and candidate genes in domestic and aquaculture animals, such as carcass traits in Nellore cattle [11], growth and carcass traits in rainbow trout [12], and reproductive traits in pigs [13]. However, to our knowledge, few wssGWASs have been performed to study growth traits in purebred Duroc pigs. Therefore, this study aims to identify genomic regions and candidate genes associated with growth traits such as days adjusted to 100 kg (AGE), average daily gain adjusted to 100 kg (ADG), backfat thickness (BF) and predicted lean meat percentage (LMP) adjusted to 100 kg in a Duroc pig population using the wssGWAS methodology. Then, gene ontology (GO) and Kyoto Encyclopedia of Genes and Genomes (KEGG) enrichment analysis facilitate further understanding of biological processes and functional terms of candidate genes for growth traits.

## 2. Materials and Methods

### 2.1. Ethics Statement

All animals used in this study were used according to the guidelines for the care and use of experimental animals established by the Ministry of Agriculture and Rural Affairs of China. The ethics committee of South China Agricultural University (SCAU, Guangzhou, China) approved the entire study. No experimental animals were anesthetized or euthanized in this study.

### 2.2. Animals, Phenotypes, and Pedigree

The animals used in this study were raised in two core farms of the Wens Foodstuff Group CO., Ltd. (Guangdong, China) with uniform standards. In brief, a total of 3945 Canadian Duroc pigs (1966 males and 1979 females) born between 2015 and 2017 were used in this study. Among them, 2084 individuals had genotypes and four growth-trait phenotypes in the pedigree, while 1843 ungenotyped individuals in the pedigree had phenotypes of AGE and ADG, and 1825 ungenotyped individuals in the pedigree had phenotypes of BF and LMP. Furthermore, the complete pedigree could be traced back 3 generations, with 5204 pigs in the full pedigree (2103 males and 3101 females).

Days to 100 kg and ADG were measured from 30 to 115 kg and then adjusted to 100 kg. AGE was adjusted to 100 kg using the formula below [14]:

$$AGE\ adjusted\ to\ 100\ \text{kg} = Measured\ age - \left(\frac{Measured\ weight - 100\ \text{kg}}{Correction\ factor\ 1}\right) \tag{1}$$

where the correction factor 1 of sire and dam are different, as follows:

$$Sire : Correction\ factor\ 1 = \frac{Measured\ weight}{Measured\ age} \times 1.826 \tag{2}$$

$$Dam : Correction\ factor\ 1 = \frac{Measured\ weight}{Measured\ age} \times 1.715 \tag{3}$$

*ADG* was adjusted to 100 kg by following formula [14]:

$$ADG\ adjusted\ to\ 100\ \text{kg} = \frac{100\ \text{kg}}{AGE\ adjusted\ to\ 100\ \text{kg}} \tag{4}$$

Adjusting LMP to 100 kg, phenotypes of BF and loin muscle depth (LMD) was measured between the last 3rd and 4th rib of Duroc pigs at the weight of 100 $\pm$ 5 kg by an Aloka 500 V SSD B ultrasound (Coromertics Medical Systems, USA) [15]. BF and LMD adjusted to 100 kg were calculated as reported by the Canadian Center for Swine Improvement (http://www.ccsi.ca/Reports/Reports_2007/Update_of_weight_adjustment_factors_for_fat_and_lean_depth.pdf):

$$BF\ adjusted\ to\ 100\ \text{kg} = Measured\ BF \times \frac{A}{A + [B \times (Measured\ Weight - 100)]} \tag{5}$$

where $A$ and $B$ are different for sire and dam, as follows:

$$Sire : A = 13.47;\ B = 0.1115 \tag{6}$$

$$Dam : A = 15.65;\ B = 0.1566 \tag{7}$$

*LMD* adjusted to 100 kg was calculated by the following equation [16]:

$$LMD\ adjusted\ to\ 100\ \text{kg} = Measured\ LMD \times \left[\frac{a}{a + b \times (Measured\ Weight - 100)}\right] \tag{8}$$

where $a$ and $b$ are gender-specific, and

$$Sire : a = 50.52;\ b = 0.228 \tag{9}$$

$$Dam : a = 52.01;\ b = 0.228 \tag{10}$$

*LMP* was adjusted to 100 kg using the formula below [16]:

$$LMP\ adjusted\ to\ 100\ \text{kg} = 61.21920 - 0.77665 \times BF + 0.15239 \times LMD \tag{11}$$

Overall, 3927 individuals were used in wssGWAS for ADG and AGE; 3909 individuals were used in wssGWAS for BF and LMP.

### 2.3. Genotyping and Quality Control (QC)

DNA was extracted from ear tissue of 2084 Duroc pigs following the standard phenol/chloroform method, then quantified and diluted to 50 ng/µL. All DNA samples were genotyped by GeneSeek porcine 50 K SNP chip from Illumina (Neogen, Lincoln, NE, USA), including 50,649 SNPs mapped to Sus scrofa11.1 (https://www.ensembl.org/biomart) in total. Quality control was performed by PLINK v1.09 (Boston, MA, USA) [17] in which

SNPs were excluded when individuals call rate was <90%, SNPs call rate was <90%, Hardy–Weinberg equilibrium $p$-value was $<10^{-6}$, minor allele frequency was <0.01, and SNPs were located in sex chromosomes and unmapped. After QC, a final set of 35,851 high-quality SNPs for 2084 Duroc pigs remained for subsequent analyses.

### 2.4. Statistical Analyses

Variance components for AGE, ADG, BF, and LMP traits were estimated with two methods using the average information restricted maximum-likelihood (AIREML), including pedigree-based Best Linear Unbiased Prediction (BLUP) and ssGBLUP. The four traits were analyzed using the same single-trait animal model, as described below:

$$Y = Xb + Za + e \tag{12}$$

where $Y$ was the vector of phenotypic values; $X$ was the incidence matrix of fix effect for relating phenotypes; $b$ was the vector of fixed effect, including birth year, sex, and farm; $Z$ was the incidence matrix of random effect; $a$ was the vector of additive genetic effects, and $e$ was the vector of residuals. Narrow sense heritability was estimated as $h^2 = \frac{\sigma_a^2}{\sigma_a^2+\sigma_e^2}$, where $\sigma_a^2$ and $\sigma_e^2$ were additive genetic variance and residual variance, respectively.

Additionally, the GEBVs of all individuals were estimated via the same single-trait model as described previously using the ssGBLUP [18] approach, and marker effects were calculated from the GEBVs. Comparing with the regular BLUP approach, ssGBLUP replaces the inverse of the pedigree relationship matrix ($A^{-1}$) with the matrix $H^{-1}$, for which the matrix $H$ combined the pedigree and the genomic relationship matrices [19]. The inverse of matrix $H$ was represented as follows:

$$H^{-1} = A^{-1} + \begin{bmatrix} 0 & 0 \\ 0 & G^{-1} - A_{22}^{-1} \end{bmatrix} \tag{13}$$

where $A_{22}^{-1}$ was the inverse matrix of the numerator relationship matrix considering genotyped animals and $G^{-1}$ was the inverse matrix of the genomic relationship matrix [20]. The genomic matrix G can be created as follows [21]:

$$G = \frac{ZDZ'}{\sum_{i=1}^{N} 2p_i(1-p_i)} \tag{14}$$

where $Z$ was a centered matrix of SNP genotypes (aa = 0, Aa = 1 and AA = 2), $D$ was a matrix of weights for SNP variances, $n$ was the number of SNPs and $p_i$ was the minor allele frequency of the i-th SNP [8].

The wssGWAS of SNP effects and weights were calculated following by Wang et al. [8]:

1. Initially, set $t = 1$, $D_{(1)} = I$;
2. Calculate $G_{(t)} = \lambda Z D_{(t)} Z'$, where $\lambda = \sum_{i=1}^{N} 2p_i(1-p_i)$;
3. Calculate GEBVs for whole data set by ssGBLUP method;
4. Calculate SNPs effects: $\hat{u}_{(t)} = \lambda D_{(t)} Z' G_{(t)}^{-1} \hat{g}$, where $\hat{g}$ was the GEBV of animals genotyped;
5. Calculate the weight of each SNP:

$$d_{i(t)} = 2\hat{u}_{i(t)}^2 p_i(1-p_i) \tag{15}$$

where $i$ was the $i$-th SNP;

6. Normalize SNP weights to keep total genetic variance constant via

$$D_{(t+1)} = \frac{tr\left(D_{(t)}\right) \times D_{(t+1)}}{tr\left(D_{(t+1)}\right)} \tag{16}$$

7. Set $t = t + 1$, then loop to step 2.

The procedure was run for three iterations, as suggested by Wang et al. [8], which reached a high accuracy of GEBVs. In this study, SNPs located within 0.8 Mb (according to the linkage disequilibrium decay of this population [22]) were grouped in a window, and the percentage of genetic variance explained by each 0.8 Mb window was calculated following as below [8]:

$$\frac{\mathrm{Var}(a_i)}{\sigma_a^2} \times 100\% = \frac{\mathrm{Var}(\sum_{j=1}^{x} Z_j g_j)}{\sigma_a^2} \quad (17)$$

where $a_i$ was the genetic value of the $i$-th region consisting of $x = 0.8$ Mb.

The procedures mentioned above were run with BLUPF90 software family programs [23] iteratively. The RENUMF90 module was used to obtain the required parameter file formats; the AIREMLF90 module was used for variance components estimation, the BLUPF90 module for GEBVs calculation, and the postGSF90 module for association analysis.

### 2.5. Identification of Candidate Genes and Functional Enrichment Analysis

Genomic windows that explained higher than 1.0% of the total genetic variance were selected as candidate QTL regions associated with growth traits in this study, which was also used in previous studies [8,13]. Since the 0.8 Mb window explained on–average 0.0473% (100% divided by 2115 genomic regions) of the genetic variance, the 1% threshold is over 20 times the expected average genetic variance explained by the 0.8 Mb window. The first three windows that explained the largest proportion of genetic variance for each trait were extended to 0.4 Mb flanking on either side of the regions. For the identified QTL regions, genes were searched using the Ensemble Sus scrofa 11.1 (https://www.ensembl.org/biomart) database within significant windows. To better understand the biological processes, GO and KEGG analyses were performed based on genes within significant regions using the database for annotation, visualization, and integrated discovery (DAVID v6.8, https://david.ncifcrf.gov/). A $p$-value of <0.05 was the threshold for significantly enriched GO terms and KEGG pathways.

## 3. Results and Discussion

### 3.1. Descriptive Statistics and Heritability for the Growth Traits

Descriptive statistics of the phenotypes are presented in Table 1. Previous studies reported that the average AGE phenotype of Duroc and other western commercial pig breeds was between 150 and 162 days, ADG was between 610 and 820 g/day, BF was between 11.69 and 18.19 mm, and LMP was between 56% and 62% [6,14,24–26]. The phenotypic averages for AGE, ADG, BF, and LMP in this study were similar to previous studies. The coefficients of variation (CV) for AGE, ADG, BF, and LMP were 7.30%, 7.25%, 17.86%, and 2.83%, respectively. The results indicated substantial phenotypic variation in these traits, except LMP. Since Duroc pigs are the terminal male parent of the Duroc × (Landrace × Yorkshire) pigs (DLY), the LMP of Duroc pigs receives long-term positive selection [27]. In other words, the lower CV of LMP indicates that the selection prior to the LMP was effective in this core Duroc population.

**Table 1.** Descriptive statistics of growth traits in the Duroc pig population.

| Traits [a] | $n$ | Mean | SD [b] | Min | Max | CV (%) [c] |
|---|---|---|---|---|---|---|
| AGE | 3927 | 163.41 | 11.93 | 125.98 | 206.32 | 7.30 |
| ADG | 3927 | 604.31 | 43.81 | 478.73 | 779.49 | 7.25 |
| BF | 3909 | 9.52 | 1.70 | 5.10 | 17.31 | 17.86 |
| LMP | 3909 | 61.08 | 1.39 | 54.93 | 65.06 | 2.28 |

[a] AGE, days to 100 kg, ADG: average daily gain adjusted to 100 kg, BF, backfat thickness adjusted to 100 kg; LMP, predicted lean meat percentage adjusted to 100 kg; [b] SD, standard deviation; [c] CV, coefficient of variation.

To better understand the genetic background of growth traits, we estimated the genetic variance ($\sigma_a^2$), residual variance ($\sigma_e^2$), and heritability ($h^2$) by different methods, including BLUP and ssGBLUP. The heritability estimated by BLUP and ssGBLUP were 0.507 and 0.343, 0.508 and 0.333, 0.512 and 0.315, and 0.554 and 0.332 for AGE, ADG, BF, and LMP, respectively (Table 2). There were differences in the heritability estimated by the two methods in this study, and the previous study showed that common environmental components lead to a possible overestimation of genetic variance in the pedigree-based BLUP method of estimating heritability [28]. Compared with the BLUP method, the ssGBLUP method has a lower standard error. The ssGBLUP method uses both pedigrees and genotyped information, and the estimated genetic parameters are theoretically more accurate [29]. Furthermore, the results from the two methods indicated that these traits were moderate heritability traits and could be genetically improved by genetic techniques.

**Table 2.** Variance components and heritability estimates of growth traits.

| Traits [a] | Models | $\sigma_a^2$ * | $\sigma_e^2$ * | $\sigma_p^2$ * | $h^2$ (SE) * |
|---|---|---|---|---|---|
| AGE | BLUP | 68.667 | 66.879 | 135.546 | 0.507 (0.0454) |
| | ssGBLUP | 44.932 | 85.981 | 130.913 | 0.343 (0.0314) |
| ADG | BLUP | 926.290 | 895.570 | 1821.860 | 0.508 (0.0453) |
| | ssGBLUP | 581.4 | 1166.3 | 1747.7 | 0.333 (0.0308) |
| BF | BLUP | 1.516 | 1.445 | 2.961 | 0.512 (0.0449) |
| | ssGBLUP | 0.877 | 1.903 | 2.780 | 0.315 (0.0289) |
| LMP | BLUP | 1.142 | 0.918 | 2.060 | 0.554 (0.0444) |
| | ssGBLUP | 0.639 | 1.283 | 1.922 | 0.332 (0.0289) |

[a] AGE, days to 100 kg, ADG: average daily gain adjusted to 100 kg, BF, backfat thickness adjusted to 100 kg; LMP, predicted lean meat percentage adjusted to 100 kg; * $\sigma_a^2$, genetic variance, $\sigma_e^2$, residual variance, $\sigma_p^2$, phenotypic variance, $h^2$, heritability; SE, standard error.

### 3.2. Summary of wssGWAS

Most important economic traits of livestock are quantitative traits with complicated genetic architectures. Therefore, uncovering the candidate genes underlying these traits has been a crucial goal in livestock breeding programs. In particular, growth rate and carcass fat content comprise the essential measuring basis of production performance in pigs, influencing the economic benefit directly. In this study, genetic variance explained by 0.8 Mb windows for each trait was achieved by wssGWAS. The first three most important QTL regions and the candidate genes are shown in Table 3. Overall, the first three QTL regions totally explained 5.96%–7.25% of the genetic variance of these traits under study. For each trait, the most significant windows explained approximately 2.56%–3.07% of the total genetic variance. Additionally, the identified windows (>1%) explained 17.07%, 18.59%, 23.87%, and 21.94% for AGE, ADG, BF, and LMP, respectively (Supplementary File, Tables S1–S5). Previous GWAS research reported that the candidate QTL regions of ADG on Sus scrofa chromosome (SSC) 1, 3, 6, 8, 13 and the candidate QTL regions of AGE on SSC 1, 3, 6, 8, 10, explaining a total of 8,09% and 4.08% of genetic variance [14], respectively. Due to LD, the wssGWAS method using the SNP window for analysis probably better identifies unknown QTL than the traditional GWAS, avoiding overestimation of the detected QTL number and false-positives [30,31]. Moreover, iterative weighting for SNPs could highlight QTL with larger effects [8]. Comparing with the results of the ssGWAS in ADG and BF by Matteo et al. [32], and our results identify the most significant QTL regions explaining greater genetic variance. Figure 1 shows the proportion of variance explained by each 0.8 Mb window for the studied traits, suggesting the polygenic genetic architecture of these traits.

**Table 3.** First three most important quantitative trait loci (QTL) regions and candidate genes for growth traits.

| Traits [a] | Chr [b] | Position (Mb) | nSNPs | gVar (%) [c] | Candidate Genes |
|---|---|---|---|---|---|
| AGE | 4 | 4.38–5.98 | 43 | 3.07 | *FAM135B* |
| | 4 | 6.75–8.35 | 43 | 1.84 | *ZFAT* |
| | 14 | 1.63–3.23 | 22 | 1.57 | *NFIL3, ROR2* |
| ADG | 4 | 4.38–5.98 | 43 | 2.56 | *FAM135B* |
| | 2 | 130.75–132.35 | 20 | 1.91 | *SLC27A6* |
| | 2 | 149.94–151.54 | 29 | 1.49 | *ADRB2* |
| BF | 7 | 29.34–30.94 | 26 | 2.97 | *DAXX, ITPR3, IP6K3, PACSIN1* |
| | 3 | 117.76–119.36 | 19 | 1.94 | *SDC1* |
| | 10 | 55.95–57.55 | 29 | 1.85 | *NRP1* |
| LMP | 2 | 8.11–9.71 | 26 | 2.68 | *NAA40, LGALS12* |
| | 3 | 117.76–119.36 | 39 | 2.08 | *SDC1* |
| | 10 | 38.67–40.27 | 15 | 2.00 | *MOB3B, RAB18, MPP7* |

[a] AGE, days to 100 kg, ADG: average daily gain adjusted to 100 kg, BF, backfat thickness adjusted to 100 kg; LMP, predicted lean meat percentage adjusted to 100 kg; [b] Chr, chromosome; [c] gVar (%) represents the proportion of genetic variance explained by 0.8 Mb. For each trait, the genomic regions are sorted in descending order according to the proportion of genetic variance explained.

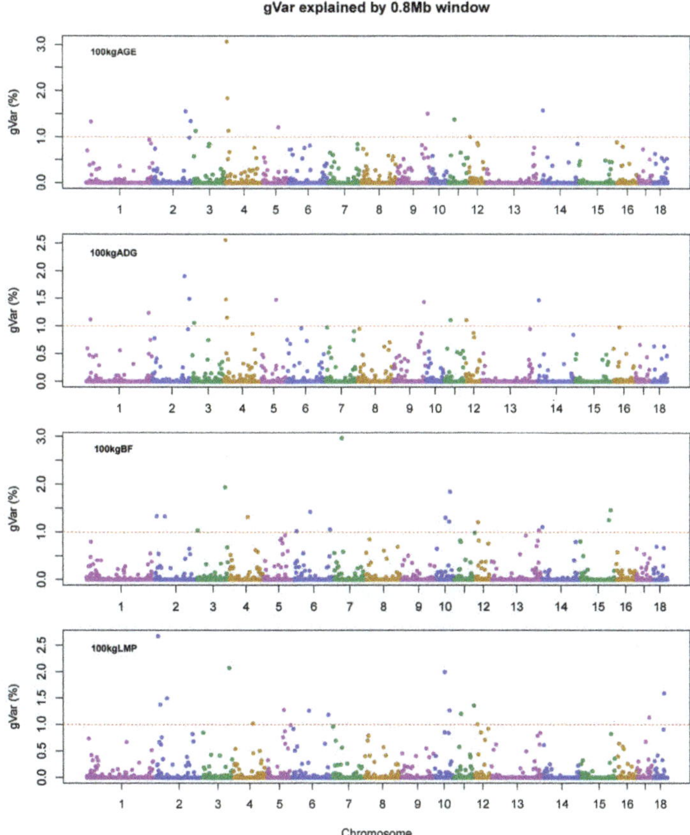


**Figure 1.** The proportion of genetic variances of the growth traits is explained by 0.8 Mb windows. gVar (%) represents the proportion of genetic variance explained by 0.8 Mb windows; 100 kg AGE, days to 100 kg; 100 kg ADG: average daily gain adjusted to 100 kg; 100 kg BF, backfat thickness adjusted to 100 kg; 100 kg LMP, predicted lean meat percentage adjusted to 100 kg.

### 3.3. *wssGWAS for AGE and ADG*

For AGE, 11 relevant QTL regions located on SSC 1, 2, 3, 4, 5, 9, 11, and 14 were identified (Supplementary Materials, Tables S1–S3). These regions explained 1.13–3.07% of total genetic variance for AGE, and 73 genes were annotated in these genomic regions. For ADG, 13 relevant QTL regions located on SSC1, 2, 3, 4, 5, 9, 11, 12, and 14 were identified, where 104 genes are located in these genomic regions (Supplementary Material, Tables S1 and S3). These regions explained total genetic variance ranged from 1.06% to 2.56% for ADG.

For the identified significant regions, there were 10 overlapped windows for AGE and ADG, which explained different proportions of genetic variance in these two traits. For complex quantitative traits, it was assumed that the linear effects of genes fitted the average of traits completely. However, the effects of genes are not always linear for the traits in practice, and the nonlinear assumption is more appropriate [14], which means that genes contributed differently and pleiotropic effects of the QTL between traits. QTLs with pleiotropic effects are common in the pig genome. For instance, Yang et al. [33] reported that a pleiotropic QTL on SSC 7 was associated with the vertebral number, carcass length, and teat number. In the present study, the region with the largest explained genetic variance for AGE and ADG, located in the region of 4.38–5.98 Mb on SSC4, seemingly had pleiotropic effects on meat and carcass traits in pigs [34]. Considering the duplication of identified windows and the strong genetic relationship of AGE and ADG, the genes identified by these two traits as common candidate genes are acceptable.

Among the significant windows of these two traits, the most important region (4.38–5.98 Mb on SSC4) harbored the *Family with Sequence Similarity 135 Member B* (*FAM135B*). The expression of *FAM135B* promotes granulin (GRN) secretion, and GRN is a secreted growth factor with high expression in epithelial, immune, chondrocytes, and neuronal cells [35]. Furthermore, *FAM135B* was reported as a candidate gene related to growth traits in beef cattle [36] and reproductive traits in Duroc pigs [37]. The *Zinc Finger And AT-Hook Domain Containing* (*ZFAT*) located in the region of 6.75–8.35 Mb on SSC4, and its mutation would lead to abnormal human body development and thyroid hormone secretion that played a key role in growth and metabolism [38].

The *Nuclear Factor, Interleukin 3 Regulated* (*NFIL3*) and the *Receptor Tyrosine Kinase-Like Orphan Receptor 2* (*ROR2*) were located in the regions of 1.63–3.23 Mb on SSC14. Wang et al. [39] reported that *NFIL3* affected the circadian lipid metabolism program, lipid–absorption, and export of intestinal epithelial through mouse experiments. The mice knocked out *ROR2* resulted in shortened or deformed bones and neurodevelopmental dysplasia [40].

The *Solute Carrier Family 27 Member 6* (*SLC27A6*) gene is located in the region of 130.75–132.35 Mb on SSC9. The *SLC27A6* gene had high expression in fat and muscle tissue and worked on lipid metabolism in pigs [41]. The *Adrenoceptor $\beta$ 2* (*ADRB2*) gene, located in the region of 149.94–151.54 Mb on SSC2, encoded the β-adrenergic receptor that played an essential role in regulating metabolic level [42]. Furthermore, Bachman et al. [43] found that the knockout mice *ADRBs* have a reduced metabolic rate and accelerated fat deposition. The members of the *Tumor Necrosis Factor Receptor Superfamily* (*TNFS*), among which *TNFS11* was identified in the region of 24.24–25.04 Mb on SSC11, were responsible for bone growth in mice [44], and the variation of *TNFS11* led to the low level of serum insulin-like growth factor 1 (*IGF1*) influencing growth rate [45].

### 3.4. *wssGWAS for BF*

A total of 17 relevant QTL regions on SSC2, 3, 4, 6, 7, 10, 12, 13, 14, and 15 were identified for BF (Supplementary Material, Tables S1 and S4), where 99 genes were targeted in these genomic regions. These genomic regions explained 1.02–2.97% of the total genetic variance for BF.

The most significant window was located in the region of 29.34–30.94 Mb on SSC7, where four genes were targeted and were related to BF. In previous studies, *Death Domain*

*Associated Protein* (*DAXX*) was reported to affect fat deposition and fatty acid synthesis via regulating the transcriptional activity of the androgen receptor negatively [46,47]. For *Inositol 1,4,5-Trisphosphate Receptor Type 3* (*ITPR3*), another gene located in the most important window, it was confirmed that mutations could cause taste disorders in mice [48]. Nonetheless, the *Inositol Hexakisphosphate Kinase 3* (*IP6K3*) gene was located in the same region. The mice without this gene resulted in a lower growth rate and metabolism and a shorter lifespan [49]. *Protein–Kinase C and Casein Kinase–Substrate In Neurons 1* (*PACSIN1*), a fourth gene located in the region of 29.34–30.94 Mb, was identified concerning the bodyweight [50] and loin muscle area [51] in pigs.

*CYP7A1*, a member of Cytochrome P450 Family 7 Subfamily A, was identified in the region of 74.12–74.92 Mb on SSC4. The *CYP7A1* gene-encoded enzyme cholesterol 7α-hydroxylase mainly catalyzes the decomposition of cholesterol and synthesis of cholic acid [52]. The *SECIS-Binding Protein 2* (*SECISBP2*) was located in the region of 0.43–1.22 Mb on SSC14, and its mutation brought about abnormal thyroid hormone metabolism in humans [53].

### 3.5. wssGWAS for LMP

Altogether, 15 relevant regions on SSC2, 3, 4, 5, 6, 10, 11, 12, 17 and 18 were identified for LMP. These regions explained 1.00–2.68% of total genetic variance for LMP and 115 genes located in these genomic regions (Supplementary Materials, Tables S1 and S5). The *N-α-Acetyltransferase 40* (*NAA40*) gene and *Galectin 12* (*LGALS12*) gene were located in the region of 8.11–9.71 Mb on SSC2 with the highest percentage of total genetic variance. Liu et al. [54] demonstrated that knockout male rats of the *NAA40* gene exhibited abnormal lipid metabolism and reduced fat mass. In addition, *NAA40* was identified to be associated with the metabolism/transport of fatty acids or lipids in pigs [55]. For *LGALS12*, this gene was preferentially expressed in adipocytes, and mice lacking *LGALS12* resulted in increased mitochondrial respiration, reduced adiposity and decreased insulin resistance/glucose tolerance [56]. Furthermore, *LGALS12* has been identified to be associated with intramuscular and subcutaneous fat in pigs [57].

The *Corticotropin-Releasing Hormone Receptor 2* (*CRHR2*) gene, located in the region of 42.05–42.83 Mb on 18, was highly expressed in adipose tissue, which was involved in the regulation of energy homeostasis and the anorexia effect of fat levels in the corticotropin-releasing hormone (CRH) system [58]. For the region of 41.40–42.12 Mb on SSC2, *Peroxisomal Biogenesis Factor 16* (*PEX16*) and *Cryptochrome Circadian Regulator 2* (*CRY2*) were associated with LMP. Hofer et al. [59] found that the silence of *PEX6* affects adipocyte differentiation and increases peroxisomal fatty acid oxidation–reduction. For the *CRY2* gene, Mármol-Sánchez et al. [60] reported that the polymorphism of *CRY2* was significantly associated with stearic acid content in the longissimus dorsi muscle in Duroc pigs. The *Acyl-CoA Thioesterase 8* (*ACOT8*) gene is located in the region of 47.67–48.83 Mb on SSC17. The protein encoded by this gene is an acyl-CoA thioesterase enzyme that influences the thyroid hormone to regulate lipid storage and utilization according to metabolic demands [61].

### 3.6. BF and LMP Overlap Regions

In the present study, six genomic regions were found to be associated with both BF and LMP, including 41.40–42.12 Mb on SSC2,117.76–119.36 Mb on SSC3, 67.38–68.18 Mb, and 155.99–156.71 Mb on SSC6, and 38.67–40.27 Mb and 55.95–57.55 Mb on SSC10. Notably, BF and LMP were used as an important indicator of carcass fat content in production. Moreover, the genetic correlation of lipid deposition with growth rate and feed efficiency traits were positively high and negatively moderate, respectively [62]. Therefore, these overlap and pleiotropic regions were valuable for growth traits in pigs.

*Potassium Inwardly Rectifying Channel Subfamily J Member 11* (*KCNJ11*), located in the region of 41.40–42.12 Mb on SSC2, was associated with type 2 diabetes in humans [63]. The region of 117.76–119.36 Mb on SSC3 was the second most important window for BF and LMP, which explained 1.94% and 2.08% of the additive genetic variance, respectively, and the *Syndecan 1* (*SDC1*) gene was detected. The *SDC1* gene has been proved to consume the intradermal fat layer, improve glucose tolerance, and significantly reduce body fat content in knockout mice [64]. Two genomic regions stood out on SSC6, which explained 1.05% and 1.42% of additive genetic variance for BF, and 1.26% and 1.19% of additive genetic variance for LMP, respectively. However, the annotated genes in one of these regions are not reported to be associated with growth traits, and no genes are described in the other region on SSC6, pending further studies.

The *Neuropilin 1* (*NRP1*) gene is located in the region of 55.95–57.55 Mb on SCC10, and several studies have exhibited its function in regulating fat cell–activity [65] and reducing dietary insulin resistance [66]. For the region of 38.67–40.27 Mb on SSC10, three genes were identified to be associated with BF and LMP. The *MOB Kinase Activator 3B* (*MOB3B*) gene was significantly associated with intramuscular fat and residual feed intake in cattle [67]. *Ras-Related Protein Rab-18* (*RAB18*), another gene located in the region of 38.67–40.27 Mb on SSC10, encoded a crucial Rab guanosine triphosphatase that controls the growth and maturation of lipid droplet, which lipid droplet was an intracellular organelle to stores triglycerides and cholesterol [68]. Still, in the same region, the *Membrane Palmitoylated Protein 7* (*MPP7*) gene was detected, and Bhoj et al. [69] reported that differences in *MPP7* gene expression affected glucose metabolism in the body.

### 3.7. GO and KEGG Analysis

In the current study, gene set enrichment analyses revealed that several terms might be related to growth traits. Among them, seven biological processes, two cellular components, one molecular function, and four KEGG pathways were targeted significantly (Table 4).

**Table 4.** Significant gene ontology (GO) terms and Kyoto Encyclopedia of Genes and Genomes (KEGG) pathways associated with growth traits in Duroc pigs ($p < 0.05$).

| Term [a] | Count | *p*-Value | Genes |
|---|---|---|---|
| GO:0003727—single-stranded RNA binding | 4 | 0.004495 | *SNRPC, NXF1, JMJD6, POLR2G* |
| GO:0032435—negative regulation of proteasomal ubiquitin-dependent protein catabolic process | 3 | 0.020544 | *WAC, UBXN1, SDCBP* |
| GO:0002924—negative regulation of humoral immune response mediated by circulating immunoglobulin | 2 | 0.029686 | *PTPN6, FOXJ1* |
| GO:0030335—positive regulation of cell migration | 5 | 0.031742 | *ROR2, SEMA4D, CSF1R, SDCBP, SPHK1* |
| GO:0008076—voltage-gated potassium channel complex | 4 | 0.035714 | *KCNC1, KCNJ11, KCNJ2, ABCC8* |
| GO:0005783—endoplasmic reticulum | 11 | 0.038199 | *GPC2, CREB3L1, VWF, P3H3, BRINP1, ATL3, PLAAT3, EEF1G, SRP68, CLDN14, GANAB* |
| GO:1904504—positive regulation of lipophagy | 2 | 0.044199 | *ADRB2, SPTLC1* |
| GO:0032651—regulation of interleukin-1 β production | 2 | 0.044199 | *S1PR3, SPHK1* |
| GO:0030501—positive regulation of bone mineralization | 3 | 0.049487 | *ADRB2, OSR1, FBN2* |
| GO:0010107—potassium ion import | 3 | 0.049487 | *KCNJ11, KCNJ16, KCNJ2* |
| ssc04742—taste transduction | 5 | 0.000381 | *TAS1R1, GRM4, ITPR3, GNB3, SCNN1A* |
| ssc04911—insulin secretion | 5 | 0.019468 | *CREB3L1, KCNJ11, CAMK2A, ITPR3, ABCC8* |
| ssc04725—cholinergic synapse | 5 | 0.045538 | *CREB3L1, CAMK2A, ITPR3, GNB3, KCNJ2* |
| ssc03320—PPAR signaling pathway | 4 | 0.047474 | *ACOX1, SLC27A6, PLTP, CYP7A1* |

[a] GO, gene ontology, KEGG, Kyoto Encyclopedia of Genes and Genomes pathway.

The positive regulation of bone mineralization (GO:0030501) is a key biological process of bone formation, which promotes the deposition of inorganic minerals in the organic–matter of the bone. Bone mineralization affects the strength and density of bone, enabling it to bear the body weight. Shim et al. [70] found that rapid weight gains were correlated with bone mineralization in broilers.

The positive regulation of lipophagy (GO:1904504) is an autophagic process that promotes cells to activate autophagy-related molecules to degrade lipids and regulate intracellular lipid content. Excessive fat deposition in pigs reduces feed conversion rate and affects growth rate, but also affects the quality of animal products [71]. Hence, the function of lipophagy in preventing excess fat deposition may improve the growth traits of pigs. Moreover, the PPAR signaling pathway (ssc03320) is the main pathway associated with lipid metabolism in pigs [72]. Free fat acid from lipophagy is a well-characterized ligand for PPAR$\gamma$ (peroxisome proliferator-activated receptor $\gamma$) [68], which activated the PPAR signaling to induce agouti-related peptide expression (AgRP). Sandoval et al. [73] found that AgRP co-expressed neuropeptide Y stimulated food intake and reduced energy expenditure.

Potassium ion import (GO:0010107) mediates the transmembrane transport of ions and plays a key role in material exchange, energy transfer, and signal transduction. In particular, resting potassium currents make sour taste cells particularly sensitive to changes in intracellular pH, thereby affecting sour taste transduction [74]. Besides this, the taste transduction (ssc04742) pathway is the biological process by which the taste receptors of the organism detect and encode taste information through various transduction mechanisms. Several studies have shown that taste affects appetite and feed intake, and leads to a decrease in growth traits, such as body weight [75]. Moreover, the taste transduction pathway stimulates cephalic phase responses [76], promoting the process of salivary, gastric acid, and cephalic insulin secretion. Moreover, the insulin secretion (ssc04911) pathway was related to feeding intake, which promotes digestive metabolism and nutrient absorption and thus improves the growth trait.

## 4. Conclusions

In conclusion, we indicated 41 genomic regions to be associated with four growth traits (AGE, ADG, BF, and LMP) in a Canadian Duroc pig population using the wssGWAS method. The identified windows explained 1.00 to 3.07% of the genetic variance. Furthermore, 21 genes with related functional validation in previous studies were highlighted as candidate genes for growth traits in pigs. Moreover, GO, and KEGG enrichment analyses implied that the identified genes took part in bone formation, the immune system, and digestion, which were associated with growth traits. Such a full use of phenotypic and genotypic data and genealogical information will further advance our understanding of the genetic architecture and accelerate the genetic improvement of these economically important traits in pigs. In addition, the SNPs within identified regions may be useful for marker-assisted selection or genomic selection in future pig breeding.

**Supplementary Materials:** The following are available online at https://www.mdpi.com/2073-4425/12/1/117/s1, Table S1: Genomic regions of 0.8 Mb explained more than 1% of genetic variance for growth traits in Duroc pigs; Table S2: The explained genetic variance of SNPs within significant windows for AGE; Table S3: The explained genetic variance of SNPs within significant windows for ADG; Table S4: The explained genetic variance of SNPs within significant windows for BF; Table S5: The explained genetic variance of SNPs within significant windows for LMP.

**Author Contributions:** J.Y. and Z.W. conceived and designed the experiment. D.R., Z.Z., and R.D. performed the experiments. Y.Q., S.Z., J.W., C.X., L.H., S.H., G.C., and E.Z. collected the samples and recorded the phenotypes. D.R., Z.Z., and J.Y. analyzed the data and wrote the manuscript. Z.W. contributed to the materials. All authors have read and agreed to the published version of the manuscript.

**Funding:** This research was funded by the Local Innovative and Research Teams Project of Guangdong Province (2019BT02N630), the National Modern Agricultural Industry Science and Technology Innovation Center Creation Project of Guangzhou (2018KCZX01), the Natural Science Foundation of Guangdong Province (2018B030315007) and the Pearl River S and T Nova Program of Guangzhou (201906010011). The funders had no role in study design, data collection and analysis, decision to publish, or preparation of the manuscript.

**Institutional Review Board Statement:** This study was approved by the Ethics Committee of South China Agricultural University (SCAU, Guangzhou, China and Approval number SCAU#0017).

**Informed Consent Statement:** Not applicable.

**Data Availability Statement:** The datasets generated and/or analyzed during the current study are not publicly available since the studied population is consisted of the nucleus herd of Wens Foodstuff Group Co., Ltd., but are available from the corresponding author on reasonable request.

**Acknowledgments:** The authors would like to thank Wens Foodstuff Group Co., Ltd. (Guangdong, China) for providing all phenotypic data, pedigree information, and ear tissue samples.

**Conflicts of Interest:** The authors declare no conflict of interest.

## References

1. Szűcs, I.; Vida, V. Global tendencies in pork meat-production, trade and consumption. *Appl. Stud. Agribus. Commer.* **2017**, *11*, 105–111. [CrossRef]
2. Fontanesi, L.; Schiavo, G.; Galimberti, G.; Calò, D.G.; Russo, V. A genomewide association study for average daily gain in Italian Large White pigs. *J. Anim. Sci.* **2014**, *92*, 1385–1394. [CrossRef] [PubMed]
3. Hoque, M.A.; Suzuki, K.; Kadowaki, H.; Shibata, T.; Oikawa, T. Genetic parameters for feed efficiency traits and their relationships with growth and carcass traits in Duroc pigs. *J. Anim. Breed. Genet.* **2007**, *124*, 108–116. [CrossRef] [PubMed]
4. Visscher, P.M.; Wray, N.R.; Zhang, Q.; Sklar, P.; McCarthy, M.I.; Brown, M.A.; Yang, J. 10 Years of GWAS Discovery: Biology, Function, and Translation. *Am. J. Hum. Genet.* **2017**, *101*, 5–22. [CrossRef] [PubMed]
5. Georges, M.; Charlier, C.; Hayes, B. Harnessing genomic information for livestock improvement. *Nat. Rev. Genet.* **2018**, *20*, 135–156. [CrossRef]
6. Jiang, Y.; Tang, S.; Wang, C.; Wang, Y.; Qin, Y.; Wang, Y.; Zhang, J.; Song, H.; Mi, S.; Yu, F.; et al. A genome-wide association study of growth and fatness traits in two pig populations with different genetic backgrounds. *J. Anim. Sci.* **2018**, *96*, 806–816. [CrossRef]
7. Qiao, R.; Gao, J.; Zhang, Z.; Li, L.; Xie, X.; Fan, Y.; Cui, L.; Ma, J.; Ai, H.; Ren, J.; et al. Genome-wide association analyses reveal significant loci and strong candidate genes for growth and fatness traits in two pig populations. *Genet. Sel. Evol.* **2015**, *47*, 17. [CrossRef]
8. Wang, H.; Misztal, I.; Aguilar, I.; Legarra, A.; Fernando, R.L.; Vitezica, Z.; Okimoto, R.; Wing, T.; Hawken, R.; Muir, W.M. Genome-wide association mapping including phenotypes from relatives without genotypes in a single-step (ssGWAS) for 6-week body weight in broiler chickens. *Front. Genet.* **2014**, *5*, 134. [CrossRef]
9. Garrick, D.J.; Taylor, J.F.; Fernando, R.L. Deregressing estimated breeding values and weighting information for genomic regression analyses. *Genet. Sel. Evol.* **2009**, *41*, 55. [CrossRef]
10. Stafuzza, N.B.; Silva, R.M.d.O.; Fragomeni, B.D.O.; Masuda, Y.; Huang, Y.J.; Gray, K.A.; Lourenco, D.A.L. A genome-wide single nucleotide polymorphism and copy number variation analysis for number of piglets born alive. *Bmc Genom.* **2019**, *20*, 321. [CrossRef]
11. Silva, R.M.D.O.; Stafuzza, N.B.; Fragomeni, B.D.O.; De Camargo, G.M.F.; Ceacero, T.M.; Cyrillo, J.N.D.S.G.; Baldi, F.; Boligon, A.A.; Mercadante, M.E.Z.; Lourenco, D.L.; et al. Genome-Wide Association Study for Carcass Traits in an Experimental Nelore Cattle Population. *PLoS ONE* **2017**, *12*, e0169860. [CrossRef]
12. Gonzalez-Pena, D.; Gao, G.; Baranski, M.; Moen, T.; Cleveland, B.M.; Kenney, P.B.; Vallejo, R.L.; Palti, Y.; Leeds, T.D. Genome-Wide Association Study for Identifying Loci that Affect Fillet Yield, Carcass, and Body Weight Traits in Rainbow Trout (Oncorhynchus mykiss). *Front. Genet.* **2016**, *7*, 203. [CrossRef] [PubMed]
13. Gao, N.; Chen, Y.; Liu, X.; Zhao, Y.; Zhu, L.; Liu, A.; Jiang, W.; Peng, X.; Zhang, C.; Tang, Z.; et al. Weighted single-step GWAS identified candidate genes associated with semen traits in a Duroc boar population. *BMC Genom.* **2019**, *20*, 797. [CrossRef]
14. Tang, Z.; Xu, J.; Yin, L.; Yin, D.; Zhu, M.; Yu, M.; Li, X.; Zhao, S.; Liu, X. Genome-Wide Association Study Reveals Candidate Genes for Growth Relevant Traits in Pigs. *Front. Genet.* **2019**, *10*, 302. [CrossRef] [PubMed]
15. Suzuki, K.; Kadowaki, H.; Shibata, T.; Uchida, H.; Nishida, A. Selection for daily gain, loin-eye area, backfat thickness and intramuscular fat based on desired gains over seven generations of Duroc pigs. *Livest Prod. Sci.* **2005**, *97*, 193–202. [CrossRef]
16. Zhao, Y.; Jin, C.; Xuan, Y.; Zhou, P.; Fang, Z.; Che, L.; Xu, S.; Feng, B.; Li, J.; Jiang, X.; et al. Effect of maternal or post-weaning methyl donor supplementation on growth performance, carcass traits, and meat quality of pig offspring. *J. Sci. Food Agric.* **2019**, *99*, 2096–2107. [CrossRef]

17. Chang, C.C.; Chow, C.C.; Tellier, L.C.; Vattikuti, S.; Purcell, S.M.; Lee, J.J. Second-generation PLINK: Rising to the challenge of larger and richer datasets. *Gigascience* **2015**, *4*, 7. [CrossRef]
18. Wang, H.; Misztal, I.; Aguilar, I.; Legarra, A.; Muir, W.M. Genome-wide association mapping including phenotypes from relatives without genotypes. *Genet. Res.* **2012**, *94*, 73–83. [CrossRef]
19. Aguilar, I.; Misztal, I.; Johnson, D.L.; Legarra, A.; Tsuruta, S.; Lawlor, T.J. Hot topic: A unified approach to utilize phenotypic, full pedigree, and genomic information for genetic evaluation of Holstein final score. *J. Dairy Sci.* **2010**, *93*, 743–752. [CrossRef]
20. VanRaden, P.M. Efficient methods to compute genomic predictions. *J. Dairy Sci.* **2008**, *91*, 4414–4423. [CrossRef]
21. VanRaden, P.M.; Van Tassell, C.P.; Wiggans, G.R.; Sonstegard, T.S.; Schnabel, R.D.; Taylor, J.F.; Schenkel, F.S. Invited review: Reliability of genomic predictions for North American Holstein bulls. *J. Dairy Sci.* **2009**, *92*, 16–24. [CrossRef] [PubMed]
22. Zhuang, Z.; Ding, R.; Peng, L.; Wu, J.; Ye, Y.; Zhou, S.; Wang, X.; Quan, J.; Zheng, E.; Cai, G.; et al. Genome-wide association analyses identify known and novel loci for teat number in Duroc pigs using single-locus and multi-locus models. *BMC Genom.* **2020**, *21*, 344. [CrossRef] [PubMed]
23. Misztal, I.; Tsuruta, S.; Strabel, T.; Auvray, B.; Druet, T.; Lee, D. BLUPF90 and related programs (BGF90). In Proceedings of the 7th World Congress on Genetics Applied to Livestock Production, Montpellier, France, 19–23 August 2002; pp. 743–744.
24. Herrera-Cáceres, W.; Sánchez, J.P. Selection for feed efficiency using the social effects animal model in growing Duroc pigs: Evaluation by simulation. *Genet. Sel. Evol.* **2020**, *52*, 53. [CrossRef] [PubMed]
25. Sanchez, M.P.; Tribout, T.; Iannuccelli, N.; Bouffaud, M.; Servin, B.; Tenghe, A.; Dehais, P.; Muller, N.; Del Schneider, M.P.; Mercat, M.J.; et al. A genome-wide association study of production traits in a commercial population of Large White pigs: Evidence of haplotypes affecting meat quality. *Genet. Sel. Evol.* **2014**, *46*, 12. [CrossRef]
26. van Kuijk, S.J.A.; Jacobs, M.; Smits, C.H.M.; Han, Y. The effect of hydroxychloride trace minerals on the growth performance and carcass quality of grower/finisher pigs: A meta-analysis. *J. Anim. Sci.* **2019**, *97*, 4619–4624. [CrossRef]
27. Howard, J.T.; Jiao, S.; Tiezzi, F.; Huang, Y.; Gray, K.A.; Maltecca, C. Genome-wide association study on legendre random regression coefficients for the growth and feed intake trajectory on Duroc Boars. *BMC Genet.* **2015**, *16*, 59. [CrossRef]
28. Ødegård, J.; Meuwissen, T.H. Estimation of heritability from limited family data using genome-wide identity-by-descent sharing. *Genet. Sel. Evol.* **2012**, *44*, 16. [CrossRef]
29. Marques, D.B.D.; Bastiaansen, J.W.M.; Broekhuijse, M.; Lopes, M.S.; Knol, E.F.; Harlizius, B.; Guimaraes, S.E.F.; Silva, F.F.; Lopes, P.S. Weighted single-step GWAS and gene network analysis reveal new candidate genes for semen traits in pigs. *Genet. Sel. Evol.* **2018**, *50*, 40. [CrossRef]
30. Peters, S.O.; Kizilkaya, K.; Garrick, D.J.; Fernando, R.L.; Reecy, J.M.; Weaber, R.L.; Silver, G.A.; Thomas, M.G. Bayesian genome-wide association analysis of growth and yearling ultrasound measures of carcass traits in Brangus heifers. *J. Anim. Sci.* **2012**, *90*, 3398–3409. [CrossRef]
31. Habier, D.; Fernando, R.L.; Kizilkaya, K.; Garrick, D.J. Extension of the bayesian alphabet for genomic selection. *BMC Bioinform.* **2011**, *12*, 186. [CrossRef]
32. Bergamaschi, M.; Maltecca, C.; Fix, J.; Schwab, C.; Tiezzi, F. Genome-wide association study for carcass quality traits and growth in purebred and crossbred pigs1. *J. Anim. Sci.* **2020**, *98*. [CrossRef]
33. Yang, J.; Huang, L.; Yang, M.; Fan, Y.; Li, L.; Fang, S.; Deng, W.; Cui, L.; Zhang, Z.; Ai, H.; et al. Possible introgression of the VRTN mutation increasing vertebral number, carcass length and teat number from Chinese pigs into European pigs. *Sci. Rep.* **2016**, *6*, 19240. [CrossRef]
34. Choi, I.; Steibel, J.P.; Bates, R.O.; Raney, N.E.; Rumph, J.M.; Ernst, C.W. Identification of Carcass and Meat Quality QTL in an F(2) Duroc × Pietrain Pig Resource Population Using Different Least-Squares Analysis Models. *Front. Genet.* **2011**, *2*, 18. [CrossRef]
35. Bateman, A.; Bennett, H.P. The granulin gene family: From cancer to dementia. *BioEssays* **2009**, *31*, 1245–1254. [CrossRef]
36. Seabury, C.M.; Oldeschulte, D.L.; Saatchi, M.; Beever, J.E.; Decker, J.E.; Halley, Y.A.; Bhattarai, E.K.; Molaei, M.; Freetly, H.C.; Hansen, S.L.; et al. Genome-wide association study for feed efficiency and growth traits in U.S. beef cattle. *BMC Genom.* **2017**, *18*, 386. [CrossRef]
37. Zhang, Z.; Chen, Z.; Ye, S.; He, Y.; Huang, S.; Yuan, X.; Chen, Z.; Zhang, H.; Li, J. Genome-Wide Association Study for Reproductive Traits in a Duroc Pig Population. *Animals* **2019**, *9*, 732. [CrossRef]
38. Bassett, J.H.; Williams, G.R. Role of Thyroid Hormones in Skeletal Development and Bone Maintenance. *Endocr. Rev.* **2016**, *37*, 135–187. [CrossRef]
39. Wang, Y.; Kuang, Z.; Yu, X.; Ruhn, K.A.; Kubo, M.; Hooper, L.V. The intestinal microbiota regulates body composition through NFIL3 and the circadian clock. *Science* **2017**, *357*, 912–916. [CrossRef]
40. DeChiara, T.M.; Kimble, R.B.; Poueymirou, W.T.; Rojas, J.; Masiakowski, P.; Valenzuela, D.M.; Yancopoulos, G.D. Ror2, encoding a receptor-like tyrosine kinase, is required for cartilage and growth plate development. *Nat. Genet.* **2000**, *24*, 271–274. [CrossRef] [PubMed]
41. Reyer, H.; Varley, P.F.; Murani, E.; Ponsuksili, S.; Wimmers, K. Genetics of body fat mass and related traits in a pig population selected for leanness. *Sci. Rep.* **2017**, *7*, 9118. [CrossRef]
42. Mei, C.G.; Gui, L.S.; Wang, H.C.; Tian, W.Q.; Li, Y.K.; Zan, L.S. Polymorphisms in adrenergic receptor genes in Qinchuan cattle show associations with selected carcass traits. *Meat Sci.* **2018**, *135*, 166–173. [CrossRef]
43. Bachman, E.S.; Dhillon, H.; Zhang, C.Y.; Cinti, S.; Bianco, A.C.; Kobilka, B.K.; Lowell, B.B. betaAR signaling required for diet-induced thermogenesis and obesity resistance. *Science* **2002**, *297*, 843–845. [CrossRef]

44. Odgren, P.R.; Kim, N.; MacKay, C.A.; Mason-Savas, A.; Choi, Y.; Marks, S.C., Jr. The role of RANKL (TRANCE/TNFSF11), a tumor necrosis factor family member, in skeletal development: Effects of gene knockout and transgenic rescue. *Connect. Tissue Res.* **2003**, *44*, 264–271. [CrossRef]
45. Whyte, M.P.; Totty, W.G.; Novack, D.V.; Zhang, X.; Wenkert, D.; Mumm, S. Camurati-Engelmann disease: Unique variant featuring a novel mutation in TGFβ1 encoding transforming growth factor β 1 and a missense change in TNFSF11 encoding RANK ligand. *J. Bone Miner. Res.* **2011**, *26*, 920–933. [CrossRef]
46. Heemers, H.; Vanderhoydonc, F.; Roskams, T.; Shechter, I.; Heyns, W.; Verhoeven, G.; Swinnen, J.V. Androgens stimulate coordinated lipogenic gene expression in normal target tissues in vivo. *Mol. Cell. Endocrinol.* **2003**, *205*, 21–31. [CrossRef]
47. Lin, D.Y.; Lai, M.Z.; Ann, D.K.; Shih, H.M. Promyelocytic leukemia protein (PML) functions as a glucocorticoid receptor co-activator by sequestering Daxx to the PML oncogenic domains (PODs) to enhance its transactivation potential. *J. Biol. Chem.* **2003**, *278*, 15958–15965. [CrossRef]
48. Tordoff, M.G.; Ellis, H.T. Taste dysfunction in BTBR mice due to a mutation of Itpr3, the inositol triphosphate receptor 3 gene. *Physiol. Genom.* **2013**, *45*, 834–855. [CrossRef]
49. Moritoh, Y.; Oka, M.; Yasuhara, Y.; Hozumi, H.; Iwachidow, K.; Fuse, H.; Tozawa, R. Inositol hexakisphosphate kinase 3 regulates metabolism and lifespan in mice. *Sci. Rep.* **2016**, *6*, 1–13. [CrossRef]
50. Sun, H.; Wang, Z.; Zhang, Z.; Xiao, Q.; Mawed, S.; Xu, Z.; Zhang, X.; Yang, H.; Zhu, M.; Xue, M.; et al. Genomic signatures reveal selection of characteristics within and between Meishan pig populations. *Anim Genet.* **2018**, *49*, 119–126. [CrossRef]
51. Zhuang, Z.; Li, S.; Ding, R.; Yang, M.; Zheng, E.; Yang, H.; Gu, T.; Xu, Z.; Cai, G.; Wu, Z.; et al. Meta-analysis of genome-wide association studies for loin muscle area and loin muscle depth in two Duroc pig populations. *PLoS ONE* **2019**, *14*, e0218263. [CrossRef]
52. Li, T.; Chanda, D.; Zhang, Y.; Choi, H.S.; Chiang, J.Y. Glucose stimulates cholesterol 7alpha-hydroxylase gene transcription in human hepatocytes. *J. Lipid Res.* **2010**, *51*, 832–842. [CrossRef] [PubMed]
53. Dumitrescu, A.M.; Liao, X.H.; Abdullah, M.S.; Lado-Abeal, J.; Majed, F.A.; Moeller, L.C.; Boran, G.; Schomburg, L.; Weiss, R.E.; Refetoff, S. Mutations in SECISBP2 result in abnormal thyroid hormone metabolism. *Nat. Genet.* **2005**, *37*, 1247–1252. [CrossRef] [PubMed]
54. Liu, Y.; Zhou, D.; Zhang, F.; Tu, Y.; Xia, Y.; Wang, H.; Zhou, B.; Zhang, Y.; Wu, J.; Gao, X.; et al. Liver Patt1 deficiency protects male mice from age-associated but not high-fat diet-induced hepatic steatosis. *J. Lipid Res.* **2012**, *53*, 358–367. [CrossRef] [PubMed]
55. Zhang, J.; Zhang, Y.; Gong, H.; Cui, L.; Ma, J.; Chen, C.; Ai, H.; Xiao, S.; Huang, L.; Yang, B. Landscape of Loci and Candidate Genes for Muscle Fatty Acid Composition in Pigs Revealed by Multiple Population Association Analysis. *Front. Genet.* **2019**, *10*, 1067. [CrossRef] [PubMed]
56. Yang, R.-Y.; Yu, L.; Graham, J.; Hsu, D.K.; Lloyd, K.C.K.; Havel, P.J.; Liu, F.-T. Ablation of a galectin preferentially expressed in adipocytes increases lipolysis, reduces adiposity, and improves insulin sensitivity in mice. *Proc. Natl. Acad. Sci. USA* **2011**, *108*, 18696–18701. [CrossRef]
57. Wu, W.; Zhang, D.; Yin, Y.; Ji, M.; Xu, K.; Huang, X.; Peng, Y.; Zhang, J. Comprehensive transcriptomic view of the role of the LGALS12 gene in porcine subcutaneous and intramuscular adipocytes. *BMC Genom.* **2019**, *20*, 509. [CrossRef]
58. Seres, J.; Bornstein, S.R.; Seres, P.; Willenberg, H.S.; Schulte, K.M.; Scherbaum, W.A.; Ehrhart-Bornstein, M. Corticotropin-releasing hormone system in human adipose tissue. *J. Clin. Endocrinol. Metab.* **2004**, *89*, 965–970. [CrossRef]
59. Hofer, D.C.; Pessentheiner, A.R.; Pelzmann, H.J.; Schlager, S.; Madreiter-Sokolowski, C.T.; Kolb, D.; Eichmann, T.O.; Rechberger, G.; Bilban, M.; Graier, W.F.; et al. Critical role of the peroxisomal protein PEX16 in white adipocyte development and lipid homeostasis. *Biochim. Et Biophys. Acta. Mol. Cell Biol. Lipids* **2017**, *1862*, 358–368. [CrossRef]
60. Mármol-Sánchez, E.; Quintanilla, R.; Cardoso, T.F.; Jordana Vidal, J.; Amills, M. Polymorphisms of the cryptochrome 2 and mitoguardin 2 genes are associated with the variation of lipid-related traits in Duroc pigs. *Sci. Rep.* **2019**, *9*, 9025. [CrossRef]
61. Krause, K.; Weiner, J.; Hönes, S.; Klöting, N.; Rijntjes, E.; Heiker, J.T.; Gebhardt, C.; Köhrle, J.; Führer, D.; Steinhoff, K.; et al. The Effects of Thyroid Hormones on Gene Expression of Acyl-Coenzyme A Thioesterases in Adipose Tissue and Liver of Mice. *Eur. Thyroid J.* **2015**, *4*, 59–66. [CrossRef]
62. Godinho, R.M.; Bergsma, R.; Silva, F.F.; Sevillano, C.A.; Knol, E.F.; Lopes, M.S.; Lopes, P.S.; Bastiaansen, J.W.M.; Guimarães, S.E.F. Genetic correlations between feed efficiency traits, and growth performance and carcass traits in purebred and crossbred pigs. *J. Anim. Sci.* **2018**, *96*, 817–829. [CrossRef] [PubMed]
63. Omori, S.; Tanaka, Y.; Takahashi, A.; Hirose, H.; Kashiwagi, A.; Kaku, K.; Kawamori, R.; Nakamura, Y.; Maeda, S. Association of CDKAL1, IGF2BP2, CDKN2A/B, HHEX, SLC30A8, and KCNJ11 with susceptibility to type 2 diabetes in a Japanese population. *Diabetes* **2008**, *57*, 791–795. [CrossRef] [PubMed]
64. Kasza, I.; Suh, Y.; Wollny, D.; Clark, R.J.; Roopra, A.; Colman, R.J.; MacDougald, O.A.; Shedd, T.A.; Nelson, D.W.; Yen, M.I.; et al. Syndecan-1 is required to maintain intradermal fat and prevent cold stress. *PLoS Genet.* **2014**, *10*, e1004514. [CrossRef]
65. Belaid, Z.; Hubint, F.; Humblet, C.; Boniver, J.; Nusgens, B.; Defresne, M.P. Differential expression of vascular endothelial growth factor and its receptors in hematopoietic and fatty bone marrow: Evidence that neuropilin-1 is produced by fat cells. *Haematologica* **2005**, *90*, 400–401.
66. Dai, X.; Okon, I.; Liu, Z.; Bedarida, T.; Wang, Q.; Ramprasath, T.; Zhang, M.; Song, P.; Zou, M.H. Ablation of Neuropilin 1 in Myeloid Cells Exacerbates High-Fat Diet-Induced Insulin Resistance Through Nlrp3 Inflammasome In Vivo. *Diabetes* **2017**, *66*, 2424–2435. [CrossRef]

67. Higgins, M.G.; Kenny, D.A.; Fitzsimons, C.; Blackshields, G.; Coyle, S.; McKenna, C.; McGee, M.; Morris, D.W.; Waters, S.M. The effect of breed and diet type on the global transcriptome of hepatic tissue in beef cattle divergent for feed efficiency. *BMC Genom.* **2019**, *20*, 525. [CrossRef]
68. Liu, K.; Czaja, M.J. Regulation of lipid stores and metabolism by lipophagy. *Cell Death Differ.* **2013**, *20*, 3–11. [CrossRef]
69. Bhoj, E.J.; Romeo, S.; Baroni, M.G.; Bartov, G.; Schultz, R.A.; Zinn, A.R. MODY-like diabetes associated with an apparently balanced translocation: Possible involvement of MPP7 gene and cell polarity in the pathogenesis of diabetes. *Mol. Cytogenet.* **2009**, *2*, 5. [CrossRef]
70. Shim, M.Y.; Karnuah, A.B.; Mitchell, A.D.; Anthony, N.B.; Pesti, G.M.; Aggrey, S.E. The effects of growth rate on leg morphology and tibia breaking strength, mineral density, mineral content, and bone ash in broilers. *Poult. Sci.* **2012**, *91*, 1790–1795. [CrossRef]
71. Cameron, N.D.; Curran, M.K. Responses in carcass composition to divergent selection for components of efficient lean growth rate in pigs. *Anim. Sci.* **2010**, *61*, 347–359. [CrossRef]
72. Wang, W.; Xue, W.; Jin, B.; Zhang, X.; Ma, F.; Xu, X. Candidate gene expression affects intramuscular fat content and fatty acid composition in pigs. *J. Appl Genet.* **2013**, *54*, 113–118. [CrossRef] [PubMed]
73. Sandoval, D.; Cota, D.; Seeley, R.J. The integrative role of CNS fuel-sensing mechanisms in energy balance and glucose regulation. *Annu. Rev. Physiol.* **2008**, *70*, 513–535. [CrossRef] [PubMed]
74. Ye, W.; Chang, R.B.; Bushman, J.D.; Tu, Y.H.; Mulhall, E.M.; Wilson, C.E.; Cooper, A.J.; Chick, W.S.; Hill-Eubanks, D.C.; Nelson, M.T.; et al. The K+ channel KIR2.1 functions in tandem with proton influx to mediate sour taste transduction. *Proc. Natl. Acad. Sci. USA* **2016**, *113*, E229–E238. [CrossRef]
75. Ribani, A.; Bertolini, F.; Schiavo, G.; Scotti, E.; Utzeri, V.J.; Dall'Olio, S.; Trevisi, P.; Bosi, P.; Fontanesi, L. Next generation semiconductor based sequencing of bitter taste receptor genes in different pig populations and association analysis using a selective DNA pool-seq approach. *Anim. Genet.* **2017**, *48*, 97–102. [CrossRef]
76. Des Gachons, C.P.; Breslin, P.A. Salivary amylase: Digestion and metabolic syndrome. *Curr. Diabetes Rep.* **2016**, *16*, 102. [CrossRef]

*Article*

# Identification of Circular RNAs in Hypothalamus of Gilts during the Onset of Puberty

**Qingnan Li [1,†], Xiangchun Pan [2,†], Nian Li [2], Wentao Gong [2], Yaosheng Chen [1,*] and Xiaolong Yuan [2,3,*]**

1 State Key Laboratory of Biocontrol, School of Life Sciences, Sun Yat-Sen University, North Third Road, Guangzhou Higher Education Mega Center, Guangzhou 510006, China; liqn27@mail2.sysu.edu.cn

2 National Engineering Research Center for Breeding Swine Industry, Guangdong Provincial Key Lab of Agro-Animal Genomics and Molecular Breeding, College of Animal Science, South China Agricultural University, Guangzhou 510642, China; 15186255977@126.com (X.P.); 17746076937@163.com (N.L.); g_w_tao@163.com (W.G.)

3 Guangdong Provincial Key Laboratory of Laboratory Animals, Guangdong Laboratory Animals Monitoring Institute, Guangzhou 510260, China

* Correspondence: chyaosh@mail.sysu.edu.cn (Y.C.); yxl@scau.edu.cn (X.Y.)

† These authors contributed equally to this article.

**Abstract:** The disorders of puberty have shown negative outcomes on health of mammals, and the hypothalamus is thought to be the main regulator of puberty by releasing GnRH. Many studies show that the circular RNAs (circRNAs) might be implicated in the timing of puberty in mammals. However, the circRNAs in the hypothalamus of gilts have not been explored. To profile the changes and biological functions of circRNAs in the hypothalamus during the onset of puberty, RNA-seq was utilized to establish pre-, in-, and post-pubertal hypothalamic circRNAs profiles. In this study, the functions of hypothalamic circRNAs were enriched in the signaling pathway of neurotrophin, progesterone-mediated oocyte maturation, oocyte meiosis, insulin, ErbB, and mTOR, which have been highly suggested to be involved in the timing of puberty. Furthermore, 53 circRNAs were identified to be putative hypothalamus-specific expressed circRNAs, and some of them were exclusively expressed in the one of three pubertal stages. Moreover, 22 differentially expressed circRNAs were identified and chosen to construct the circRNA-miRNA-gene network. Moreover, 10 circRNAs were found to be driven by six puberty-related genes (*ESR1*, *NF1*, *APP*, *ENPP2*, *ARNT*, and *DICER1*). Subsequently, the expression changes of several circRNAs were confirmed by RT-qPCR. Collectively, the preliminary results of hypothalamic circRNAs provided useful information for the investigation of the molecular mechanism for the timing of puberty in gilts.

**Keywords:** hypothalamus; puberty; circRNAs; pubertal genes

**Citation:** Li, Q.; Pan, X.; Li, N.; Gong, W.; Chen, Y.; Yuan, X. Identification of Circular RNAs in Hypothalamus of Gilts during the Onset of Puberty. *Genes* **2021**, *12*, 84. https://doi.org/10.3390/genes12010084

Received: 1 December 2020
Accepted: 8 January 2021
Published: 12 January 2021

**Publisher's Note:** MDPI stays neutral with regard to jurisdictional claims in published maps and institutional affiliations.



## 1. Introduction

In female pigs, puberty is widely defined as the emergence of the first estrous and capable of reproduction [1]. There are more evidences demonstrated that gilts having an earlier age at puberty can shorten the generation interval of livestock [2,3] and farrow multiple litters [4]. Nevertheless, the basic molecular mechanisms that regulate the onset of puberty have not been largely explored in gilts. Generally, the onset of puberty is controlled and driven by hypothalamic-pituitary-gonadal (HPG) axis. The release of gonadotropin-releasing hormone (GnRH) from the hypothalamus leads to the release of FSH and LH from the pituitary [5], and the FSH and LH act on the folliculogenesis, oogenesis, and sex steroid of the gonads to arouse the timing of puberty in mammals [6]. Lomniczi, A. et al. showed that disrupting the release of pulsatile GnRH in hypothalamus delayed puberty [7]. Pandolfi, E.C. et al. demonstrated that the deletion of homeodomain protein sine oculis-related homeobox 6 (Six6) in hypothalamic GnRH neuron can leads to infertility [8]. These demonstrations indicate that the hypothalamus plays an essential role in the onset of puberty.

Circular RNAs (circRNAs) are covalently closed transcript generated by the back-splicing. This back-splicing jointed a canonical 5′ splice site sequence to an upstream 3′ splice site sequence to produce the only region of a circRNA (BMJ, back-spliced junction) [9]. Multiple circRNAs have been showed to be generated by a single gene through alternative splicing [10]. Recently, next-generation sequencing has shown that circRNAs are widespread expression in mammals [11–13]. Moreover, circRNAs have been suggested to be stage-specific, cell- specific, and tissue-specific in the development of mammals [10,14]. Furthermore, most circRNAs are consisted of exons, while a few numbers of circRNAs are formed by the exon-intronic or intronic RNA in mammals [13]. It has been shown that the exonic circRNA may induce active DNA methylation through recruit specific protein, such as the FLI1 exonic circRNA recruits the methylcytosine dioxygenase TET1 to the promoter region of its parental gene [15].

Recently, several studies have showed that circRNAs could regulate the transcription of genes [16–18]. Specifically, circRNAs can be used as sponge for microRNAs (miRNAs). For example, Hall et al. showed that circ_Lrp6 was the sponge for circ_Lrp6 to counterbalance functions of the miRNA in functions of the miRNA in VSMCs [19]. Jost et al. produced the artificial circRNAs to inhibit the viral protein production by acting as sponges for the miRNA relevant in human disease [20]. Moreover, increasing evidence has shown that circRNAs are significantly enriched in mammalian brain and are related to physiological development of the brain [21]. Study has shown that the unique patterns of circRNAs across tissues and development stages seemed to reflect the reproductively capable individuals [22]. In addition, recent research indicated that circRNAs were closely connected to development in pig's brain. For instance, M.T. et al. identified large amounts of circRNAs in fetal brain of pig, and indicated that circRNA was significant impacted gilts' brain development [23]. These results indicated that circRNAs might play an indispensable role in multiple critical biological process in pigs. However, circRNAs has rarely been studies in onset of puberty of gilts.

In this study, the hypothalamus of pre-, in-, and post-pubertal gilts were utilized for RNA-seq analysis to explore the expression of circRNAs driven by the pubertal genes, and then investigate a circRNA-miRNA-gene network. It is hoped that the results of this study will provide insight into the potential function of circRNA in gilts during the onset of puberty and help in identifying circRNAs that play pivotal role in this process.

## 2. Materials and Methods

### 2.1. Ethics Statement

All animal experiments were approved by the Animal Care and Use Committee of the South China Agricultural University, Guangzhou, China (permit number: SCAU#2013-10), and conducted with the Regulations for the Administration of Affairs Concerning Experimental Animals (Ministry of Science and Technology, China; revised in June 2004).

### 2.2. Animals

All the experimental Landrace × Yorkshire crossbred gilts were monitored periodically for signs of puberty, including body weight, days of age, change of vulva, and the reaction to the boars, and the onset of puberty was identified by looking at this information. After that, three stages during the onset of puberty (pre-, in-, and post-puberty) were used. Thereinto, three gilts were designated as pre-puberty gilts (160 days old) without any pubertal signs (weight = 81.38 ± 2.40 kg); three gilts were selected as the in-pubertal gilts which exhibited first pubertal signs (weight = 110.00 ± 2.00 kg); three gilts (14 days old) were served as the post-pubertal gilts beyond the pubertal phase (weight = 122.82 ± 9.11 kg). After euthanasia, the hypothalamuses of gilts were immediately removed, placed in the liquid nitrogen, and then stored it at −80 °C until further use. In addition, refer to other researchers' experimental studies on circRNAs, three replicates in each group were used in this study [24].

### 2.3. RNA Sequencing and the Transcriptome Assembly

Total RNA from the pre-, in- and post-pubertal hypothalamuses of gilts was isolated with the Trizol agent (Invitrogen, Carlsbad, CA, USA). After quality testing of total RNA using the Agilent Bioanalyzer 2100 System (Agilent Technologies, Santa Clara, CA, USA), RNA samples with RNA integrity value of greater than 7.0 were left behind. Subsequently, rRNA was removed using the Epicentre Ribo-zero rRNA removal kits (Epicentre, Madison, WI, USA). Then we used the rRNA-depleted RNAs to compose double-stranded cDNA with the mRNA-Seq Sample Preparation Kit (Illumina, San Diego, CA, USA). Each sample was sequenced using the HiSeq 3000 sequencer according to the manufacturer's instructions for 5 μg cDNA and generated 150 bp paired-end reads. These raw reads were subjected to quality control using the Cutadapt software [25] to remove the 3′ adaptor-trimming, the low-quality reads which had >10% of unknown bases or >50% of the low-mass bases. Remaining reads after quality control were clean reads, which will be mapped onto the pig reference genome Sus scrofa11.1 by BWA [26] and bowtie2 [27] software.

### 2.4. circRNA Identification and Data Analysis

CIRI2 [28] was used to identify circRNA after BWA, and find_circ [29] was used to identify circRNAs after bowtie2, which based on the reference genome alignment. We screened the number of unique junctions read to be at least 2, removed RNA with unclear breakpoints, and filtered out RNA with a length greater than 100 kb (genome length) as potential circRNA. Analysis included three replicates for each stage. Finally, circRNAs in pre-, in- and post-pubertal hypothalamuses of gilts were identified. Subsequently, the two software identified the intersection of circRNAs as the candidate circRNAs, and the annotation of circRNAs was proposed by CIRI for further study, which was based on the annotation file from Ensembl release 95. Furthermore, circRNAs originating from exons was used for further analyses. Length of circRNA is the sum of the lengths of the exons that form the circRNA. Besides, we obtained the circRNAs expression with BSJ reads, and used EBSeq package to calculate RPM [30]. In addition, the screening criteria for differential expression were FDR < 0.05, $\log2 - \text{fold} - \text{change} - \geq 1$. Furthermore, the screening criteria for stage-specific circRNAs were as follows: circRNA detecting only in a unique stage was judged as stage-specific circRNA. The tissue-specific screening criteria were as follows: identification of circRNAs in this study were matched with the known pig' circRNAs through the starting and ending genomic positions of circRNAs, and the new circRNAs that were not matched in the database were regarded as the presumed tissue-specific circRNAs. In this study, the ggsignif package was used to perform statistical tests for differences between groups (Welch two-sample *t*-test).

### 2.5. Pathway Analysis and circRNA-miRNA-mRNA Network Construction

The parental genes of circRNA were used for Gene Ontology (GO) analysis and Kyoto Encyclopedia of Genes and Genomes (KEGG) pathway enrichment analysis that the cutoff criterion was $p < 0.05$ and the results were conducted with KOBAS 3.0 online software ( http://kobas.cbi.pku.edu.cn/) [31]. In addition, the differentially expressed genes were screened under the condition of FDR < 0.05, $\log2 - \text{fold} - \text{change} - \geq 3$. Furthermore, miRanda software [32] was used to predict circRNA-miRNA connections, miRanda match score ≥120. Then miRanda was as well as used to predict differentially expressed target genes of these miRNA, miRanda match score ≥200. Finally, cytoscape software [33] was used to draw a network interaction map between circRNA-miRNA-gene. Moreover, this analysis is based on the part of the transcript containing only exons.

### 2.6. circRNA Validation by RT-qPCR

We used RT and quantitative PCR (RT-qPCR) assays to validate the reliability of the high-throughput RNA sequencing data with divergent primers flanking the BSJ [9]. Used primeScript RT Reagent Kit (TaKaRa, Osaka, Japan) in a Mx3005P real-time PCR System (Stratagene, La Jolla, CA, USA) for qPCR according to the manufacturer's protocol.

Furthermore, the divergent primers of 5 circRNAs were designed to verify the accuracy of the RNA-seq. In order to normalize the expression of circRNAs, GAPDH was served as an internal reference. The PCR standard procedure were denaturation 94 °C (5 min), 40 cycles at 94 °C (10 s), 52 to 62 °C (15 s), and 72 °C (30 s). We used the $2^{-\Delta\Delta Ct}$ method to analyze the RT-qPCR data. The pre-, in- and post-pubertal hypothalamuses were come from three gilts. Moreover, three biological replicates were carried out in each qRT-PCR. The Student's t test was used to assess the differences between any two pubertal groups of gilts, and the screening criteria for statistically significant were $p < 0.05$.

## 3. Results

### 3.1. Identification of Hypothalamus-Derivced circRNAs during the Onset of Puberty

Totally, 2582 circRNAs candidates were identified by CIRI2 and find_circ software (Figure 1a, Supplementary Table S1). Respectively, 1619, 1273, and 1936 circRNAs were identified during the pre-, in- and post-puberty stages (Figure 1b). And the average circRNAs expression were highest in in-puberty compared with other two stages (Figure 1c). Moreover, circRNAs split into three categories: 2388 exonic circRNAs, 65 intronic circRNAs and 129 intergenic region circRNAs. Furthermore, these 2582 circRNAs were derived from 1487 genes, of which 1461 genes identified as able to produce 2388 exonic circRNA and 57 genes identified as able to produce 65 intronic circRNA (Figure 1d). Furthermore, the 2388 exonic circRNAs were used for subsequent analyses.

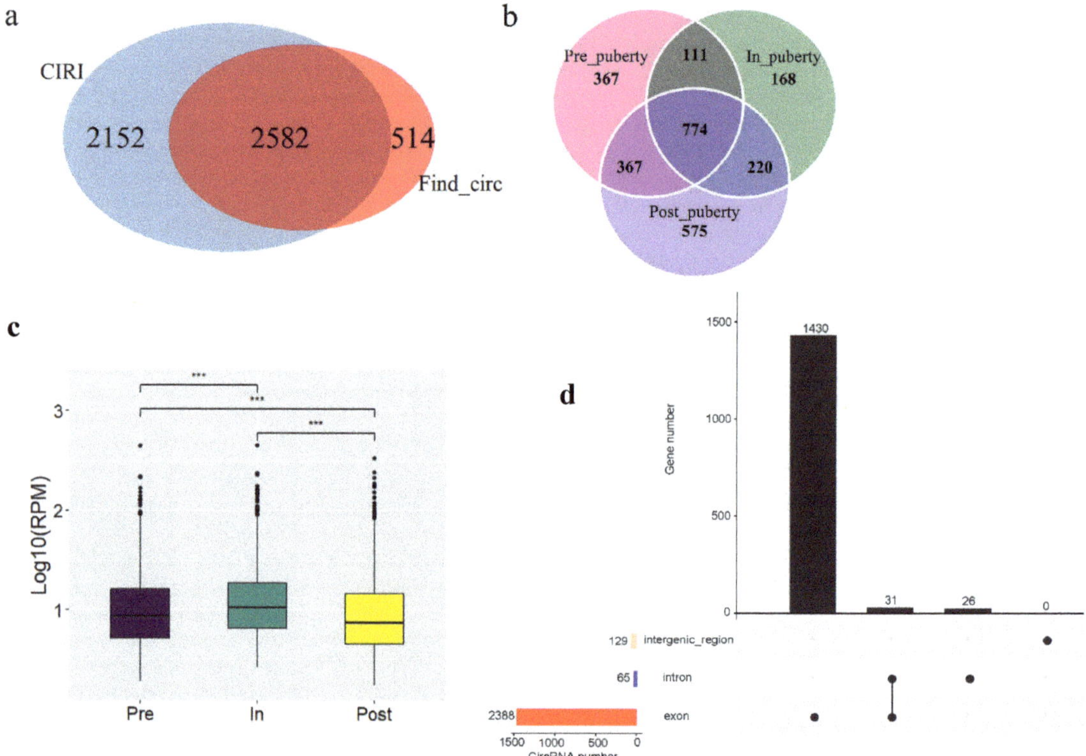


**Figure 1.** Overview of the identified circRNAs by RNA-seq analyses in ovaries of gilts. (**a**) The circRNAs were identified by two algorithms (CIRI and find_circ). (**b**) The number of unique and common circRNAs during pre-, in- and post-puberty. (**c**) Expression level of circRNAs in the pre-, in- and post-puberty stages. (**d**) The number of three types of circRNA and the number of corresponding parental genes. *** $p < 0.001$.

### 3.2. Key Pathways of cirRNAs in Pubertal Transition

To further explored the circRNA involved in pubertal hypothalamus, the KEGG analysis was used to perform the parental genes of all circRNAs (Supplementary Table S2). Notably, the functional pathways that were significantly overrepresented in pubertal hypothalamus included ras signaling pathway, insulin signaling pathway, ErbB signaling pathway, mTOR signaling pathway, neurotrophin signaling pathway, progesterone-mediated oocyte maturation and oocyte meiosis signaling pathway (Figure 2a). For the ras signaling pathway, *NF1* drive to "circ 12:43516178-43526438" which was uniquely expressed in the in-puberty, and drive to "circ 12:43673069-43691261" which was expressed in pre- and post-puberty; *EXOC2* drive to "circ 7:229160-252991" which was uniquely expressed in in-puberty and post-puberty, and drive to "circ 7:306094-311618" which was expressed in pre-puberty (Figure 2b, Supplementary Table S2). For the insulin signaling pathway, *PPP1CB* drive to "circ 3:110442415-110455791" which was expressed in pre- and post-puberty, but not expressed in post-puberty (Figure 2b, Supplementary Table S2). For the ErbB signaling pathway, *MAP2K4* drive to "circ 12:56438413-56490424" which was expressed in pre- and post-puberty, and drive to "circ 12:56464310-56490424" which was uniquely expressed in post-puberty (Figure 2b, Supplementary Table S2). For the mTOR signaling pathway, *RICTOR* drive to "circ 16:24175667-24179590" which was expressed in pre- and post-puberty, and drive to "circ 16:24193110-24213797" which was expressed in in- and post-puberty (Figure 2b, Supplementary Table S2). For the Neurotrophin signaling pathway, *PRDM4* drive to "circ 5:12644891-12657528" which was uniquely expressed in in-puberty (Figure 2b, Supplementary Table S2). For the progesterone-mediated oocyte maturation, *MAPK10* drive to "circ 8:132670606-132709741" which was expressed in in- and post-puberty, drive to "circ 8:132704592-132733488" which was uniquely expressed in in-puberty (Figure 2b, Supplementary Table S2). For the oocyte meiosis signaling pathway, *PPP3CB* drive to "circ 14:76278036-76289229" which was uniquely expressed in-puberty (Figure 2b, Supplementary Table S2). The detailed information of these circRNA is shown in Supplementary Table S3.

### 3.3. The Stage-Specific circRNAs in the Pubertal Transition

To explore the expression changes of circRNA expressed in all stages, circRNAs were used for differentially expressed analysis except for the stage-specific circRNAs. 367, 168, and 575 putative stage-specific circRNAs were exclusively identified during the pre-, in- and post-puberty stages, respectively (Figure 1b). Furthermore, the expression levels of specific post-puberty circRNAs were significantly lower than specific pre-puberty circRNAs (*t*-test, $p$-value $< 9.7 \times 10^{-5}$), as well as significantly lower than specific in-puberty circRNAs (*t*-test, $p$-value $< 0.025$) (Figure 3a). In addition, the KEGG enrichment with the parental genes of stage-specific circRNAs were showed in Figure 3b, neurotrophin signaling pathway and ErbB signaling pathway were enriched in the pre- and post-puberty stages; axon guidance pathway was enriched in the in-puberty stage; insulin signaling pathway as well as progesterone-mediated oocyte maturation were enriched in the post-puberty stage (Supplementary Table S4). Moreover, 111 genes generated stage-specific and non-specific circRNAs, of which 100 genes generated pre-specific and non-specific circRNAs, and 11 genes generated in-specific and non-specific circRNAs (Supplementary Table S5).

### 3.4. Potentially Regulated Network of Differentially Expressed circRNAs

In order to explore the putative functions of differentially expressed circRNAs, we identified a total of 22 differentially expressed circRNAs (Supplementary Table S6) and showed the expression in Figure 4a. Thereinto, 11 differentially up-regulated circRNAs and three differentially down-regulated circRNAs were identified in the pre- vs. in-puberty group; two differentially up-regulated circRNAs and four differentially down-regulated circRNA were identified in the pre- vs. post-puberty group; two differentially up-regulated circRNAs and five differentially down-regulated circRNA were identified in the in- vs. post-puberty group (Supplementary Table S6). Later, circRNAs mentioned above were used to predict the binging sites, and the top three possible miRNA targets were listed in Table 1. After that, differentially expressed genes were used to predict the circRNA-miRNA-gene regulatory network (Figure 4b). Noticeably, we highlight *FSTL4*, *TSHR*, *SULT1E1*, *NPFFR2*, *RGCC*, and *ADAMTS4* genes, which were associated with puberty [34–39] (Supplementary Table S7). Interestingly, one of these differentially expressed circRNAs, "circ 11:4104218-4118265" that interacted with *FSTL4* via ssc-miR-34a, was down-regulated in the pre- vs. in-puberty groups, as well as down-regulated in the pre- vs. post-puberty groups (Supplementary Tables S6 and S7). In addition, "circ 3:103726106-103773127" was up-regulated in the pre- vs. in-puberty groups but down-regulated in the in- vs. post-puberty groups (Supplementary Table S6), and this circRNAs interacted with *SULT1E1* via ssc-miR-4331-3p (Supplementary Table S7). According to this result, we found that some differential expression of circRNAs interacted with differentially expressed genes via miRNAs, whereafter they potentially regulate the onset of puberty.

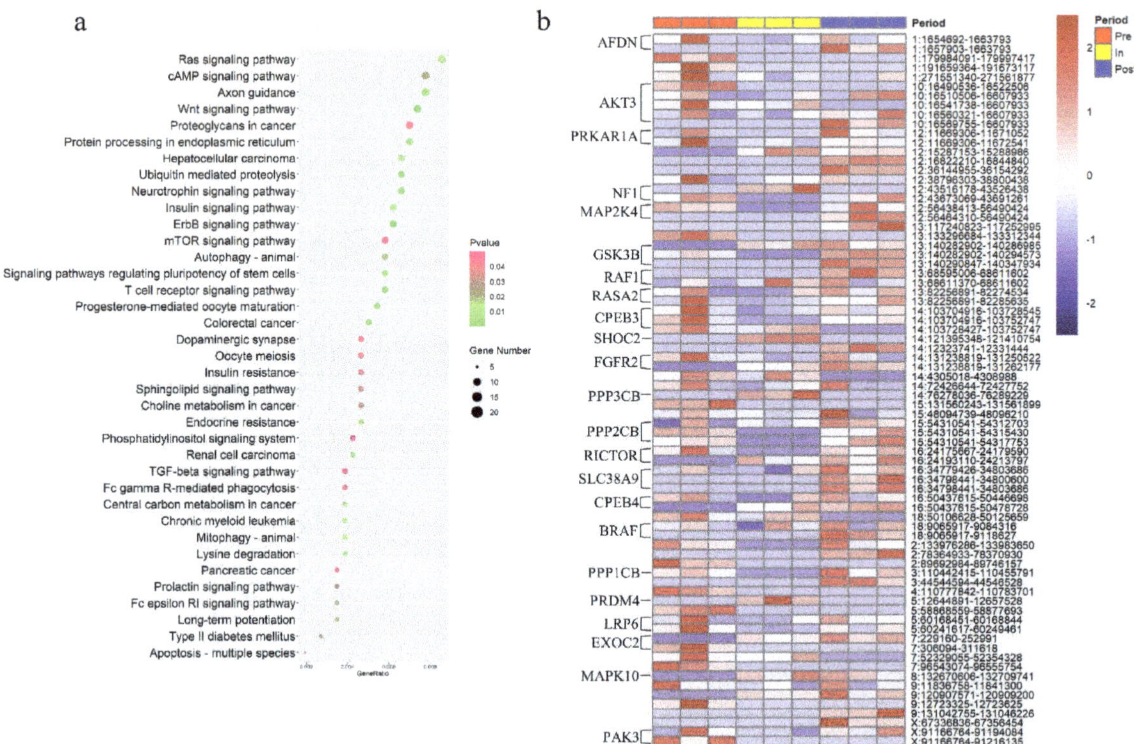


**Figure 2.** The Key signaling pathway of CirRNAs in pubertal transition. (**a**) KEGG analysis of all identified circRNAs ($p < 0.05$). (**b**) Expression level of circRNAs involved in pubertal key pathways in pre-, in- and post-puberty.

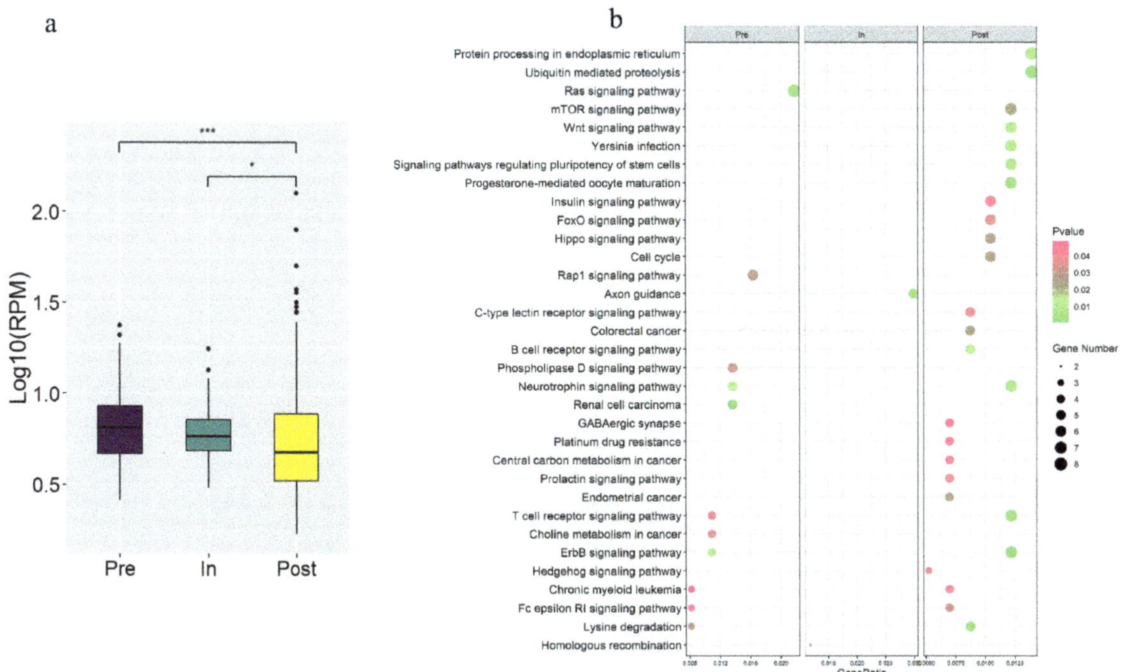


**Figure 3.** Analysis results of stage-specific circRNAs. (**a**) Expression level of stage-specific circRNAs during Pre-, In-, Post-puberty. (**b**) KEGG analysis with parental genes of stage-specific circRNAs during pre-, in-, post-puberty ($p < 0.05$). * $p < 0.05$, *** $p < 0.001$.

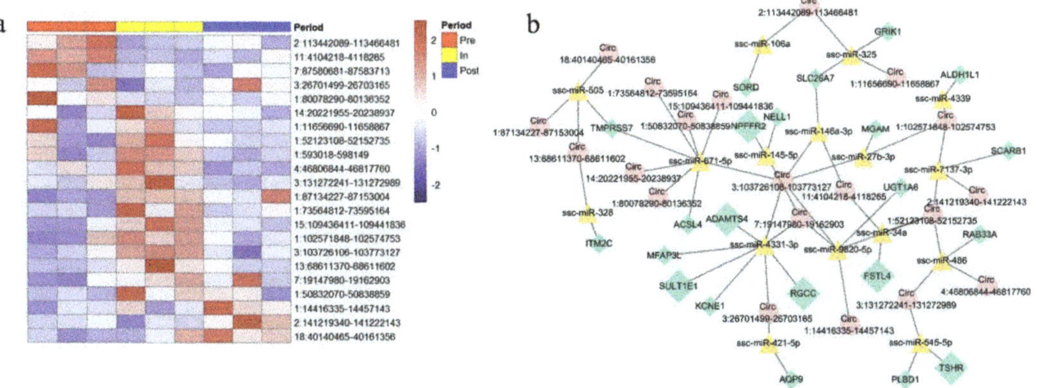


**Figure 4.** Analysis of differentially regulated circRNAs. (**a**) Heatmap of differentially expressed circRNAs in pubertal transition. (**b**) Differentially expressed circRNAs interact with differentially expressed genes via miRNAs and the differentially regulated status was show in Tables S6 and S7. The red circle represented circRNAs, the yellow triangle represented miRNAs, the green diamond represented genes.

**Table 1.** The differentially regulated circRNAs in this study of hypothalamus of gilts.

| circRNA ID | Position | Strand | circRNA Type | Parental Gene | Top 3 miRNA Targets |
|---|---|---|---|---|---|
| 1:102571848-102574753 | Chr1:102571848-102574753 | + | exon | *DCC* | ssc-miR-9814-3p, ssc-miR-15b, ssc-miR-144 |
| 1:11656690-11658867 | Chr1:11656690-11658867 | − | exon | *TIAM2* | ssc-miR-383, ssc-miR-7857-3p, ssc-miR-185 |
| 1:14416335-14457143 | Chr1:14416335-14457143 | − | exon | *ESR1* | ssc-miR-4331-3p, ssc-miR-16, ssc-miR-7143-3p |
| 1:50832070-50838859 | Chr1:50832070-50838859 | + | exon | *SMAP1* | ssc-miR-4331-3p, ssc-miR-9825-5p, ssc-miR-383 |
| 1:52123108-52152735 | Chr1:52123108-52152735 | + | exon | *RIMS1* | ssc-miR-4331-3p, ssc-miR-574-5p, ssc-miR-34a |
| 1:593018-598149 | Chr1:593018-598149 | + | exon | *WDR27* | ssc-miR-4331-3p, ssc-miR-7135-3p, ssc-miR-181b |
| 1:73564812-73595164 | Chr1:73564812-73595164 | + | exon | *SOBP* | ssc-miR-4331-3p, ssc-miR-27b-5p, ssc-miR-491 |
| 1:80078290-80136352 | Chr1:80078290-80136352 | − | exon | *HS3ST5* | ssc-miR-4331-3p, ssc-miR-9820-5p, ssc-miR-9-1 |
| 1:87134227-87153004 | Chr1:87134227-87153004 | + | exon | *PHIP* | ssc-miR-545-3p, ssc-miR-4331-3p, ssc-miR-148b-5p |
| 11:4104218-4118265 | Chr11:4104218-4118265 | + | exon | *CDK8* | ssc-miR-4331-3p, ssc-miR-9822-3p, ssc-miR-424-5p |
| 13:68611370-68611602 | Chr13:68611370-68611602 | − | exon | *RAF1* | ssc-miR-208b |
| 14:20221955-20238937 | Chr14:20221955-20238937 | + | exon | *NEK1* | ssc-miR-4331-3p, ssc-miR-505, ssc-miR-2483 |
| 15:109436411-109441836 | Chr15:109436411-109441836 | − | exon | *NDUFS1* | ssc-miR-30a-3p, ssc-miR-30e-3p, ssc-miR-676-5p |
| 18:40140465-40161356 | Chr18:40140465-40161356 | − | exon | *BBS9* | ssc-miR-199b-5p, ssc-miR-186-5p, ssc-miR-186-3p |
| 2:113442089-113466481 | Chr2:113442089-113466481 | − | exon | *FBXL17* | ssc-miR-4331-3p, ssc-miR-145-5p, ssc-miR-320 |
| 2:141219340-141222143 | Chr2:141219340-141222143 | + | exon | *MATR3* | ssc-miR-10390, ssc-miR-186-5p, ssc-miR-421-5p |
| 3:103726106-103773127 | Chr3:103726106-103773127 | − | exon | *CRIM1* | ssc-miR-574-5p, ssc-miR-4331-3p, ssc-miR-146a-3p |
| 3:131272241-131272989 | Chr3:131272241-131272989 | + | exon | *RNASEH1* | ssc-miR-149 |
| 3:26701499-26703165 | Chr3:26701499-26703165 | + | exon | *SMG1* | ssc-miR-4331-3p, ssc-miR-106a, ssc-miR-20a-5p |
| 4:46806844-46817760 | Chr4:46806844-46817760 | + | exon | *NBN* | ssc-miR-4331-3p, ssc-miR-9820-5p, ssc-miR-7140-3p |

+ Positive strand; − Negative strand.

### 3.5. The Hypothalamus-Specific circRNAs in Puberty

The known circRNA of pig come from circAtlas 2.0, which was includes thousands of known circRNAs in nine porcine tissue types (brain, heart, kidney, liver, lung, skeletal muscle, spleen, testis, and retina) [40]. In order to investigate the specific circRNAs in hypothalamus tissue, 2518 circRNAs which overlapped in known circAtlas 2.0 were excluded, and leaving another 53 circRNAs as being putative hypothalamus-specific circRNAs. Moreover, the 53 putative hypothalamus-specific circRNAs were significantly shorter than that of the known circRNAs (*t*-test, *p*-value = 0.00082) (Figure 5a). Furthermore, the expression of these putative hypothalamus-specific circRNAs was significantly lower than that of the known circRNAs during the onset of puberty (*t*-test, *p*-value $< 0.001$) (Figure 5b). In addition, the expression of these 53 hypothalamus-specific circRNAs was shown in Figure 5c, some of which were only expressed in one of three stages, and 5 genes were able to produce circRNAs at all stages without variation. Interestingly, the circRNA "circ 1:68439845-68491357" was only expressed in pre-puberty and "circ 1:68645763-68678869" was only expressed in post-puberty (Supplementary Table S8), both of which were derived from *GRIK2* associated with excitatory neurotransmission in the mammalian central nervous system [41]. The parental genes of these putative hypothalamus-specific circRNAs were enriched in "ssc04360: axon guidance" and "ssc04015: rap1 signaling pathway" pathways (Supplementary Table S9). Meanwhile, these parental genes of hypothalamus-specific circRNAs were related to "GO 0048666: neuron development" and "GO 0030182 neuron differentiation" terms (Supplementary Table S9).

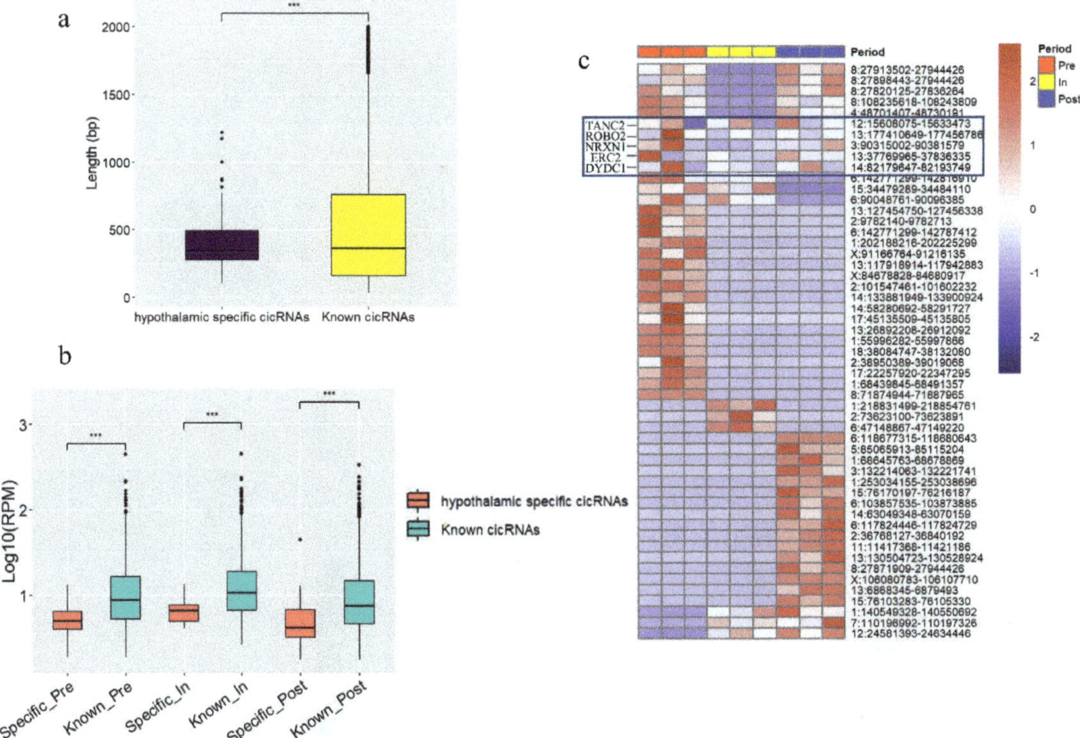


**Figure 5.** Analysis results of hypothalamus-specific circRNAs. (**a**) Length of tissue-specific circRNAs and known circRNAs. (**b**) Significant difference analysis between hypothalamus-specific circRNAs and known circRNAs. (**c**) The expression of hypothylamus-specific circRNAs in three stages, the blue box represents circRNAs and its parental genes without variation. *** $p < 0.001$.

### 3.6. circRNAs in Pubertal Genes

To further explore the function of circRNAs in puberty, the 20 pubertal genes were selected and investigated through reviewing the literature and databases by hand (Supplementary Table S10). Subsequently, we found that 16 circRNAs were driven by 6 pubertal genes (Supplementary Table S10). Thereinto, *ESR1* drive to circRNAs "circ 1:14416335-14457143", which was differentially expressed during pubertal hypothalamus; *APP* drive to four circRNAs ("circ 13:189505179-189528484", "circ13:189505179-189544139", "circ 13:189523366-189528484" and "circ 13:189597995-189600156"); *NF1* drive to two circRNAs ("circ 12:43516178-43526438" and "circ 12:43673069-43691261"); *ENPP2* and *ARNT* respectively drive to circRNAs "circ 4:19360870-19367922" and circRNAs "4:98369520-98372553", which were uniquely expressed in pre-pubertal hypothalamus; *DICER1* drive to circRNAs "circ 1:14416335-14457143", which was always expressed during pubertal hypothalamus (Supplementary Table S10). These results will become the focus for further analysis.

### 3.7. Validation of circRNAs by RT-qPCR

In order to verify the accuracy of RNA-seq data, five circRNAs were randomly selected for validation experiments. Thereinto, circRNA "circ 1:11656690-11658857", "circ 1:87134227-87153004", "circ 2:141219340-141222143", "circ 3:26701499-87153004" were differential expression and circRNA "circ 6:9159375-91605991" was no differential expression. First, the divergent primers were used to this study, then the RT-qPCR were used to verified (see Section 2 for detail). Accordingly, the RT-qPCR assay results showed a similar tendency of expression with our RNA-seq data (Figure 6), further confirming the reliability of sequencing.

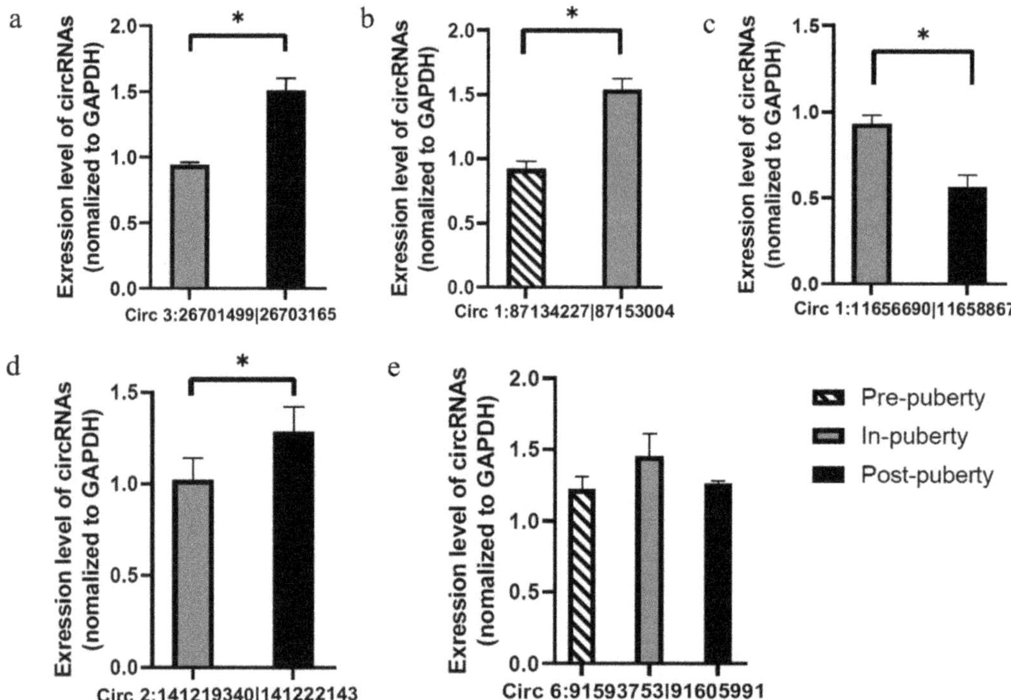


**Figure 6.** RT-qPCR validation of circRNAs. Five circRNAs were randomly selected for RT-qPCR validation, of which four circRNAs (**a**–**d**) were differential expression and one circRNA (**e**) was no significant difference. The primer informations were listed in Table S11. * $p < 0.05$.

## 4. Discussion

Puberty is a complex physiological process regulated by multiple pathways. Hypothalamus, the main female puberty organs, directly mediate the pulsatile release of GnRH, which play crucial roles during onset of puberty [7,42]. Due to the delay in puberty, about 30% of gilts has been culled, which has obviously harmed the financial stake of the modern commercial farms [43]. CircRNAs were found to be function in many biological processes and widely expressed in mammal [44,45]. With the development of next-generation sequencing technology, research on the regulation of circRNA in animal puberty has made small progress year by year. However, puberty-associated circRNA expression remains unclear in gilts. Thus, our study focused on exploring the potential role of circRNAs in pubertal hypothalamus of gilts. A total of 2582 circRNAs were identified, of which 1110 were putative stage-specific circRNAs, 53 were putative hypothalamus-specific circRNAs and 22 were differentially expressed circRNAs.

For all identified circRNAs, 2388 exonic circRNAs were generated from 1461 parental genes (Figure 1d). This result may be explained by the fact that one gene could produce different circRNAs through different splicing forms. Furthermore, we uncovered several genes in some of the key pathways associated with the timing of puberty that could drive expression of differential circRNAs in differential pubertal stage. Neurofibromin 1 (*NF1*) is the main Ras regulator and plays an important role in neurons [46]. For the ras signaling pathway, "circ 12:43516178-43526438" driven by *NF1* was only expressed in in-puberty, while "circ 12:43516178-43526438", driven by *NF1*, was not expressed in in-puberty but expressed in pre- and post-puberty, indicating that these two *NF1*-driven circRNAs have different roles during the onset of puberty. When entering in-puberty, *NF1* might specifically splicing "circ 12:43516178-43526438" to play a crucial role. *MAP2K4* as the direct upstream activator of NH2-terminal kinase pathway, which plays an important role in regulating neuron survival and apoptosis in response to cerebral ischemia [47,48]. Similarly, for the ErbB signaling pathway, the two circRNAs driven by *MAP2K4* might have different effects. When entering pre-puberty, *MAP2K4* spliced "circ 12:56438413-56490424", and after in-puberty, *MAP2K4* re-spliced "circ 12:56438413-56490424" and spliced "circ 12:56464310-56490424". In addition, previous report has shown that *MAPK10* could block the hypothalamic-pituitary-thyroid axis, thereby reducing energy expenditure and promoting obesity [49]. In this study, *MAPK10* drive to the specific expression of circRNA "circ 8:132670606-132709741" during in- and post-puberty in the progesterone-mediated oocyte maturation signaling pathway. These results suggested that the circRNAs identified on the relevant signaling pathway might play a crucial role in pubertal transition.

Moreover, 367, 168, and 575 circRNAs were uniquely expressed in pre-, in-, post-puberty, respectively (Figure 3b), among which the uniquely expressed circRNAs in post-puberty had the highest average expression (Figure 3a). Importantly, parental genes of stage-specific circRNAs were involved in neurotrophin signaling pathway, ErbB signaling pathway, axon guidance pathway, insulin signaling pathway as well as progesterone-mediated oocyte maturation; these processed have been reported to regulate the puberty [50–55]. In addition, circRNA-miRNA-gene network was used to predict the relationship between differential circRNA and differential genes. Interestingly, differentially expressed circRNAs "circ 11:4104218-4118265" that downregulated in the pre- vs. post-puberty groups interacted with *FSTL4* that downregulated in the pre- vs. post-puberty groups (Tables S6 and S7). This result suggested that circRNAs "circ 11:4104218-4118265" might be the sponge for ssc-miR-34, thereby promote the expression of *FSTL4*. The formation of circRNA is affected by alternative splicing and methylation [9]. Moreover, our previous studies have found that there had differential methylation pattern in genes during the onset of puberty in gilts [56]. It is possible, therefore, that the parental genes might be influenced by other epigenetic regulation to produce stage-specific exonic circRNA.

In addition, the circRNA "circ 7:19147980-19162903" that upregulated in the pre- vs. in-puberty groups interacted with *SULT1E1* and *NPFFR2* that up-regulated in the pre- vs. in-puberty groups (Supplementary Tables S6 and S7). This result suggested that circRNAs "circ 7:19147980-19162903" might be the sponge for ssc-miR-4331-3p and ssc-miR-145-5p, thereby promote the expression of *SULT1E1* and *NPFFR2*, respectively. These results showed that there might be a competitive binding of miRNA by circRNA to affect gene expression in pubertal hypothalamus of gilts. Subsequently, we identified 53 hypothetical hypothalamic-specific circRNAs, which were involved in axon guidance and rap1 signaling pathway pathways, neuron development and neuron differentiation. Previous study reported that there were complex changes in the central nervous system in the pubertal hypothalamus [57]. Other study showed that proper axon guidance is essential for the migration of GnRH neurons in the brain [53]. Another study demonstrated that the rap1 plays a crucial role in mediating cAMP-induced MAPK activation of specific cell types [58]. It may be the case therefore that hypothalamus-specific circRNAs affect the development of neurons in hypothalamus and subsequently affect the onset of puberty. Interestingly, we found that two tissue-specific circRNA (circ 1:68439845-68491357, circ 1:68645763-68678869) were derived from the gene associated with excitatory neurotransmission in the mammalian central nervous system, suggesting that these two circRNAs might play a vital role in the specific differentiation of the hypothalamus.

Moreover, 10 pubertal genes-driven circRNAs were found in this study (Supplementary Table S10). Thereinto, *APP*, which implicated in neural development and reproduction [59], drive to four circRNAs ("circ 13:189505179-189528484", "circ 13:189505179-189544139", "circ 13:189523366-189528484" and "circ 13:189597995-189600156"); *ESR1*, which associated with the timing of puberty [60], drive to circRNAs "circ 1:14416335-14457143", of which "circ 1:14416335-14457143" was up-regulated in pre- vs. post-puberty group (Supplementary Table S6). Moreover, *DICER1*, which was essential for the normal development of the reproductive system [61], drive to circRNA "circ 7:116399251-116408577". In addition, these circRNAs was constituted in multiple exons except "circ 4:47041713-47042593" and "circ 7:116399251-116408577". Our results provide new insight into the existence of hypothalamus-derived circRNAs in gilts. However, the underlying mechanism of these circRNAs during gilts' pubertal onset still requires carefully elucidation and verification.

## 5. Conclusions

During pubertal transition, 2582 circRNAs were identified hypothalamus, of which 1110 circRNAs were putative stagce-specific circRNAs, 53 circRNAs were putative hypothalamus-specific expressed circRNAs, and 22 circRNAs were significantly differentially expressed. These cirRNA were mostly enriched in neurotrophin signaling pathway, progesterone-mediated oocyte maturation, ras signaling pathway, insulin signaling pathway, ErbB signaling pathway, mTOR signaling pathway and oocyte meiosis signaling pathway, which had been highly implicated in puberty. Moreover, 16 circRNAs were driven by six genes, i.e., *ESR1*, *NF1*, *APP*, *ENPP2*, *ARNT*, and *DICER1*. These preliminary results indicated circRNAs involved in the timing of puberty at the hypothalamus level in gilts, and provided useful information for the investigation of the molecular mechanism of pubertal onset in mammals.

**Supplementary Materials:** The following are available online at https://www.mdpi.com/2073-4425/12/1/84/s1, Table S1: Information of all identified circRNAs. Table S2: The KEGG pathways enriched using parental genes of all CircRNAs. Table S3: Key pathways related to the timing of puberty in parental genes of circRNAs. Table S4: The KEGG pathways enriched using parental genes of stage-specific CircRNAs. Table S5: Parental Genes That Are Capable of Producing Stage-specific And Non-specific CircRNAss. Table S6: The differentially regulated circRNAs. Table S7: The differentially expressed genes associated with puberty our development of ovary. Table S8: Tissue-specific CircRNAs. Table S9: The KEGG pathways and GO term enriched using parental genes of tissue-specific CircRNAs. Table S10: CircRNAs in Pubertal Genes. Table S11: Primers used for qRT-PCR.

**Author Contributions:** Data curation: Q.L., X.P., N.L., and W.G.; funding acquisition: Y.C. and X.Y.; supervision: Y.C. and X.Y.; writing—original draft: Q.L. and X.P.; writing—review and editing: Y.C. and X.Y. All authors have read and agreed to the published version of the manuscript.

**Funding:** This research was funded by the National Key R&D Program of China (grant number: 2018YFD0501200), the Special Fund for Science and Technology Innovation of Guangdong Province (grant number: 2018B020203003), the National Natural Science Foundation of China (grant number: 31902131), the earmarked fund for the China Agriculture Research System (grant number: CARS-35), the National Natural Science Foundation of Guangdong Province (grant number: 2019A1515010676), the Youth Innovative fund of Guangdong Education Department (grant number: 2018KQNCX019), and China Postdoctoral Science Foundation (grant number: 2020M672556).

**Institutional Review Board Statement:** The animal study was reviewed and approved by The Animal Care and Use Committee of the South China Agricultural University, Guangzhou, China (permit number: SCAU#2013-10).

**Informed Consent Statement:** Informed consent was obtained from all subjects involved in the study.

**Data Availability Statement:** The datasets used in this study have been submitted to the European Nucleotide Archive under accession number PRJEB39729.

**Conflicts of Interest:** The authors claim that there is no conflict of interest.

## References

1. Martinat-Botte, F.; Royer, E.; Venturi, E.; Boisseau, C.; Guillouet, P.; Furstoss, V.; Terqui, M. Determination by echography of uterine changes around puberty in gilts and evaluation of a diagnosis of puberty. *Reprod. Nutr. Dev.* **2003**, *43*, 225–236. [CrossRef] [PubMed]
2. Luo, L.; Yao, Z.; Ye, J.; Tian, Y.; Yang, C.; Gao, X.; Song, M.; Liu, Y.; Zhang, Y.; Li, Y.; et al. Identification of differential genomic DNA Methylation in the hypothalamus of pubertal rat using reduced representation Bisulfite sequencing. *Reprod. Biol. Endocrinol.* **2017**, *15*, 81. [CrossRef] [PubMed]
3. Nonneman, D.J.; Schneider, J.F.; Lents, C.A.; Wiedmann, R.T.; Vallet, J.L.; Rohrer, G.A. Genome-wide association and identification of candidate genes for age at puberty in swine. *BMC Genet.* **2016**, *17*, 50. [CrossRef] [PubMed]
4. Tummaruk, P.; Tantasuparuk, W.; Techakumphu, M.; Kunavongkrit, A. Age, body weight and backfat thickness at first observed oestrus in crossbred Landrace x Yorkshire gilts, seasonal variations and their influence on subsequence reproductive performance. *Anim. Reprod. Sci.* **2007**, *99*, 167–181. [CrossRef]
5. Root, A.W. Hormonal changes in puberty. *Pediatr. Ann.* **1980**, *9*, 365–375. [CrossRef] [PubMed]
6. Dutta, S.; Mark-Kappeler, C.J.; Hoyer, P.B.; Pepling, M.E. The steroid hormone environment during primordial follicle formation in perinatal mouse ovaries. *Biol. Reprod.* **2014**, *91*, 68. [CrossRef] [PubMed]
7. Lomniczi, A.; Wright, H.; Ojeda, S.R. Epigenetic regulation of female puberty. *Front. Neuroendocrinol.* **2015**, *36*, 90–107. [CrossRef]
8. Pandolfi, E.C.; Tonsfeldt, K.J.; Hoffmann, H.M.; Mellon, P.L. Deletion of the Homeodomain Protein Six6 From GnRH Neurons Decreases GnRH Gene Expression, Resulting in Infertility. *Endocrinology* **2019**, *160*, 2151–2164. [CrossRef] [PubMed]
9. Kristensen, L.S.; Andersen, M.S.; Stagsted, L.; Ebbesen, K.K.; Hansen, T.B.; Kjems, J. The biogenesis, biology and characterization of circular RNAs. *Nat. Rev. Genet.* **2019**, *20*, 675–691. [CrossRef]
10. Salzman, J.; Chen, R.E.; Olsen, M.N.; Wang, P.L.; Brown, P.O. Cell-type specific features of circular RNA expression. *PLoS Genet.* **2013**, *9*, e1003777. [CrossRef]
11. Zhang, X.O.; Dong, R.; Zhang, Y.; Zhang, J.L.; Luo, Z.; Zhang, J.; Chen, L.L.; Yang, L. Diverse alternative back-splicing and alternative splicing landscape of circular RNAs. *Genome Res.* **2016**, *26*, 1277–1287. [CrossRef] [PubMed]
12. Jeck, W.R.; Sorrentino, J.A.; Wang, K.; Slevin, M.K.; Burd, C.E.; Liu, J.; Marzluff, W.F.; Sharpless, N.E. Circular RNAs are abundant, conserved, and associated with ALU repeats. *RNA* **2013**, *19*, 141–157. [CrossRef] [PubMed]
13. Guo, J.U.; Agarwal, V.; Guo, H.; Bartel, D.P. Expanded identification and characterization of mammalian circular RNAs. *Genome Biol.* **2014**, *15*, 409. [CrossRef]
14. Xia, S.; Feng, J.; Lei, L.; Hu, J.; Xia, L.; Wang, J.; Xiang, Y.; Liu, L.; Zhong, S.; Han, L.; et al. Comprehensive characterization of tissue-specific circular RNAs in the human and mouse genomes. *Brief. Bioinform.* **2017**, *18*, 984–992. [CrossRef] [PubMed]
15. Chen, N.; Zhao, G.; Yan, X.; Lv, Z.; Yin, H.; Zhang, S.; Song, W.; Li, X.; Li, L.; Du, Z.; et al. A novel FLI1 exonic circular RNA promotes metastasis in breast cancer by coordinately regulating TET1 and DNMT1. *Genome Biol.* **2018**, *19*, 218. [CrossRef]
16. Hansen, T.B.; Jensen, T.I.; Clausen, B.H.; Bramsen, J.B.; Finsen, B.; Damgaard, C.K.; Kjems, J. Natural RNA circles function as efficient microRNA sponges. *Nature* **2013**, *495*, 384–388. [CrossRef]
17. Conn, S.J.; Pillman, K.A.; Toubia, J.; Conn, V.M.; Salmanidis, M.; Phillips, C.A.; Roslan, S.; Schreiber, A.W.; Gregory, P.A.; Goodall, G.J. The RNA binding protein quaking regulates formation of circRNAs. *Cell* **2015**, *160*, 1125–1134. [CrossRef]

18. Zheng, Q.; Bao, C.; Guo, W.; Li, S.; Chen, J.; Chen, B.; Luo, Y.; Lyu, D.; Li, Y.; Shi, G.; et al. Circular RNA profiling reveals an abundant circHIPK3 that regulates cell growth by sponging multiple miRNAs. *Nat. Commun.* **2016**, *7*, 11215. [CrossRef]
19. Hall, I.F.; Climent, M.; Quintavalle, M.; Farina, F.M.; Schorn, T.; Zani, S.; Carullo, P.; Kunderfranco, P.; Civilini, E.; Condorelli, G.; et al. Circ_Lrp6, a Circular RNA Enriched in Vascular Smooth Muscle Cells, Acts as a Sponge Regulating miRNA-145 Function. *Circ. Res.* **2019**, *124*, 498–510. [CrossRef]
20. Jost, I.; Shalamova, L.A.; Gerresheim, G.K.; Niepmann, M.; Bindereif, A.; Rossbach, O. Functional sequestration of microRNA-122 from Hepatitis C Virus by circular RNA sponges. *RNA Biol.* **2018**, *15*, 1032–1039. [CrossRef]
21. You, X.; Vlatkovic, I.; Babic, A.; Will, T.; Epstein, I.; Tushev, G.; Akbalik, G.; Wang, M.; Glock, C.; Quedenau, C.; et al. Neural circular RNAs are derived from synaptic genes and regulated by development and plasticity. *Nat. Neurosci.* **2015**, *18*, 603–610. [CrossRef] [PubMed]
22. Zhou, T.; Xie, X.; Li, M.; Shi, J.; Zhou, J.J.; Knox, K.S.; Wang, T.; Chen, Q.; Gu, W. Rat BodyMap transcriptomes reveal unique circular RNA features across tissue types and developmental stages. *RNA* **2018**, *24*, 1443–1456. [CrossRef] [PubMed]
23. Veno, M.T.; Hansen, T.B.; Veno, S.T.; Clausen, B.H.; Grebing, M.; Finsen, B.; Holm, I.E.; Kjems, J. Spatio-temporal regulation of circular RNA expression during porcine embryonic brain development. *Genome Biol.* **2015**, *16*, 245. [CrossRef] [PubMed]
24. Shen, Y.; Guo, X.; Wang, W. Identification and characterization of circular RNAs in zebrafish. *FEBS Lett.* **2017**, *591*, 213–220. [CrossRef]
25. Chen, C.; Khaleel, S.S.; Huang, H.; Wu, C.H. Software for pre-processing Illumina next-generation sequencing short read sequences. *Source Code Biol. Med.* **2014**, *9*, 8. [CrossRef]
26. Li, H.; Durbin, R. Fast and accurate long-read alignment with Burrows-Wheeler transform. *Bioinformatics* **2010**, *26*, 589–595. [CrossRef]
27. Langmead, B.; Salzberg, S.L. Fast gapped-read alignment with Bowtie 2. *Nat. Methods* **2012**, *9*, 357–359. [CrossRef]
28. Gao, Y.; Wang, J.; Zhao, F. CIRI: An efficient and unbiased algorithm for de novo circular RNA identification. *Genome Biol.* **2015**, *16*, 4. [CrossRef]
29. Hansen, T.B.; Veno, M.T.; Damgaard, C.K.; Kjems, J. Comparison of circular RNA prediction tools. *Nucleic Acids Res.* **2016**, *44*, e58. [CrossRef]
30. Leng, N.; Dawson, J.A.; Thomson, J.A.; Ruotti, V.; Rissman, A.I.; Smits, B.M.; Haag, J.D.; Gould, M.N.; Stewart, R.M.; Kendziorski, C. EBSeq: An empirical Bayes hierarchical model for inference in RNA-seq experiments. *Bioinformatics* **2013**, *29*, 1035–1043. [CrossRef]
31. Wu, J.; Mao, X.; Cai, T.; Luo, J.; Wei, L. KOBAS server: A web-based platform for automated annotation and pathway identification. *Nucleic Acids Res.* **2006**, *34*, W720–W724. [CrossRef] [PubMed]
32. John, B.; Enright, A.J.; Aravin, A.; Tuschl, T.; Sander, C.; Marks, D.S. Human MicroRNA targets. *PLoS Biol.* **2004**, *2*, e363. [CrossRef] [PubMed]
33. Su, G.; Morris, J.H.; Demchak, B.; Bader, G.D. Biological network exploration with Cytoscape 3. *Curr. Protoc. Bioinform.* **2014**, *47*, 8–13. [CrossRef] [PubMed]
34. Wang, Y.; Cheng, T.; Lu, M.; Mu, Y.; Li, B.; Li, X.; Zhan, X. TMT-based quantitative proteomics revealed follicle-stimulating hormone (FSH)-related molecular characterizations for potentially prognostic assessment and personalized treatment of FSH-positive non-functional pituitary adenomas. *EPMA J.* **2019**, *10*, 395–414. [CrossRef] [PubMed]
35. Marx, S.J. Hyperplasia in glands with hormone excess. *Endocr. Relat. Cancer* **2016**, *23*, R1–R14. [CrossRef] [PubMed]
36. Braun, B.C.; Okuyama, M.W.; Muller, K.; Dehnhard, M.; Jewgenow, K. Steroidogenic enzymes, their products and sex steroid receptors during testis development and spermatogenesis in the domestic cat (*Felis catus*). *J. Steroid Biochem. Mol. Biol.* **2018**, *178*, 135–149. [CrossRef] [PubMed]
37. Thorson, J.F.; Heidorn, N.L.; Ryu, V.; Czaja, K.; Nonneman, D.J.; Barb, C.R.; Hausman, G.J.; Rohrer, G.A.; Prezotto, L.D.; McCosh, R.B.; et al. Relationship of neuropeptide FF receptors with pubertal maturation of gilts. *Biol. Reprod.* **2017**, *96*, 617–634. [CrossRef] [PubMed]
38. Nowicki, A.; Skupin-Mrugalska, P.; Jozkowiak, M.; Wierzchowski, M.; Rucinski, M.; Ramlau, P.; Krajka-Kuzniak, V.; Jodynis-Liebert, J.; Piotrowska-Kempisty, H. The Effect of 3′-Hydroxy-3,4,5,4′-Tetramethoxy -stilbene, the Metabolite of the Resveratrol Analogue DMU-212, on the Motility and Proliferation of Ovarian Cancer Cells. *Int. J. Mol. Sci.* **2020**, *21*, 1100. [CrossRef]
39. Richards, J.S.; Hernandez-Gonzalez, I.; Gonzalez-Robayna, I.; Teuling, E.; Lo, Y.; Boerboom, D.; Falender, A.E.; Doyle, K.H.; LeBaron, R.G.; Thompson, V.; et al. Regulated expression of ADAMTS family members in follicles and cumulus oocyte complexes: Evidence for specific and redundant patterns during ovulation. *Biol. Reprod.* **2005**, *72*, 1241–1255. [CrossRef]
40. Wu, W.; Ji, P.; Zhao, F. CircAtlas: An integrated resource of one million highly accurate circular RNAs from 1070 vertebrate transcriptomes. *Genome Biol.* **2020**, *21*, 101. [CrossRef]
41. Barbon, A.; Vallini, I.; Barlati, S. Genomic organization of the human GRIK2 gene and evidence for multiple splicing variants. *Gene* **2001**, *274*, 187–197. [CrossRef]
42. Trout, W.E.; Diekman, M.A.; Parfet, J.R.; Moss, G.E. Pituitary responsiveness to GnRH, hypothalamic content of GnRH and pituitary LH and FSH concentrations immediately preceding puberty in gilts. *J. Anim. Sci.* **1984**, *58*, 1423–1431. [CrossRef] [PubMed]
43. Stancic, I.; Stancic, B.; Bozic, A.; Anderson, R.; Harvey, R.; Gvozdic, D. Ovarian activity and uterus organometry in delayed puberty gilts. *Theriogenology* **2011**, *76*, 1022–1026. [CrossRef] [PubMed]

44. Memczak, S.; Jens, M.; Elefsinioti, A.; Torti, F.; Krueger, J.; Rybak, A.; Maier, L.; Mackowiak, S.D.; Gregersen, L.H.; Munschauer, M.; et al. Circular RNAs are a large class of animal RNAs with regulatory potency. *Nature* **2013**, *495*, 333–338. [CrossRef] [PubMed]
45. Li, Y.; Zheng, Q.; Bao, C.; Li, S.; Guo, W.; Zhao, J.; Chen, D.; Gu, J.; He, X.; Huang, S. Circular RNA is enriched and stable in exosomes: A promising biomarker for cancer diagnosis. *Cell Res.* **2015**, *25*, 981–984. [CrossRef]
46. Xie, K.; Colgan, L.A.; Dao, M.T.; Muntean, B.S.; Sutton, L.P.; Orlandi, C.; Boye, S.L.; Boye, S.E.; Shih, C.C.; Li, Y.; et al. NF1 Is a Direct G Protein Effector Essential for Opioid Signaling to Ras in the Striatum. *Curr. Biol.* **2016**, *26*, 2992–3003. [CrossRef]
47. Gu, L.; Wu, Y.; Hu, S.; Chen, Q.; Tan, J.; Yan, Y.; Liang, B.; Tang, N. Analysis of Association between MAP2K4 Gene Polymorphism rs3826392 and IL-1b Serum Level in Southern Chinese Han Ischemic Stroke Patients. *J. Stroke Cerebrovasc. Dis.* **2016**, *25*, 1096–1101. [CrossRef]
48. Kiddle, S.J.; Steves, C.J.; Mehta, M.; Simmons, A.; Xu, X.; Newhouse, S.; Sattlecker, M.; Ashton, N.J.; Bazenet, C.; Killick, R.; et al. Plasma protein biomarkers of Alzheimer's disease endophenotypes in asymptomatic older twins: Early cognitive decline and regional brain volumes. *Transl. Psychiatry* **2015**, *5*, e584. [CrossRef]
49. Nogueiras, R.; Sabio, G. Brain JNK and metabolic disease. *Diabetologia* **2020**. [CrossRef]
50. Calabrese, F.; Richetto, J.; Racagni, G.; Feldon, J.; Meyer, U.; Riva, M.A. Effects of withdrawal from repeated amphetamine exposure in peri-puberty on neuroplasticity-related genes in mice. *Neuroscience* **2013**, *250*, 222–231. [CrossRef]
51. Ojeda, S.R.; Lomniczi, A.; Sandau, U. Contribution of glial-neuronal interactions to the neuroendocrine control of female puberty. *Eur. J. Neurosci.* **2010**, *32*, 2003–2010. [CrossRef] [PubMed]
52. Dziedzic, B.; Prevot, V.; Lomniczi, A.; Jung, H.; Cornea, A.; Ojeda, S.R. Neuron-to-glia signaling mediated by excitatory amino acid receptors regulates ErbB receptor function in astroglial cells of the neuroendocrine brain. *J. Neurosci.* **2003**, *23*, 915–926. [CrossRef] [PubMed]
53. Oleari, R.; Caramello, A.; Campinoti, S.; Lettieri, A.; Ioannou, E.; Paganoni, A.; Fantin, A.; Cariboni, A.; Ruhrberg, C. PLXNA1 and PLXNA3 cooperate to pattern the nasal axons that guide gonadotropin-releasing hormone neurons. *Development* **2019**, *146*. [CrossRef] [PubMed]
54. Qiu, X.; Dao, H.; Wang, M.; Heston, A.; Garcia, K.M.; Sangal, A.; Dowling, A.R.; Faulkner, L.D.; Molitor, S.C.; Elias, C.F.; et al. Insulin and Leptin Signaling Interact in the Mouse Kiss1 Neuron during the Peripubertal Period. *PLoS ONE* **2015**, *10*, e121974. [CrossRef]
55. Ye, J.; Yao, Z.; Si, W.; Gao, X.; Yang, C.; Liu, Y.; Ding, J.; Huang, W.; Fang, F.; Zhou, J. Identification and characterization of microRNAs in the pituitary of pubescent goats. *Reprod. Biol. Endocrinol.* **2018**, *16*, 51. [CrossRef]
56. Yuan, X.; Zhou, X.; Chen, Z.; He, Y.; Kong, Y.; Ye, S.; Gao, N.; Zhang, Z.; Zhang, H.; Li, J. Genome-Wide DNA Methylation Analysis of Hypothalamus During the Onset of Puberty in Gilts. *Front. Genet.* **2019**, *10*, 228. [CrossRef]
57. Naule, L.; Maione, L.; Kaiser, U.B. Puberty, a sensitive window of hypothalamic development and plasticity. *Endocrinology* **2020**. [CrossRef]
58. Vossler, M.R.; Yao, H.; York, R.D.; Pan, M.G.; Rim, C.S.; Stork, P.J. cAMP activates MAP kinase and Elk-1 through a B-Raf- and Rap1-dependent pathway. *Cell* **1997**, *89*, 73–82. [CrossRef]
59. Coronel, R.; Lachgar, M.; Bernabeu-Zornoza, A.; Palmer, C.; Dominguez-Alvaro, M.; Revilla, A.; Ocana, I.; Fernandez, A.; Martinez-Serrano, A.; Cano, E.; et al. Neuronal and Glial Differentiation of Human Neural Stem Cells Is Regulated by Amyloid Precursor Protein (APP) Levels. *Mol. Neurobiol.* **2019**, *56*, 1248–1261. [CrossRef]
60. Kugelberg, E. Reproductive endocrinology: ESR1 mutation causes estrogen resistance and puberty delay in women. *Nat. Rev. Endocrinol.* **2013**, *9*, 565. [CrossRef]
61. Hong, X.; Luense, L.J.; McGinnis, L.K.; Nothnick, W.B.; Christenson, L.K. Dicer1 is essential for female fertility and normal development of the female reproductive system. *Endocrinology* **2008**, *149*, 6207–6212. [CrossRef] [PubMed]

*Article*

# Profiling Novel Alternative Splicing within Multiple Tissues Provides Useful Insights into Porcine Genome Annotation

**Wen Feng, Pengju Zhao, Xianrui Zheng, Zhengzheng Hu and Jianfeng Liu** *

Key Laboratory of Animal Genetics, Breeding and Reproduction, Ministry of Agriculture, College of Animal Science and Technology, China Agricultural University, Beijing 100193, China; wfeng@cau.edu.cn (W.F.); zhaopengju2014@gmail.com (P.Z.); zxr07sk1@163.com (X.Z.); zhengzheng0517@163.com (Z.H.)

* Correspondence: liujf@cau.edu.cn; Tel.: +86-(010)-62731921

Received: 13 November 2020; Accepted: 24 November 2020; Published: 26 November 2020

**Abstract:** Alternative splicing (AS) is a process during gene expression that results in a single gene coding for different protein variants. AS contributes to transcriptome and proteome diversity. In order to characterize AS in pigs, genome-wide transcripts and AS events were detected using RNA sequencing of 34 different tissues in Duroc pigs. In total, 138,403 AS events and 29,270 expressed genes were identified. An alternative donor site was the most common AS form and accounted for 44% of the total AS events. The percentage of the other three AS forms (exon skipping, alternative acceptor site, and intron retention) was approximately 19%. The results showed that the most common AS events involving alternative donor sites could produce different transcripts or proteins that affect the biological processes. The expression of genes with tissue-specific AS events showed that gene functions were consistent with tissue functions. AS increased proteome diversity and resulted in novel proteins that gained or lost important functional domains. In summary, these findings extend porcine genome annotation and highlight roles that AS could play in determining tissue identity.

**Keywords:** alternative splicing; transcript; protein; domain; single nucleotide polymorphism

---

## 1. Introduction

Alternative splicing (AS) is a regulated process that generates multiple transcripts from a single gene. It therefore plays an important role in expanding protein diversity. AS affects 95% of multi-exonic genes in humans, and occurs in high proportions in other animals [1–3]. AS has four basic modes, including exon skipping, alternative donor sites, alternative acceptor sites, and intron retention (IR) [4]. In addition to these, multiple promoters [5] and multiple polyadenylation sites are two other mechanisms by which different mRNAs may be generated from the same gene [6].

Many alternatively spliced isoforms play important roles in the biological timing and development of tissues [7–10]. It is becoming clear that a large number of AS events contributes to the acquisition of adult tissue functions and identity in human tissue development [11]. It has been revealed that RNA mis-splicing underlies a growing number of human diseases [12]. Protein coding genes always have several alternatively spliced isoforms, emphasizing the importance of AS in gene expression [13]. Proteins generated by different AS also have the potential to gain or lose domains. This can substantially change the protein products, which results in similar or even opposite biological functions, which in turn can affect phenotypes [14]. Point mutations, such as single nucleotide polymorphisms (SNPs), have been shown to have substantial phenotypic variation and affect pre-mRNA splicing [15–20]. AS leads to the early termination of translation by the introducing premature stop codons [21]. Regulation of AS plays roles in multiple eukaryotic biological processes, including cell growth [22], chromatin modification [23], and tissue development [24].

As the best medical models for human diseases, pigs have similar anatomical and physiological structures as humans [25]. AS in pigs plays an essential role in the regulation of gene expression in genital tissues [26,27]. However, compared to the relevant studies in humans, it has still left a gap in our knowledge of the features of pig AS in multiple pig tissues. Various research groups have attempted to understand the role of AS in pigs by using expressed sequence tags (ESTs). A total of 1223 genes, with an average of 2.8 splicing variants per gene, have been detected among 16,540 unique genes using the EST data [28]. However, it has been reported that more AS could be detected using RNA sequencing (RNA-seq) than via EST technology in humans and plants [1,29]. However, to date, only one study has cataloged AS in a cross-breed at the pig genome-wide level, using RNA-seq [30]. In this previous study by Beiki et al., they mainly focus on transcripts and transcript structures. In our study, we further detect what affects the generation of AS, and how AS affects downstream progress.

In this study, a profile of AS events within multiple tissues of the Duroc pigs was constructed to extend the porcine AS genome annotation. Genome-wide transcript identification was performed in 34 normal tissues of the pig, using over 340 Gb of sequence data generated from 116 RNA sequencing libraries. In total, 2,486,239 known transcripts and 2,430,911 novel transcripts were identified. Tissue-specific expression patterns of novel and known transcripts were examined, and then the effects of the divergence of novel isoforms on their translation were analyzed. AS variations regulated by SNPs were also analyzed. The data reported here provide a valuable resource for enhancing the understanding and utilization of pig AS.

## 2. Materials and Methods

### 2.1. Sample Collection and Sequencing

PBMCs (peripheral blood mononuclear cells) and 33 different tissues were removed from nine unrelated, healthy Duroc pigs from Shenzhen Jinxinnong Technology Co., Ltd. (Shenzhen, China) in this study (Table S1). These nine Durocs consist of three infant pigs at the age of 3 days, three adult pigs at the age of 3 months, and three adult pigs at the age of 1 year old. For tissues collected from infant pigs, the samples are referred to tissuename_I (e.g., Brain_I), while from adult pigs they are called tissuename_A (e.g., Brain_A). The sample collection and treatment were fully conducted in strict accordance with the protocol approved by the Institutional Animal Care and Use Committee (IACUC) of China Agricultural University (no. DK2017/163) and Shenzhen Jinxinnong Technology Co., Ltd. The tissues and cells were then used for RNA and protein extraction.

### 2.2. Separation of RNA from Tissues

According to the standard protocols of Trizol method (Invitrogen, Carlsbad, CA, United States), the total RNA was isolated from mixture of equally unrelated pig pool tissues. RNA degradation and contamination were monitored on 1% agarose gels. The purity and contamination of total RNA was measured using a NanoPhotometer trophotometer (IMPLEN, Westlake Village, CA, United States) and Qubit RNA Assay Kit in a Qubit 2.0 Flurometer (Life Technologies, Carlsbad, CA, United States). The RNA's integrity was measured using the RNA Nano 6000 Assay Kit of the Bioanalyzer 2100 system (Agilent Technologies, Santa Clara, CA, USA). RNA samples that met the criteria of having an RNA integrity number (RIN) value of 7.0 or higher and a total RNA amount of 5 μg or higher were included and batched for RNA sequencing.

### 2.3. Library Construction and RNA Sequencing

The total RNA of the samples meeting the quality control (QC) criteria were ribosomal RNA-depleted and depleted QC, using the RiboMinus Eukaryote System v2 (Thermo Fisher Scientific, Waltham, MA, USA) and RNA 6000 Pico chip (Agilent Technologies) according to the manufacturer's protocol. RNA sequencing libraries were constructed using the NEBNext Ultra RNA Library Prep Kit (Illumina, Santiago, CA, United States), with 3 μg rRNA-depleted RNA, according to the manufacturer's recommendation. RNA-seq library preparations were clustered on a cBot Cluster Generation System using a HiSeq PE Cluster Kit v4 cBot (Illumina), and sequenced using the Illumina Hiseq 2500 platform according to the manufacturer's instructions, with a data size of per sample of a minimum of 10 G clean reads (corresponding to 126 bp paired-end reads). The sequenced RNA-Seq raw data for 34 pig tissues is available from NCBI Sequences Read Archive with the BioProject number PRJNA392949.

### 2.4. *Read Alignment to the Reference Sus Scrofa 11.1 Genome*

The RNA-Seq raw data were trimmed based on the quality control for downstream analyses by following steps: BBmap [31] automatically detected the adapter sequence of reads and removed those reads containing Illumina adapters; the Q20, Q30, and GC content of the clean data were calculated by FASTQC [32] for quality control and filtering; homopolymer trimming to 3′ end of fragments and removal of the N bases of 3′ end were carried out by Fastx toolkit v0.014 [33]. The resulting sequences then were mapped to a reference *Sus scrofa* 11.1 genome by Hisat 0.1.6-beta 64-bit [34]. Ensemble *Sus scrofa* 11.1 version 91 [35] annotation was used as the transcript model reference for the alignment and splice junction findings, as well as for all protein-coding genes and isoform expression-level quantifications. In addition, StringTie 1.0.4 [36] calculated the FPKM (fragments per kilobase of exon model per million mapped reads (controlling for fragment length and sequencing depth)) values. A gene or transcript was defined as expressed when its expression was measured above 0.1 FPKM in all tissues [37].

### 2.5. *Alternative Splicing Events in Pig Transcriptome*

Asprofile (v1.0.4) software (https://ccb.jhu.edu/software/ASprofile/) was used to classify and count the AS events in each sample [38]. Asprofile counts twelve AS events types in total. In our analysis, only exon skipping, alternative donor site, alternative acceptor site, and intron retention were included. Tissue-specific AS events in pig transcriptome were detected by Splicing Express software [39].

### 2.6. *Comparison of Novel and Known Protein Domains*

Translated DNA sequences of novel transcripts were aligned to UniProt database by DIAMOND software (v4.4.0) (https://www.crystalimpact.com/diamond/) [39]. HMMER3 [40] was used to determine conserved protein domains for novel isoforms and their most similar known transcripts. The domain changes in novel and known proteins were carried out in a pairwise manner. Protein conformation was predicted by the Swiss model [41].

### 2.7. *SNP Compared with Alternative Splicing Variations*

Here we mainly used Pvaas software [42] to detect the single nucleotide1 variant (SNV) mutation associated with the novel AS, which is mainly to determine the correlation between the SNV and AS by Fisher exact test. Its reliability was assessed by the *p*-value after correcting by the Benjamini–Hochberg procedure. Further filtering was performed in order to ensure the accuracy of SNV: the minor mutation frequency should be greater than 5% (the minor mutation reads number accounts for more than 5% of AS); the number of mutant reads is ≥5; the *p*-value after correction <0.001; and the reads number of the new AS is ≥10.

### 2.8. *Known and Novel AS Validation Using Quantitative Reverse Transcription PCR*

RNAs from six pig tissues, including heart, liver, lung, kidney, brain, and lymph were transcribed into cDNA using PrimeScript RT reagent Kit with gDNA Eraser (TaKaRa Bio, Beijing, China) for PCR reaction. Two reverse-transcribed reaction systems were conducted. For reverse-transcribed reaction system I, a DNA-free master mix of each reaction was prepared, with 2.0 µL 5× gDNAEraser buffer, 1.0 µL gDNA Eraser, 7 µL RNase-free dH2O, and 1.0 µg total RNA. Reverse-transcribed reaction system I was standing for 5 min. Reverse-transcribed reaction system II consisted of 4.0 µL 5× PrimeScript

Buffer 2, 1.0 μL PrimeScript RT Enzyme Mix I, 1.0 μL RT Primer Mix, 4.0 μL RNase-free dH2O, and 10.0 μL reverse-transcribed reaction system I. Reverse-transcribed reaction system II reacting was performed at 37 °C for 15 min, followed by 85 °C for 5 s.

The primers for qPCR amplification were designed by primer-blast and confirmed by Oligo 7.0. The details about the primers are listed in Table S2. The selected internal reference gene was β-actin, which is commonly used in swine tissue as an internal reference gene. RT-qPCR was performed by the LightCycler 480 SYBR Green I Master kit (Roche Applied Sciences, Indianapolis, IN, United States). A total of 20 μL volumes consisted of 2 μL cDNA, 10 μL SYBR green master mix, 1 μL forward primer and 1 μL reverse primer (10 μmol/L), and 6 μL nuclease water. RT-qPCR conditions were as follows: pre-incubation for 5 min at 95 °C, and amplification for 40 cycles of 10 s at 95 °C, 20 s at 60 °C, and 20 s at 72 °C. After this, a high-resolution melting curve was generated, using the following protocol: 5 s at 95 °C and 1 min at 65 °C, followed by a gradual increase in temperature from 60 °C to 97 °C, using a ramp rate of 0.02 °C per second. Results were analyzed with the standard Light- Cycler 480 software, version 1.5 (Roche), using the 2-ΔΔCt method [43] to calculate the relative expression level of the target gene for each sample.

### 2.9. Protein Extraction and Western Blotting

Proteins were isolated from the heart and kidney of the Durocs. Fresh frozen tissue was thawed, cut into small pieces, and extensively washed with pre-cooled PBS (Gibco, Rockville, MD, USA). Tissues were suspended in 100 μL RIPA Lysis Buffer (Beyotime, Nanjing, China) and supplemented with 1% proteinase inhibitor (PMSF; Beyotime, Nanjing, China). The supernatant was collected and determined with a BCA (Bicinchoninic acid) assay kit (Thermo Fisher Scientific). The protein concentration of all samples were adjusted to 2 ug/ul with RIPA buffer (Beyotime, Nanjing, China). Samples containing 30 μg protein were separated on 9% sodium dodecyl sulfate–polyacrylamide gels (SDS-PAGE), and then electrotransferred onto a nitrocellulose membrane for 1 h using Bio-Rad Trans-Blot. The membrane was blocked with 5% non-fat milk in Tris-buffered saline (20 mM Tris-HCl, pH 7.6, 137 mM NaCl) containing 0.1% tris-buffered saline +Tween-20 (TBST, Gibco) for 30 min at room temperature, and incubated at 4 °C overnight with the following primary antibodies: Anti- Immunoglobulin Binding Protein 1(IGBP1) antibodies (ab70545, Abcam, Cambrige, UK). The membranes were washed three times with TBST for 10 min and incubated with goat anti-rabbit IgG (Heavy + Light) secondary antibody, an Horseradish peroxidase (HRP) conjugate (Thermo Fisher Scientific), for 1 h at room temperature. The antibody–antigen complexes were detected using Western blotting (WB) luminal reagent. The bands on the developed film were quantified with Quantity One v4.6.2 software (Bio-Rad, Hercules, CA, United States). The β-actin was used as a loading control for normalization.

## 3. Results

### 3.1. Identification of Novel Transcripts in 34 Different Pig Tissues

To discover and map novel transcripts, RNA-seq of 34 different pig tissues was performed. On average, 48.48 million reads per tissue were sequenced from 116 strand-specific and paired-end cDNA libraries (Table S3). Of these sequences, an average of 43.97 million reads (90.7%) per sample passed the strict quality control (QC). A total of 1495 million high-quality reads (376.7 Gb, 135-fold genome coverage) were aligned to the porcine reference genome (*Sus scrofa* 11.1), and 1223 million mapped fragments (310.1 Gb, 110.8-fold genome coverage) with an average alignment rate of 88.29% were recovered (Table S1). These mapped reads were then assembled and quantified as candidate transcripts using Stringtie software. This step produced a total of 2,486,239 transcripts from 29,270 genes across all tissues; of these, 60,578 transcripts (23,887 loci) were annotated in the pig Ensemble database (https://www.ensembl.org/), and 144,134 transcripts from 2424 non-coding genes, as well as 15,385 coding genes, were considered as potential AS. The remaining 2,281,529 novel transcripts were from 26,493 loci without any annotated information. Details for each tissue are available in Table S4. These novel transcripts enhanced the pig genome annotation and increased the number of average transcripts for each gene in pigs.

### 3.2. Classification of Alternative Splicing Types

AS events were classified into 12 types, including the four basic types of AS events (exon skipping, alternative donor site, IR, and alternative acceptor site) (Figure 1A). A total of 138,403 AS events and 29,270 genes across all 34 tissues were detected. A gene had 4.73 AS events on an average. Alternative donor sites accounted for the most common AS event type (60,851; 44%), whereas the other three types shared similar percentages (exon skipping, 19%; alternative acceptor site, 19%; IR, 18%) (Figure 1A). The most significant increase was evident in the number of novel alternative donor sites, which accounted for 44% of all novel AS events (Figure 1B). The average length of IR was found to be the longest, with exon skipping the shortest, which is consistent with the definition of each AS type (Figure 1B).

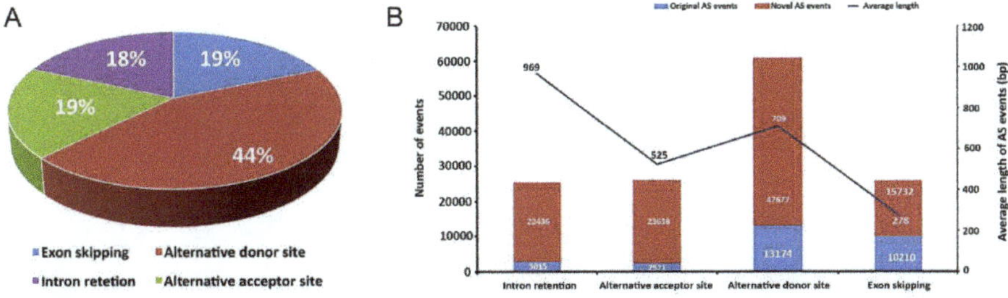


**Figure 1.** Four modes of alternative splicing (AS). (**A**) The proportion of exon skipping, alternative donor sites, alternative acceptor sites, and intron retention. (**B**) Left *y*-axis: the number of known (blue bars) and novel (red bars) alternative splicing events. Right *y*-axis: average length of alternative splicing events (black polyline).

The novel AS events were categorized into two types, novel transcripts of known genes and novel transcripts of unknown genes. The number of novel transcripts in different tissues is shown in Tables S5 and S6. More novel transcripts of unknown genes than known genes were detected. The correlation between the number of previously annotated isoforms and novel isoforms of annotated genes was analyzed (Figure 2A). Transcripts with ensemble IDs were considered to be annotated: the greater the number of annotated isoforms per gene, the more the novel isoforms were detected (Figure 2A). Comparing to annotated isoforms, novel isoforms did not increase significantly, perhaps due to the limited isoforms per gene, which would not change even with improved annotation methods.

The number of AS events related to different genes varied. The number of AS events of the AHNAK gene was 391. However, thousands of genes only had a single AS event. To explain the variation in the number of AS events of the genes, a cluster analysis was performed. Genes with more than 10 AS events were grouped together; those with between 1 and 10 events were grouped together, and those with one AS event were grouped together. A Kyoto Encyclopedia of Genes and Genomes (KEGG) analysis of these three groups (Figure 2B, Table S7) determined that for genes with one AS event, a total of 24 pathways were significantly enriched, including the phosphatidylinositol signaling system ($p$-value = $6.30 \times 10^{-9}$; number of genes = 59), inositol phosphate metabolism ($p$-value = $2.80 \times 10^{-6}$; number of genes = 43), glycerophospholipid metabolism ($p$-value = $1.30 \times 10^{-4}$; number of genes = 49). These genes were mostly involved in lipid metabolism. Other substantial pathways were mainly related to cancer and reproduction. For genes with $1 <$ AS events $\leq 10$, altogether 70 pathways were significantly enriched. Forty-six genes were involved in the measles pathway ($p$-value = $3.9 \times 10^{-8}$; number of genes = 46). Genes with $1 <$ AS events $\leq 10$ were mostly involved in diseases and signaling pathways, such as the forkhead box O (FoxO) and hypoxia-inducible factors (HIF)-1 signaling pathways. However, for genes with >10 AS events, only seven pathways were significantly enriched. These pathways were related to neuroactive and immune-related diseases. The second significant pathway of the seven was the olfactory transduction pathway ($p$-value = $4.9 \times 10^{-145}$), enriched with 639 genes.

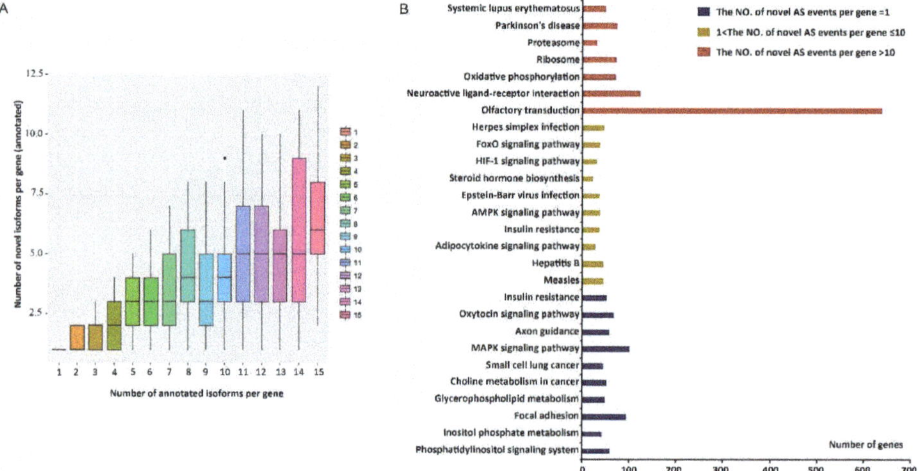


**Figure 2.** The difference in the number of alternative splicing-related genes. (**A**) The relationship between the number of annotated isoforms and novel isoforms per gene. (**B**) Top 10 enriched Kyoto Encyclopedia of Genes and Genome (KEGG) pathways of genes with different numbers of AS events (ASEs). Blue bars represent genes with one ASE, yellow bars represent genes with 1 < ASEs ≤ 10, and red bars represent genes with >10 ASEs.

### 3.3. Tissue Specificity of Alternative Splicing in Different Tissues

The tissue-specific AS events in the pig transcriptome were detected with Jekroll splicing express [39]. As reported in a previous study [37], transcripts with an expression level above 0.1 FPKM in only one tissue and an expression level less than 0.1 FPKM in all other tissues were defined to be tissue-specific transcripts. The number of tissue-specific transcripts varied substantially in different tissues, with the peripheral blood mononuclear cells possessing the most unique isoforms (1633) and the pancreas possessing the least (133; Table 1). Further examination of these tissue-specific transcripts showed that 86% of the genes that encode them were expressed in a single tissue at levels above 0.1 FPKM, revealing that the tissue specificity of these transcripts occurs at the transcriptional level. The remaining 14% of the tissue-specific transcripts had other isoforms expressed in other tissues, indicating that their tissue specificity likely stems from AS (Figure 3A).

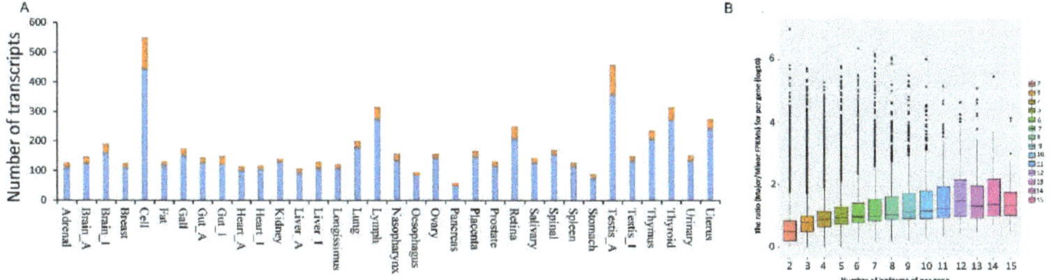

**Figure 3.** (**A**) Single tissue-specific transcripts. The number of known and novel transcripts that are expressed at or above 0.1 FPKM (fragments per kilobase of exon model per million mapped reads) in only one tissue and less than FPKM in all others. Blue areas represent the transcripts and related genes expressed only in one tissue (gene expression dependent). Red areas represent the transcripts expressed in one tissue, with other isoforms present in other tissues (alternative splicing dependent). (**B**) The expression of a gene with different numbers of alternative splicing. The *y*-axis represents the ratio = log10* (the highest FPKM/lowest FPKM) of a gene with multiple AS events, which means the greater the ratio, the greater the difference between the highest and lowest expression of a different transcript per gene. The plot represents this ratio for a gene.

**Table 1.** Distribution of tissue ASE types in different tissues.

| Tissue | AS with Gene | Exon Skipping | Alternative 5′ Splice Site | Alternative 3′ Splice Site | Intron Retention |
|---|---|---|---|---|---|
| Brain_A | 147 | 113 | 60 | 52 | 58 |
| Breast | 126 | 92 | 80 | 71 | 63 |
| Testis_I | 150 | 80 | 71 | 70 | 99 |
| Lung | 202 | 94 | 95 | 75 | 225 |
| Placenta | 167 | 141 | 102 | 83 | 99 |
| Stomach | 89 | 48 | 56 | 48 | 34 |
| Gut_A | 144 | 76 | 111 | 54 | 86 |
| PBMC | 551 | 402 | 303 | 212 | 716 |
| Fat | 132 | 78 | 72 | 63 | 81 |
| Lymph | 317 | 133 | 497 | 116 | 479 |
| Prostate | 132 | 61 | 58 | 75 | 105 |
| Thymus | 238 | 215 | 194 | 118 | 224 |
| Heart_A | 116 | 89 | 55 | 72 | 56 |
| Brain_I | 192 | 101 | 54 | 81 | 116 |
| Gall | 175 | 78 | 70 | 64 | 146 |
| Nasopharynx | 157 | 84 | 87 | 82 | 92 |
| Retina | 252 | 192 | 113 | 104 | 228 |
| Thyroid | 316 | 114 | 82 | 79 | 350 |
| Liver_A | 108 | 97 | 69 | 66 | 45 |
| Gut_I | 150 | 77 | 69 | 77 | 107 |
| Oesophagus | 95 | 86 | 63 | 32 | 57 |
| Salivary | 143 | 67 | 128 | 48 | 117 |
| Urinary | 152 | 70 | 40 | 61 | 136 |
| Testis_A | 459 | 359 | 322 | 172 | 313 |
| Heart_I | 119 | 102 | 38 | 52 | 56 |
| Kidney | 140 | 156 | 113 | 88 | 78 |
| Ovary | 157 | 91 | 69 | 65 | 127 |
| Spinal | 170 | 84 | 84 | 80 | 102 |
| Uterus | 276 | 102 | 62 | 101 | 286 |
| Adrenal | 127 | 99 | 56 | 49 | 84 |
| Liver_I | 131 | 131 | 85 | 69 | 68 |
| Longissimus | 122 | 91 | 90 | 53 | 58 |
| Pancreas | 58 | 41 | 33 | 19 | 40 |
| Spleen | 128 | 117 | 56 | 64 | 75 |

To identify developmentally regulated changes in AS, StringTie [36] was used to quantify the expression of novel and known transcripts in different tissues (Table S8). The average level of expression of novel transcripts in the pancreas, liver, longissimus dorsi, spleen, prostate, adrenal gland, and breast was higher (FPKM > 10) than that of known transcripts. The difference in the expression level of transcripts between the highest expression transcript and the lowest expression transcript was compared (the ratio of major FPKM and minor FPKM) to the gene whose number of AS events was greater than 1. This revealed that more isoforms of a gene are associated with a greater difference in its highest and lowest expression (Figure 3B). A total of 116,439 genes were tissue-specific, of which 109,773 genes were newly detected. A total of 5,683 genes were expressed across all 34 tissues, suggesting that many genes are expressed in multiple tissues simultaneously (Table S9). These results are useful resources for improving the transcript annotations of the pig genome.

### 3.4. Potential Effect of Novel Alternative Splicing on the Pig Proteome

Translated DNA sequences of novel transcripts were aligned to the UniProt database using DIAMOND software [44]. To determine the effect of AS on the pig proteome, computationally predicted proteins encoded by the novel isoforms of known genes were compared to their respective known annotated proteins. A total of 1184 alternatively spliced transcripts were mapped to 361 new proteins (Table S10). These results, therefore, improved the annotation of unknown proteins in the pig genome.

HMMER3 (biosequence analysis using profile hidden Markov models) was used to identify conserved domains within the novel and known proteins, which were then compared in a pairwise manner [40]. A novel transcript of apolipoprotein B MRNA editing enzyme catalytic subunit 3B (APOBEC3B) in uterine tissue was mapped to a protein with the UniProt identifier D3U1S2_PIG. Compared to a similar known transcript (UniProt identifier: F1SNY0_PIG), the novel transcript (3072 bp) was 11 bp shorter. The annotated protein contained two APOBEC-like N-terminal domains, whereas the new isoform lacked an APOBEC-like N-terminal domain, potentially altering its function substantially (Figure 4A). The novel transcript had three coding exons, which was fewer than the known transcript (Figure 4B). It has been shown that DNA ligase 3 (LIG3) proteins activate leukemia via a transcriptional error [45]. The current results reveal a novel transcript (UniProt identifier: I3L9T2_PIG) of LIG3 that had gained an additional domain (DNA ligase 3 BRCA1 C-terminal domain) not present in the known transcript (UniProt identifier: I3LN18_PIG; Figure 4C). This novel transcript had one more coding exon than the known transcript (Figure 4D). The protein conformations of these four transcripts were predicted (Figure 4E,F), and the loss of the APOBEC3B domain and the gain of the LIG3 domain were clearly demonstrated. Although AS may cause changes in protein conformation, further experiments are needed to understand expression and function of the predicted proteins.

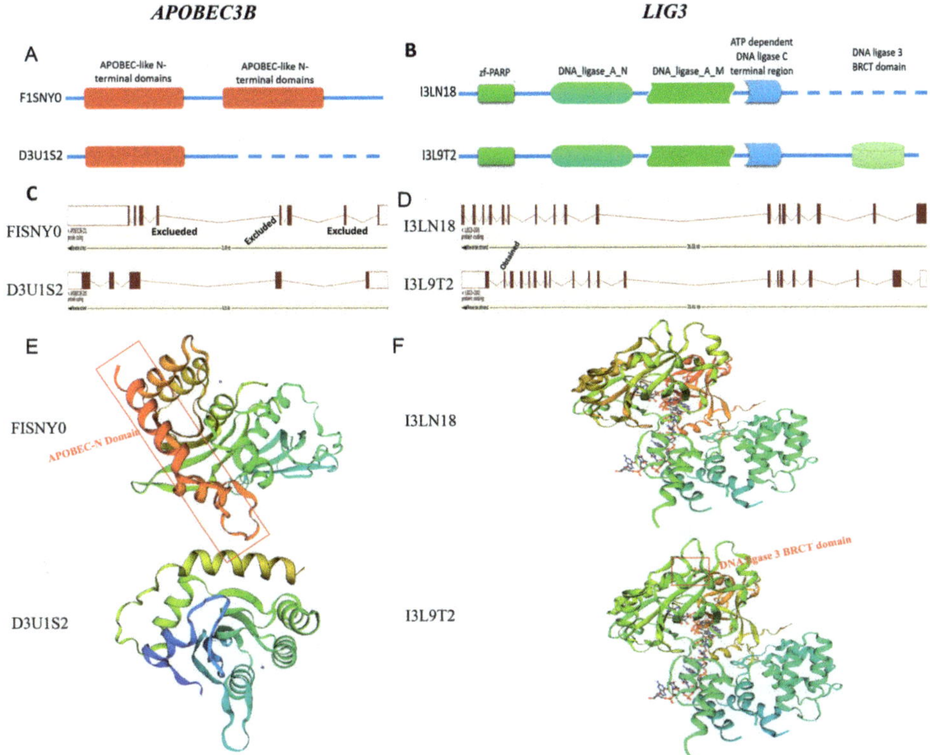


**Figure 4.** Examples of proteins encoded by known transcripts compared with those encoded by novel transcripts. (**A**) Apolipoprotein B MRNA editing enzyme catalytic subunit 3b (APOBEC3B)'s loss of an APOBEC-like N-terminal domain. (**B**) DNA ligase 3 (LIG3)'s gain of a DNA ligase 3 BRCA1 C-terminal (BRCT) domain. (**C**,**D**) The number of coding exons of APOBEC3B and LIG3. (**E**,**F**) The predicted protein conformations of APOBEC3B and LIG3.

### *3.5. Conserved Alternative Splicing between Pig and Human*

Pigs are generally considered a promising medical model for human diseases, and pig orthologs for many human disease-associated genes have been identified [25]. To investigate the orthologous relationship between humans and pigs at the protein level, pig proteins were aligned with human proteins using Basic Local Alignment Search Tool (BLAST; https://blast.ncbi.nlm.nih.gov/), with the criteria of minimum length ≥ 50 bp and identity ≥ 80%. As such, 4168 (56.97%) pig proteins that are orthologous to human proteins were identified.

A similar study reported a global analysis of AS transitions between human infants and adult hearts [46]. Infant and adult pig hearts were included in the current study. Similar to human infant hearts, IR and exon skipping occurred more frequently in the infant pig. In contrast, exon skipping occurred more frequently in the adult pig hearts, but least frequently in the adult human hearts. The specific AS genes that were differentially expressed in infant human and pig hearts were compared. Two genes, sperm-associated antigen 5 (SPAG5) and LIG3, were found to be shared between human and pig. Pig proteins of SPAG5 and LIG3 were aligned to their human proteins (Figure 5A). For two proteins, the identity between the pig and human was 100% and 99.5%. These conserved proteins indicate genome evolution and facilitate further exploration of the potential functional similarity between human and pig genomes.

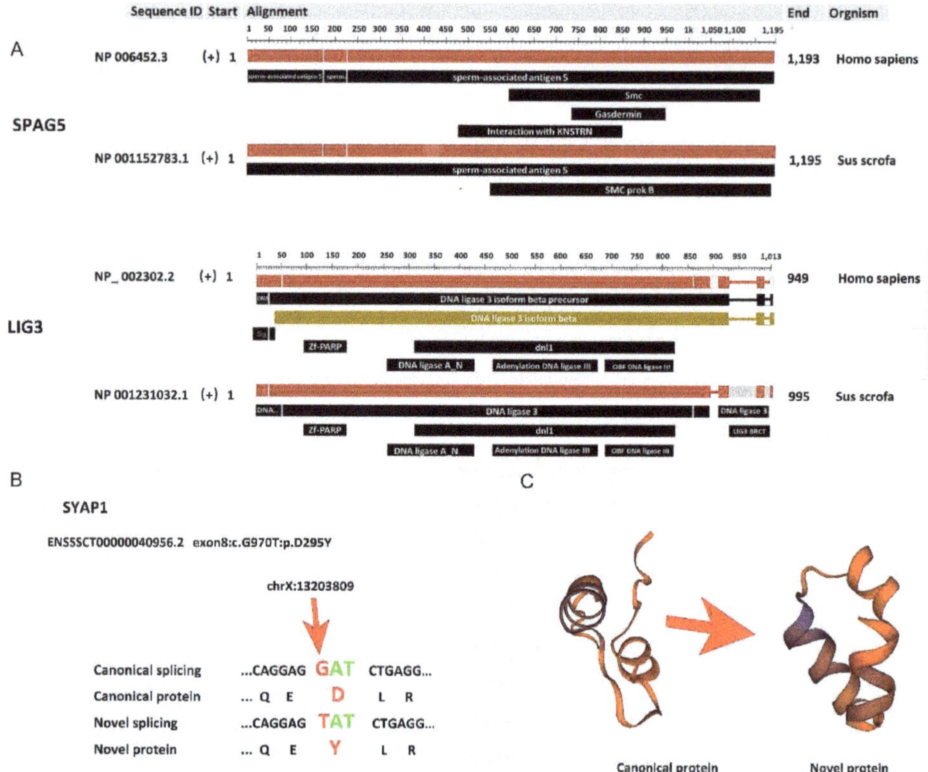


**Figure 5.** (**A**) Example of a homologous protein related to an AS gene between human and pig (downloaded from NCBI Multiple Sequence Alignment Viewer, Version 1.15.0). (**B**) Effect of single nucleotide polymorphism (SNP) mutation of synapse-associated protein 1 (SYAP1) on alternative splicing. (**C**) Protein conformation of canonical and novel isoforms as predicted by the Swiss model.

### 3.6. Effect of Single Nucleotide Polymorphism on Pig Alternative Splicing

Deep RNA sequencing can detect sequence variations associated with AS. AS events were grouped into three types: canonical splicing with a GT–AG intron (i.e., GT and AG splicing signals at donor and acceptor sites, respectively), semi-canonical splicing with GC–AG or AT–AC introns, and novel splicing without GT–AG introns [47]. Semi-canonical and novel splicing are also called aberrant splicing. SNPs occur in different regions of a gene. They change the expression of transcripts and therefore the protein abundance, especially when present in the exonic regions (Table S11). The changes in exons owning to AS may cause synonymous or non-synonymous mutation in the proteins. Some aberrant splicing of the exons led to non-synonymous mutations in the codons, and caused the premature termination of translation in some tissues (Table S12). Related SNPs, genes, and codons when an aberrant splicing occurred in more than 20 tissues are described in Table 2. Six novel splicing events were shown to cause non-synonymous mutations in genes (Table 2)—in particular, a missense SNV (exon8:c.G970T:p.D295Y) within the synapse-associated protein 1 (SYAP1) gene was found that generated a serine to threonine substitution (Figure 5B), leading to a change in the SYAP1 protein conformation (Figure 5C).

**Table 2.** Statistics of different SNP mutation types in different tissues.

| Position | Type | Canonical | Aberrant | Gene | Mutation | Detail |
|---|---|---|---|---|---|---|
| chr7: 22889860 | Novel | T28228<br>C188 | T2132<br>C17845 | TMP-CH242-74M17.2<br>ENSSSCG00000001229 | Non-synonymous | ENSSSCT00000001330<br>exon3:c.A395G:p.H132R |
| chr7: 22889835 | Novel | A27034<br>C0 | A1996<br>C2429 | TMP-CH242-74M17.2<br>ENSSSCG00000001229 | Non-synonymous | ENSSSCT00000001329<br>exon3:c.T370G |
| chr7: 22825399 | Novel | T2495<br>C0 | T2956<br>C2075 | SLA-1<br>ENSSSCG00000001231 | Non-synonymous | ENSSSCT00000036412<br>exon3:c.A404G:p.D135G |
| chr1: 236442561 | Novel | G2635<br>T0 | G1216<br>T722 | TLN1<br>ENSSSCG00000005317 | Non-synonymous | ENSSSCT00000005856<br>exon8:c.C745A:p.H249N |
| chr9: 67827403 | Novel | G1128<br>T593 | G7<br>T7237 | C4BPA<br>ENSSSCG00000015663 | Non-synonymous | ENSSSCT00000017061<br>exon4:c.C605T:p.T202I |
| chr9: 67827383 | Novel | G1506<br>A520 | G16<br>A8310 | C4BPA<br>ENSSSCG00000015663 | Synonymous | ENSSSCT00000017061<br>exon4:c.G585A:p.K195K |
| chr4: 90572872 | Semi-canonical | C17145<br>T7513 | C27<br>T785 | TAGLN2<br>ENSSSCG00000006395 | Synonymous | ENSSSCT00000007008<br>exon3:c.G246A:p.G82G |
| chr7: 1955628 | Semi-canonical | A6919<br>G5 | A26<br>G4623 | TUBB2B<br>ENSSSCG00000001006 | Synonymous | ENSSSCT00000001098<br>exon1:c.T42C:p.N14N |
| chr6: 103414552 | Novel | A6090<br>G14 | A79<br>G325 | LOC733637<br>ENSSSCG00000003692 | Synonymous | ENSSSCT00000025362<br>exon4:c.T495C:p.F165F |
| chrX: 13203809 | Novel | T21856<br>G0 | G0<br>T4538 | SYAP1<br>ENSSSCG00000012148 | Non-synonymous | ENSSSCT00000040956.2<br>exon8:c.G970T:p.D295Y |
| chr7: 24913889 | Semi-canonical | C17224<br>T61 | C14<br>T1190 | SLA-DRB1<br>ENSSSCG00000001455 | Synonymous | ENSSSCT00000001612<br>exon1:c.G39A:p.A13A |
| chr6: 103414681 | Novel | A19996<br>G481 | A114<br>G332 | LOC733637<br>ENSSSCG00000003692 | Synonymous | ENSSSCT00000025362<br>exon4:c.T366C:p.G122G |
| chr7: 23636509 | Novel | C9015<br>T2779 | C3066<br>T2870 | SLA-7<br>ENSSSCG00000024161 | Synonymous | ENSSSCT00000035331<br>exon4:c.C852T:p.H284H |

### 3.7. Validation of Known and Novel Transcripts Using RT-qPCR and Western Blotting

The reliability of transcript identification in this study was verified with RT-qPCR of one known and five novel transcripts from six tissues. Of these transcripts, TCONS_01698330 and TCONS_01698333 are known and novel transcripts of DDX17 gene, respectively. TCONS_00393548, TCONS_00028775, TCONS_01245051, and TCONS_01413087 are novel transcripts of MICU2, IGBP1, TCIRG1, and NFATC2IP genes, respectively. These six transcripts that were detected with RT-qPCR and RNA-seq showed consistent expression patterns (Figure 6A). Gene expression levels measured using these two methods were highly correlated ($R^2 = 0.92$). These results confirm the accuracy of the identification and characterization of pig transcripts.

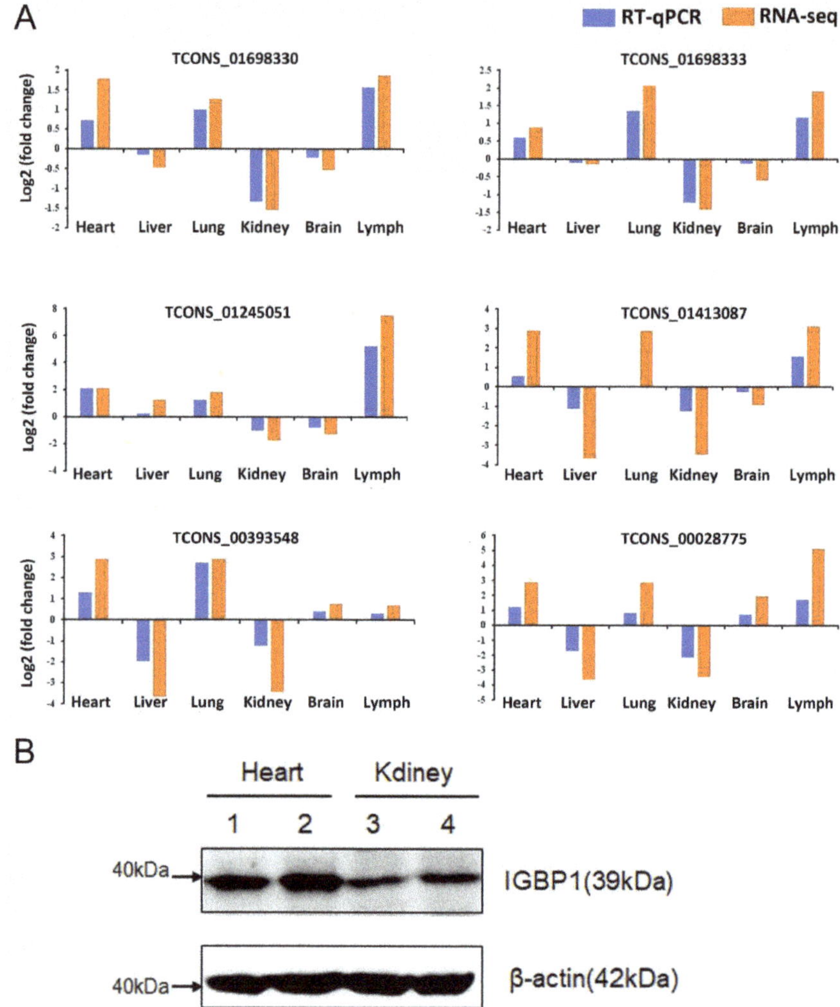


**Figure 6.** (**A**) Validation of transcripts using RT-qPCR. Blue bars and red bars represent gene expression measured with RT-qPCR and RNA-seq, respectively. (**B**) The protein level of Immunoglobulin Binding Protein 1 (IGBP) using western blotting.

To determine whether AS was also reflected at the protein level, Western blotting was performed. Among the five genes used for RT-qPCR, only the antibody of IGBP1 (Abcam) for pig was available. β-actin was used as house-keeping gene. Fortunately, only one AS type and transcript of the IGBP1 gene was detected. We can see that the mRNA of IGBP1 is translated to similar changes at protein levels.

## 4. Discussion

In the present study, novel AS from 34 different tissues in nine Duroc pigs was profiled via deep RNA sequencing, providing important insights into porcine genome annotation. A total of 138,403 AS events and 29,270 expressed genes were detected, with 4.73 AS events per gene on average. The average number of AS events per gene has improved to 1.93 splicing variants in this study compared with the 2.8 splicing variants per gene by EST data [28,30]. Previous studies have shown that exon skipping, alternative donor sites, alternative acceptor splice sites, and IR are the four basic AS types. In the current study, the alternative donor site was the most common AS form, accounting for 44% of the total AS events. The percentage of the other three AS forms was approximately 19%. These results suggest that AS events involving different alternative splicing types can produce different transcripts or proteins that affect biological processes. In a previous study to detect AS events in nine pig tissues using Iso-seq and RNA-seq data, alternative acceptor splice site was shown to be the most prevalent AS type in each tissue [30]. The study by Beiki et al. [30] shared four tissues in common with the current study, including the brain, liver, spleen, and thymus. In the current study, exon skipping was the predominant splicing event in the brain, liver, and spleen (Table 1), which is consistent with a study where exon skipping is the most prevalent AS type in animals [4]. In addition, IR was the most frequent of AS events in the thymus, which is similar to that reported in humans [48]. Nine Duroc pigs, three infants and six adults, were used to detect AS events in the present study, whereas a single cross-bred pig was used for AS analysis in the previous study [30]. Compared to results in humans, we can see it is common that AS show different trends in different situation. Even we do not share the exact same results, both of our results make sense. The different breeds used in these two studies could have produced different results, but regardless of the splicing mechanism, this large increase in new, alternatively spliced transcripts improves our knowledge of the diversity of the pig proteome dramatically.

Genes with only one AS event were predominantly located in the pathway of lipid metabolism. Genes with more than 10 AS events were associated with pregnancy or disease-related signaling pathways. Whether genes belonging to a pathway are related in terms of their AS events requires further investigation. The functions of genes with tissue-specific AS were found to be nearly consistent with their tissue of expression. In addition, the expression of a transcript in different tissues could be different. For example, the expression of the gene glutathione peroxidase 4 (GPX4) ranges from 20 to over 400 FPKM, and its transcript expression in the adult testes was much higher than in other tissues (Figure S1). The GPX4 gene translates into specific enzymes in the testes of adult rats, which is an important structural protein in mature sperm [49,50].

Genes expressed in specific tissues are reported as tissue-enriched genes with tissue-specific functions in multicellular organisms (Figure S2). Results shows that 8482 well-annotated protein isoforms demonstrated tissue-specific expression characteristics. The expression level of 16,356 well-annotated protein isoforms in a specific tissue is at least five times higher FPKM compared to its second-highest expression level in the tissue. In the present study, tissues with complex biological processes usually had more tissue-enriched genes that are closely related to the function of the corresponding tissue. For example, the rhodopsin (RHO) gene, enriched in the retina, plays an important role in the deposition of retinal pigments [51]. Therefore, these specific, tissue-enriched genes can not only confirm the biological properties of known genes, but also predict the potential function of unknown genes in the pig genome. Besides these tissue-specific genes, there are still some genes that can be highly expressed in some functionally related tissues/organs, and these genes usually have similar functions. A total of 1318 well-annotated protein isoforms could be divided into seven types (Figure S3). These results show that most of the enriched genes belong to the brain system (72.7%), followed by the muscle system, adrenal and thymic system (6.6%), and liver and gallbladder system (4.5%).

In general, the majority of novel splicing events resulted in new proteins. Novel alternatively spliced transcripts often contain domain losses or gains relative to their most similar known isoform, and even led to opposite functions [14]. In the current analysis, proteins encoded by novel transcripts were predicted in silico. The 1184 alternatively spliced transcripts mapped to 361 new proteins. Novel and known transcripts of APOBEC3B and LIG3 were compared as examples and showed changes in transcript length. The novel transcripts of APOBEC3B and LIG3 showed the loss and gain of a domain, respectively, as predicted by HMMER3 (Figure 4A,D). However, additional experiments are still necessary to confirm the expression and function of new proteins.

SNPs affect protein-binding sites (or cause mutations in the binding proteins themselves) and contribute to aberrant splicing. Some non-synonymous mutations also cause premature translational termination. A missense SNV (exon3:c.T376A:p.S126T) within SYAP1 was found that generated a serine to threonine substitution, which caused changes in protein conformation of these two isoforms, as predicted by the Swiss model.

## 5. Conclusions

In summary, the current analysis greatly expands the pig transcriptome to include more than 2,281,529 newly annotated transcripts. It was found that if genes had more AS events, the expression difference between the highest and lowest expressions would be greater. The expression of genes with tissue-specific AS events is consistent with their tissue-specific functions. This expansion of isoforms increases the known proteome diversity and results in novel proteins that gain and lose important functional domains. Point mutations, such as SNPs, in exons could lead to new AS. Taken together, these findings extend pig genome annotation and highlight the roles that AS plays in tissue identity in organisms.

**Supplementary Materials:** The following are available online at http://www.mdpi.com/2073-4425/11/12/1405/s1, Figure S1: The specific AS events of the GPX4 gene in the testis of adult pigs. Figure S2: Numbers of tissue-enriched isoforms for known and unknown proteins, Figure S3: Numbers of group-enriched isoforms in different tissue groups, Table S1: Alignment information of 34 pig tissues, Table S2: The details about the primers, Table S3: Information of sequence data from 34 pig tissues, Table S4: Statistics of predicted transcripts and related genes (FPKM > 0.1), Table S5: Statistics of novel AS that are known genes, Table S6: Statistics of novel transcripts of unknown genes, Table S7: Enriched KEGG terms of genes with different number of AS events, Table S8: The average expression of novel and known transcripts (/FPKM), Table S9: Statistic of novel and known genes expressed in different number tissues, Table S10: Statistic of new proteins that AS were alligned successfully, Table S11: Specific SNP in different Tissues (Gene level), Table S12: Specific SNP indifferent Tissues (Exonic level).

**Author Contributions:** W.F. and J.L. designed the experiments. P.Z. performed population analyses. W.F. contributed to computational analyses. X.Z. collected samples and prepared them for sequencing. Z.H. designed the primers for RT-qPCR validation. W.F. wrote, J.L. revised the paper. All authors have read and agreed to the published version of the manuscript.

**Funding:** This work was supported by the National Natural Science Foundations of China (31661143013, 31790414), National Key R&D Program of China (2018YFD0501200),the Beijing Natural Science Foundation (65199011). The funders had no role in study design, data collection and analysis, and interpretation of decision to publish, nor the preparation of the manuscript.

**Acknowledgments:** We thank Shenzhen Jinxinnong Technology Co., Ltd. for offering nine Duroc pigs.

**Conflicts of Interest:** The authors declare no conflict of interest. The funders had no role in the design of the study; in the collection, analyses, or interpretation of data; in the writing of the manuscript, or in the decision to publish the results.

## References

1. Pan, Q.; Shai, O.; Lee, L.J.; Frey, B.J.; Blencowe, B.J. Deep surveying of alternative splicing complexity in the human transcriptome by high-throughput sequencing. *Nat. Genet.* **2008**, *40*, 1413–1415. [CrossRef] [PubMed]

2. Chacko, E.; Ranganathan, S. Genome-wide analysis of alternative splicing in cow: Implications in bovine as a model for human diseases. *BMC Genom.* **2009**, *10*, S11. [CrossRef]
3. Liu, S.; Zhou, X.; Hao, L.; Piao, X.; Hou, N.; Chen, Q. Genome-Wide Transcriptome Analysis Reveals Extensive Alternative Splicing Events in the Protoscoleces of Echinococcus granulosus and Echinococcus multilocularis. *Front. Microbiol.* **2017**, *8*, 929. [CrossRef] [PubMed]
4. Sammeth, M.; Foissac, S.; Guigo, R. A General Definition and Nomenclature for Alternative Splicing Events. *PLoS Comput. Biol.* **2008**, *4*, e1000147. [CrossRef]
5. Koenigsberger, C.; Chicca, J.J.; Amoureux, M.C.; Edelman, G.M.; Jones, F.S. Differential regulation by multiple promoters of the gene encoding the neuron-restrictive silencer factor. *Proc. Natl. Acad. Sci. USA* **2000**, *97*, 2291–2296. [CrossRef]
6. Di Giammartino, D.C.; Nishida, K.; Manley, J.L. Mechanisms and Consequences of Alternative Polyadenylation. *Mol. Cell* **2011**, *43*, 853–866. [CrossRef]
7. Emrich, S.J.; Barbazuk, W.B.; Li, L.; Schnable, P.S. Gene discovery and annotation using LCM-454 transcriptome sequencing. *Genome Res.* **2007**, *17*, 69–73. [CrossRef]
8. Ruhl, C.; Stauffer, E.; Kahles, A.; Wagner, G.; Drechsel, G.; Ratsch, G.; Wachter, A. Polypyrimidine Tract Binding Protein Homologs from Arabidopsis Are Key Regulators of Alternative Splicing with Implications in Fundamental Developmental Processes. *Plant. Cell* **2012**, *24*, 4360–4375. [CrossRef]
9. Staiger, D.; Brown, J.W.S. Alternative Splicing at the Intersection of Biological Timing, Development, and Stress Responses. *Plant. Cell* **2013**, *25*, 3640–3656. [CrossRef]
10. Gutierrez-Arcelus, M.; Ongen, H.; Lappalainen, T.; Montgomery, S.B.; Buil, A.; Yurovsky, A.; Bryois, J.; Padioleau, I.; Romano, L.; Planchon, A.; et al. Tissue-Specific Effects of Genetic and Epigenetic Variation on Gene Regulation and Splicing. *PLoS Genet.* **2015**, *11*, e1004958. [CrossRef]
11. Baralle, F.E.; Giudice, J. Alternative splicing as a regulator of development and tissue identity. *Nat. Rev. Mol. Cell Biol.* **2017**, *18*, 437–451. [CrossRef]
12. Scotti, M.M.; Swanson, M.S. RNA mis-splicing in disease. *Nat. Rev. Genet.* **2016**, *17*, 19–32. [CrossRef]
13. Duque, P. A role for SR proteins in plant stress responses. *Plant Signal. Behav.* **2011**, *6*, 49–54. [CrossRef]
14. Sun, X.Y.; Yang, Q.Y.; Deng, Z.P.; Ye, X.F. Digital Inventory of Arabidopsis Transcripts Revealed by 61 RNA Sequencing Samples. *Plant Physiol.* **2014**, *166*, 869–878. [CrossRef]
15. Cheng, J.H.; Zhou, T.; Liu, C.D.; Shapiro, J.P.; Brauer, M.J.; Kiefer, M.C.; Barr, P.J.; Mountz, J.D. Protection from Fas-Mediated Apoptosis by a Soluble Form of the Fas Molecule. *Science* **1994**, *263*, 1759–1762. [CrossRef]
16. Jiang, F.K.; Guo, M.; Yang, F.; Duncan, K.; Jackson, D.; Rafalski, A.; Wang, S.C.; Li, B.L. Mutations in an AP2 Transcription Factor-Like Gene Affect Internode Length and Leaf Shape in Maize. *PLoS ONE* **2012**, *7*, e37040. [CrossRef]
17. Li, X.R.; Zhu, C.S.; Yeh, C.T.; Wu, W.; Takacs, E.M.; Petsch, K.A.; Tian, F.; Bai, G.H.; Buckler, E.S.; Muehlbauer, G.J.; et al. Genic and nongenic contributions to natural variation of quantitative traits in maize. *Genome Res.* **2012**, *22*, 2436–2444. [CrossRef]
18. Xing, A.Q.; Williams, M.E.; Bourett, T.M.; Hu, W.N.; Hou, Z.L.; Meeley, R.B.; Jaqueth, J.; Dam, T.; Li, B.L. A pair of homoeolog ClpP5 genes underlies a virescent yellow-like mutant and its modifier in maize. *Plant J.* **2014**, *79*, 192–205. [CrossRef]
19. Tejedor, J.R.; Papasaikas, P.; Valcarcel, J. Genome-Wide Identification of Fas/CD95 Alternative Splicing Regulators Reveals Links with Iron Homeostasis. *Mol. Cell* **2015**, *57*, 23–38. [CrossRef] [PubMed]
20. Julien, P.; Minana, B.; Baeza-Centurion, P.; Valcarcel, J.; Lehner, B. The complete local genotype-phenotype landscape for the alternative splicing of a human exon. *Nat. Commun.* **2016**, *7*, 11558. [CrossRef]
21. Lewis, B.P.; Green, R.E.; Brenner, S.E. Evidence for the widespread coupling of alternative splicing and nonsense-mediated mRNA decay in humans. *Proc. Natl. Acad. Sci. USA* **2003**, *100*, 189–192. [CrossRef] [PubMed]
22. Kalsotra, A.; Cooper, T.A. Functional consequences of developmentally regulated alternative splicing. *Nat. Rev. Genet.* **2011**, *12*, 715–729. [CrossRef] [PubMed]
23. Kornblihtt, A.R.; Schor, I.E.; Allo, M.; Dujardin, G.; Petrillo, E.; Munoz, M.J. Alternative splicing: A pivotal step between eukaryotic transcription and translation. *Nat. Rev. Mol. Cell Biol.* **2013**, *14*, 153–165. [CrossRef] [PubMed]

24. Wang, E.T.; Sandberg, R.; Luo, S.J.; Khrebtukova, I.; Zhang, L.; Mayr, C.; Kingsmore, S.F.; Schroth, G.P.; Burge, C.B. Alternative isoform regulation in human tissue transcriptomes. *Nature* **2008**, *456*, 470–476. [CrossRef] [PubMed]
25. Bassols, A.; Costa, C.; Eckersall, P.D.; Osada, J.; Sabria, J.; Tibau, J. The pig as an animal model for human pathologies: A proteomics perspective. *Proteom. Clin. Appl.* **2014**, *8*, 715–731. [CrossRef]
26. Zhang, L.; Tao, L.; Ye, L.; He, L.; Zhu, Y.Z.; Zhu, Y.D.; Zhou, Y. Alternative splicing and expression profile analysis of expressed sequence tags in domestic pig. *Genom. Proteom. Bioinform.* **2007**, *5*, 25–34. [CrossRef]
27. Thatcher, S.R.; Zhou, W.G.; Leonard, A.; Wang, B.B.; Beatty, M.; Zastrow-Hayes, G.; Zhao, X.Y.; Baumgarten, A.; Li, B.L. Genome-Wide Analysis of Alternative Splicing in Zea mays: Landscape and Genetic Regulation. *Plant Cell* **2014**, *26*, 3472–3487. [CrossRef]
28. Tang, L.T.; Ran, X.Q.; Mao, N.; Zhang, F.P.; Niu, X.; Ruan, Y.Q.; Yi, F.L.; Li, S.; Wang, J.F. Analysis of alternative splicing events by RNA sequencing in the ovaries of Xiang pig at estrous and diestrous. *Theriogenology* **2018**, *119*, 60–68. [CrossRef]
29. Zhang, Y.H.; Cui, Y.; Zhang, X.L.; Wang, Y.M.; Gao, J.Y.; Yu, T.; Lv, X.Y.; Pan, C.Y. Pig StAR: mRNA expression and alternative splicing in testis and Leydig cells, and association analyses with testicular morphology traits. *Theriogenology* **2018**, *118*, 46–56. [CrossRef]
30. Beiki, H.; Liu, H.; Huang, J.; Manchanda, N.; Nonneman, D.; Smith, T.P.L.; Reecy, J.M.; Tuggle, C.K. Improved annotation of the domestic pig genome through integration of Iso-Seq and RNA-seq data. *BMC Genom.* **2019**, *20*, 344. [CrossRef]
31. Bushnell, B. BBMap. Available online: https://sourceforge.net/projects/bbmap/ (accessed on 3 June 2016).
32. Andrews, S. FastQC: A Quality Control Tool for High Throughput Sequence Data. Available online: http://www.bioinformatics.babraham.ac.uk/projects/fastqc (accessed on 6 August 2016).
33. Fastx_toolkit. Available online: http://hannonlab.cshl.edu/fastx_toolkit/download.html (accessed on 13 June 2016).
34. Kim, D.; Langmead, B.; Salzberg, S.L. HISAT: A fast spliced aligner with low memory requirements. *Nat. Methods* **2015**, *12*, 357–360. [CrossRef]
35. Ensemble. Available online: http://ftp.ensembl.org/pub/release-91/gff3/sus_scrofa/ (accessed on 19 June 2016).
36. Pertea, M.; Pertea, G.M.; Antonescu, C.M.; Chang, T.-C.; Mendell, J.T.; Salzberg, S.L. StringTie enables improved reconstruction of a transcriptome from RNA-seq reads. *Nat. Biotechnol.* **2015**, *33*, 290–295. [CrossRef]
37. Mele, M.; Ferreira, P.G.; Reverter, F.; DeLuca, D.S.; Monlong, J.; Sammeth, M.; Young, T.R.; Goldmann, J.M.; Pervouchine, D.D.; Sullivan, T.J.; et al. Human genomics. The human transcriptome across tissues and individuals. *Science* **2015**, *348*, 660–665. [CrossRef]
38. Florea, L.; Song, L.; Salzberg, S.L. Thousands of exon skipping events differentiate among splicing patterns in sixteen human tissues. *F1000Research* **2013**, *2*, 188. [CrossRef]
39. Kroll, J.E.; Kim, J.; Ohno-Machado, L.; de Souza, S.J. Splicing Express: A software suite for alternative splicing analysis using next-generation sequencing data. *PeerJ* **2015**, *3*, e1419. [CrossRef]
40. HMMER3. Available online: https://www.ebi.ac.uk/Tools/hmmer/ (accessed on 9 August 2018).
41. Swiss-Model. Available online: https://swissmodel.expasy.org/ (accessed on 2 July 2020).
42. Wang, L.; Nie, J.J.; Kocher, J.P. PVAAS: Identify variants associated with aberrant splicing from RNA-seq. *Bioinformatics* **2015**, *31*, 1668–1670. [CrossRef]
43. Livak, K.J.; Schmittgen, T.D. Analysis of Relative Gene Expression Data Using Real-Time Quantitative PCR and the 2−ΔΔCT Method. *Methods* **2001**, *25*, 402–408. [CrossRef]
44. Buchfink, B.; Xie, C.; Huson, D.H. Fast and sensitive protein alignment using DIAMOND. *Nat. Methods* **2015**, *12*, 59–60. [CrossRef]
45. Muvarak, N.; Kelley, S.; Robert, C.; Baer, M.R.; Perrotti, D.; Gambacorti-Passerini, C.; Civin, C.; Scheibner, K.; Rassool, F.V. c-MYC Generates Repair Errors via Increased Transcription of Alternative-NHEJ Factors, LIG3 and PARP1, in Tyrosine Kinase–Activated Leukemias. *Mol. Cancer Res.* **2015**, *13*, 699–712. [CrossRef]
46. Wang, H.; Chen, Y.; Li, X.; Chen, G.; Zhong, L.; Chen, G.; Liao, Y.; Liao, W.; Bin, J. Genome-wide analysis of alternative splicing during human heart development. *Sci. Rep.* **2016**, *6*, 35520. [CrossRef]
47. Sheth, N.; Roca, X.; Hastings, M.L.; Roeder, T.; Krainer, A.R.; Sachidanandam, R. Comprehensive splice-site analysis using comparative genomics. *Nucleic Acids Res.* **2006**, *34*, 3955–3967. [CrossRef] [PubMed]

48. Zhang, J.; Jiang, H.; Xie, T.; Zheng, J.; Tian, Y.; Li, R.; Wang, B.; Lin, J.; Xu, A.; Huang, X.; et al. Differential Expression and Alternative Splicing of Transcripts Associated With Cisplatin-Induced Chemoresistance in Nasopharyngeal Carcinoma. *Front. Genet.* **2020**, *11*, 52. [CrossRef] [PubMed]
49. Conrad, M. Transgenic mouse models for the vital selenoenzymes cytosolic thioredoxin reductase, mitochondrial thioredoxin reductase and glutathione peroxidase 4. *BBA Gen. Subj.* **2009**, *1790*, 1575–1585. [CrossRef] [PubMed]
50. Michaelis, M.; Gralla, O.; Behrends, T.; Scharpf, M.; Endermann, T.; Rijntjes, E.; Pietschmann, N.; Hollenbach, B.; Schomburg, L. Selenoprotein P in seminal fluid is a novel biomarker of sperm quality. *Biochem. Biophys. Res. Commun.* **2014**, *443*, 905–910. [CrossRef]
51. Yu, X.; Shi, W.; Cheng, L.; Wang, Y.; Chen, D.; Hu, X.; Xu, J.; Xu, L.; Wu, Y.; Qu, J.; et al. Identification of a rhodopsin gene mutation in a large family with autosomal dominant retinitis pigmentosa. *Sci. Rep.* **2016**, *6*, 19759. [CrossRef]

*Article*

# Prolonged Effect of Seminal Plasma on Global Gene Expression in Porcine Endometrium

**Marek Bogacki *, Beenu Moza Jalali, Anna Wieckowska and Monika M. Kaczmarek**

Institute of Animal Reproduction and Food Research of Polish Academy of Sciences in Olsztyn, 10-748 Olsztyn, Poland; beenu.jalali@pan.olsztyn.pl (B.M.J.); anna.kitewska@gmail.com (A.W.); m.kaczmarek@pan.olsztyn.pl (M.M.K.)

* Correspondence: m.bogacki@pan.olsztyn.pl; Tel.: +48-89-5391-3131

Received: 8 October 2020; Accepted: 29 October 2020; Published: 3 November 2020

**Abstract:** Seminal plasma (SP) deposited in the porcine uterine tract at the time of mating is known to elicit an initial response that is beneficial for pregnancy outcome. However, whether SP has any long-term effect on alterations in endometrial molecular and cellular processes is not known. In this study, using microarray analyses, differential changes in endometrial transcriptome were evaluated after Day 6 of SP-infusion (6DPI) or Day 6 of pregnancy as compared to corresponding day of estrous cycle. Both, pregnancy and SP induced significant changes in the endometrial transcriptome and most of these changes were specific for a particular group. Functional analysis of differentially expressed genes (DEGs) using Ingenuity Pathway Analysis revealed that inhibition in immune response was affected by both pregnancy and SP infusion. Long-term effects of SP included differential expression of genes involved in inhibition of apoptosis, production of reactive oxygen species and steroid biosynthesis, and activation of processes such as proliferation of connective tissue cells and microvascular endothelial cells. Moreover, interleukin-2 and interferon-$\gamma$ was identified to be responsible for regulating expression of many DEGs identified on 6DPI. The present study provides evidence for the long-term effects of SP on porcine endometrium that can be beneficial for pregnancy success.

**Keywords:** seminal plasma; endometrium; global gene expression; microarray; pig

## 1. Introduction

The high rate of pregnancy failure in human and livestock has been attributed mainly to the unsynchronized development of the embryos with the proper preparation of the female reproductive tract and the impaired communication between the developing embryos and uterus [1–3].

Understanding of the molecular embryo-maternal cross-talk is crucial for solving infertility problems, reducing pregnancy loss and identifying hormonal, paracrine, and autocrine factors regulating the developmental potential of the offspring. Effective recognition of the embryo in the maternal tract is crucial for the preparation of an appropriate environment in the uterus for the embryo's development, implantation, and final establishment of pregnancy [4]. However, exactly when the oviduct and uterus recognize the presence of embryos and how the maternal pathway changes its environment in response to embryos is not fully understood.

In pigs, transcriptomic profiling of pregnant and non-pregnant animals has been conducted and pointed major differences in endometrial genes activities in the post-conception period of pregnancy [5], pre-attachment phase [6,7] and during onset of implantation [8]. Identified alterations in uterine transcriptome lead to morphological, biochemical and immunological changes and are reflection of action of para- and autocrine signals released by maternal tract as well as developing embryos.

The application of artificial insemination (AI) and other reproductive technologies shows that pregnancy can be maintained without any semen being deposited in the uterus (embryo transfer outcome) or with highly diluted semen for AI [9]. However, studies conducted in pigs and other livestock species show reduction of early fetal loss due to exposure to seminal constituents. Seminal plasma (SP) is a mixture of various components and serves not only as a vehicle for sperm transport, protection, and nutrition but also affects gamete interaction and successful establishment of pregnancy. Biologically active molecules present in SP (estrogens, testosterone, prostaglandins, cytokines, and growth factors) interact with uterine epithelium to induce a series of changes in the maternal immune response and modify cellular composition, structure, and function of the reproductive tract, directly in local tissues or indirectly in tissues distal to the uterus, for example ovary [10]. In pigs, it was shown that introduction of SP before natural service or/and AI leads to the increase in farrowing rate and litter size [11]. Additionally, increased litter size was also reported after pre-sensitization of the uterus to sperm and seminal antigens in a previous estrous cycle [12] and an increased embryo survival was noted after addiction of leukocyte antigens to boar semen at breeding [13].

The immediate response to full semen insemination in pigs is a rapid influx of inflammatory cells such as neutrophils into the uterine lumen [14] and macrophages, granulocytes, and lymphocytes into the endometrial stroma [15]. The activation of maternal immune system does not cause the rejection of seminal antigens due to the presence of several immunoregulatory molecules in boar SP, such as prostaglandin E (PGE) and tumor growth factor β (TGF-β) [16]. It has been suggested that constituents of SP deposited in lower reproductive tract can easily access the upper reproductive tract and induce biologically relevant changes in the endometrium [17]. Concomitant with this observation, SP interacts with uterine cells in pigs to induce expression of several cytokines: granulocyte-macrophage colony-stimulating factor (GM-CSF), interleukin 6 (IL-6), and macrophage chemotactic protein-1 (MCP1) 34 h after infusion [18]. Moreover, components of SP can alter expression of prostaglandin synthesis enzymes in the porcine oviduct and uterus, acting in favor of PGE2 action as a critical element of early embryo transport and development but also can modify endometrial vascularity up to 10 days after infusion [19,20]. Additionally, in our previous study, prolonged modulatory effects of SP infusion at least for 6 days, demonstrated by induction of T helper (Th) and T regulatory (Treg) cells, increased interleukin 10 (IL10), and decreased expression granulocyte-macrophage colony stimulating factor (GM-CSF), was observed [21].

Though, the immune modulatory effects of SP on porcine endometrium are well documented, to our knowledge there is no published microarray data showing long term effect of SP on molecular changes in porcine endometrium that might be relevant for hatching blastocysts on Day 6 of pregnancy when they reach the uterine horn from the oviduct. That is why, we hypothesized that SP infusion can induce not only immediate but also prolonged transcriptome changes required for endometrial receptivity and modulation of immune response in the uterine environment.

## 2. Materials and Methods

### 2.1. Animals and Treatments

All procedures involving the use of animals were conducted in accordance with the national guidelines for agricultural animal care and approved by the Animal Ethics Committee, University of Warmia and Mazury in Olsztyn, Poland; Decision 86/2011. Estrous induction and synchronization, insemination, and SP infusion were performed as previously described [21]. Cross-bred gilts (*Sus scrofa domesticus*) of similar age and weight were subjected to a hormonal treatment with an intramuscular injection of 750 I.U equine chorionic gonadotropin (eCG) and 500 I.U. human chorionic gonadotropin (hCG) given after 72 h. Gilts randomly divided into three groups ($n$ = 5), 24 h after hCG injection were treated as following: (1) artificially inseminated twice within an interval of 12 h, (2) received two intrauterine infusions of 100 mL SP with an interval of 12 h or (3) received two intrauterine infusions of 100 mL PBS within 12 h (reference group). All the treatments were given at two time points within an interval of 12 h to mimic regular procedure of artificial insemination. Artificial insemination was performed with 100 mL of 2.5 × 109 spermatozoa diluted from the fresh semen collected from a boar with semen extender Safe Cell + (IMV technologies, L'Aigle, France). All the treatments were deposited into the uterus using post cervical artificial insemination methods (PC Blue, SafeBlue Foamtip® with PC Cannula-Minitube) to facilitate interaction of treatments with the endometrium. SP for intrauterine infusion was prepared from whole semen collected and pooled together from four fertile boars, which were used for AI. SP was separated by centrifugation of the whole semen at 1200× *g* at 4 °C for 20 min, divided into 100 mL aliquots and frozen at −20 °C until needed for intrauterine infusion. Animals were slaughtered in a local abattoir at day 6 of pregnancy or at day 6 after SP or PBS infusion. Uterine horns were flushed with PBS and opened longitudinally along the anti-mesometrial surface. Endometrial explants were collected from the upper part of uterine horns and snap-frozen in liquid nitrogen. In the group of artificially inseminated animals, only those animals were included in the study in which pregnancy was confirmed by the presence of blastocysts in the uterine flushings.

### 2.2. RNA Isolation

Total RNA was isolated from 30 mg of grinded in liquid nitrogen and homogenized endometrial tissue with the use of a Qiagen RNeasy Mini Kit (Qiagen, Valencia, CA, USA) and genomic DNA contamination was removed by DNAse treatment (RNase free DNAse Kit, Qiagen, Valencia, CA, USA). The initial RNA quality and quantity were determined spectrophotometry using NanoDrop ND-1000 (Thermo Scientific, Pittsburgh, PA, USA). Subsequently, RNA integrity was evaluated with microfluidic electrophoresis by 2100 Bioanalyzer (Agilent Technologies, Santa Clara, CA, USA) and RNA integrity number (RIN) was calculated for each sample using Agilent 2100 Expert software (Agilent Technologies, Inc., Santa Clara, CA, USA). Only samples with a RIN above 8.0 were processed further.

### 2.3. Microarrays

The Porcine (V2) Gene Expression Microarray 4_44 (Agilent Technologies, Santa Clara, CA, USA) was used for differential gene expression analysis. As positive internal controls of microarrays performance an RNA Agilent Spike-In Kit, One Color was used (Agilent Technologies, Santa Clara, CA, USA). Total RNA obtained from cyclic and pregnant gilts was amplified and labeled with Cy3 dye using Quick Amp Kit, One Color (Agilent Technologies, Santa Clara, CA, USA). After purification of labeled RNA (Qiagen RNeasy Kit), the yield (ng of cRNA) and specific activity (pmol of Cy3/mg of cRNA) were quantified using NanoDrop ND-1000. Labeled cRNA was fragmented, mixed with hybridization buffer, and added to the microarray slide. On each array ($n = 4$, one slide) a combination of samples from all three groups were hybridized for 17 h at 65 °C in an Agilent hybridization oven. Afterwards, arrays were dissociated from the hybridization chamber and washed twice in GE wash buffers. After washing, slides were scanned using Agilent G2565CA Microarray Scanner at settings recommended for the 4_44 K array format. Images obtained after scanning were analyzed using Agilent Feature Extraction software v. 10.5.1.1 (Agilent Technologies Inc., Santa Clara, CA, USA). A detailed analysis including filtering of outlier spots, background subtraction from features, and dye normalization was performed.

### 2.4. Data Analysis

Data obtained after extraction was further analyzed using GeneSpring GX 11.0.2 (Agilent Technologies, Santa Clara, CA, USA). To determine differentially expressed genes (DEGs) data were normalized with quantile method and afterwards moderated *t*-test (Benjamini–Hochberg false discovery rate (FDR) < 0.05, absolute fold change $|Fc| > 1.5$) was performed to compare endometrial transcriptomes between: (1) pregnant and cyclic ($n = 5$) as well as (2) SP infused and cyclic animals ($n = 4$, data from one array were not correlating with other arrays after principal component analysis (PCA)). For identification of differentially expressed probe sets with unknown target sequence the annotation was done manually using NCBI blast algorithm [22]. When porcine sequence for particular mRNA was not available the annotation was performed for human, murine, and cattle transcript, and the date was included only if the query cover and percent identity was equal or higher than 70%. To identify biological processes, pathways and upstream regulators Ingenuity Pathway Analysis (IP Ingenuity Systems-Qiagen, Aarhus, Denmark). Biological processes, pathways, and upstream regulators were considered statistically significant if Fishers exact tests $p$-value $\geq 0.05$ and each process associated with at least four DEGs. Biological processes and pathways connected to cancer, diseases and disorders, nervous, respiratory, digestive, renal and urological system, organismal survival and functions, drug metabolism, organ development and behavior were not taken into consideration while examining IPA results.

### 2.5. Quantitative Real-Time PCR

For validation of microarray results 2 µg of total RNA were transcribed to cDNA with the use of High Capacity RNA to cDNA kit (Applied Biosystems, Foster City, CA, USA). Real-time PCRs were performed using 7900 HT Real-Time PCR System (Applied Biosystems) using 15 ng cDNA, TaqMan Universal MasterMix II, no UNG. TaqMan assays are listed in Table 1. The initial denaturation was carried out at 95 °C

**Table 1.** TaqMan assays used for real-time PCR validation of microarray results.

| Gene Symbol | Gene Name | Test ID | Entrez Gene ID |
|---|---|---|---|
| *ACTB* * | *Actin, β* | Ss03376081_ u1 | 397653 |
| *TGFA* | *Transforming growth factor, α* | Ss03383643_u1 | 397484 |
| *S100A12* | *S100 calcium binding protein A12* | Ss04246259_g1 | 100301483 |
| *S100A8* | *S100 calcium binding protein A8* | Ss04246257_g1 | 100127488 |
| *CCR3* | *Chemokine (C-C motif) receptor 3* | Ss03378176_u1 | 414373 |
| *CXCL11* | *Chemokine (C-X-C motif) ligand 11* | Ss03648934_m1 | 100169744 |
| *HPRT* * | *Hypoxanthine phosphoribosyltransferase 1* | Ss03388273_m1 | 397351 |
| *SLA-DQA1* | *MHC class II histocompatibility antigen SLA-DQA* | Ss03389952_m1 | 100153387 |
| *IL18* | *Interleukin 18* | Ss03391204_m1 | 397057 |
| *LGALS1* | *Galectin 1* | Ss03388270_m1 | 396491 |
| *PDCD10* | *Programmed cell death 10* | Ss03820202_s1 | 100157978 |
| *LY96* | *Lymphocyte antigen 96* | Ss03389453_m1 | 100125555 |

* Reference genes.

For 15 min and was followed by 40 cycles of denaturation at 95 °C (15 s) and primer annealing and elongation at 60 °C. Non template controls and non-enzyme controls were included in the experiment. Gene expression levels were calculated with the use of Real-Time PCR Miner (http://ewindup.info/miner/) and normalized using the geometric mean of expression levels of two reference genes-hypoxanthine guanine phosphoribosyl transferase (HPRT) and β-actin (ACTB). The statistical differences in gene expression between the endometrium from pregnant and cyclic as well as seminal plasma infused and cyclic animals was analyzed with GraphPad PRISM v. 5.0 Software (GraphPad Software, Inc., San Diego, CA, USA) by Student's $t$-test. Confirmed differences in gene expression ($p < 0.05$) were expressed as fold changes.

## 3. Results

### 3.1. Differential Changes in Endometrial Transcriptome

Pairwise comparisons of endometrial samples collected from pigs on Day 6 of pregnancy (6DP) and Day 6 after SP infusion (6DPI) with PBS-infused cycling control pigs on corresponding day of estrous cycle (6DC) were performed to identify endometrial transcriptome changes in response to pregnancy and SP constituents, respectively. Statistical analysis revealed a pregnancy-induced change (fold change > 1.5 or < −1.5; $p < 0.05$; false discovery rate (FDR) = 5%) in 444 probes representing 281 differentially regulated (DEGs) of which 225 were downregulated and 56 were upregulated, on Day 6 as compared to controls on 6DC (Table S1). Whereas, genes showing the most downregulation included *S1008*, *CGA*, and *HLA-DQA1*, genes with highest upregulation were *GPR116*, *COL4A1*, and *CADPS2*. On the other hand, SP-infusion resulted in statistically significant alterations in 342 probes representing 255 DEGs. A downregulation of 118 genes and upregulation of 137 genes was observed on Day 6 after SP infusion as compared to 6DC (Table S2). SP infusion decreased the expression of *ATP6V1C2*, *NMU*, *S100A8*, *S100A12*, *ANGPTL3*, *NOS1*, *CCR3* and upregulated *PCDHB15*, *KLHL5*, *RASGEF1A*, *NMB*, and *CAPZB*.

### 3.2. Comparison between Pregnancy-Induced and Seminal Plasma-Induced Changes in Endometrial Transcriptome

The list of pregnancy and SP-induced DEGs were uploaded to jvenn software (http://jvenn.toulouse.inra.fr/app/example.html) to visualize common DEGs across both presented comparisons and DEGs that were identified as a result of either pregnancy or SP-infusion (Figure 1). This comparison identified 19 common genes (Table S3) that were differentially regulated by pregnancy and SP-infusion. Whereas, in both the groups, 15 genes were found to be downregulated only three genes were upregulated. The expression of transcript coding for chloride channel, voltage sensitive 5 (CLCN5) was lower during Day 6 of pregnancy (Fc = −2.66) and higher on Day 6 after SP infusion (Fc = 1.53). Most of the DEGs common between two groups were involved with immune regulation (*IL15*, *IL18*, *LGALS1*, *FKBP3*) or were DEGs related with molecular transport (*S100A8*, *S100A12*, *CLCN5*) and structural organization (*FN1*, *COL7A1*, *TUBA1B*).

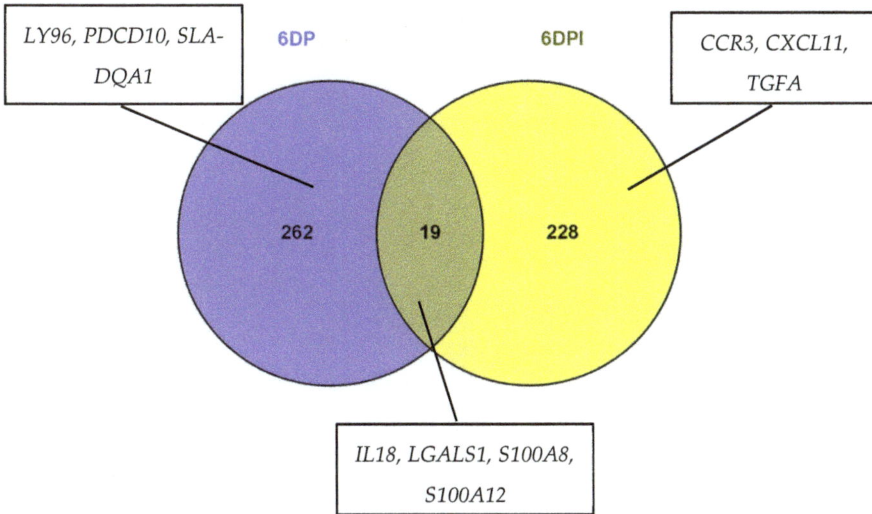


**Figure 1.** Venn diagram showing number of shared and unique DEGs altered on Day 6 of pregnancy (6DP) and Day 6 after SP infusion (6DPI) in comparison to 6 Day of estrous cycle. Genes chosen for qPCR validation are listed in boxes.

### 3.3. Analysis of Biological Processes, Pathways, and Upstream Regulators of Identified DEGs

To classify identified DEGs altered on Day 6 of pregnancy and Day 6 after SP infusion under functional categories, tools available in IPA database were employed. Analysis of DEGs using core analysis module of IPA revealed many altered functions in porcine endometrium as a result of pregnancy or SP-infusion. Furthermore, a comparison analysis, comparing disease and biological functions, canonical pathways, and upstream regulators of DEGs was also performed to evaluate the differences in activated and inhibited functions between pregnant and SP-infused animals (Figure 2). Whereas, most of the identified functions were specific to either pregnancy status or SP-infusion, inhibition of immune functions was common to both the groups. A pregnancy specific activation was observed in processes related to organization of cytoskeleton, organization of cytoplasm, and transmigration of leukocytes ($Z > 2.0$; Figure 3A). The molecular functions inhibited by pregnancy included chemoattraction of leukocytes, homing of leukocytes, and cytotoxicity of lymphocytes ($Z < -2.0$) (Figure 3B, Table S4). On the other hand, SP infusion, besides inhibiting cytotoxicity of lymphocytes, also inhibited molecular functions such as apoptosis, lipid metabolism, senescence of fibroblasts, and production of reactive oxygen species ($Z < -2.0$; Figure 3C, Table S5) and activated only function related to proliferation of microvascular endothelial cells and connective tissue cells ($Z > 2.0$; Figure 3D). Interestingly, alterations in molecular signaling pathways associated HIF1$\alpha$ signaling were specific to Day 6 of pregnancy and activation of Wnt/$\beta$-catenin signaling was specific to SP-infused group. Genes associated with PPAR signaling were differentially altered in both the groups (Figure 2B).

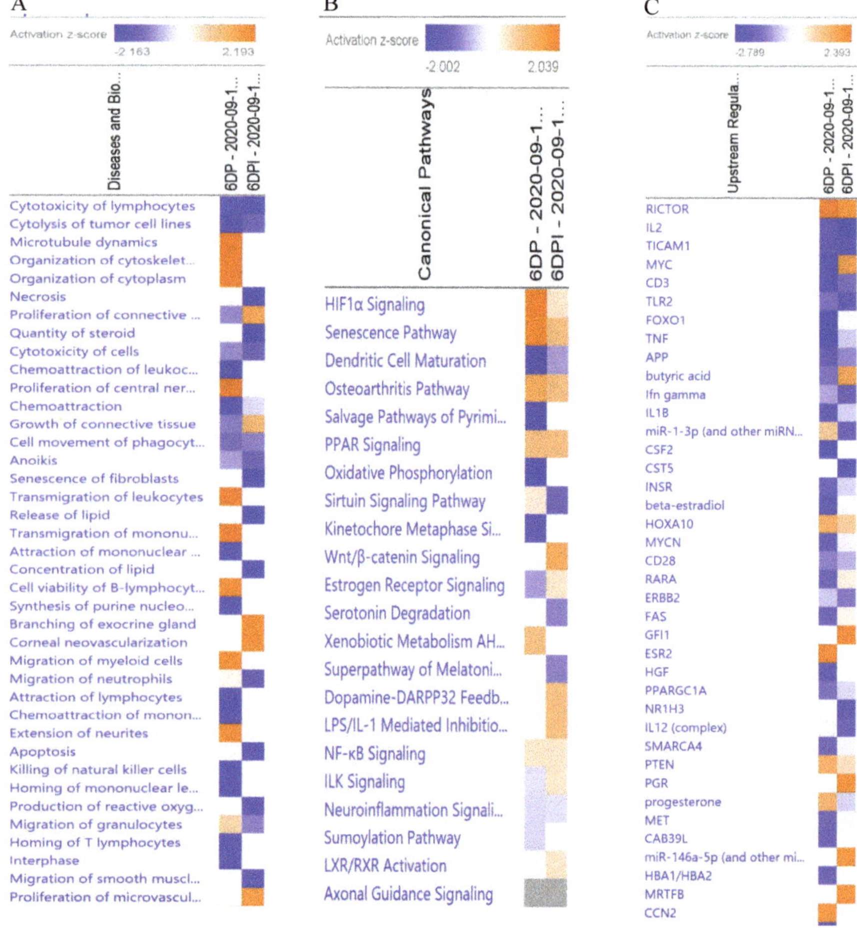


**Figure 2.** Comparison analysis of (**A**) diseases and bio functions, (**B**) canonical pathways, and (**C**) upstream regulators associated with DEGs identified on 6DP and 6DPI. A Z-score > 2 (orange color) is associated with activated functions, pathways, or upstream regulators and a Z-score < 2 (blue color) is associated with inhibited functions, pathways, or upstream regulators.

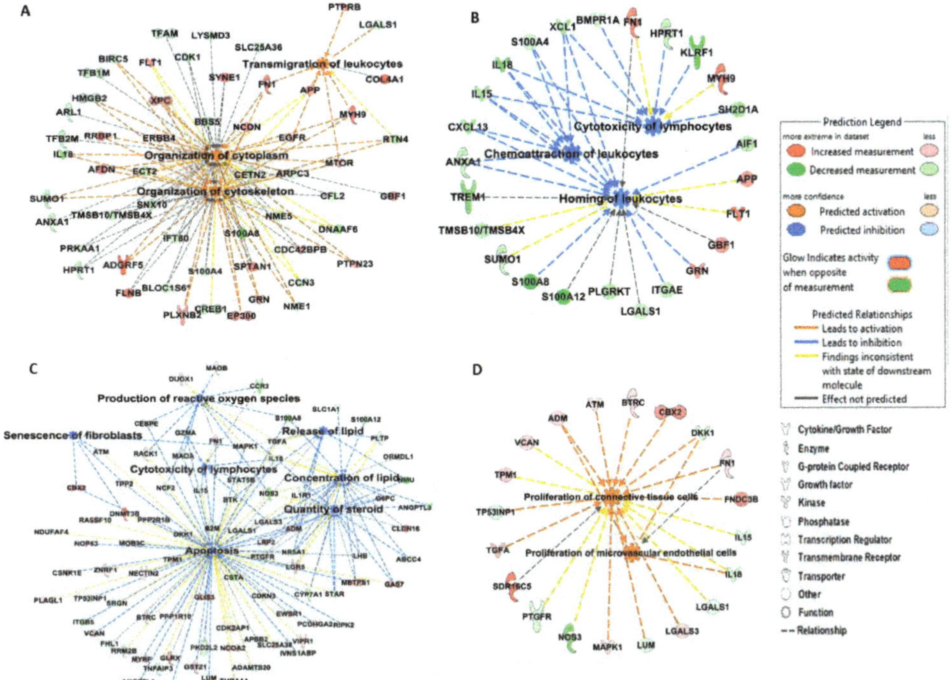


**Figure 3.** IPA depicting networks integrating DEGs identified on 6DP as compared to 6DC in (**A**) activated functions: organization of cytoskeleton, organization of cytoplasm, and transmigration of leukocytes, and (**B**) inhibited immunological functions: cytotoxicity of lymphocytes, chemoattraction of leukocytes, and homing of leukocytes. DEGs identified that 6DPI as compared to 6DC were found to be associated with (**C**) activated functions: proliferation of connective tissue cells and proliferation of microvascular endothelial cells and (**D**) inhibited functions: cytotoxicity of lymphocytes, apoptosis, production of reactive oxygen species, concentration and release of lipid and quantity of steroid, on Day 6 after SP infusion in the porcine endometrium. Red and green colors depict an increase or decrease, respectively, in the expression of a gene. The color intensity of nodes indicates a fold change increase or decrease associated with a particular DEG.

In our study, upstream analysis function of IPA was used to identify the molecules including cytokines, transcription factors, or hormones (2.0 < Z score < −2.0) that might be upstream regulators of altered gene expression in porcine endometrium as a result of pregnancy or SP-infusion. Many cytokines including interleukin (IL-1β), IL-2, TNF, transcription regulators such as FOXO1, NUPR1 and nuclear receptor ESR2, PPARA, and AHR were found to affect the gene expression in endometrium on Day 6 of pregnancy (Figure 4A, Table S6). Upstream analysis of SP-induced DEGs revealed that upstream regulators such as IL-2 and RICTOR were the only common regulators among two groups. Other molecules affecting the expression of genes in SP-infused endometrium on Day 6 included *IFNγ, GFI1, HSF1, TNFRSF1A, TLR2, and NR1H3* (Figure 4B, Table S7). Interestingly, we observed PGR to be an upstream regulator of SP-induced DEGs (Figure 4B).

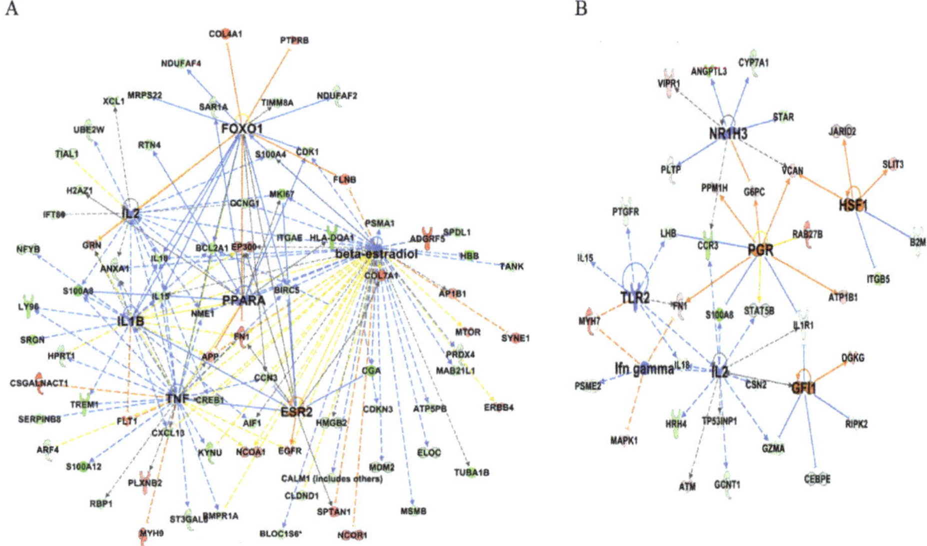


**Figure 4.** Networks generated for identified upstream regulators of DEGs in porcine endometrium on (**A**) Day 6 of pregnancy or (**B**) Day 6 after SP infusion. Red and green colors depict an increase or decrease, respectively, in the expression of a gene. The color intensity of nodes indicates a fold change increase or decrease associated with a particular DEG.

### 3.4. qRT-PCR Validation of Microarray Results

For q PCR validation of microarray data, 10 DEGs, shown in Figure 1, were chosen. These genes were associated with immune function, molecular transport, and cell proliferation. Most of the assessed DEGs showed similar expression profiles when comparing microarray and qPCR data (Figure 5). However, qPCR data revealed a pregnancy induced upregulation of TGFA expression that was not observed in the microarray data. A comparison of fold change and $p$ value of DEGs obtained after qPCR and microarray data analysis is presented in Table 2.

**Table 2.** Results of microarray experiment validation with qPCR. 6DP—6 Day of pregnancy vs. 6 Day of estrous cycle, 6DPI—6 Day after SP infusion vs. 6 Day of estrous cycle, Fc—fold change.

| Gene Symbol | qPCR | | Microarray | |
|---|---|---|---|---|
| | Fc | *p*-Value | Fc | *p* Corr |
| | | 6DP | | |
| *IL18* | −3.52 | 0.002 | −3.7 | 0.036 |
| *LGALS1* | −1.75 | 0.020 | −2.89 | 0.046 |
| *LY96* | −2.32 | 0.023 | −3.22 | 0.034 |
| *PDCD10* | −1.85 | 0.026 | −2.16 | 0.047 |
| *S100A12* | −4.69 | 0.026 | −5.43 | 0.049 |
| *S100A8* | −5.29 | 0.014 | −9.45 | 0.041 |
| *SLA-DQA1* | −6.56 | 0.002 | −5.16 | 0.041 |
| | | 6DPI | | |
| *CCR3* | −6.80 | 0.011 | −8.59 | 0.003 |
| *CXCL11* | −2.38 | 0.009 | −2.22 | 0.003 |
| *TGFA* | 2.25 | 0.007 | 2.64 | 0.006 |
| *IL18* | −3.64 | 0.002 | −3.54 | 0.003 |
| *S100A8* | −6.71 | 0.004 | −12.18 | 0.003 |
| *S100A12* | −6.80 | 0.011 | −5.98 | 0.002 |

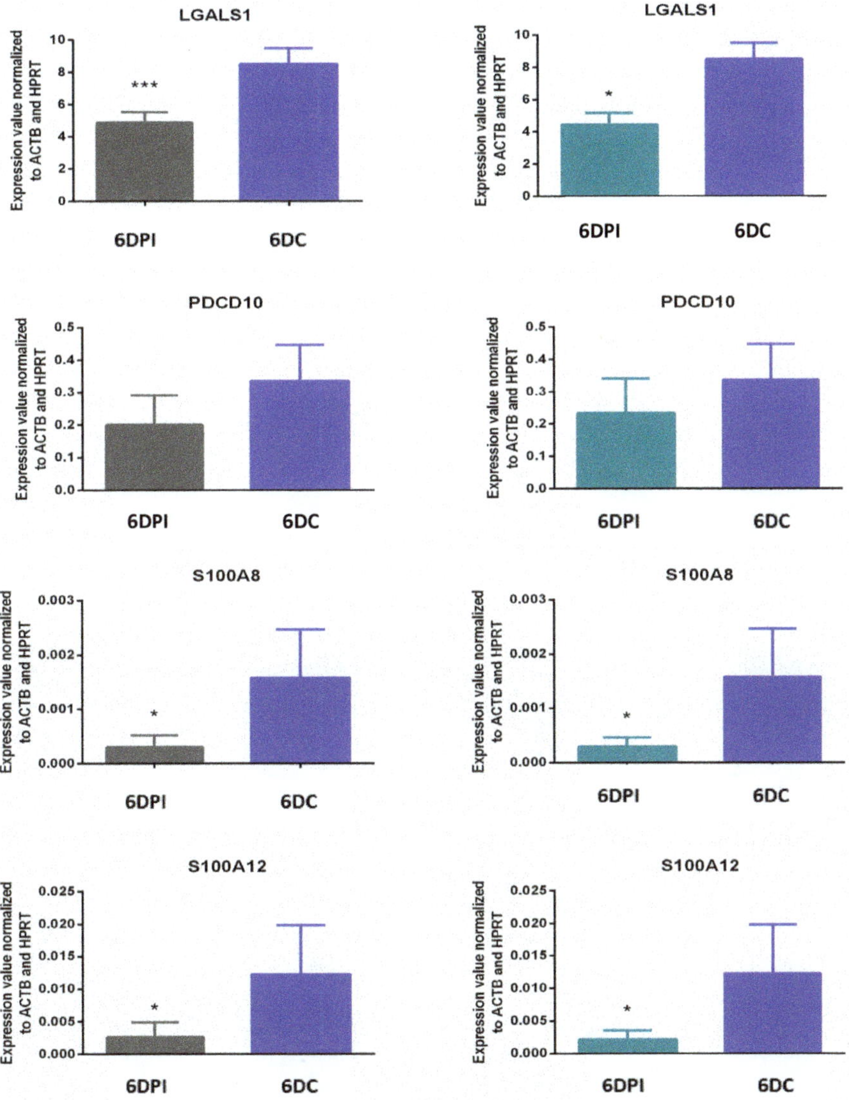


**Figure 5.** *Cont.*

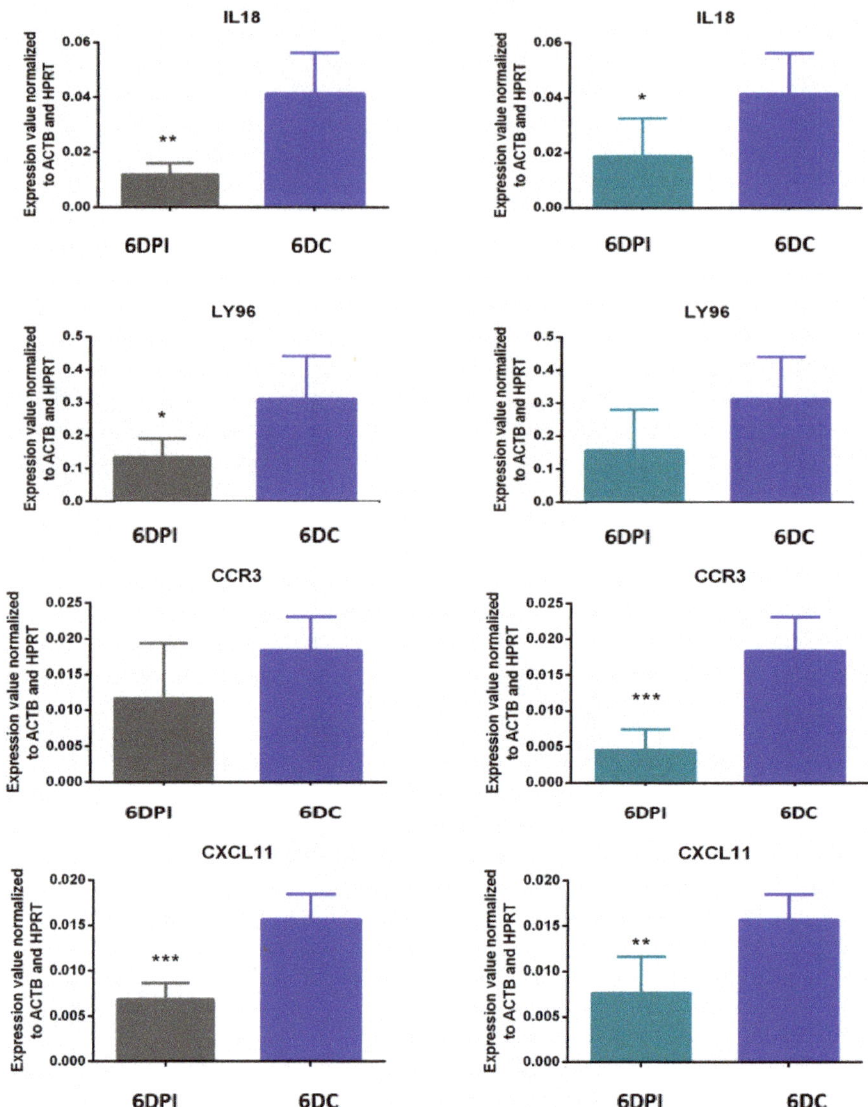


**Figure 5.** *Cont.*

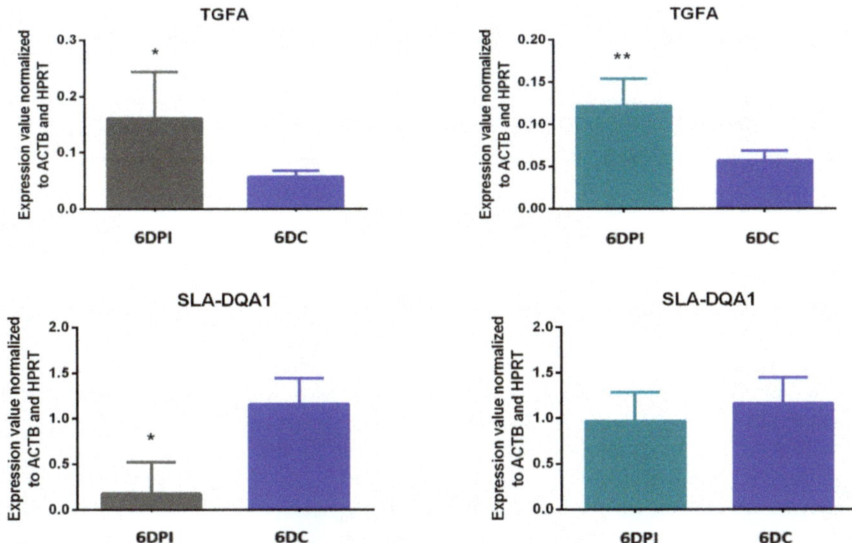


**Figure 5.** Validation of microarray results using qPCR. Expression of *CCR3* (*chemokine (C-C motif) receptor 3*), *CXCL11* (*chemokine (C-X-C motif) ligand 11*), *TGFA* (*transforming growth factor, α*), *LGALS1* (*galectin 1*), *IL-18* (*interleukin 18*), *LY96* (*lymphocyte antigen 96*), *PDCD10* (*programmed cell death 10*), *SLA-DQA1* (*MHC class II histocompatibility antigen SLA-DQA*), *S100A8* (*S100 calcium binding protein A8*), and *S100A12* (*S100 calcium binding protein A12*) in the Day 6 of pregnant or SP infused animals as compared to Day 6 of cycling control animals. Expression values were normalized to expression of *ACTB* (*Actin, β*) and *HPRT* (*Hypoxanthine phosphoribosyltransferase1*). Data are presented as mean ± standard error. * $p \leq 0.05$, ** $p \leq 0.01$, *** $p \leq 0.001$ by *t*-test. The fold change and *p* values are shown in Table 2.

## 4. Discussion

Although, nowadays, pregnancy in pigs is a result of AI with diluted semen or the result of embryo transfer techniques during which only the residual SP enters the reproductive tract, there is documented evidence that SP affects the biological functions of the uterus and evidence that interaction between male SP and female tissues promotes fertility, pregnancy, and finally health of offspring [23]. Many transcriptomic studies in humans, cattle, and pigs have been carried out to evaluate the effects of SP on endometrium [17,24,25]. These reports support the results that SP itself modifies the transcriptome, although semen either after mating or AI results in the maximum changes in the molecular profiles in peri-ovulatory uterine tract of pigs [25,26]. However, in pigs, these studies either reported the immediate effects (after 24 h) of SP on uterine tract or effect of SP followed by AI [27]. Our previous studies have shown that SP can induce long term changes in the endometrium that can be observed at least till Day 6 of its infusion, therefore, in this study, we evaluated SP-induced long-term changes in endometrial transcriptome to identify significantly altered molecular and cellular processes that might prepare endometrium for a possible pregnancy. We also compared these changes with the list of DEGs obtained on Day 6 of pregnancy to evaluate distinct and shared pathways between the two treatments.

Our data demonstrated that as many as 255 and 281 genes are differentially regulated after 6 days of SP infusion and on Day 6 of pregnancy as compared to corresponding day of estrous cycle with only 19 being common to both the groups. A comparison of the biological, molecular, and cellular functions altered by SP-infusion or pregnancy revealed that most of these processes are specific to either SP-infused or pregnant groups of animals, highlighting specific actions of SP constituents. Many DEGs found in both the groups were responsible for inhibition of immune function. Processes such as organization of cytoskeleton and transmigration of leukocytes were specific for pregnancy induced DEGs. Treatment with SP inhibited processes such as apoptosis, necrosis, production of reactive oxygen species and steroid transport. On the other hand, connective tissue cell and microvascular endothelial cell proliferation was activated by SP. Whereas, pathways affected by SP, such as endometrial immune response and steroid biosynthesis were inhibited after Day 6 of its infusion, these responses were activated immediately after SP infusion [25].

### 4.1. Immune Regulation

Consistent with the literature reports, modulation of immune responses was one of the topmost processes altered by both the treatments, i.e., AI and SP. Blastocysts and SP are known to differentially regulate genes involved in immune response on Day 6 [5,21]. It is well known that immediate effects of SP on endometrium include an inflammatory reaction, a response to paternal antigens and mainly to clear reproductive tract of any pathogen deposited at the time of mating [13]. A recent study reported minimal effect of SP treatments on the differential expression of genes in the porcine upper reproductive tract [25]. Consistent with these reports, a very small number of immune related genes were differentially regulated in our study. However, there was no difference between the genes regulated either as a result of pregnancy after AI (six DEGs) or SP treatment (eight DEGs) showing the similarities between both the treatments. In the present study genes involved with immune functioning such as *IL15*, *IL18*, and *LGALS1* were found to be downregulated in both the groups and additionally, *STAT5* and *GZMA* was downregulated by SP infusion. Interleukins 15 and 18 are pro-inflammatory cytokines, these cytokines are not only the regulators of innate immune response but also enhance the cytotoxicity of natural killer cells (NK cells) [28,29]. Furthermore, a decrease in cytotoxicity of lymphocytes was also evident from the downregulation of *GZMA*, a factor secreted by the cytotoxic T cells and natural killer cells that induces apoptosis [30]. A downregulation of these factors might result in a dampened innate immune response at the time of blastocyst hatching and in turn may affect endometrial immune tolerance to paternal antigens. Downregulation of many of the genes associated with immune regulation in SP-infused animals was found to be a result of inhibition of toll-like receptor (TLR) 2 and IL-2 signaling. The activation of TLR-signaling is indeed found to be detrimental to the success of pregnancy [5]. In the endometrium of pregnant pigs, negative regulation of immune responses could be a result of inhibition TNF or IL-1 signaling ($Z > -2.0$). Both SP and pregnancy status are able to generate moderate changes in endometrial immune response. Current data closely corresponded with our previous results suggesting that the immunomodulatory effects of SP, also observed at the protein level, last up to at least 6 days after its infusion at which time blastocysts enter the uterus from the oviduct [21,23].

### 4.2. Cell Death and Survival

A large number of endometrial genes associated with the category cell death and survival were downregulated in SP-infused animals as compared to PBS-infused controls. These DEGs resulted in an inhibition of the apoptotic signaling and promotion of microvascular and connective tissue cell proliferation. Corresponding to the proliferation of connective tissue, an inhibition in senescence of fibroblasts was also observed, confirming an earlier observation that SP constituents can also have an effect on stromal layer of endometrium [17]. Our data emphasize the utility of SP in suppressing apoptosis in endometrial cells during early pregnancy. Increase in proliferation and inhibition of apoptosis of endometrial cells during early pregnancy period is an important step for generation of receptivity to implanting embryos in pigs [31]. A recent report reveals pro-survival effect of SP on endometrial epithelial and stromal cells [17]. In vitro treatment of human endometrial cells resulted in inhibition of pathways promoting apoptosis [17]. In this study we observed downregulation of many apoptotic genes such as *TNFAIP3*, *CXCL11*, *ABCC4*, and *B2M* and upregulation of genes inhibiting apoptosis including *FN1*, *GLRX1*, *CDK2AP1*, and *LGALS3*. Though direct participation of all of these genes has not been evaluated in endometrial cell proliferation or in apoptosis, their role in apoptosis is established. The tumor necrosis factor-α-induced-protein 3 (TNFAIP3) and chemokines including CXCL11 are inducers of epithelial apoptosis [32,33] which is important for species with invasive implantation, however, in pigs, conceptuses do not breach the epithelial layer during implantation. An inhibition in endometrial apoptosis resulting from downregulation of these genes will be favorable for generation of receptive endometrium in pigs.

Many genes that were upregulated in the endometrium of SP-infused pigs were found to participate in proliferation of microvascular endothelial cells and proliferation of connective tissue cells that consists mostly of fibroblast. The genes associated with these categories included *TGFA, MAPK1, ADM, LGALS3, DNMT3B* and *DKK*. Galectins, including LGALS3, are multifunctional proteins associated with immune regulation, cell proliferation, and angiogenesis. Growth factor, TGFA, that activates epidermal growth factor receptor (EGFR) and LGALS3 has been shown to mediate angiogenesis through VEGF and bFGF-mediated angiogenic response [34–36]. Increased endometrial angiogenesis is a hallmark of successful pregnancy, it ensures proper growth and development of the embryo. A SP-induced upregulation of these factors suggests its possible effect on endometrial vasculogenesis.

### 4.3. Oxidation Stress

A downregulation of genes associated with production of reactive oxygen species (ROS) such as *MAOA, NCF2, NOS3* and *CCR3* was observed in endometrial transcriptome of SP-infused pigs. A balance in production ROS is important as at moderate concentrations it has important signaling roles under physiological conditions, but sustained ROS production can have detrimental effects. The expression of *MAOA* which is induced by TNFα has been reported in human and rat endometrium where it is upregulated during the implantation period. The role of this gene in endometrium is not clear yet. However, as *MAOA* is identified as a source of ROS generation, we speculate that a decreased activity of this gene during early pregnancy might create a decrease in detrimental ROS species which in turn might act on downregulation of *CCR3* expression. We also observed an increase in the expression of gene coding antioxidative enzyme, *GLRX*. Glutaredoxins were reported to be expressed in human endometrium where they are associated with the regulation of the cellular redox state and antioxidation defense mechanisms [37].

### 4.4. Lipid Metabolism

Processes related to lipid metabolism, such as steroid and lipid quantity and lipid release were found to be inhibited in the endometrium of SP-infused animals (Table S5). Some of the DEGs related to this process included downregulated expression of STAR, NR5A1, and Cyp7A1. Steroid biosynthesis has previously been reported to be regulated by the SP-infusion. However, the process was shown to be activated after 24 h of the SP treatment [25]. In our study, the one probable explanation for downregulation of these genes could be SP-induced upregulation of COX-2 [18]. Though, regulation of steroid biosynthesis by COX-2 has not been evaluated in porcine endometrium, it was shown to be a negative modulator of steroid biosynthesis in Leydig and bovine luteal cells [38,39]. More importantly, it has been suggested that COX-2 inhibits steroidogenesis through MAPK-signaling [40] and MAPK was found to be upregulated after SP-infusion in our study. As we did not measure systemic progesterone concentrations, whether a result of downregulation of these genes was observed in systemic progesterone concentration needs to be evaluated. Interestingly, it has been reported that early progesterone treatment decreases uterine capacity at 105 days of gestation due to accelerated fetal growth [41]. Moreover, a slower rate of conceptus development is attributed to greater fertility as in Meishan breed of pigs. Consistent with these findings, SP has also been shown to increase embryo viability and at the same time slow the growth (size) of embryos on day 9 of gestation [18]. Our observations concerning a possible long-term effect of SP on steroid biosynthesis are worth further exploration in terms of effect of SP on fertility rate in *Sus scrofa domesticus* through moderate changes in progesterone synthesis.

In conclusion, our results clearly show that SP can induce long term effects on the gene expression in the porcine endometrium. Long-term effects of SP on endometrium include inhibition of processes related to immune response, apoptosis, and steroid biosynthesis and activation of processes such as proliferation of cells. In modern pig breeding, use of SP has been neglected. Our study paves the way for further research on the benefits of addition of SP or its constituents to the semen extenders during AI. The effects of SP on endometrium might prove to be advantageous for blastocyst development, preparing uterus for the conceptus attachment and finally for improving the fertility rate in pigs.

**Supplementary Materials:** The following are available online at http://www.mdpi.com/2073-4425/11/11/1302/s1, Table S1: Microarray analysis data listing differentially regulated genes that were up- or downregulated on day 6 of pregnancy as compared to day 6 of the estrous cycle, Table S2: Microarray analysis data listing differentially regulated genes that were up- or downregulated on day 6 after seminal plasma infusion as compared to day 6 of the estrous cycle, Table S3: List of 19 common differentially regulated genes between day 6 of pregnancy and day 6 after seminal plasma infusion, Table S4: Disease and function analysis (IPA analysis) of DEGs identified on day 6 of pregnancy as compared to day 6 of the estrous cycle, Table S5: Disease and function analysis (IPA analysis) of DEGs identified on day 6 after seminal plasma infusion as compared to day 6 of the estrous cycle, Table S6: List of upstream regulators that might regulate the expression of DEGs identified on day 6 of pregnancy, Table S7: List of upstream regulators that might regulate the expression of DEGs identified on Tabelday 6 after seminal plasma infusion.

**Author Contributions:** Conceptualization, M.B.; methodology, A.W., M.B. and M.M.K.; software, A.W., M.M.K.; validation, A.W.; investigation, A.W., M.B.; resources, M.B.; data curation, B.M.J.; writing—original draft preparation, A.W., M.B., B.M.J.; supervision, M.B.; project administration, M.B.; funding acquisition, M.B. All authors have read and agreed to the published version of the manuscript.

**Funding:** This research was funded by National Science Center Grant No 2011/01/B/NZ4/03603.

**Acknowledgments:** We would like to thank Michal Blitek and Piotr Grezlikowski for helping with the animal preparation and handling as well as Paula Wojnicz and Marta Romaniewicz for their technical assistance.

**Conflicts of Interest:** The authors declare no conflict of interest.

## References

1. Geisert, R.D.; Johnson, G.A.; Burghardt, R.C. Implantation and Establishment of Pregnancy in the Pig. *Adv. Anat. Embryol. Cell Biol.* **2015**, *216*, 137–163. [PubMed]
2. Simon, C.; Moreno, C.; Remohi, J.; Pellicer, A. Molecular interactions between embryo and uterus in the adhesion phase of human implantation. *Human Reprod.* **1998**, *13*, 219–232. [CrossRef] [PubMed]
3. Wolf, E.; Arnold, G.; Bauersachs, S.; Beier, H.; Blum, H.; Einspanier, R.; Frohlich, T.; Herrler, A.; Hiendleder, S.; Kolle, S.; et al. Embryo-maternal communication in bovine—Strategies for deciphering a complex cross-talk. *Reprod. Domest. Anim.* **2003**, *38*, 276–289. [CrossRef] [PubMed]
4. Spencer, T.; Johnson, G.; Bazer, F.; Burghardt, R.; Palmarini, M. Pregnancy recognition and conceptus implantation in domestic ruminants: Roles of progesterone, interferons and endogenous retroviruses. *Reprod. Fert. Dev.* **2007**, *19*, 65–78. [CrossRef] [PubMed]
5. Alminana, C.; Heath, P.; Wilkinson, S.; Sanchez-Osorio, J.; Cuello, C.; Parrilla, I.; Gil, M.; Vazquez, J.; Vazquez, J.; Roca, J.; et al. Early developing pig embryos mediate their own environment in the maternal tract. *PLoS ONE* **2012**, *7*, e33625. [CrossRef] [PubMed]
6. Ostrup, E.; Bauersachs, S.; Blum, H.; Wolf, E.; Hyttel, P. Differential endometrial gene expression in pregnant and nonpregnant sows. *Biol. Reprod.* **2010**, *83*, 277–285. [CrossRef]
7. Samborski, A.; Graf, A.; Krebs, S.; Kessler, B.; Reichenbach, M.; Reichenbach, H.; Ulbrich, S.; Bauersachs, S.B. Transcriptome changes in the porcine endometrium during the preattachment phase. *Biol. Reprod.* **2013**, *89*, 1–16. [CrossRef]
8. Samborski, A.; Graf, A.; Krebs, S.; Kessler, B.; Bauersachs, S. Deep sequencing of the porcine endometrial transcriptome on day 14 of pregnancy. *Biol. Reprod.* **2013**, *88*, 1–13. [CrossRef]
9. Ziecik, A.; Waclawik, A.; Kaczmarek, M.; Blitek, A.; Jalali, B.; Andronowska, A. Mechanisms for the establishment of pregnancy in the pig. *Reprod. Domest. Anim.* **2011**, *46*, 31–41. [CrossRef] [PubMed]
10. Robertson, S. Seminal fluid signaling in the female reproductive tract: Lessons from rodents and pigs. *J. Anim. Sci.* **2007**, *85*, E36–E44. [CrossRef]

11. Flowers, W.; Esbenshade, K. Optimizing management of natural and artificial mating in Swine. *J. Reprod. Fertil.* **1993**, *52*, 217–228.
12. Murray, F.; Grifo, A.; Parker, C. Increased litter size in gilts by intrauterine infusion of seminal and sperm antigens before breeding. *J. Anim. Sci.* **1983**, *56*, 895–900. [CrossRef] [PubMed]
13. Almlid, T. Does enhanced entigenicity of semen increase the litter size in pigs. *Z. Fur Tierz. Und Zucht. J. Anim. Breed. Genet.* **1981**, *98*, 1–10.
14. Rozeboom, K.; Troedsson, M.; Crabo, B. Characterization of uterine leukocyte infiltration in gilts after artificial insemination. *J. Reprod. Fert.* **1998**, *114*, 195–199. [CrossRef]
15. Engelhardt, H.; Croy, B.; King, G. Role of uterine immune cells in early pregnancy in pigs. *J. Reprod. Fertil.* **1997**, *52*, 115–131. [CrossRef]
16. Letterio, J.; Roberts, A. Regulation of immune responses by TGF-β. *Ann. Rev. Immunol.* **1998**, *16*, 137–161. [CrossRef]
17. Chen, J.C.; Johnson, B.A.; Erikson, D.W.; Piltonen, T.T.; Barragan, F.; Chu, S.; Kohgadai, N.; Irwin, J.C.; Greene, W.C.; Giudice., L.C.; et al. Seminal plasma induces global transcriptomic changes associated with cell migration, proliferation and viability in endometrial epithelial cells and stromal fibroblasts. *Hum. Reprod.* **2014**, *29*, 1255–1270. [CrossRef]
18. O'Leary, S.; Jasper, M.; Warnes, G.; Armstrong, D.; Robertson, S. Seminal plasma regulates endometrial cytokine expression, leukocyte recruitment and embryo development in the pig. *Reproduction* **2004**, *128*, 237–247. [CrossRef]
19. Kaczmarek, M.; Krawczynski, K.; Blitek, A.; Kiewisz, J.; Schams, D.; Ziecik, A. Seminal plasma affects prostaglandin synthesis in the porcine oviduct. *Theriogenology* **2010**, *74*, 1207–1220. [CrossRef]
20. Kaczmarek, M.; Krawczynski, K.; Filant, J. Seminal Plasma Affects Prostaglandin Synthesis and Angiogenesis in the Porcine Uterus. *Biol. Reprod.* **2013**, *88*, 1–11. [CrossRef] [PubMed]
21. Jalali, B.; Kitewska, A.; Wasielak, M.; Bodek, G.; Bogacki, M. Effects of seminal plasma and the presence of a conceptus on regulation of lymphocyte- cytokine network in porcine endometrium. *Mol. Reprod. Dev.* **2014**, *81*, 270–281. [CrossRef]
22. Altschul, S.; Gish, W.; Miller, W.; Myers, E.; Lipman, D. Basic local alignment search tool. *J. Mol. Biol.* **1990**, *215*, 403–410. [CrossRef]
23. Waberski, D.; Schäfer, J.; Bölling, A.; Scheld, M.; Henning, H.; Hambruch, N.; Schuberth, H.J.; Pfarrer, C.; Wrenzycki, C.; Hunter, R.H.F. Seminal plasma modulates the immunecytokine network in the porcine uterine tissue and pre-ovulatory follicles. *PLoS ONE* **2018**, *28*, e0202654.
24. Mateo-Otero, Y.; Fernández-López, P.; Gil-Caballero, S.; Fernandez-Fuertes, B.; Bonet, S.; Barranco, I.; Yeste, M. 1H Nuclear magnetic resonance of pig seminal plasma reveals intra-ejaculate variation in metabolites. *Biomolecules* **2020**, *15*, 906. [CrossRef]
25. Álvarez-Rodríguez, M.; Martinez, C.A.; Wright, D.; Rodríguez-Martinez, H. The role of semen and seminal plasma in inducing large-scale genomic changes in the female porcine peri-ovulatory tract. *Sci. Rep.* **2020**, *5061*, 1–16. [CrossRef]
26. Alvarez-Rodriguez, M.; Atikuzzaman, M.; Venhoranta, H.; Wright, D.; Rodriguez-Martinez, H. Expression of immune regulatory genes in the porcine internal genital tract is differentially triggered by spermatozoa and seminal plasma. *Int. J. Mol. Sci.* **2019**, *20*, 513. [CrossRef]
27. Martinez, C.A.; Cambra, J.M.; Parrilla, I.; Roca, J.; Ferreira-Dias, G.; Pallares, F.J.; Lucas, X.; Vazquez, J.M.; Martinez, E.A.; Gil, M.A.; et al. Seminal Plasma Modifies the Transcriptional Pattern of the Endometrium and Advances Embryo Development in Pigs. *Front. Vet Sci.* **2019**, *6*, 465. [CrossRef]
28. Liew, F.Y.; McInnes, I.B. Role of interleukin 15 and interleukin 18 in inflammatory response. *Ann. Rheum. Dis.* **2002**, *61*, ii100–ii102. [CrossRef]
29. Chaix, J.; Tessmer, M.S.; Hoebe, K.; Fuse´ri, N.; Ryffel, B.; Dalod, M.; Alexopoulou, L.; Beutler, B.; Brossay, L.; Vivier, E.; et al. Priming of natural killer cells by interleukin-18. *J. Immunol.* **2008**, *181*, 1627–1631. [CrossRef]
30. Lieberman, J. Granzyme A activates another way to die. *Immunol. Rev.* **2010**, *235*, 93–104. [CrossRef] [PubMed]
31. Lim, W.; Bae, H.; Bazer, F.W.; Song, G. Stimulatory effects of fibroblast growth factor 2 on proliferation and migration of uterine luminal epithelial cells during early pregnancy. *Biol. Reprod.* **2017**, *96*, 185–198. [CrossRef]

32. Elbaz, M.; Hadas, R.; Bilezikjian, L.M.; Gershon, E. Uterine Foxl2 regulates the adherence of the Trophectoderm cells to the endometrial epithelium. *Reprod. Biol. Endocrinol.* **2018**, *16*, 12. [CrossRef]
33. Park, D.W.; Yang, K.M. Hormonal regulation of uterine chemokines and immune cells. *Clin. Experim. Reprod. Med.* **2011**, *38*, 179–185. [CrossRef]
34. Hofer, E.; Schweighofer, B. Signal transduction induced in endothelial cells by growth factor receptors involved in angiogenesis. *Thromb. Haemost.* **2007**, *97*, 355–363.
35. Funasaka, T.; Raz, A.; Nangia-Makker, P. Galectin-3 in angiogenesis and metastasis. *Glycobiology* **2014**, *24*, 886–891. [CrossRef]
36. Blois, S.M.; Conrad, M.L.; Freitag, N.; Barrientos, G. Galectins in angiogenesis: Consequences for gestation. *J. Reprod. Immunol.* **2015**, *108*, 33–41. [CrossRef]
37. Stavréus-Evers, A.; Masironi, B.; Landgren, B.M.; Holmgren, A.; Eriksson, H.; Sahlin, L. Immunohistochemical localization of glutaredoxin andthioredoxin in human endometrium: A possible association with pinnipeds. *Mol. Hum. Reprod.* **2002**, *8*, 546–551. [CrossRef]
38. Nakamura, T.; Sakamoto, K. Reactive oxygen species up-regulates cyclooxygenase-2, p53, and Bax mRNA expression in bovine luteal cells. *Bioch. Biophys. Res. Commun.* **2001**, *284*, 203–210. [CrossRef]
39. Wang, X.; Dyson, M.T.; Jo, Y.; Stocco, D.M. Inhibition of cyclooxygenase-2 activity enhances steroidogenesis and steroidogenic acute regulatory gene expression in MA-10 mouse Leydig cells. *Endocrinology* **2003**, *144*, 3368–3375. [CrossRef]
40. Abidi, P.; Zhang, H.; Zaidi, S.M.; Shen, W.J.; Leers-Sucheta, S.; Cortez, Y.; Han, J.; Azhar, S. Oxidative stress-induced inhibition of adrenal steroidogenesis requires participation of p38 mitogen-activated protein kinase signaling pathway. *J. Endocrinol.* **2008**, *198*, 193–207. [CrossRef]
41. Vallet, J.L.; Christenson, R.K. Effect of progesterone, mifepristone, and estrogen treatment during early pregnancy on conceptus development and uterine capacity in Swine. *Biol. Reprod.* **2004**, *70*, 92–98. [CrossRef]

*Article*

# Expression Profile of Porcine TRIM26 and Its Inhibitory Effect on Interferon-β Production and Antiviral Response

**Hui Huang [1,2,†], Mona Sharma [1,†], Yanbing Zhang [1], Chenxi Li [1], Ke Liu [1], Jianchao Wei [1], Donghua Shao [1], Beibei Li [1], Zhiyong Ma [1], Ruibing Cao [2,*] and Yafeng Qiu [1,*]**

1 Shanghai Veterinary Research Institute, Chinese Academy of Agricultural Sciences, Shanghai 200241, China; huanghui0406@outlook.com (H.H.); monasharma1990@yahoo.com (M.S.); zhangyanbing129@outlook.com (Y.Z.); lichenxihsy@outlook.com (C.L.); liuke@shvri.ac.cn (K.L.); jianchaowei@shvri.ac.cn (J.W.); shaodonghua@shvri.ac.cn (D.S.); lbb@shvri.ac.cn (B.L.); zhiyongma@shvri.ac.cn (Z.M.)

2 College of Veterinary Medicine, Nanjing Agricultural University, Nanjing 210095, China

* Correspondence: crb@njau.edu.cn (R.C.); yafengq@shvri.ac.cn (Y.Q.); Tel.: +86-258-439-6028 (R.C.); +86-213-468-0292 (Y.Q.)

† These authors contributed equally to this work.

Received: 29 August 2020; Accepted: 15 October 2020; Published: 19 October 2020

**Abstract:** TRIM26, a member of the tripartite motif (TRIM) family has been shown to be involved in modulation of innate antiviral response. However, the functional characteristics of porcine TRIM26 (porTRIM26) are unclear. In this study, we used a synthesized antigen peptide to generate a polyclonal antibody against porTRIM26 with which to study the expression and function of porTRIM26. We demonstrated that polyinosinic:polycytidylic acid (poly (I:C)) stimulation and viral infection (vesicular stomatitis (VSV) or porcine reproductive and respiratory syndrome virus (PRRSV)) induce expression of porTRIM26, whereas knock-down expression of porTRIM26 promotes interferon (IFN)-β production after poly (I:C) stimulation and virus infection (VSV or PRRSV). The importance of the porTRIM26-mediated modulation of the antiviral response was also shown in VSV- or PRRSV-infected cells. In summary, these findings show that porTRIM26 has an inhibitory role in IFN-β expression and the antiviral response.

**Keywords:** TRIM26; antiviral response; IFN-β; poly (I:C); VSV; PRRSV

---

## 1. Introduction

Tripartite motif (TRIM) proteins, a large family of ubiquitin E3 ligase, include more than 80 and 60 members in human and mouse, respectively [1]. Moreover, more than 50 members of porcine *TRIM* genes have been annotated in the GenBank database, although most of them were predicted with computational analyses [2,3]. Recent studies have shown that many members of the TRIM family are expressed in response to interferons (IFNs) and are involved in the processes of innate immune response, especially during viral infections [4–6]. These reports strongly suggest that understanding the molecular functions of porcine TRIM proteins could offer insights into the regulation of innate immune response in swine species.

TRIM26 is a member of the TRIM family, and has a structure similar to that of many other members of the family. They are all structurally characterized by a RING finger domain (E3 ligase with ubiquitin ligase activity) with two B-box domains, followed by a coiled-coil (CC) region and a C-terminal protein binding domain [7–9]. In human, *TRIM26* gene is located in the MHC class I region [10,11]. Likewise, the porcine MHC (swine leukocyte antigen (SLA)) region contains *TRIM26* gene according to a sequencing analysis. However, although porcine *TRIM26* (*porTRIM26*) gene has been identified [2,12,13], the biological functions of the protein, especially in the immune response, remain to be determined.

Previous studies in human and mouse cell lines have shown that TRIM26 plays a controversial role in the regulation of IFN-β production and innate antiviral response, which may be attributable to the different experimental systems used [14,15]. However, how porTRIM26 affects the regulation of IFN-β expression and the innate antiviral response is unknown. In this study, for the first time, we determined the expression profile of porTRIM26 in different pig tissues and clarified its effect on the modulation of IFN-β expression and innate antiviral response.

## 2. Materials and Methods

### 2.1. Tissue Sample Collection

Piglets (Shanghai great white pig strain; ~30 days old) were purchased from the Shanghai Academy of Agricultural Sciences (Shanghai, China), euthanized, and dissected to obtain various tissue samples (liver, spleen, lung, kidney, submandibular lymph node, hilar lymph node, mesenteric lymph node, inguinal lymph node, and thymus). The tissue samples were immediately snap-frozen in liquid nitrogen and stored at −80 °C. All animal experiments were approved by the Institutional Animal Care and Use Committee of Shanghai Veterinary Research Institute, Chinese Academy of Agricultural Sciences, Shanghai, China (IACUC No: Shvri-po-2016060501) and were performed in compliance with the Guidelines on the Humane Treatment of Laboratory Animals (Ministry of Science and Technology of the People's Republic of China, Policy No. 2006398).

### 2.2. Cells, Viruses and Infections

Porcine alveolar macrophages (PAM) were generated as shown in our previous study [16]. HEK 293T cells, porcine iliac artery endothelial cells (PIEC), ST cells, PK-15 cells, BHK-21 cells, and Marc-145 cells were maintained in Dulbecco modified Eagle medium (Thermo Fisher Scientific, Shanghai, China) supplemented with 10% fetal bovine serum (Thermo Fisher Scientific, Shanghai, China) at 37 °C in a 5% $CO_2$ atmosphere. Vesicular stomatitis virus (VSV) and highly pathogenic porcine reproductive and respiratory syndrome virus (HP-PRRSV) were maintained in our lab and were propagated on BHK-21 cells and Marc-145 cells, respectively. The viruses were titrated with the median tissue culture infective dose ($TCID_{50}$) methods, as described in a previous study [17]. According to the progress of virus infection and expression of porTRIM26, we chose 24 h after infection for collecting the viral infected samples: PIEC cells were infected with VSV at a multiplicity of infection (MOI) of 1 for 24 h, at which point peak titer was reached with the induction of porTRIM26; PAM were infected with PRRSV at a multiplicity of infection (MOI) of 1 for 24 h, at which point peak titer was reached with the induction of porTRIM26.

### 2.3. Cloning and Sequence Analysis of porTRIM26

The primers for cloning *porTRIM26* gene (shown in Table S1) were designed based on the *porTRIM26* gene sequence (GenBank accession number, NM_001123209.1). The amplified sequence was confirmed with DNA sequencing and inserted into the p3×Flag-CMV-14 vector (Sigma, St. Louis, MO, USA) to generate a recombinant plasmid expressing FLAG-tagged porTRIM26 (pFlag-porTRIM26). An amino acid sequence alignment of the deduced protein sequence by this construct and the TRIM26 proteins of another three species, *Homo sapiens* (NP_001229712.1), *Mus musculus* (NP_001020770.2), and *Rattus norvegicus* (XP_008770914.1), was conducted using the software Lasergene version 7.1 (Madison, WI, USA). A phylogenetic tree based on the sequences of different species was constructed by the neighbor-joining method using the MEGA software (version 6.06).

### 2.4. Generation of Polyclonal Antibody Against porTRIM26

A polyclonal antibody directed against porTRIM26 was generated as described in a previous study [18]. Briefly, a peptide of 20 amino acids corresponding to residues 364–383 of the porTRIM26 sequence was synthesized chemically and conjugated with keyhole limpet hemocyanin (KLH) as the carrier protein. Rabbits were immunized five times with the peptide-KLH conjugate combined with complete or incomplete Freund's adjuvants. HEK 293T cells were then transfected with the porTRIM26-expressing recombinant plasmid described in Section 2.3 and a Western blotting analysis was performed to confirm the specificity of the polyclonal antibody. All of the animal experiments were approved by IACUC in Shanghai Veterinary Research Institute, CAAS (No: Shvri-po-201606 0501) and followed the guidelines described in Section 2.1.

### 2.5. Plasmid Transfection, Small Interfering RNA (siRNA), and Polyinosinic:polycytidylic Acid (poly (I:C)) Stimulation

Cells were grown to 70–80% confluence and transfected with the plasmids expressing FLAG-porTRIM26 or the empty vector (p3×Flag-CMV-14 vector, as a negative control) with Lipofectamine 2000 (Thermo Fisher Scientific, Shanghai, China), according to the manufacturer's instructions. After 24 h of transfection, PIEC cells were transfected with poly (I:C) at a final concentration of 3 μg/mL (InvivoGen, Toulouse, France) and stimulated for 6 h, at which *porTRIM26* and *IFN-β* was obviously induced according to a multiple time-point sample analysis by qPCR. At 6 h after stimulation, the samples were collected for further analysis.

To investigate the role of porTRIM26, one porTRIM26 specific siRNA out of 4 (the targeted sequence: GCCTGTACCAGAGCTCTTA) was selected and transfected into PIEC cells or PAM by Lipofectamine RNAiMAX (Thermo Fisher Scientific, Shanghai, China) following the manufacture's instructions. Likewise, a scrambled siRNA (sequence: TTCTCCGAACGTGTCACGT) was transfected as the negative control (NC). After 72 h of transfection, cells were treated with poly (I:C): the PIEC cells were treated as described above, and PAM were stimulated with poly (I:C) at a final concentration of 3 μg/mL without transfection. At 6 h after stimulation, the samples were collected for further analysis.

### 2.6. Western Blottting Analysis

The protein samples were prepared as previously described [19]. In brief, the membranes transferred with the protein samples were blocked with 5% skim milk for 1 h at room temperature. The membrane was incubated overnight at 4 °C with the individual primary antibodies (anti-Flag (1:1000, M2, Sigma), anti-porTrim26 (1:1000, generated in this study), anti-VSV G (1:1000, Abcam), anti-PRRSV N (1:1000, generated by a synthetic peptide of N), and anti-actin (1:10,000, Sigma)). The membrane was then incubated for 1 h at room temperature with the secondary antibodies: horseradish-peroxidase-conjugated goat anti-mouse IgG (1:5000, Abcam) or goat anti-rabbit IgG (1:10,000, Abcam).

### 2.7. Enzyme-Linked Immunosorbent Assay (ELISA)

The concentrations of porcine IFN-β in supernatants from PIEC cell culture or PAM culture were measured by ELISA kit (Lengton, Shanghai, China).

### 2.8. Reverse Transcription (RT)-Quantitative PCR (qPCR) Analysis

Total RNA was extracted with RNAiso Reagent (Takara, Dalian, China) and the cDNA was prepared with PrimeScript RT Reagent Kit (Takara, Dalian, China). Gene expression was analyzed by RT-qPCR using SYBR Green qPCR Master Mix (Takara, Dalian, China). The specific primers are shown in Table S1. The expression of the glyceraldehyde 3-phosphate dehydrogenase gene (*GAPDH*) was used as the reference. Expression was calculated relative to that of *GAPDH* ($2^{-\Delta Ct}$).

### 2.9. Statistical Analysis

All data were analyzed with GraphPad Prism software (Graphpad Software, Inc, La Jolla, CA, USA). An unpaired Student's *t*-test was used to determine significant differences. Values were considered statistically significant when $p < 0.05$. Data were given as mean ± SEM as indicated; 'n' refers to the sample size.

## 3. Results

### 3.1. Sequence Analysis and Generation of Polyclonal Antibody against porTRIM26

Based on the porcine *TRIM26* gene sequence (GenBank accession number NM_001123209.1), the porcine *TRIM26* gene was amplified with RT-PCR from the total RNA extracted from pig lungs (Shanghai great white pig strain). A sequence analysis with BLAST showed that the cloned gene sequence from Shanghai great white pig strain was identical to the porcine *TRIM26* gene sequence in GenBank. The full-length cDNA of porcine *TRIM26* contained 1638 bp and encoded a protein of 546 amino acid residues including one Ring domain, two B-box domains, one coiled-coil domain, and one C-terminal domain, according to the protein BLAST information. A BLASTp analysis in the National Center for Biotechnology Information (NCBI) nonredundant database detected more than 200 TRIM26 orthologues (>50% identity) from more than 100 species. Notably, among the mammalian sequences, human and mouse TRIM26 shared 91% identity and 84% identity with porTRIM26, respectively, which is consistent with the data generated with the Clustal W method (Figure 1A). A phylogenetic analysis was performed to investigate the difference in the identities with TRIM26 of the other species. The *porTRIM26* gene clustered on a separate branch of the dendrogram within the sequences of Mammalia and was phylogenetically closest to the human sequence than to the mouse sequence (Figure 1B).

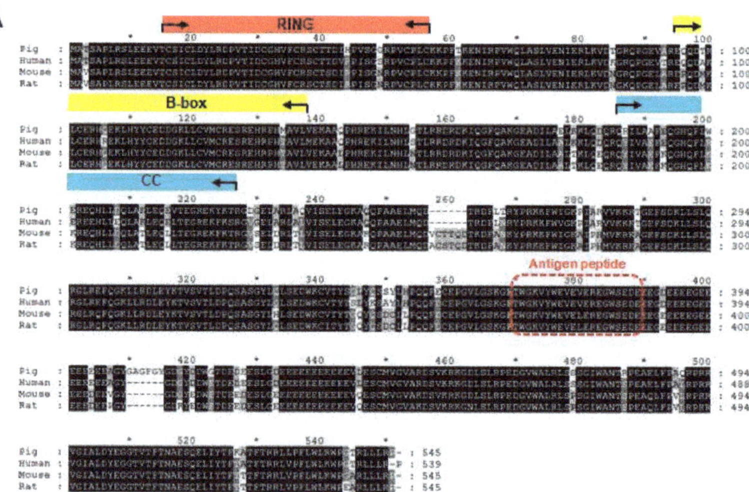


**Figure 1.** *Cont.*

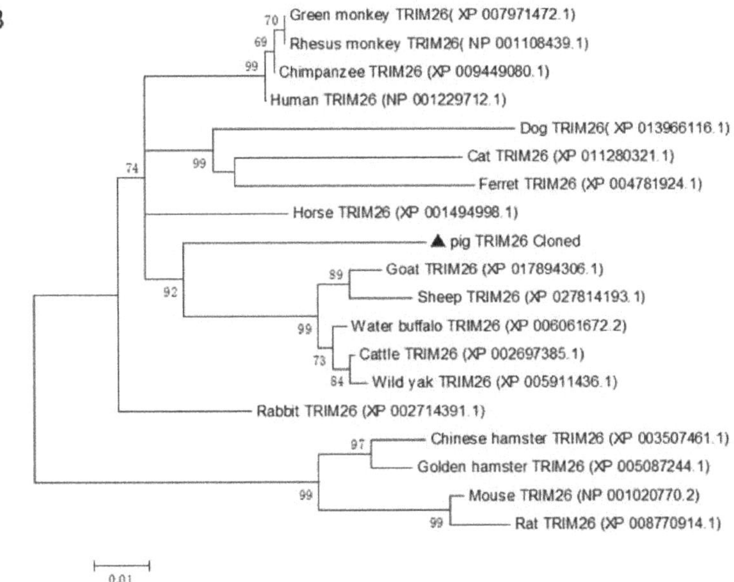


**Figure 1.** Multiple alignment and phylogenetic analysis of TRIM26. (**A**) Polypeptide of porcine TRIM26 was aligned with that of human, mouse, and rat. The RING finger domain (RING), the B-box domains (B-box), and the coiled-coil region (CC) are labeled in reference to the human TRIM26. The antigenic peptide is shown in the red box. (**B**) The phylogenetic tree of TRIM26, constructed by neighbor-joining method using MEGA 6.06.

To study porTRIM26 expression in different tissues and cells of pigs, we developed an anti-porTRIM26 antibody after synthesizing an antigen peptide (Figure 1A). We tested the specificity of the antibody with a Western blotting analysis. Our result indicated that the anti-porTRIM26 antibody specifically detected the expression of porTRIM26 in HEK 293T cells (Figure 2A), and that the signal was equal to that detected with an anti-Flag antibody.

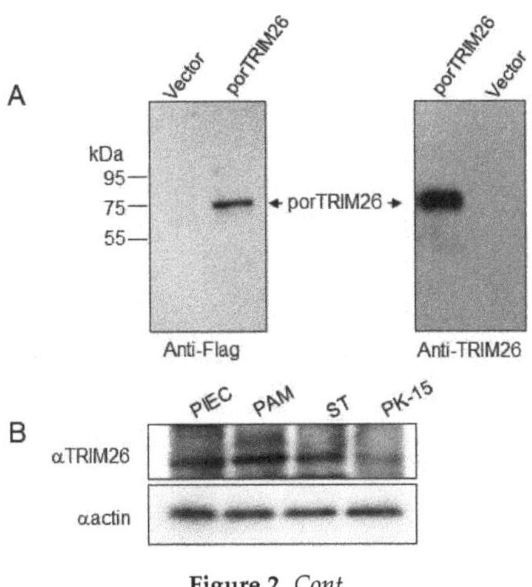


**Figure 2.** *Cont.*

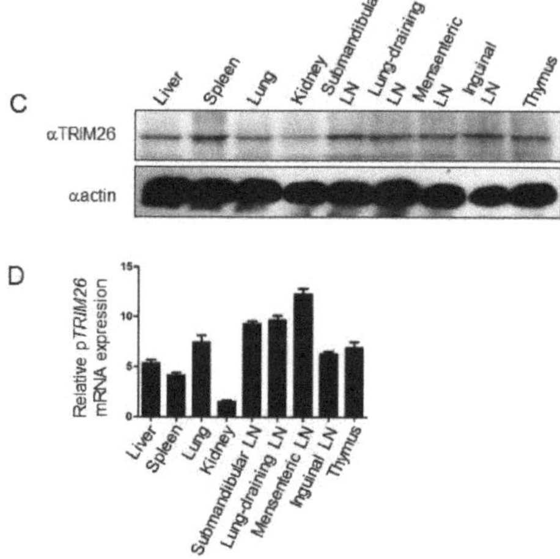


**Figure 2.** (**A**) Specificity of polyclonal antibody against porcine TRIM26 was analyzed with Western blotting. (**B**) Expression profiles of TRIM26 in different porcine cell lines, analyzed with Western blotting with anti-TRIM26 antibody. (**C**) Expression profiles of TRIM26 in different pig tissues ($n$ = 3), analyzed with Western blotting and an anti-TRIM26 polyclonal antibody. Approximately 100 mg of each tissue was homogenized with lysis buffer as per the Materials and Method section for protein sample preparation. (**D**) Expression profiles of *TRIM26* in different pig tissues ($n$ = 3) analyzed by reverse transcription-quantitative PCR.

### *3.2. Expression Profiles of porTRIM26 in Different Cell Lines and Tissues*

The polyclonal antibody was used to determine the expression profiles of porTRIM26 in different porcine cell lines and tissues in a Western blotting analysis. First, the expression of porTRIM26 in different porcine cell lines was determined by Western blotting. The PK15 cell line had a lower level of porTRIM26 in comparison to that in other cell lines (Figure 2B). The expression of porTRIM26 in different tissues was also determined with Western blotting. In almost all examined tissues porTRIM26 was differentially expressed, including in liver, spleen, lungs, kidneys, lymph nodes, and thymus (Figure 2C). The lowest expression of *porTRIM26* was observed in the kidneys in RNA level (Figure 2D), consistent with its expression in protein level (Figure 2C).

### *3.3. porTRIM26 Negatively Regulates Expression of IFN-β*

The role of porTRIM26 in the regulation of IFN-β expression is controversial, which may be attributable to the different experimental systems, such as different cell types used in different studies. Whether porTRIM26 affects IFN-β production, either positively or negatively, remains unclear. We first investigated its effect on poly (I:C)-induced IFN-β expression using two kinds of porcine cells, PIEC cells and PAM. Notable, poly (I:C) stimulation induced the expression of porTRIM26 in both porcine cell lines (Figure 3A), consistent with the previous report in human and mouse cell lines [15]. To investigate the effect of porTRIM26 on IFN-β expression, a porTRIM26-specific siRNA was designed and used to transfect PIEC cells and PAM. A Western blotting analysis showed that the transfection of the siRNA reduced the expression of porTRIM26 (Figure 3B,C). The transfection of porTRIM26 siRNA promoted poly (I:C)-induced IFN-β expression in both PAM and PIEC cells (Figure 3B,C). In contrast, the overexpression of porTRIM26 in PIEC attenuated the poly (I:C)-induced IFN-β expression (Figure 3D). Collectively, these results reveal that porTRIM26 modulates IFN-β expression downstream of poly (I:C)-stimulated innate signaling.

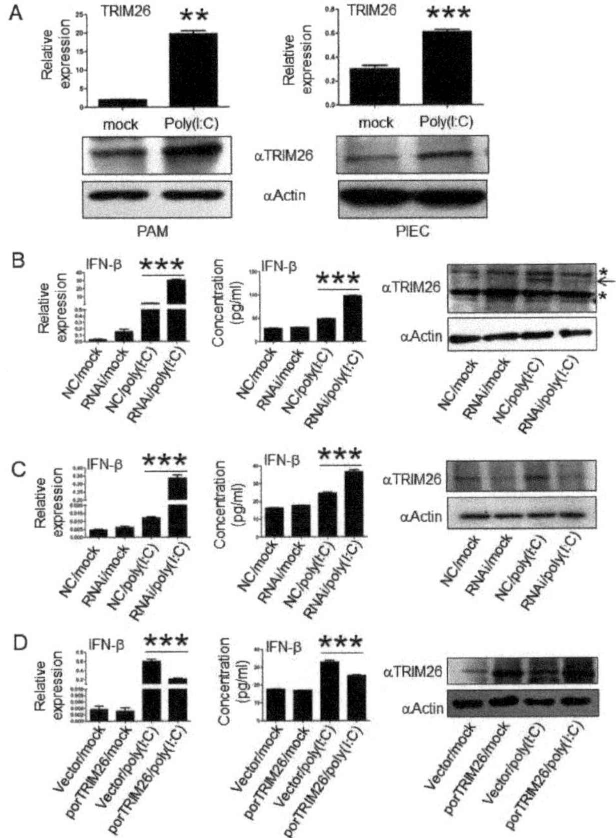


**Figure 3.** Effect of porTRIM26 on poly (I:C)-induced interferon (IFN)-β. (**A**) Expression of TRIM26 in porcine alveolar macrophages (PAM) or porcine iliac artery endothelial cells (PIEC) with or without poly(I:C) treatment was measured with qPCR and Western blotting. Poly (I:C) treatment induced the expression of TRIM26 in both PAM and PIEC cells. (**B**) PAM were transfected with porTRIM26 siRNA or negative control (NC) siRNA for 72 h and then stimulated with poly (I:C) for 6 h. Supernatant and cells were harvested and analyzed by enzyme-linked immunosorbent assay (ELISA) and qPCR, respectively. (**C**) PIEC cells were transfected with porTRIM26 siRNA or NC siRNA for 72 h, and then stimulated with poly (I:C) for 6 h. Supernatant and cells were harvested and analyzed by ELISA and qPCR, respectively. (**D**) PIEC cells were transfected with plasmid encoding porTRIM26 or the empty vector. After 24 h, the cells were treated with poly (I·C) or vehicle and incubated for another 6 h. Level of *IFN-β* mRNA was determined with RT-qPCR. Data present are mean ± SEM pooled from one independent experiment; $n \geq 3$ for each of the analyzed parameters. *, $p < 0.05$; **, $p < 0.01$; ***, $p < 0.001$ in comparison between mock and poly (I:C) group (**A**); between NC and RNAi with poly (I:C) stimulation (**B**,**C**); and between the empty vector and pFlag-porTRIM26 with poly (I:C) stimulation (**D**).

### *3.4. porTRIM26 Negatively Regulates Antiviral Response to VSV Infection*

Although we had shown that porTRIM26 plays a role in poly (I:C)-stimulated innate immune signaling, whether it plays a role in virus-triggered signaling remained to be determined. In previous studies, VSV has been used to study the effect of TRIM26 on IFN-β expression and antiviral response. Because PAM are resistant to VSV infection according to our preliminary experiments, we next investigated the effect of porTRIM26 (either positive or negative) on VSV infection in PIEC cells. Similar to poly (I:C) stimulation of PIEC cells, VSV infection also induced the expression of porTRIM26 in PIEC cells (Figure 4A). To identify the role of porTRIM26 in VSV infection we transfected PIEC cells with porTRIM26 siRNA, as described above. The knock-down of porTRIM26 expression significantly increased the expression of IFN-β during VSV infection compared with that in cells infected with the negative control. The viral titers were higher in the VSV infected-negative control cells than in VSV infected cells in which the expression of TRIM26 was knocked down (Figure 4B). To confirm these results, we transfected PIEC cells with the plasmid, as described above, to overexpress porTRIM26. ELISA data showed that the overexpression of porTRIM26 significantly reduced the expression of IFN-β relative to that in the VSV-infected empty-vector-transfected group. Notable, VSV infection was significantly greater in the porTRIM26-overexpressed cells than in the VSV-infected empty-vector-transfected cells. Taken together, these data suggest that porTRIM26 promotes viral infection by inhibiting IFN-β expression.

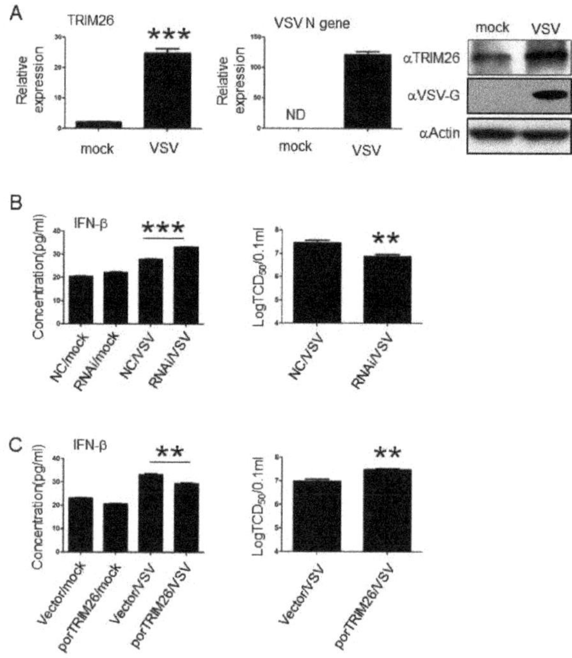


**Figure 4.** Effect of porTRIM26 on IFN-β expression and vesicular stomatitis virus (VSV) infection in PIEC cells. (**A**) PIEC cells were infected with VSV at a multiplicity of infection (MOI) of 1 for 24 h. The porTRIM26 mRNA and protein level was measured by RT-qPCR and Western blotting, respectively. (**B**) PIEC cells were transfected with porTRIM26 siRNA or NC siRNA. After 72 h, the cells were mock infected or infected with VSV at a MOI of 1. Supernatants were collected at 24 h post infection (hpi) for tissue culture infective dose ($TCID_{50}$) assay. (**C**) PIEC cells were transfected with plasmid encoding porTRIM26 or empty vector for 24 h and then infected with VSV at an MOI of 1. Supernatants were collected at 24 hpi for either a $TCID_{50}$ assay or ELISA. Data are mean ± SEM pooled from one independent experiment; $n \geq 3$ for each of the analyzed parameters. ND, no detected. **, $p < 0.01$; ***, $p < 0.001$ in comparison: mock-infected vs. VSV-infected group (**A**); NC-treated vs. siRNA-treated groups after VSV infection (**B**); empty-vector-transfected vs. pFlag-pTRIM26-transfected after VSV infection (**C**).

### 3.5. porTRIM26 Negatively Regulates Antiviral Response to PRRSV Infection

Our data showed that porTRIM26 inhibits IFN-β production and innate antiviral response. Therefore, we examined whether this also occurs during other viral infections. A previous study showed with an RNA-sequencing (RNA-seq) analysis that PRRSV infection could induce the expression of porTRIM26 [20], so we examined the biological functions of porTRIM26 using PRRSV, to which PAM are susceptible. Our data show that PRRSV infection induced the expression of TRIM26 in PAM (Figure 5A), consistent with the previous RNA-seq data. We also investigated the role of porTRIM26 in IFN-β production and the antiviral response in PRRSV-infected PAM. Our ELISA data showed that knocking down the expression of porTRIM26 increased the expression of IFN-β compared with that in the PRRSV-infected negative control (NC) cells (Figure 5B). We also noted that the viral titer was significantly lower in the PRRSV-infected porTRIM26-siRNA-transfected cells than in the PRRSV-infected negative control (NC) cells. These data suggest that PRRSV may inhibit IFN-β production and the antiviral response by inducing porTRIM26. These results confirm that porTRIM26 plays an inhibitory role in IFN-β expression and the antiviral response.

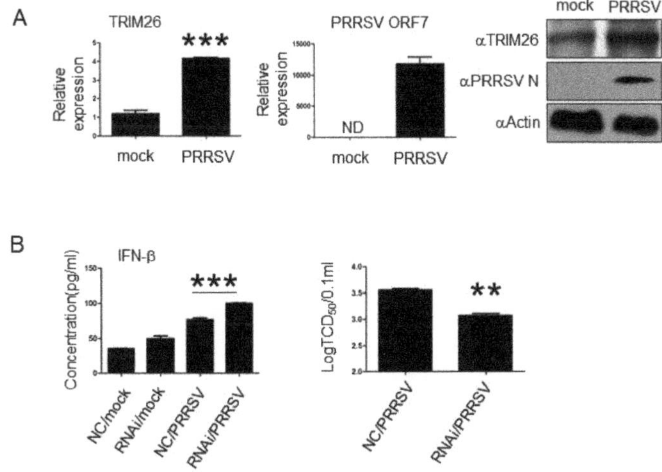


**Figure 5.** Effect of porTRIM26 on IFN-β expression and porcine reproductive and respiratory syndrome virus (PRRSV) infection in PAM. (**A**) PAM were infected with PRRSV at a MOI of 1 for 24 h. The mRNA levels of *TRIM26* and PRRSV *ORF7* were measured with RT-qPCR. The protein levels of TRIM26 and PRRSV N were measured with Western blotting. PRRSV infection induced expression of TRIM26. (**B**) PAM were transfected with the porTRIM26 siRNA or NC siRNA. After 72 h, cells were mock-infected or infected with PRRSV at an MOI of 1. Supernatants were collected at 24 hpi for either a $TCID_{50}$ assay or ELISA. The knock-down of porTRIM26 expression increased the production of IFN-b and inhibited PRRSV infection. Data are mean ± SEM pooled from one independent experiment; $n \geq 3$ for each of the analyzed parameters. ND: not detected. **, $p < 0.01$; ***, $p < 0.001$ in comparison: mock-infected vs. PRRSV-infected group (**A**); NC-siRNA-transfected vs. *porTRIM26*-siRNA-transfected cells after infection with PRRSV (**B**).

## 4. Discussion

Several studies have reported the gene sequence of *porTRIM26* based on genomic sequencing analyses [12,13]. However, the expression profiles of porTRIM26 in tissues and its biological functions have been unclear until now. In this study, we confirmed that porTRIM26 inhibits IFN-β production after poly (I:C) stimulation or viral infection. Using a rabbit anti-pTRIM26 polyclonal antibody generated with synthesized antigen peptide, we showed that poly (I:C) stimulation or viral infection (VSV or PRRSV) induces the expression of porTRIM26. By overexpressing or knocking down the expression of porTRIM26, we demonstrated that the induction of porTRIM26 modulates the IFN-β expression induced by poly (I:C) stimulation. Our data demonstrate that the induction of porTRIM26 negatively regulates IFN-β production and the antiviral response to VSV or PRRSV infection. Collectively, these data provide novel evidence that porTRIM26 modulates the innate antiviral response by inhibiting IFN-β production.

Although more than 50 *TRIM* genes of the pig have been annotated, only TRIM21 has yet been functionally analyzed. Swine TRIM21 restricts foot-and-mouth disease virus (FMDV) infection with an intracellular neutralization mechanism [21]. Our data demonstrate that porTRIM26 negatively regulates IFN-β production and the antiviral response to PRRSV infection. These findings not only verify the role of porTRIM26 in IFN-β expression and the antiviral response but also extend our understanding of how PRRSV uses host proteins, such as porTRIM26, to interfere with the innate antiviral response.

A previous study has shown that TRIM26 negatively regulates IFN-β production and the innate antiviral response by inhibiting activation of interferon regulatory factor 3 (IRF3) [15]. Because PRRSV infection interferes with the activation of IRF3, it cannot be excluded that PRRSV infection inhibits the activation of IRF3 by inducing porTRIM26. The exact mechanism by which PRRSV inhibits IFN-β production via induction of porTRIM26 needs to be clarified in future research.

Although TRIM26 negatively regulates IFN-β production and the innate antiviral response, one study has reported that TRIM26 has the opposite function, promoting IFN-β production and the innate antiviral response [14]. It is possible that the different experimental systems used, including different methods and viruses, have generated different results. For example, TRIM21 is reported to degrade IRF3, thus limiting the type I IFN response after Japanese encephalitis virus (JEV) infection [22]. In contrast, another report suggested that TRIM21 acts as a positive regulator of the IRF3 pathway during Sendai virus infection [23]. Whether or not porTRIM26 positively affects IFN-β production and the innate antiviral response needs to be investigated in other swine viruses. However, our results suggest that porTRIM26 modulates IFN-β production and the innate antiviral response.

In summary, this is the first study to identify the biological functions of porTRIM26. We have demonstrated that porTRIM26 plays an inhibitory role in IFN-β expression and the innate antiviral response. These findings extend our understanding of how some swine viruses, such as PRRSV, inhibit IFN-β production to evade the host's innate immune response.

**Supplementary Materials:** The following are available online at http://www.mdpi.com/2073-4425/11/10/1226/s1, Table S1: PCR primer used in the study.

**Author Contributions:** H.H. and M.S.: Conceptualization, methodology, writing–original daft, visualization. Y.Z. and C.L.: Conceptualization, visualization, methodology. K.L., J.W., D.S. and B.L.: Methodology, visualization. R.C. and Z.M.: Supervision. Y.Q.: Conceptualization, supervision, writing—review and editing. All authors have read and agreed to the published version of the manuscript.

**Funding:** This study was supported, in part, by the national key Research and Development program of China (2018YFD0500100), the National Natural Science Foundation of China (31972693), the Chinese Special Fund for Ago-scientific Research in the Public Interest (No. 2020JB04), and the Elite program of CAAS (to YQ).

**Conflicts of Interest:** The authors declare that they have no conflict of interest.

## References

1. Rajsbaum, R.; Garcia-Sastre, A.; Versteeg, G.A. TRIMmunity: The roles of the TRIM E3-ubiquitin ligase family in innate antiviral immunity. *J. Mol. Biol.* **2014**, *426*, 1265–1284. [CrossRef] [PubMed]
2. Renard, C.; Vaiman, M.; Chiannilkulchai, N.; Cattolico, L.; Robert, C.; Chardon, P. Sequence of the pig major histocompatibility region containing the classical class I genes. *Immunogenetics* **2001**, *53*, 490–500. [CrossRef]
3. Kimsa, M.W.; Strzalka-Mrozik, B.; Kimsa, M.C.; Mazurek, U.; Kruszniewska-Rajs, C.; Gola, J.; Adamska, J.; Twardoch, M. Differential expression of tripartite motif-containing family in normal human dermal fibroblasts in response to porcine endogenous retrovirus infection. *Folia Biol.* **2014**, *60*, 144–151.
4. van Tol, S.; Hage, A.; Giraldo, M.I.; Bharaj, P.; Rajsbaum, R. The TRIMendous Role of TRIMs in Virus-Host Interactions. *Vaccines* **2017**, *5*, 23. [CrossRef] [PubMed]
5. Hu, H.; Sun, S.C. Ubiquitin signaling in immune responses. *Cell Res.* **2016**, *26*, 457–483. [CrossRef] [PubMed]

6. Nisole, S.; Stoye, J.P.; Saib, A. TRIM family proteins: Retroviral restriction and antiviral defence. *Nat. Rev. Microbiol.* **2005**, *3*, 799–808. [CrossRef] [PubMed]
7. McNab, F.W.; Rajsbaum, R.; Stoye, J.P.; O'Garra, A. Tripartite-motif proteins and innate immune regulation. *Curr. Opin. Immunol.* **2011**, *23*, 46–56. [CrossRef] [PubMed]
8. Han, K.; Lou, D.I.; Sawyer, S.L. Identification of a genomic reservoir for new TRIM genes in primate genomes. *PLoS Genet.* **2011**, *7*, e1002388. [CrossRef]
9. Short, K.M.; Cox, T.C. Subclassification of the RBCC/TRIM superfamily reveals a novel motif necessary for microtubule binding. *J. Biol. Chem.* **2006**, *281*, 8970–8980. [CrossRef]
10. Chu, T.W.; Capossela, A.; Coleman, R.; Goei, V.L.; Nallur, G.; Gruen, J.R. Cloning of a new "finger" protein gene (ZNF173) within the class I region of the human MHC. *Genomics* **1995**, *29*, 229–239. [CrossRef]
11. Genome-wide association study implicates HLA-C*01:02 as a risk factor at the major histocompatibility complex locus in schizophrenia. *Biol. Psychiatry* **2012**, *72*, 620–628. [CrossRef]
12. Ando, A.; Shigenari, A.; Kulski, J.K.; Renard, C.; Chardon, P.; Shiina, T.; Inoko, H. Genomic sequence analysis of the 238-kb swine segment with a cluster of TRIM and olfactory receptor genes located, but with no class I genes, at the distal end of the SLA class I region. *Immunogenetics* **2005**, *57*, 864–873. [CrossRef] [PubMed]
13. Tanaka-Matsuda, M.; Ando, A.; Rogel-Gaillard, C.; Chardon, P.; Uenishi, H. Difference in number of loci of swine leukocyte antigen classical class I genes among haplotypes. *Genomics* **2009**, *93*, 261–273. [CrossRef]
14. Ran, Y.; Zhang, J.; Liu, L.L.; Pan, Z.Y.; Nie, Y.; Zhang, H.Y.; Wang, Y.Y. Autoubiquitination of TRIM26 links TBK1 to NEMO in RLR-mediated innate antiviral immune response. *J. Mol. Cell Biol.* **2016**, *8*, 31–43. [CrossRef] [PubMed]
15. Wang, P.; Zhao, W.; Zhao, K.; Zhang, L.; Gao, C. TRIM26 negatively regulates interferon-beta production and antiviral response through polyubiquitination and degradation of nuclear IRF3. *PLoS Pathog.* **2015**, *11*, e1004726. [CrossRef] [PubMed]
16. Qi, P.; Liu, K.; Wei, J.; Li, Y.; Li, B.; Shao, D.; Wu, Z.; Shi, Y.; Tong, G.; Qiu, Y.; et al. Nonstructural Protein 4 of Porcine Reproductive and Respiratory Syndrome Virus Modulates Cell Surface Swine Leukocyte Antigen Class I Expression by Downregulating beta2-Microglobulin Transcription. *J. Virol.* **2017**, *91*. [CrossRef] [PubMed]
17. Shi, Z.; Wei, J.; Deng, X.; Li, S.; Qiu, Y.; Shao, D.; Li, B.; Zhang, K.; Xue, F.; Wang, X.; et al. Nitazoxanide inhibits the replication of Japanese encephalitis virus in cultured cells and in a mouse model. *Virol. J.* **2014**, *11*, 10. [CrossRef] [PubMed]
18. Lateef, S.S.; Gupta, S.; Jayathilaka, L.P.; Krishnanchettiar, S.; Huang, J.S.; Lee, B.S. An improved protocol for coupling synthetic peptides to carrier proteins for antibody production using DMF to solubilize peptides. *J. Biomol. Tech. JBT* **2007**, *18*, 173–176.
19. Qiu, Y.; Shen, Y.; Li, X.; Ding, C.; Ma, Z. Molecular cloning and functional characterization of a novel isoform of chicken myeloid differentiation factor 88 (MyD88). *Dev. Comp. Immunol.* **2008**, *32*, 1522–1530. [CrossRef]
20. Badaoui, B.; Rutigliano, T.; Anselmo, A.; Vanhee, M.; Nauwynck, H.; Giuffra, E.; Botti, S. RNA-sequence analysis of primary alveolar macrophages after in vitro infection with porcine reproductive and respiratory syndrome virus strains of differing virulence. *PLoS ONE* **2014**, *9*, e91918. [CrossRef]
21. Fan, W.; Zhang, D.; Qian, P.; Qian, S.; Wu, M.; Chen, H.; Li, X. Swine TRIM21 restricts FMDV infection via an intracellular neutralization mechanism. *Antivir. Res.* **2016**, *127*, 32–40. [CrossRef] [PubMed]
22. Manocha, G.D.; Mishra, R.; Sharma, N.; Kumawat, K.L.; Basu, A.; Singh, S.K. Regulatory role of TRIM21 in the type-I interferon pathway in Japanese encephalitis virus-infected human microglial cells. *J. Neuroinflamm.* **2014**, *11*, 24. [CrossRef] [PubMed]
23. Yang, K.; Shi, H.X.; Liu, X.Y.; Shan, Y.F.; Wei, B.; Chen, S.; Wang, C. TRIM21 is essential to sustain IFN regulatory factor 3 activation during antiviral response. *J. Immunol. (Baltim. Md. 1950)* **2009**, *182*, 3782–3792. [CrossRef] [PubMed]

*Article*

# 3′quant mRNA-Seq of Porcine Liver Reveals Alterations in UPR, Acute Phase Response, and Cholesterol and Bile Acid Metabolism in Response to Different Dietary Fats

**Maria Oczkowicz [1,*], Tomasz Szmatoła [1,2], Małgorzata Świątkiewicz [3], Anna Koseniuk [1], Grzegorz Smołucha [1], Wojciech Witarski [1] and Alicja Wierzbicka [1]**

1 Department of Animal Molecular Biology, National Research Institute of Animal Production, ul Krakowska 1, 32-083 Balice, Poland; tomasz.szmatola@izoo.krakow.pl (T.S.); anna.koseniuk@izoo.krakow.pl (A.K.); grzegorz.smolucha@izoo.krakow.pl (G.S.); wojciech.witarski@izoo.krakow.pl (W.W.); alicja.wierzbicka@izoo.krakow.pl (A.W.)

2 Centre of Experimental and Innovative Medicine, University of Agriculture in Kraków, Al. Mickiewicza 24/28, 30-059 Kraków, Poland

3 Department of Animal Nutrition and Feed Science, National Research Institute of Animal Production, ul Krakowska 1, 32-083 Balice, Poland; m.swiatkiewicz@izoo.krakow.pl

* Correspondence: maria.oczkowicz@izoo.krakow.pl; Tel.: +48666081109

Received: 20 August 2020; Accepted: 16 September 2020; Published: 18 September 2020

**Abstract:** Animal fats are considered to be unhealthy, in contrast to vegetable fats, which are rich in unsaturated fatty acids. However, the use of some fats, such as coconut oil, is still controversial. In our experiment, we divided experimental animals (domestic pigs) into three groups differing only in the type of fat used in the diet: group R: rapeseed oil ($n$ = 5); group B: beef tallow ($n$ = 5); group C: coconut oil ($n$ = 6). After transcriptomic analysis of liver samples, we identified 188, 93, and 53 DEGs (differentially expressed genes) in R vs. B, R vs. C, and B vs. C comparisons, respectively. Next, we performed a functional analysis of identified DEGs with String and IPA software. We observed the enrichment of genes engaged in the unfolded protein response (UPR) and the acute phase response among genes upregulated in B compared to R. In contrast, cholesterol biosynthesis and cholesterol efflux enrichments were observed among genes downregulated in B when compared to R. Moreover, activation of the UPR and inhibition of the sirtuin signaling pathway were noted in C when compared to R. The most striking difference in liver transcriptomic response between C and B was the activation of the acute phase response and inhibition of bile acid synthesis in the latest group. Our results suggest that excessive consumption of animal fats leads to the activation of a cascade of mutually propelling processes harmful to the liver: inflammation, UPR, and imbalances in the biosynthesis of cholesterol and bile acids via altered organelle membrane composition. Nevertheless, these studies should be extended with analysis at the level of proteins and their function.

**Keywords:** pigs; fatty acids; 3′quant mRNA-seq; nutrigenomics

## 1. Introduction

The study of lipid metabolism is becoming increasingly important in the context of the growing incidence of human metabolic diseases like obesity, NAFLD (Non-Alcoholic Fatty Liver Disease), and type 2 diabetes, as well as neurodegenerative disorders and cancer [1,2]. Dietary fats are one of the most critical determinants of the vulnerability of organisms to the development of diseases. Fatty acids are essential components of cell membranes and function as signaling molecules, regulating enzyme activity and preserving homoeostasis [3].

The recent development of advanced sequencing techniques accelerated research on the interactions between nutrients and diet. The RNA-seq results have given insight into the lipid metabolism of many species, including mice, humans, and pigs. The domestic pig is widely used as a model animal in biomedical research. It seems to be perfect for nutrigenomic research since, as an omnivore, it allows researchers to test various diets. Additionally, genetically, it is much more diverse than laboratory rats or mice, and thus better reflects human biological processes. Thus, this species is commonly used as a model in studies of the molecular background of some human diseases, including cancer and atherogenesis [4,5].

Recently, several studies investigated the effect of omega-3 and omega-6 in the diet of pigs on the transcriptome in different tissues [6–8], showing that different fatty acid compositions have a significant effect on the expression of genes engaged in fatty acid synthesis and metabolism. We also performed a preliminary study on the impact of different sources of fat in the diet on the liver transcriptome in pigs [9]; however, due to the limited number of samples and identified differentially expressed genes, we were not able to recognize all aspects of the observed effects. In the present paper, we described the comprehensive functional analysis of Quant-seq 3′mRNA-seq quantitative profiling (Lexogen, Vienna, Austria) of liver samples collected from animals receiving rapeseed oil, beef tallow, or coconut oil in the diet.

Although dietary recommendations point to the harmful effects of excessive consumption of saturated fat, there is still a lot of doubt about the health effects of consuming fats from various sources, e.g., coconut or rapeseed oil. Our investigation aimed to identify the possible molecular mechanism of metabolic changes that occur after receiving different fats in the diet and how they contribute to the pathogenesis of diseases.

## 2. Materials and Methods

### 2.1. Animals and Diets

Animals for the study were kept at the Testing Station of the National Research Institute of Animal Production in Grodziec Śląski. The experiment described in this manuscript was conducted on animals used in previously published studies [10–13]. All procedures included in this study relating to the use of live animals were in agreement with the local Ethics Committee for Experiments with Animals in Cracow (Resolution No. 912 dated 26 April 2012). In this study, we used 16 samples of liver collected from crossbred fatteners divided into three dietary groups, in which the diets differed among each other in terms of fodder fat: 3% rapeseed oil (group R−$n$ = 5, 3 gilts and 2 barrows), 3% beef tallow (group B−$n$ = 5, 2 gilts and 3 barrows), and 3% coconut oil (group C−$n$ = 6, 3 gilts and 3 barrows). Two samples were excluded from the initial number of 18 samples, due to the low quality of RNA-sequencing. All animals were kept in individual straw-bedded pens in uniform conditions. The animals were healthy, and as equal as possible in regards to body weight. The diets of all of the groups were isonitrogenous and isoenergetic (metabolized energy (ME): $R$ = 13.4, $B$ = 13.3, $C$ = 13.4 MJ/kg feed), and were formulated to cover the nutritional requirements of the pigs. The ingredient composition and nutritive value of the diets are presented elsewhere [11], while the fatty acid compositions of the feed mixtures are shown in Figure 1. Briefly, the group I feed mix contained 80% UFA content (44% MUFA and 36% PUFA), group B contained 67% UFA (32% MUFA and 35% PUFA), and group C contained 45% UFA (16% MUFA and 29% PUFA). The experimental fattening lasted from 60 to 118 kg of live weight of the animals. At the end of the experiment, all the pigs were slaughtered by stunning with high-voltage electric tongs (voltage 240–400 V), and samples of subcutaneous adipose tissue from the area between the last thoracic and the first lumbar vertebrae were collected for transcriptome analysis. All samples were stored in a freezer (−85 °C) until analysis.

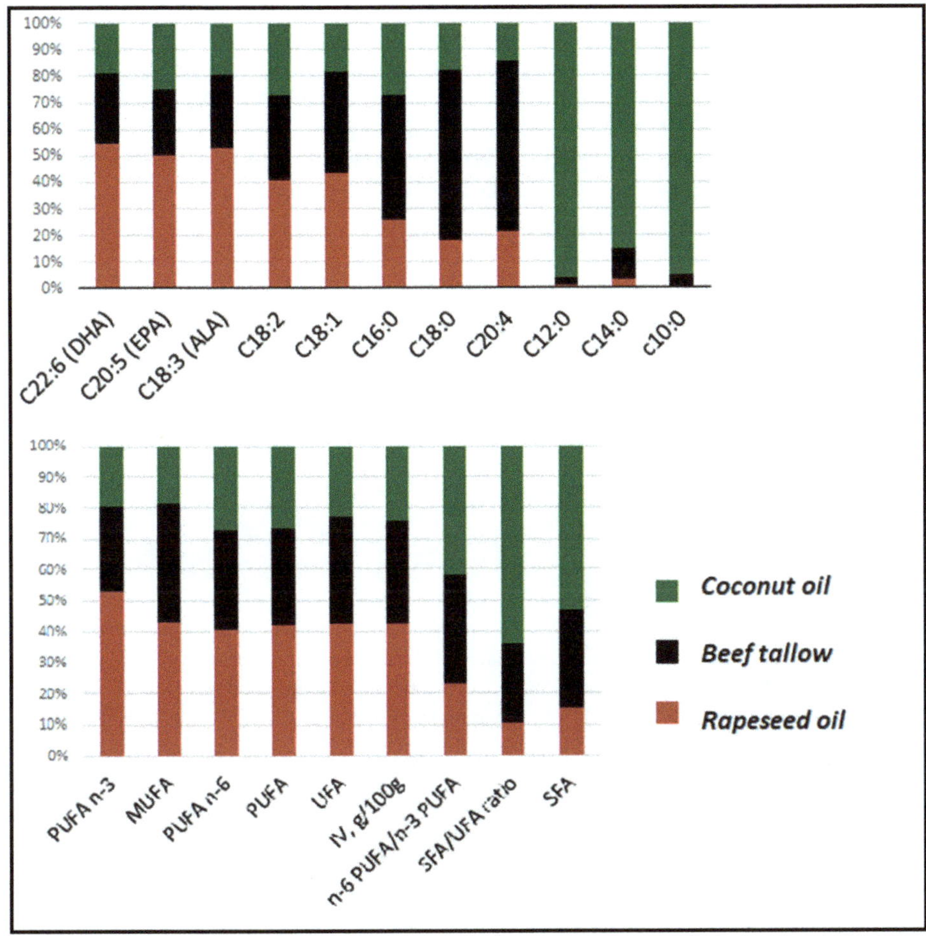


**Figure 1.** Percentage of the most abundant fatty acids in the diets.

### 2.2. RNA Isolation and 3′Quant mRNA Library Construction and Sequencing

RNA isolation was performed using the SPLIT RNA Extraction Kit (Lexogen, Vienna, Austria) according to the manufacturer's recommendations. Quality of RNA was assessed using a Tapestation 2200 (Agilent, Santa Clara, CA, USA), while quantity was evaluated by nanodrop. A sample of 200 ng of RNA was used for library preparation using the QuantSeq 3′ mRNA-Seq Library Prep Kit FWD for Illumina (Lexogen, Vienna, Austria) according to the manufacturer's protocol. Assessment of library quantity and quality was performed using a Qubit (Thermofisher Scientific, Foster City, CA, USA) and Tapestation 2200 (Agilent, Santa Clara, CA, USA), respectively. The sequencing of the pooled libraries (50 bp single-read) was performed on an Illumina Hiseq 2500 instrument (Illumina, San Diego, CA, USA), using the High-Output v4—SR 50 Cycle kit (Illumina, San Diego, CA, USA) at the DNA Sequencing Center at Brigham Young University (Provo, UT, USA).

### 2.3. Bioinformatic Analysis

Demultiplexed fastq files were downloaded from the sequencing provider server. Next, the quality check, trimming of reads, and mapping of reads were conducted with FastQC 11.8, FLEXBAR 3.5.0, and TopHat 2.1.1 software, respectively. Samtools 1.9, RSeQC, and HTSeq-count 0.11.1 software, and Gtf-Ensembl annotation 96 were used for evaluation of the mapping statistics and read counts. Differential expression analysis was performed using DEseq 2 software. Genes with $p$-adjusted < 0.05 (FDR-False Discovery Rate) Benjamini–Hochberg (BH) adjustment and no fold-change threshold were regarded as differentially expressed. Functional analysis with STRING software was performed separately for upregulated and downregulated genes using the *Sus scrofa* database. Functional analysis with IPA software was performed using porcine gene names with databases for all available species (human, rat, mouse). Figures were made with the Biorender.com and Paint programs.

### 2.4. Quantitative PCR

RNA was reverse transcribed using a high capacity cDNA archive kit (Thermo Fisher Scientific). Next, qPCR was performed using TaqMan gene expression assays and TaqMan gene expression master mix on a QuantStudio 7-flex instrument. Relative quantity data were analyzed on the Thermo Fisher cloud. Statistical significance was assessed using the Relative Quantification application on Thermo Fisher Connect.

## 3. Results

### 3.1. Fatty Acids Profiles of Diets Used in the Experiment

In our experiment, we compared the expression of genes in the liver of pigs fed with three isoenergetic diets differing only in the type of fat. The detailed information about the diets are presented elsewhere [10]. The proportions of the most important fatty acids in each diet are presented in Figure 1. Group R (rapeseed oil) obtained feed with the highest amounts of unsaturated fatty acids (UFA) and the lowest quantity of saturated fatty acids (SFA). The predominant fatty acids in this group were behenic acid (C22:0), eicosapentaenoic acid (EPA) (C20:5), docosahexaenoic acid (DHA) (C22:6), and α-linolenic acid (C18:3). The second group had an intermediate ratio of SFA/UFA and high content of palmitic acid (C16:0), stearic acid (C18:0), arachidonic acid (C20:4), and palmitoleic acid (C16:1). Group C displayed the highest SFA/UFA ratio and a high content of short/medium-chain saturated fatty acids: myristic acid (C14:0), capric acid (C10:0), and lauric acid (C12:0).

Diets used in the study did not change the phenotype of animals dramatically [10]. We observed no differences in the weight of animals, the weight of the liver and main muscles, or backfat thickness. Different diets did not affect average feed utilization or growth of animals. However, substantial differences were observed in the fatty acid composition of the adipose tissue. We found a high correlation between the dietary fatty acid composition of the diet and adipose tissue fatty acid composition [10–12].

### 3.2. 3′Quant mRNA Statistics and DEGs Identified in the Liver after Different Dietary Treatments

After 3′quant mRNA-sequencing, we obtained, on average, 2,776,234 raw reads per sample. More than 79% of them were mapped to *Sus scrofa* 11.1. (Table 1). On average, 11,500 genes with read counts >1 were identified in each sample. All gene expression data were submitted to the GEO NCBI database (accession number: GSE144247). Using DESeq2 software, we performed comparisons of gene expression between all three dietary groups (rapeseed oil vs. beef tallow, rapeseed vs. coconut oil, beef tallow vs. coconut oil, the first group in the comparison is the reference group). MA plots and PCA obtained after DESeq2 analysis are presented in Supplementary Figure S1. We identified 188 DEGs in the group R vs. group B comparison (77 downregulated in group B and 111 upregulated in group B (adjusted $p$-value < 0.05). In the comparison of group R with group C, 93 DEGs were identified (45 downregulated in group C and 48 upregulated in group C). The lowest number of DEGs (53) was noted in the group B versus group C comparison (32 downregulated in group C and 21 upregulated in group C) (Supplementary Table S1. Twenty genes with the highest adjusted $p$-value from each comparison are listed in Table 2.

**Table 1.** RNA-seq statistics of Quant-seq 3′mRNA-sequencing.

| Sample Name | No of Raw Reads | No of Mapped Reads | Percentage of Mapped Reads (%) |
|---|---|---|---|
| 17w | 3,355,869 | 2,635,495 | 78.5 |
| 19w | 2,460,130 | 1,927,026 | 78.3 |
| 20w | 2,029,705 | 1,615,350 | 79.6 |
| 21w | 2,819,316 | 2,187,454 | 77.6 |
| 22w | 2,657,611 | 2,111,384 | 79.4 |
| 24w | 2,298,911 | 1,803,582 | 78.5 |
| 41w | 3,469,934 | 2,760,471 | 79.6 |
| 42w | 2,223,698 | 1,817,882 | 81.8 |
| 43w | 4,124,251 | 3,197,675 | 77.5 |
| 45w | 2,157,086 | 1,710,536 | 79.3 |
| 46w | 2,415,641 | 1,932,621 | 80.0 |
| 48w | 2,470,229 | 1,981,470 | 80.2 |
| 49w | 2,715,605 | 2,203,190 | 81.1 |
| 52w | 2,631,687 | 2,119,487 | 80.5 |
| 53w | 2,110,920 | 1,648,309 | 78.1 |
| 54w | 2,031,614 | 1,625,581 | 80.0 |
| 55w | 4,000,000 | 3,266,253 | 81.7 |
| 56w | 4,000,000 | 3,262,080 | 81.6 |

**Table 2.** Top 20 DEGs with the highest adjusted *p*-value for each comparison.

| Ensembl ID | Base Mean | log2 Fold Change | *p*-Adjusted | Gene Description | Gene Symbol |
|---|---|---|---|---|---|
| | | | **Beef Tallow vs. Coconut Oil** | | |
| ENSSSCG00000021443 | 109 | −1.46 | 0.0005 | serum/glucocorticoid regulated kinase 1 | *SGK1* |
| ENSSSCG00000032381 | 221 | −1.42 | 0.0000 | lipopolysaccharide induced TNF factor | *LITAF* |
| ENSSSCG00000004700 | 288 | −1.34 | 0.0007 | protein disulfide isomerase family A member 3 | *PDIA3* |
| ENSSSCG00000028758 | 177 | −1.28 | 0.0000 | lipopolysaccharide binding protein | *LBP* |
| ENSSSCG00000000892 | 127 | −1.23 | 0.0008 | histidine ammonia-lyase | *HAL* |
| ENSSSCG00000033214 | 460 | −1.16 | 0.0098 | glycine N-methyltransferase | *GNMT* |
| ENSSSCG00000016174 | 361 | −1.16 | 0.0185 | fibronectin 1 | *FN1* |
| ENSSSCG00000008998 | 23136 | −1.01 | 0.0001 | fibrinogen α chain | *FGA* |
| ENSSSCG00000013514 | 256 | −0.94 | 0.0001 | LRRCT domain-containing protein | ***PLIN5*** |
| ENSSSCG00000008550 | 317 | −0.86 | 0.0124 | solute carrier family 5 member 6 | *SLC5A6* |
| ENSSSCG00000022126 | 122 | −0.79 | 0.0025 | epidermal growth factor receptor | *EGFR* |
| ENSSSCG00000002487 | 1830 | −0.78 | 0.0027 | α-1-antichymotrypsin 2 | ***SERPINA3-2*** |
| ENSSSCG00000002749 | 37957 | −0.59 | 0.0162 | haptoglobin | *HP* |
| ENSSSCG00000011453 | 1755 | −1.43 | 0.0000 | inter-α-trypsin inhibitor heavy chain 4 | *ITIH4* |
| ENSSSCG00000011741 | 340 | 0.45 | 0.0185 | golgi integral membrane protein 4 | *GOLIM4* |
| ENSSSCG00000027072 | 172 | 0.56 | 0.0112 | ATP synthase inhibitory factor subunit 1 | *ATP5IF1* |
| ENSSSCG00000002529 | 355 | 0.65 | 0.0036 | 40S ribosomal protein S21 | *RPS21* |
| ENSSSCG00000008829 | 182 | 0.80 | 0.0076 | OCIA domain containing 2 | *OCIAD2* |

**Table 2.** *Cont.*

| Ensembl ID | Base Mean | log2 Fold Change | *p*-Adjusted | Gene Description | Gene Symbol |
|---|---|---|---|---|---|
| ENSSSCG00000026044 | 221 | 1.34 | 0.0005 | farnesyl-diphosphate farnesyltransferase 1 | *FDFT1* |
| ENSSSCG00000006238 | 84 | 2.11 | 0.0007 | cytochrome P450 family 7 subfamily A member 1 | *CYP7A1* |
| | | | **Rapeseed Oil vs. Beef Tallow** | | |
| ENSSSCG00000006238 | 100 | −2.63 | $4 \times 10^{-8}$ | cytochrome P450 family 7 subfamily A member 1 | *CYP7A1* |
| ENSSSCG00000028821 | 27 | −2.14 | $1.6 \times 10^{5}$ | SAS-6 centriolar assembly protein | *SASS6* |
| ENSSSCG00000004586 | 19 | −2.02 | $6.5 \times 10^{5}$ | family with sequence similarity 81 member A | *FAM81A* |
| ENSSSCG00000033822 | 110 | −1.91 | $6.6 \times 10^{5}$ | thyroid hormone responsive | *THRSP* |
| ENSSSCG00000026044 | 252 | −1.80 | $1.6 \times 10^{5}$ | farnesyl-diphosphate farnesyltransferase 1 | *FDFT1* |
| ENSSSCG00000006719 | 94 | −1.41 | $1.6 \times 10^{7}$ | hydroxy-delta-5-steroid dehydrogenase, 3 β- and steroid delta-isomerase 1 | *HSD3B1* |
| ENSSSCG00000006040 | 186 | −1.20 | $2.6 \times 10^{6}$ | dihydropyrimidinase | *DPYS* |
| ENSSSCG00000001849 | 153 | −0.89 | $1.7 \times 10^{4}$ | alanyl aminopeptidase, membrane | *ANPEP* |
| ENSSSCG00000039388 | 448 | −0.84 | $8.3 \times 10^{5}$ | | |
| ENSSSCG00000023686 | 5283 | −0.78 | $3.4 \times 10^{4}$ | transthyretin | *TTR* |
| ENSSSCG00000015106 | 124 | 1.23 | $1.5 \times 10^{4}$ | hypoxia up-regulated 1 | *HYOU1* |
| ENSSSCG00000032381 | 218 | 1.35 | $3.4 \times 10^{4}$ | lipopolysaccharide induced TNF factor | *LITAF* |
| ENSSSCG00000011297 | 222 | 1.36 | $6.6 \times 10^{5}$ | abhydrolase domain containing 5, lysophosphatidic acid acyltransferase | *ABHD5* |
| ENSSSCG00000015140 | 293 | 1.38 | $8.7 \times 10^{9}$ | heat shock protein family A (Hsp70) member 8 | *HSPA8* |
| ENSSSCG00000030095 | 78 | 1.49 | $2.4 \times 10^{4}$ | zinc finger and BTB domain containing 16 | *ZBTB16* |
| ENSSSCG00000020754 | 32 | 1.51 | $6.5 \times 10^{5}$ | RNA polymerase I subunit G | *CD3EAP* |
| ENSSSCG00000005601 | 1136 | 1.54 | $1.8 \times 10^{13}$ | heat shock protein family A (Hsp70) member 5 | *HSPA5* |
| ENSSSCG00000022126 | 97 | 1.73 | $4.0 \times 10^{13}$ | epidermal growth factor receptor | *EGFR* |
| ENSSSCG00000010686 | 369 | 1.79 | $8.3 \times 10^{6}$ | BAG cochaperone 3 | *BAG3* |
| ENSSSCG00000035058 | 43 | 1.82 | $6.3 \times 10^{6}$ | phosphotyrosine interaction domain containing 1 | *PID1* |
| | | | **Rapeseed Oil vs. Coconut Oil** | | |
| ENSSSCG00000023331 | 67 | −1.35 | 0.0000 | ubiquitin like 5 | *UBL5* |
| ENSSSCG00000007710 | 75 | −1.14 | 0.0033 | MLX interacting protein lik | *MLXIPL* |
| ENSSSCG00000027926 | 94 | −1.14 | 0.0062 | formimidoyltransferase cyclodeaminase | *FTCD* |
| ENSSSCG00000003302 | 193 | −1.01 | 0.0004 | Lysoplasmalogenase | **TMEM86B** |
| ENSSSCG00000031881 | 48 | −1.01 | 0.0003 | CDC42 small effector 1 | *CDC42SE1* |
| ENSSSCG00000035790 | 110 | −0.93 | 0.0048 | BTG anti-proliferation factor 1 | *BTG1* |
| ENSSSCG00000006719 | 105 | −0.92 | 0.0013 | hydroxy-delta-5-steroid dehydrogenase, 3 β- and steroid delta-isomerase 1 | ***HSD3B1*** |

**Table 2.** *Cont.*

| Ensembl ID | Base Mean | log2 Fold Change | *p*-Adjusted | Gene Description | Gene Symbol |
|---|---|---|---|---|---|
| ENSSSCG00000031302 | 46 | −0.91 | 0.0015 | C-terminal binding protein 1 | *CTBP1* |
| ENSSSCG00000010627 | 45 | −0.88 | 0.0064 | programmed cell death 4 | *PDCD4* |
| ENSSSCG00000013514 | 225 | −0.73 | 0.0064 | LRRCT domain-containing protein | ***PLIN5*** |
| ENSSSCG00000001849 | 163 | −0.65 | 0.0062 | alanyl aminopeptidase, membrane | *ANPEP* |
| ENSSSCG00000014855 | 870 | −0.62 | 0.0002 | ribosomal protein S3 | *RPS3* |
| ENSSSCG00000011000 | 253 | 0.86 | 0.0013 | | |
| ENSSSCG00000022126 | 69 | 0.94 | 0.0007 | epidermal growth factor receptor | *EGFR* |
| ENSSSCG00000039147 | 108 | 0.96 | 0.0014 | | |
| ENSSSCG00000015140 | 278 | 1.14 | 0.0000 | heat shock protein family A (Hsp70) member 8 | *HSPA8* |
| ENSSSCG00000004093 | 290 | 1.14 | 0.0000 | iodotyrosine deiodinase | *IYD* |
| ENSSSCG00000010686 | 339 | 1.49 | 0.0000 | BAG cochaperone 3 | *BAG3* |
| ENSSSCG00000011437 | 132 | 1.50 | 0.0005 | 5′-aminolevulinate synthase 1 | *ALAS1* |
| ENSSSCG00000024596 | 69 | 1.62 | 0.0051 | nocturnin | *NOCT* |

Bold genes were annotated with the Uniprot database.

Among the identified DEGs, there were 30 genes in common between comparison groups R vs. B and groups R vs. group C, 25 common genes between comparison groups R vs. B and groups B vs. C and eight genes in common between the group R vs. C comparison and the B vs. C comparison (Figure 2). Two differentially expressed genes (EGFR, HSPA5) were common for all three comparisons. Interestingly, expression of all common genes was altered in the same direction in both groups: in B and C when compared to R, and in R and C when compared to B. All identified genes in common between specific comparisons are listed in Supplementary Table S2.

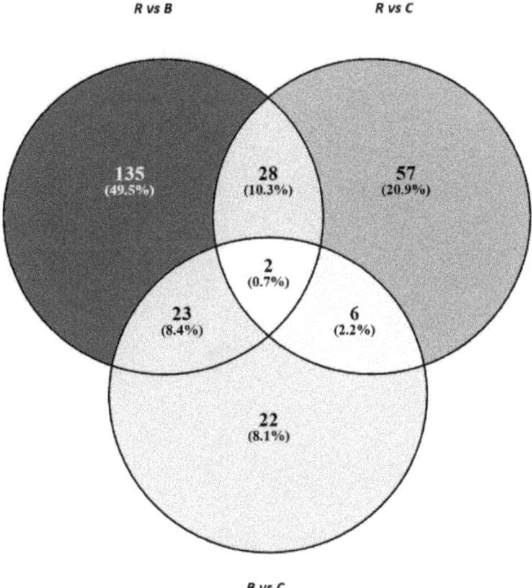


**Figure 2.** Venn diagram illustrating differentially expressed genes that are common between comparisons.

Since there was a sex imbalance in some groups in our study, we performed a comparison between females and males with DEseq2 software, regardless of dietary group. We observed only 11 differentially expressed genes between female and males, and all of them were located on sex chromosomes (Supplementary Table S3). What is more, none of the genes differentially expressed between males and females were present among the genes identified as differentially expressed between dietary groups. We concluded that the effect of sex is negligible in our study.

### 3.3. Functional Analysis of Identified DEGs with STRING

To get insights into the biological processes that are activated by specific dietary fats, we performed functional analysis of identified DEGs with String software using *Sus scrofa* genes as a background. We separately analyzed upregulated and downregulated genes in each comparison. We observed 86, 82, and 18 enrichments in rapeseed oil vs. beef tallow, rapeseed oil vs. coconut oil, and the beef tallow vs. coconut oil comparisons, respectively (Supplementary Table S4). In R vs. B and R vs. C comparisons, we observed a higher number of enrichments among downregulated genes (Supplementary Table S4). Among genes downregulated by beef tallow in the R vs. B comparison, we observed overrepresentation of genes engaged in oxidoreductase activity (*CAT*, *MSMO*, *GLUD1*), cholesterol biosynthesis (e.g., *HSD3B1*, *HMGCS1*, *MVK*), metabolism of steroids (*FDFT1*, *MVK*, *HMGCS1*), and bile acid secretion (*CYP7A1*, *HMGCR*) (Figure 3, Supplementary Table S4). Furthermore, among genes upregulated by beef tallow in this comparison, enrichment of genes associated with the metabolism of RNA (*GNL3*, *NOP58*, *NOPB*), cellular response to stress (*HSPA5*, *HSPA8*, *BAG3*), the urea cycle (*ENSSSCG00000016159*, *SLC25A15*), and metabolism of polyamines (*AMD1*, *ENSSSCG00000016159*, *SLC25A15*) were observed (Figure 4, Supplementary Table S4).

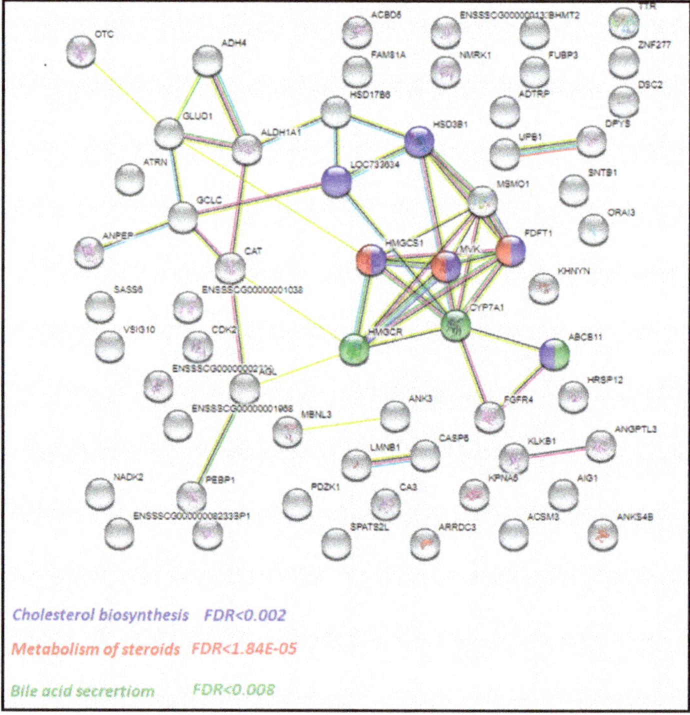


**Figure 3.** The network of DEGs downregulated in the beef tallow group when compared to the rapeseed oil group according to STRING software analysis.

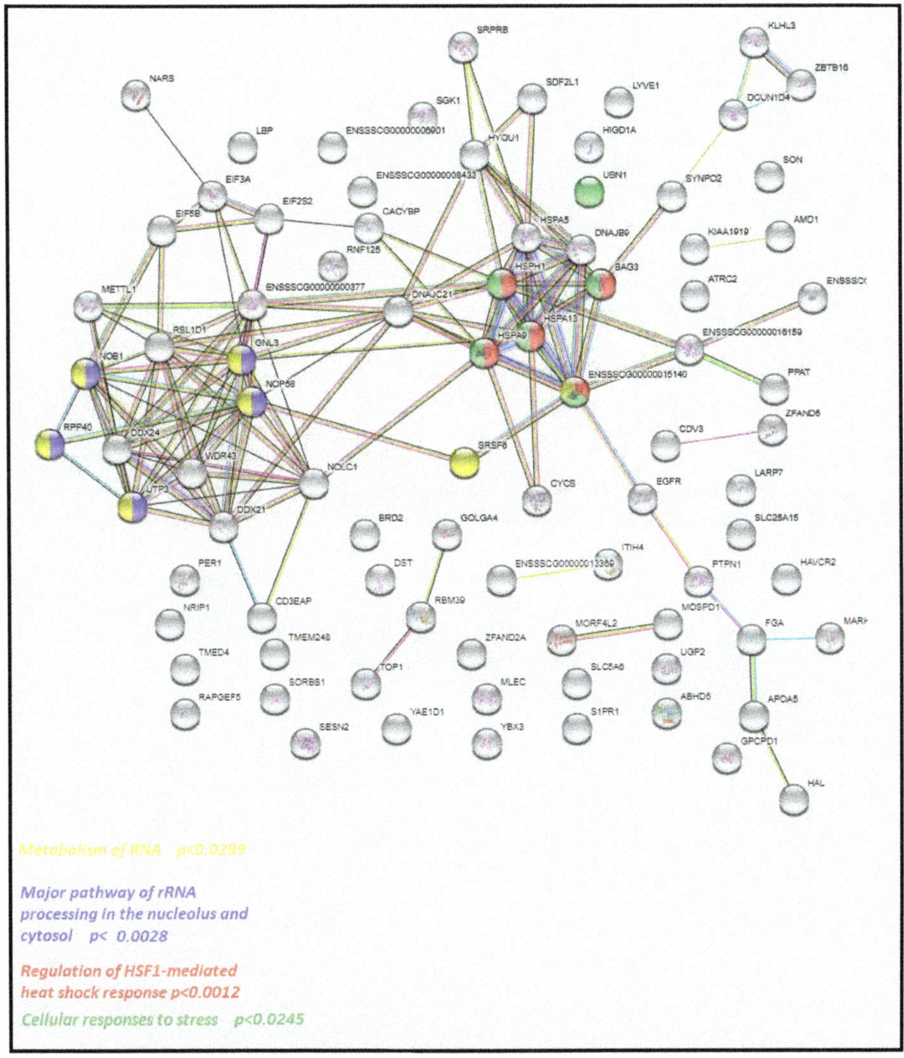


**Figure 4.** The network of DEGs upregulated in the beef tallow group when compared to the rapeseed oil group according to STRING software analysis.

In the second comparison (R vs. C), we observed that genes upregulated by coconut oil are overrepresented in pathways connected to stress response and cellular signaling: the regulation of HSF1-mediated heat shock response (*BAG3*, *ENSSSCG00000015140*, *HSPA9*), and NOTCH3 activation and transmission of signal to the nucleus (*EGFR*, *PSEN2*). Moreover, downregulated genes were enriched in many metabolism-related pathways and terms (e.g., primary metabolic process, organic substance metabolic process, cellular metabolic process, *ANPEP*, *C1QA*, *ERH*, *FTCD*, *HSD3B1*, *MAN1A1*, *RPS3*, *SDHD*, *TMEM86B*, *UOX*).

In the comparison between beef tallow and coconut oil, we observed downregulation of genes engaged in immunity (defense response (*C1QA*, *HP*, *IL4R*, *ITIH4*), and genes engaged in protein processing in the endoplasmic reticulum (*HSPA5*, *HYOU1*, *PDIA3*). Genes upregulated by coconut oil were associated with PPAR signaling pathways (*CYP7A1*, *FABP1)* and bile secretion and cholesterol metabolism (*ABCB11, CYP7A1).* All enrichments with adequate FDR values and matching proteins are presented in Supplementary Table S4.

### *3.4. Functional Analysis of Identified DEGs with IPA*

To further analyze differentially expressed gene sets for their association with human diseases and to assess to what extent the relationships observed in pigs can be translated into humans, we performed functional analysis of identified DEGs from all comparisons with ingenuity pathway analysis (IPA) (Qiagen). Sixty-six significant canonical pathways were observed in the comparison between group R vs. B (log2 $p$-value > 1.5) (Supplementary Figure S2). Among them, the superpathway of cholesterol biosynthesis (Z-score = −2.646, $p$-value < $2.34 \times 10^9$, mevalonate pathway I (Z-score = −2, $p$-value < $3.57 \times 10^6$), the superpathway of geranylgeranyldiphosphate biosynthesis I (Z-score = −2, $p$-value < $1.07 \times 10^5$), and LXR/RXR activation (Z-score = −0.447, $p$-value < $5.12 \times 10^5$) were inhibited, while LPS/IL-1-mediated inhibition of RXR function (Z-score = 1.342, $p$-value < $1.25 \times 10^5$), acute phase response signaling (Z-score = 2.236, $p$-value < $5.74 \times 10^4$), Huntington's disease signaling (Z-score = 1, $p$-value < $2.88 \times 10^3$), and several pathways of inositol metabolism were activated by the beef tallow diet when compared to the rapeseed oil diet (Figure 5) (Supplementary Figure S2). In the second comparison (rapeseed oil vs. coconut oil), 30 canonical pathways were noted, with the sirtulin signaling pathway (Z-score = −2, $p$-value < $1.1 \times 10^4$) being inhibited and the BAG2 signaling pathway (Z-score = 1, $p$-value < $2.01 \times 10^5$) and unfolded protein response (Z-score = 2, $p$-value < $5.76 \times 10^5$) being activated by coconut oil (Figure 5). The beef tallow vs. coconut oil comparison revealed 43 enriched canonical pathways. Beef tallow activated acute phase response signaling (Z-score = −2.236, $p$-value < $4.32 \times 10^9$), while coconut oil stimulated RXR/LXR activation (Z-score = 0.816, $p$-value < $7.24 \times 10^9$) in this comparison (Figure 5). All identified canonical pathways and connections between them are presented as networks in Supplementary Figure S2.

### *3.5. Identification of Hub Genes with Cytohubba*

Our next step was to identify critical genes responsible for changes observed in the transcriptome under the influence of different types of fat. For this purpose, we used the cytoHubba–Cytoscape plugin for ranking nodes in a network by their network features. Among eleven available methods, we chose MCC, which is the most precise in predicting essential proteins from the yeast PPI network [13]. In the first comparison, R vs. B, *GNL3* was ranked number one, followed by *RSL1D1*, *UTP3*, and *DDX24*. The essential genes in the B vs. C comparison were *ORM1* and *TTR*, while in the rapeseed oil vs. coconut oil they were *RPS3* and *HSPA9* (Figure 6).

### *3.6. Validation of Quant 3′mRNA Profiling by qPCR*

To validate the 3′quant mRNA-seq results, we performed qPCR analysis of several DEGs identified in the study. We observed high concordance between RNA-seq and qPCR results. As expected, we found strong overexpression of genes engaged in the unfolded protein response (*BAG3*, *HSPA5*) and inflammation (*PID1*, *IHIT4*, *ALPL*, *LITAF*) in the beef tallow group compared to the rapeseed oil and coconut oil groups. In contrast, genes involved in bile acid secretion (*CYP7A1*) and cholesterol biosynthesis *(HSD3B1)* were downregulated in the beef tallow group when compared to both groups or the rapeseed oil group only (Figure 7). Furthermore, none of the genes analyzed by qPCR showed significantly different expression between gilts and barrows.

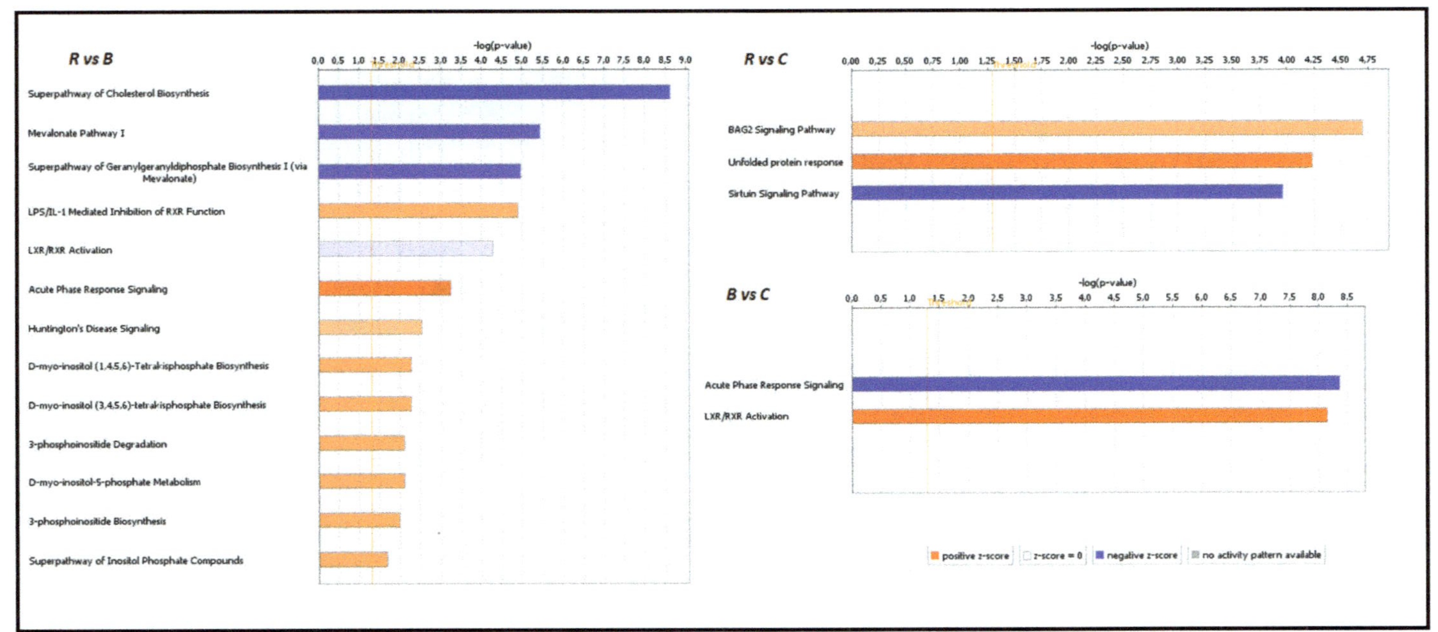


**Figure 5.** Canonical pathways with Z score ≠ 0 and log ($p$-value) > 1.3, identified in each comparison using IPA software.

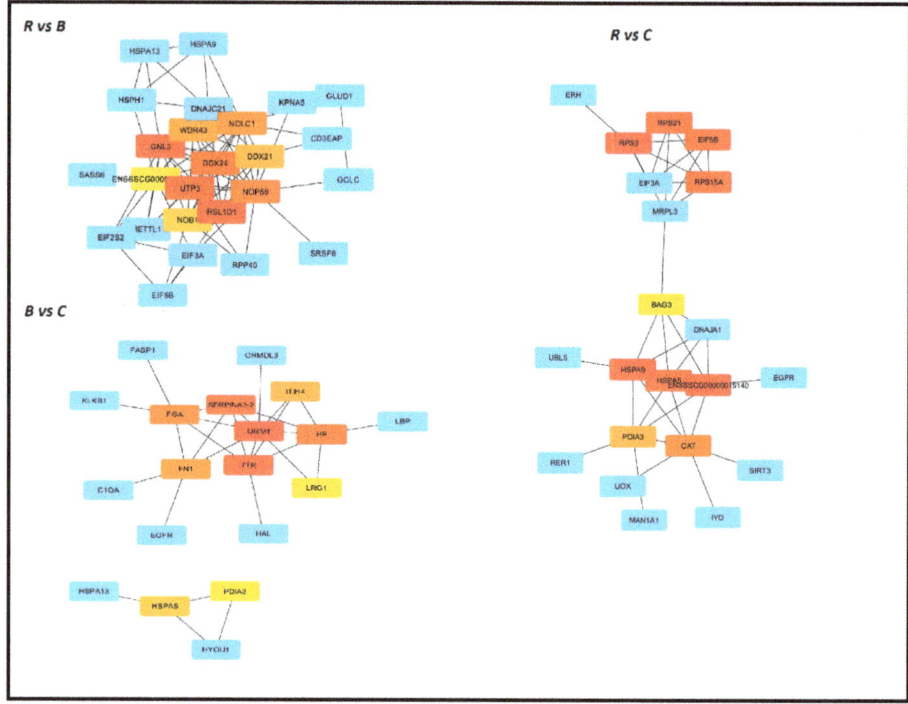


**Figure 6.** Top 10 Hub genes identified in the R vs. B comparison, R vs. C comparison, and B vs. C comparison using Cytohubba, ranked by MCC (Maximal Clique Centrality). The more intense the red color, the higher the position in the rank.

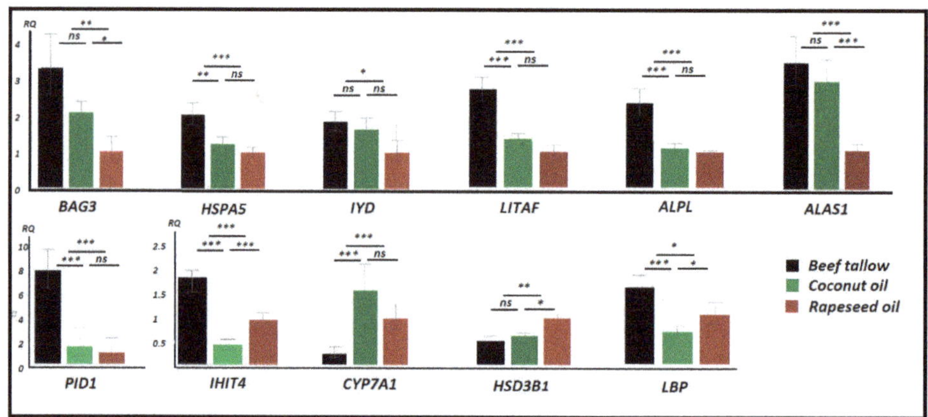


**Figure 7.** Results of the qPCR analysis of selected DEGs. RQ—Relative Quantity of mRNA, *** $p < 0.01$, ** $p < 0.05$, * $p < 0.1$, ns—not significant.

## 4. Discussion

The progress of civilization has initiated many changes in the human environment. This applies not only to nature that surrounds us but also to our way of life, especially our nutrition. First of all, fat consumption and the amount of saturated fatty acids consumed increased significantly during the last century. Moreover, the ratio of omega-6 to omega-3 in food decreased, resulting in worsening of the dietary fatty acid profile [14]. At the same time, there was a sharp increase in the incidence of diseases related to metabolism, cardiovascular diseases, neurodegenerative diseases, and cancer. These diseases are also called civilization diseases because their occurrence is closely related to the modern lifestyle.

In our experiment, we used a domestic pig (*Sus scrofa*) to illustrate the changes that occur in the liver transcriptome as a result of using various fats in the diet. Using isoenergetic and isonitrogenous diets made it possible to observe transcriptome changes at a specific stage, before drastic changes at the phenotype level occurred. For quantitative analysis of the transcriptome, we chose the 3′quant mRNA method, which allows transcriptome profiling based on the 3′UTR ends of the gene and is a cost-efficient alternative to whole transcriptome RNA-seq. By this method, we identified 11,500 genes with >1 read number, of which 308 (2.67%) were differentially expressed between samples of animals fed different diets, despite a very low sequencing coverage—approximately two million reads per sample. We observed that many of these genes are engaged in the pathogenesis of human civilization diseases and that depending on the source of dietary fat, different pathways are activated or inhibited in liver at the gene expression level.

### *4.1. Biosynthesis and Catabolism of Cholesterol are Inhibited in Animals Obtaining Beef Tallow in the Diet*

The opinion that the consumption of saturated fats increases blood cholesterol levels is generally approved and well documented [15]. Maintaining adequate blood cholesterol levels is crucial for the body since its excess results in the development of atherosclerosis, heart disease, and neurodegenerative diseases. On the other hand, cholesterol is an essential component of cell membranes and has many important functions in the organism. Cholesterol homeostasis in the body is directed by the interaction between absorption, synthesis, and excretion or conversion of cholesterol into bile acids. A reciprocal relationship between these processes is known to regulate circulating cholesterol levels in response to dietary or therapeutic interventions. Cholesterol biosynthesis is self-regulating; a high cholesterol level in the blood forces the inhibition of its synthesis in the liver. One of the mechanisms involved in this process is inhibition of the expression of the gene encoding HMGCR, a rate-limiting enzyme in cholesterol biosynthesis. This is exactly what we observed in animals fed beef tallow—the expression of *HMGCR* was ~3 fold lower than in animals fed rapeseed oil. We observed inhibition of several other genes engaged in cholesterol biosynthesis (*FDFT1*, *MVD*, *EBP*, *HSDB31*) in the group receiving beef tallow in the diet when compared to the group receiving rapeseed oil. Two of these genes (*FDFT1*, *HSD3B1*) were also downregulated in the coconut oil group compared to the rapeseed oil group but to a lesser extent, suggesting a different response of the mechanisms responsible for cholesterol homeostasis to medium-chain saturated fatty acids. The results of the functional analysis using the STRING and IPA programs indicate the existence of an additional mechanism affecting the level of cholesterol synthesis. According to the IPA results (Figure 5), RXR function was inhibited through LPS/IL-1 mediation, and simultaneously the super pathway of cholesterol biosynthesis was repressed in the beef tallow group. We suppose that it may have occurred as a result of a series of changes triggered by gut microbiome misbalance under the influence of dietary beef tallow. A diet rich in saturated fatty acid affects gut microbiota composition by enhancing overflow of dietary fat to the distal intestine in mice [15]. In the pigs fed beef tallow, an increase in pathogenic LPS-secreting bacteria appeared, which resulted in an increase in expression of *LBP* and other acute-phase response genes in the liver. Moreover, beef tallow contains substantial amounts of arachidonic acid, which is known for its pro-inflammatory properties. As a consequence of inflammation, the RXR function was inhibited, which affected the level of expression of genes engaged in cholesterol metabolism (Figure 8).

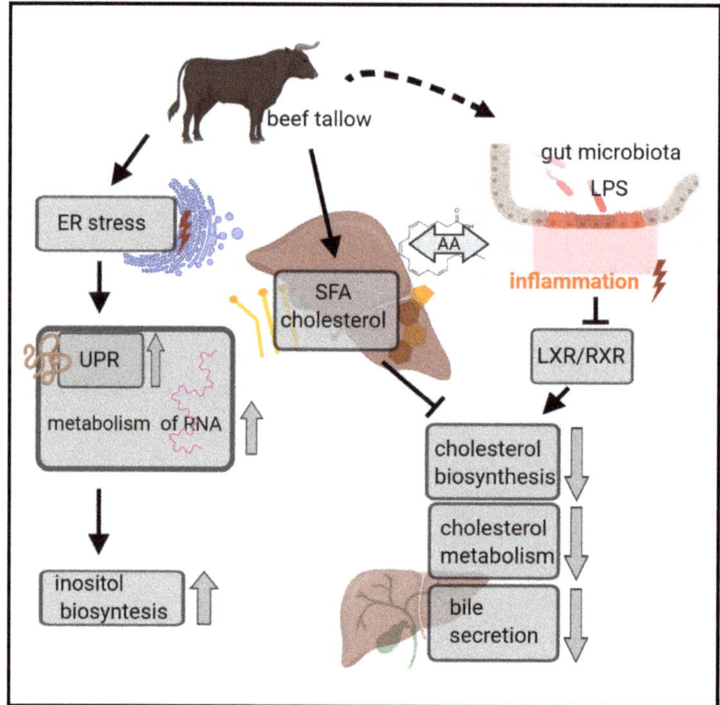


**Figure 8.** Graphical illustration of selected biological processes and canonical pathways altered in the beef tallow group when compared to rapeseed oil. Beef tallow contains pro-inflammatory ingredients (SFA, AA—arachidonic acid) that could change gut microbiota and promote inflammation. SFA and cholesterol from beef tallow decrease cholesterol biosynthesis directly in the liver and indirectly through inhibition of LXR/RXR by LPS/IL-1.

The use of coconut oil in the diet still causes a lot of controversies. As a fat containing many saturated fatty acids, and thus causing an increase in blood cholesterol levels, it is not recommended for people at risk of cardiovascular disease. On the other hand, it was shown recently that it may improve intestinal microbiota, antioxidant status, and immunity of growing rabbits [16]. Moreover, the latest meta-analysis showed that MCFA (medium-chain fatty acids), which predominate in coconut oil, increase HDLcholesterol—responsible for cholesterol efflux—content in comparison with long-chain fatty acids. *CYP7A1* is a gene coding for a rate limiting enzyme in the cholesterol catabolic pathway in the liver, which converts cholesterol to bile acids. This reaction is the major site of regulation of bile acid synthesis, which is the primary mechanism for the removal of cholesterol from the body. We observed a 6-fold decrease in *CYP7A1* gene expression in the beef tallow group compared to coconut oil, emphasizing the advantage of coconut oil over beef tallow in cholesterol efflux. It was shown in vitro that arachidonic acid is a potent inhibitor of *CYP7A1* expression [17], which is in accordance with our results.

### 4.2. Acute-Phase Response Signaling is Activated in the Beef Tallow Group when Compared to both Rapeseed Oil and Coconut Oil

Low-grade inflammation is one of the leading causes of NAFLD, cardiovascular diseases, diabetes, and neurodegenerative diseases. We observed a significant increase in the expression level of many genes connected to the immune system response and inflammation markers in liver tissue collected from animals fed with beef tallow (*LITAF, ALPL, PID1, IHIT4, LBP, HP*) (Figure 7, Supplementary

Table S4). Strikingly, the *LBP* gene, which codes for lipopolysaccharide binding protein, and *CD14* were both upregulated in the beef tallow group. LBP and CD14 drive ternary complex formation and TLR activation and as a consequence trigger the whole cascade of immune response stimulated by *NFKBiB* [18]. Interestingly, the expression of some genes (*LBP*, *IHIT4*) was also upregulated in rapeseed oil when compared to coconut oil, supporting information about the antibacterial properties of coconut oil [16]. Ingenuity pathway analysis revealed that the acute phase response canonical pathway is highly significantly activated in animals obtaining beef tallow when compared to both rapeseed oil and coconut oil. The pro-inflammatory effects of long-chain saturated fatty acids have been known for some time [19]. It has been observed that long-chain saturated fatty acids increase haptoglobin gene expression (inflammation marker) in mice adipose tissue [19]. Another study showed that the composition and metabolic activity of the gut microbiota change as a result of a steatohepatitis-inducing high-fat diet in mice [20]. Moreover, in these animals, the level of saturated fatty acids (palmitic acid) in the gut increased significantly, activated macrophages in the liver, and promoted TNF-$\alpha$ expression. The consequence of these changes was the development of NASH, which was reversible under the influence of antibiotics [20]. On the other hand, lauric acid—the main component of coconut oil—was shown previously to alleviate neuroinflammatory responses by LPS-activated microglia, supporting its beneficial effect on neurodegenerative diseases [21]. Our results, in agreement with previous studies, underline the difference in response of the immune system to dietary long-chain SFA (beef tallow) and medium-chain SFA (coconut oil), (Figure 9).

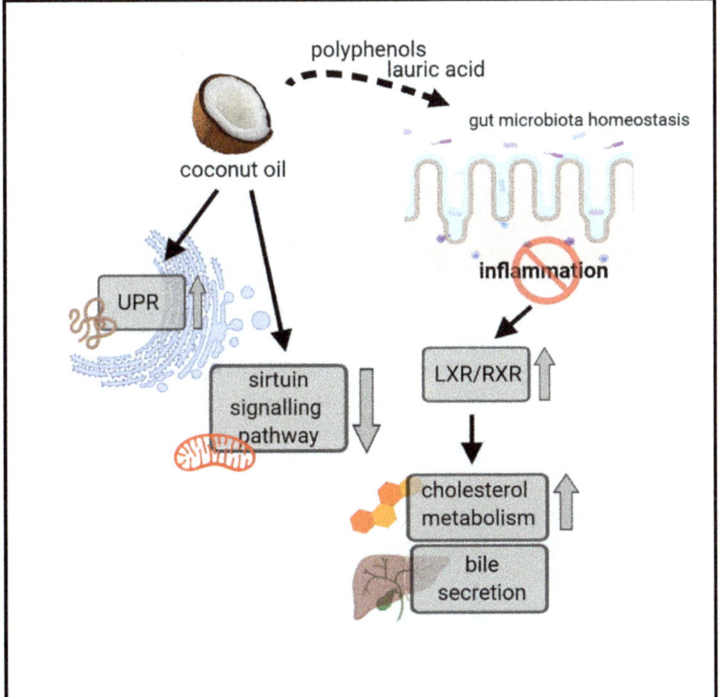


**Figure 9.** Graphical illustration of selected biological processes and canonical pathways altered in the coconut oil group when compared to the rapeseed oil (UPR, sirtuin signaling pathway) and beef tallow groups (LXR/RXR activation, cholesterol metabolism, bile secretion). Coconut oil contains antibacterial and anti-inflammatory ingredients (lauric acids, polyphenols), and cholesterol metabolism and bile acid secretion are not reduced as a result of increased inflammation, which is observed in the beef tallow group.

### 4.3. Unfolded Protein Response is Activated in both the Beef Tallow and Coconut Oil Groups when Compared to the Rapeseed Oil Group

The unfolded protein response (UPR) is a highly conserved pathway that allows the cell to manage endoplasmic reticulum (ER) stress that is imposed by the secretory demands associated with environmental forces [22]. Our results show that upon dietary beef tallow or coconut oil uptake, expression of genes engaged in UPR—*BAG3, HSP5, HSP7*—increases when compared to rapeseed oil. Consequently, functional analysis with IPA and STRING revealed that UPR is strongly activated in the coconut oil and beef tallow groups when compared to rapeseed oil. qPCR analysis confirmed overexpression of *HSPA5* and *BAG3* in beef tallow and both saturated fatty acids rich groups, respectively, when compared to rapeseed oil. All these results support the opinion that UPR is regulated by lipid-dependent mechanisms [23,24]. It is assumed that in an environment rich in saturated fatty acids, the composition of membranes in the endoplasmic reticulum changes, leads to disturbances in protein folding and results in activation of the unfolded protein response pathway [24]. Recent in vitro studies [25] indicate that palmitic acid induces ER stress and simultaneously increases inflammatory indices, while oleic acid ameliorates this action in exocrine pancreas cells. In our experiment, the ratio of palmitic to oleic acid was about twice as high in coconut oil and beef tallow as compared to rapeseed oil (Figure 1), supporting in vivo associations previously observed in vitro in pancreas cells [25].

When the proteins present in mitochondria are damaged, and their accumulation threatens the maintenance of homeostasis, mitochondrial specific UPR ($UPR^{mt}$) is activated. The central $UPR^{mt}$ coordinator is *SIRT3*, which encodes a member of the sirtuin family of class III histone deacetylases. In our study, we observed lower expression of *SIRT3* in animals fed coconut oil in the diet when compared to the rapeseed oil group. Moreover, according to IPA analysis, the sirtuin signaling pathway (with *SIRT3, SDHD, UOX* genes) was downregulated in the coconut oil group. This result may have an important clinical significance since the loss of *SIRT3* leads to deregulation of several mitochondrial pathways, which contributes to the accelerated development of the disease of ageing [26]. In general, lowering the *SIRT3* level is associated with adverse health effects. It is considered a mitochondria-localized tumor suppressor, which opposes reprogramming of cancer cell metabolism through HIF1$\alpha$ destabilization [27]. Moreover, it was shown that *SIRT3* deficiency accelerates the development of metabolic syndrome. On the other hand, in vivo experiments show that a chronic high-fat diet decreases expression of *SIRT3* in liver tissue [28]. In contrast, in vitro experiments suggest that palmitic acids increase *SIRT3* expression, contrary to oleic acids [29]. Thus, the effect of coconut oil consumption on the sirtuin signaling pathway should be further analyzed on the protein and protein function levels, especially in the context of using it to prevent neurodegenerative diseases.

During UPR activation, a decrease in RNA synthesis is often observed to protect cells against excessive accumulation of misfolded proteins [30]. According to our results, dietary beef tallow-activated genes (*DDX21, NOLC1*) suppress pre-rRNA transcription by forming ring-shaped structures surrounding multiple Pol I complexes [31,32]. Additionally, we observed upregulation of *NOB1*, which blocks the recruitment of mRNAs to the nascent ribosome [33]. Thus, it suggests the launching of repair mechanisms to stop the excessive production of misfolded proteins in the beef tallow group. On the other hand, several genes (*EIF3A, EIF5B*) engaged in translation initiation were upregulated in the beef tallow and coconut groups. It was recently found that eIF3a regulates HIF1$\alpha$ protein synthesis through the internal ribosomal entry site (IRES)-dependent translation. Therefore, it was concluded that eIF3a might be a potential therapeutic target for hepatic carcinoma (HCC) since it acts as a regulator for glycolysis—a process that is central to cancerous reprogramming of metabolism [34].

### 4.4. UPR, Inflammation, and Cholesterol and Bile Acid Metabolism are the Main Processes Affected by Dietary Fatty Acids

In our experiment, we observed that UPR, inflammation, and cholesterol and bile acid metabolism are the most altered processes under different dietary fat treatments. Interaction between endoplasmic reticulum stress—a trigger for UPR and inflammation—is involved in a variety of human pathologies [22,35]. Our results show that cholesterol and bile acid metabolism are additional components of this complex. They also support the hypothesis that a key point in these interactions is the composition of lipid membranes, since "organelle membrane" is the common cellular component enriched in all comparisons of our study (Supplementary Table S2). Considering the results of functional analysis of the identified DEGs as well as the qPCR results, it can be stated that the UPR level is low in the rapeseed oil group, while it is high in the beef tallow and coconut oil groups. The level of inflammation and cholesterol efflux is high only in the group receiving beef tallow. In contrast, cholesterol biosynthesis is low in the group receiving beef tallow and to a lesser extent in the coconut oil group. Our next step was the analysis of the causal relations between these processes and the identification of key genes regulating these interactions. The effect of bile acid and cholesterol metabolism on inflammatory processes has been well known for a long time. Numerous studies indicate that reduced flow of bile acids leads to their accumulation in the liver, which in turn causes inflammation. It was shown that bile acids act as an inflammagen, and directly activate signaling pathways in hepatocytes that stimulate the production of the proinflammatory mediators. However, other mechanisms were considered; the inflammatory response is triggered by activation of Toll-like receptor 4 (TLR4), either by bacterial lipopolysaccharide or by damage-associated molecular pattern molecules released from dead hepatocytes [36]. When we compared the effect of beef tallow and coconut oil in relation to rapeseed oil, we observed the downregulation of genes responsible for the synthesis and transport of bile acids (*CYP7A1*, *ABCB11*) only in the group receiving beef tallow. Similarly, only in this group, we observed an increase in expression of genes coding for acute-phase proteins. The exception here was the *ORM1* gene (identified as a hub gene by cytohubba software), which was upregulated in the coconut oil group. The product of this gene is classified as an acute phase protein, but it also has immunosuppressive activity. More than 25 years ago, it was found that this protein protects mice against lethality shock induced by tumor necrosis factor (TNF) or endotoxins [37]. More recently, a decrease of ORMDL protein without decreases in ORMDL mRNA levels was observed in HepG2 liver cells treated with the pro-inflammatory stimulus, and this observation was extended to in vivo models of inflammation [38]. In contrast, we observed a decrease in ORM expression accompanied by an increase in acute phase response in the beef tallow group at the mRNA level. The situation is further complicated by the fact that the *ORM1* gene is activated by bile acid through FXR—the nuclear bile acid receptor in mice [39]. This may indicate that the main difference in the action of coconut oil compared to beef tallow is that coconut oil does not reduce cholesterol catabolism and its disposal with bile acids. Thanks to this, FXR is activated, and thus ORM1's protective effect is maintained.

The interaction between inflammation and cholesterol and bile acid metabolism also acts in the opposite direction; the activation of immune response proteins affects the level of gene expression associated with cholesterol and bile acid metabolism [40]. In the liver, LPS markedly decreases the mRNA expression and activity of CYP7A1 and cholesterol transporters ABCG5 and ABCG8, which mediate cholesterol excretion into the bile in Syrian hamsters [41]. The relationship discussed here has a significant clinical implication. Until recently, hypercholesterolemia was considered the main cause of heart disease, but some scientists indicate that inflammation may play a more important role in the pathogenesis of CVD. Our results underline the significance of inflammation in this context. In essence, findings from epidemiological studies report low rates of cardiovascular disease among populations who consume coconut oil as part of their traditional diets (in India, the Philippines, and Polynesia) even though this fat is cholesterol-raising. Considering that in our study the animals receiving coconut oil did not show an increase in activation of acute-phase response, the limited influence of coconut oil consumption for the occurrence of heart disease may be due to the anti-inflammatory properties of this oil (Figure 9).

Inflammatory processes are strictly connected to endoplasmic reticulum stress, and UPR and the *ORM* genes link these two processes. Yeast cells lacking the *ORM1* and *ORM2* genes show a constitutive unfolded protein response, sensitivity to stress, and slow ER-to-Golgi transport [42]. ORM proteins function as the main regulators of the enzyme serine palmitoyltransferase—necessary for the process of sphingolipid biosynthesis, which is one of the main components of the cytoplasmic membranes of the endoplasmic reticulum, and disturbances in its homeostasis lead to ER stress [43]. The unfolded protein response can also alter sphingolipid metabolism. Bi-directional interactions between sphingolipids and the UPR have now been observed in a range of diseases, including cancer, diabetes, and liver disease [43]. Other studies have shown that ER stress can directly initiate the proinflammatory pathways, while proinflammatory agents such as ROS, TLR ligands, and cytokines can induce ER stress. As a result, activated UPR may further enhance the pro-inflammatory response [44].

As in the case of relationships discussed earlier, the connection between UPR and biosynthesis of cholesterol and bile acids is well documented and mutual. It was shown that within the ER, there are numerous membrane receptors detecting changes in cholesterol levels and cholesterol overload causes severe dysregulation of the ER [45]. To counteract these abnormalities, transcription factors (e.g., *SREBF2*, *LXR*) regulating cholesterol biosynthesis and immunological responses are triggered by the NRF2 protein [46]. In our experiment, we did not observe changes in the level of expression of genes coding for these molecules; however, they are largely regulated at the level of translation or post-translational modifications [47]. Among the DEGs identified in our report, the two genes *S1PR1* (upregulated in the beef tallow and coconut oil groups when compared to the rapeseed oil group) and *ORAI3* (downregulated only in the beef tallow group when compared to the rapeseed oil group) are potential links between cholesterol metabolism and ER stress and UPR. The protein encoded by *S1PR1* is structurally similar to G protein-coupled receptors and binds the ligand sphingosine-1-phosphate (S1P) with high affinity and high specificity. S1P, in turn, is a signaling sphingolipid acting as a bioactive lipid mediator. It is transported mainly by HDL and activates one of the S1PR1-mediated biological functions: calcium flux [48]. It was recently demonstrated that S1PR1-mediated calcium efflux is achieved through ORAI-membrane calcium channels. Thus, our data suggest that both the dietary fats beef tallow and coconut oil activate *S1PR1*, but only beef tallow downregulates *ORAI3* transcript expression. Interestingly, store-operated calcium ($Ca^{2+}$) entry (SOCE) is mediated by *Orai3* only in breast cancer cells that express the estrogen receptor, contrary to estrogen receptor-negative cancer cells, which suggests a relationship between estrogen concentration and *ORAI3* expression [49]. In our experiment, probably as a consequence of reduced cholesterol catabolism in the beef tallow group, we observed a decrease in steroid hormone biosynthesis, which theoretically could lead to a decrease in *ORAI3* expression and dysfunctional calcium channels.

Due to the limited amount of space, we are not able to discuss all the interesting relationships observed in our experiment. To mention only a few: the "inositol metabolism", "urea cycle", and "glutathione metabolism" pathways deserve additional detailed analysis in terms of the effect of dietary fats on cancer and the development of metabolic, neurodegenerative, and cardiovascular diseases. Moreover, expression of a number genes—potential therapeutic targets in liver diseases—including *EGFR*, *GLUD1*, *GNL3*, and *RGS5*, has been shown to be modified by dietary fats and should be further investigated in this regard. Although our work provides a large amount of new information on the impact of consuming different sources of fat on gene expression in the liver, we are aware that these studies should be extended with analysis at the level of proteins and their function. Furthermore, analysis of bile acid species content in the intestine and liver and histological examination of these organs would give a full view of changes introduced by consumption of different sources of fat. Even though the material was collected from all animals from the same part of the liver, our samples most likely contained a mixture of different cells (hepatocytes, parenchymal cells, immune cells). Therefore, differences in expression observed between the groups of animals tested may partly result from the proportion of the content of these cells in the samples. To accurately assess what processes occur in the specific cells of the liver under the influence of dietary fat, experiments using single-cell methods are necessary.

## 5. Conclusions

To sum up, our experiment showed that the type of consumed fat has a very significant impact on changes in gene expression in the liver of pigs. We observed that these changes are most intensively visible in three related processes: cholesterol and bile acid biosynthesis, UPR, and inflammation, playing a key role in the pathogenesis of civilization diseases. If one of these processes is dysregulated, repair mechanisms are triggered by activation of connected pathways. Therefore, when the body is excessively stimulated by improper nutrition, a vicious circle starts, in which the dysregulation of one process results in the dysregulation of the next. In our experiment, this situation most likely arose as a result of a diet containing beef tallow. In this group of animals, we observed deregulation of cholesterol and bile acid metabolism, activation of genes coding for acute phase proteins, and activation of the UPR when compared to animals fed with rapeseed oil. In contrast, in the coconut oil group, no activation of inflammation genes was observed, suggesting that some ingredients of coconut oil (lauric acid, polyphenols) can stop this vicious cycle and prevent the development of civilization diseases.

**Supplementary Materials:** The following are available online at http://www.mdpi.com/2073-4425/11/9/1087/s1, Supplementary Table S1. List of DEGs identified in each comparison. Supplementary Table S2. List of DEGs and Gene ontology terms, KEGG and Reactome pathways, common for different comparisons. Supplementary Table S3. Results of the DEseq2 comparisons between males and females. Supplementary Table S4. List of Gene ontology terms, KEGG and Reactome pathways overrepresented in each comparison after STRING functional analysis. Supplementary Figure S1. MA and PCA plots after analysis of RNA-seq results with DESeq2 software. Supplementary Figure S2. Networks of canonical pathways identified after IPA software analysis in each comparison.

**Author Contributions:** Conceptualization, M.O. and M.Ś.; methodology, A.W., A.K., M.O.; software, T.S.; validation, G.S., M.O.; formal analysis, M.O.; investigation M.O.; resources, M.Ś.; data curation, T.S.; writing—Original draft preparation, M.O.; writing—Review and editing A.W., A.K., W.W., M.O., M.Ś. and T.S.; visualization, M.O.; W.W., supervision, M.O.; project administration, M.O.; funding acquisition, M.O. All authors have read and agreed to the published version of the manuscript.

**Funding:** This research was funded by National Science Centre, Poland (grant no 2014/13/B/NZ9/02134).

**Acknowledgments:** We thank Justyna Mrozowicz for technical assistance.

**Conflicts of Interest:** The authors declare no conflict of interest.

## References

1. Othman, R. Dietary lipids and cancer. *Libyan J. Med.* **2007**, *2*, 180–184. [CrossRef] [PubMed]
2. Zanoaga, O.; Jurj, A.; Raduly, L.; Cojocneanu-Petric, R.; Fuentes-Mattei, E.; Wu, O.; Braicu, C.; Gherman, C.D.; Berindan-Neagoe, I. Implications of dietary $\omega$-3 and $\omega$-6 polyunsaturated fatty acids in breast cancer. *Exp. Ther. Med.* **2018**, *15*, 1167–1176. [CrossRef] [PubMed]
3. Halade, G.V.; Black, L.M.; Verma, M.K. Paradigm shift-Metabolic transformation of docosahexaenoic and eicosapentaenoic acids to bioactives exemplify the promise of fatty acid drug discovery. *Biotechnol. Adv.* **2018**, *36*, 935–953. [CrossRef] [PubMed]
4. Watson, A.L.; Carlson, D.F.; Largaespada, D.A.; Hackett, P.B.; Fahrenkrug, S.C. Engineered Swine Models of Cancer. *Front Genet.* **2016**, *7*, 78. [CrossRef] [PubMed]
5. Cevallos, W.H.; Holmes, W.L.; Myers, R.N.; Smink, R.D. Swine in atherosclerosis research: Development of an experimental animal model and study of the effect of dietary fats on cholesterol metabolism. *Atherosclerosis* **1979**, *34*, 303–317. [CrossRef]
6. Vitali, M.; DiMauro, C.; Sirri, R.; Zappaterra, M.; Zambonelli, P.; Manca, E.; Sami, D.; Fiego, D.P.L.; Davoli, R. Effect of dietary polyunsaturated fatty acid and antioxidant supplementation on the transcriptional level of genes involved in lipid and energy metabolism in swine. *PLoS ONE* **2018**, *13*, e0204869. [CrossRef] [PubMed]
7. Szostak, A.; Ogłuszka, M.; Pas, M.F.W.T.; Poławska, E.; Urbański, P.; Juszczuk-Kubiak, E.; Blicharski, T.; Pareek, C.S.; Dunkelberger, J.R.; Horbańczuk, J.O.; et al. Effect of a diet enriched with omega-6 and omega-3 fatty acids on the pig liver transcriptome. *Genes Nutr.* **2016**, *11*, 9. [CrossRef]

8. Ogłuszka, M.; Szostak, A.; Pas, M.F.T.; Poławska, E.; Urbański, P.; Blicharski, T.; Pareek, C.; Juszczuk-Kubiak, E.; Dunkelberger, J.R.; Horbańczuk, J.O.; et al. A porcine gluteus medius muscle genome-wide transcriptome analysis: Dietary effects of omega-6 and omega-3 fatty acids on biological mechanisms. *Genes Nutr.* **2017**, *12*, 4. [CrossRef]
9. Oczkowicz, M.; Świątkiewicz, M.; Ropka-Molik, K.; Gurgul, A.; Żukowski, K. Effects of Different Sources of Fat in the Diet of Pigs on the Liver Transcriptome Estimated by RNA-Seq. *Ann. Anim. Sci.* **2016**, *16*, 1073–1090. [CrossRef]
10. Świątkiewicz, M.; Oczkowicz, M.; Ropka-Molik, K.; Hanczakowska, E. The effect of dietary fatty acid composition on adipose tissue quality and expression of genes related to lipid metabolism in porcine livers. *Anim. Feed Sci. Technol.* **2016**, *216*, 204–215. [CrossRef]
11. Oczkowicz, M.; Szmatoła, T.; Świątkiewicz, M.; Pawlina-Tyszko, K.; Gurgul, A.; Ząbek, T. Corn dried distillers grains with solubles (cDDGS) in the diet of pigs change the expression of adipose genes that are potential therapeutic targets in metabolic and cardiovascular diseases. *BMC Genom.* **2018**, *19*, 864. [CrossRef] [PubMed]
12. Oczkowicz, M.; Szmatoła, T.; Świątkiewicz, M. Source of Dietary Fat in Pig Diet Affects Adipose Expression of Genes Related to Cancer, Cardiovascular, and Neurodegenerative Diseases. *Genes* **2019**, *10*, 948. [CrossRef] [PubMed]
13. Chin, C.H.; Chen, S.H.; Wu, H.H.; Ho, C.W.; Ko, M.T.; Lin, C.Y. cytoHubba: Identifying hub objects and sub-networks from complex interactome. *BMC Syst. Biol.* **2014**, *8* (Suppl. 4), S11. [CrossRef]
14. Simopoulos, A.P. An Increase in the Omega-6/Omega-3 Fatty Acid Ratio Increases the Risk for Obesity. *Nutrients* **2016**, *8*, 128. [CrossRef]
15. De Wit, N.; Derrien, M.; Bosch-Vermeulen, H.; Oosterink, E.; Keshtkar, S.; Duval, C.; Bosch, J.D.V.-V.D.; Kleerebezem, M.; Muller, M.; Van Der Meer, R. Saturated fat stimulates obesity and hepatic steatosis and affects gut microbiota composition by an enhanced overflow of dietary fat to the distal intestine. *Am. J. Physiol. Gastrointest. Liver. Physiol.* **2012**, *303*, G589–G599. [CrossRef] [PubMed]
16. Alagawany, M.; Abd El-Hack, M.E.; Al-Sagheer, A.A.; Naiel, M.A.; Saadeldin, I.M.; Swelum, A.A. Dietary Cold Pressed Watercress and Coconut Oil Mixture Enhances Growth Performance, Intestinal Microbiota, Antioxidant Status, and Immunity of Growing Rabbits. *Animals* **2018**, *8*, 212. [CrossRef]
17. Zhang, J.-M.; Wang, X.-H.; Hao, L.-H.; Wang, H.; Zhang, X.; Muhammad, I.; Qi, Y.; Li, G.-L.; Sun, X.-Q. Nrf2 is crucial for the down-regulation of Cyp7a1 induced by arachidonic acid in Hepg2 cells. *Environ. Toxicol. Pharmacol.* **2017**, *52*, 21–26. [CrossRef]
18. Ranoa, D.R.; Kelley, S.L.; Tapping, R.I. Human lipopolysaccharide-binding protein (LBP) and CD14 independently deliver triacylated lipoproteins to Toll-like receptor 1 (TLR1) and TLR2 and enhance formation of the ternary signaling complex. *J. Biol. Chem.* **2013**, *288*, 9729–9741. [CrossRef]
19. Bueno, A.A.; Oyama, L.M.; Motoyama, C.S.D.M.; Biz, C.R.D.S.; Silveira, V.L.; Ribeiro, E.B.; Nascimento, C.M.O.D. Long chain saturated fatty acids increase haptoglobin gene expression in C57BL/6J mice adipose tissue and 3T3-L1 cells. *Eur. J. Nutr.* **2010**, *49*, 235–241. [CrossRef]
20. Yamada, S.; Kamada, N.; Amiya, T.; Nakamoto, N.; Nakaoka, T.; Kimura, M.; Saito, Y.; Ejima, C.; Kanai, T.; Saito, Y. Gut microbiota-mediated generation of saturated fatty acids elicits inflammation in the liver in murine high-fat diet-induced steatohepatitis. *BMC Gastroenterol.* **2017**, *17*, 136. [CrossRef]
21. Nishimura, Y.; Moriyama, M.; Kawabe, K.; Satoh, H.; Takano, K.; Azuma, Y.-T.; Nakamura, Y. Lauric Acid Alleviates Neuroinflammatory Responses by Activated Microglia: Involvement of the GPR40-Dependent Pathway. *Neurochem. Res.* **2018**, *43*, 1723–1735. [CrossRef] [PubMed]
22. Grootjans, J.; Kaser, A.; Kaufman, R.J.; Blumberg, R.S. The unfolded protein response in immunity and inflammation. *Nat. Rev. Immunol.* **2016**, *16*, 469–484. [CrossRef] [PubMed]
23. Volmer, R.; van der Ploeg, K.; Ron, D. Membrane lipid saturation activates endoplasmic reticulum unfolded protein response transducers through their transmembrane domains. *Proc. Natl. Acad. Sci. USA* **2013**, *110*, 4628–4633. [CrossRef]
24. Volmer, R.; Ron, D. Lipid-dependent regulation of the unfolded protein response. *Curr. Opin. Cell Biol.* **2015**, *33*, 67–73. [CrossRef] [PubMed]
25. Ben-Dror, K.; Birk, R. Oleic acid ameliorates palmitic acid-induced ER stress and inflammation markers in naive and cerulein-treated exocrine pancreas cells. *Biosci. Rep.* **2019**, *39*. [CrossRef]

26. McDonnell, E.; Peterson, B.S.; Bomze, H.M.; Hirschey, M.D. SIRT3 regulates progression and development of diseases of aging. *Trends Endocrinol. Metab.* **2015**, *26*, 486–492. [CrossRef]
27. Finley, L.W.; Carracedo, A.; Lee, J.J.; Souza, A.; Egia, A.; Zhang, J.; Teruya-Feldstein, J.; Moreira, P.; Cardoso, S.M.; Clish, C.B.; et al. SIRT3 opposes reprogramming of cancer cell metabolism through HIF1α destabilization. *Cancer Cell.* **2011**, *19*, 416–428. [CrossRef]
28. Li, R.; Xin, T.; Li, D.; Wang, C.; Zhu, H.; Zhou, H. Therapeutic effect of Sirtuin 3 on ameliorating nonalcoholic fatty liver disease: The role of the ERK-CREB pathway and Bnip3-mediated mitophagy. *Redox Biol.* **2018**, *18*, 229–243. [CrossRef]
29. Li, S.; Dou, X.; Ning, H.; Song, Q.; Wei, W.; Zhang, X.; Shen, C.; Li, J.; Sun, C.; Song, Z. Sirtuin 3 acts as a negative regulator of autophagy dictating hepatocyte susceptibility to lipotoxicity. *Hepatology* **2017**, *66*, 936–952. [CrossRef]
30. Pizzinga, M.; Harvey, R.F.; Garland, G.D.; Mordue, R.; Dezi, V.; Ramakrishna, M.; Sfakianos, A.; Monti, M.; Mulroney, T.E.; Poyry, T.; et al. The cell stress response: Extreme times call for post-transcriptional measures. *Wiley Interdiscip. Rev. RNA* **2020**, *11*, e1578. [CrossRef]
31. Xing, Y.-H.; Yao, R.-W.; Zhang, Y.; Guo, C.-J.; Jiang, S.; Xu, G.; Dong, R.; Yang, L.; Chen, L.-L. SLERT Regulates DDX21 Rings Associated with Pol I Transcription. *Cell* **2017**, *169*, 664–678.e16. [CrossRef] [PubMed]
32. Yuan, F.; Zhang, Y.; Ma, L.; Cheng, Q.; Li, G.; Tong, T. Enhanced NOLC1 promotes cell senescence and represses hepatocellular carcinoma cell proliferation by disturbing the organization of nucleolus. *Aging Cell.* **2017**, *16*, 726–737. [CrossRef] [PubMed]
33. Parker, M.D.; Collins, J.C.; Korona, B.; Ghalei, H.; Karbstein, K. A kinase-dependent checkpoint prevents escape of immature ribosomes into the translating pool. *PLoS Biol.* **2019**, *17*, e3000329. [CrossRef] [PubMed]
34. Miao, B.; Wei, C.; Qiao, Z.; Han, W.; Chai, X.; Lu, J.; Gao, C.; Dong, R.; Gao, D.; Huang, C.; et al. eIF3a mediates HIF1α-dependent glycolytic metabolism in hepatocellular carcinoma cells through translational regulation. *Am. J. Cancer Res.* **2019**, *9*, 1079–1090.
35. Cao, S.S.; Luo, K.L.; Shi, L. Endoplasmic Reticulum Stress Interacts With Inflammation in Human Diseases. *J. Cell Physiol.* **2016**, *231*, 288–294. [CrossRef]
36. Allen, K.; Jaeschke, H.; Copple, B.L. Bile acids induce inflammatory genes in hepatocytes: A novel mechanism of inflammation during obstructive cholestasis. *Am. J. Pathol.* **2011**, *178*, 175–186. [CrossRef]
37. Libert, C.; Brouckaert, P.; Fiers, W. Protection by α 1-acid glycoprotein against tumor necrosis factor-induced lethality. *J. Exp. Med.* **1994**, *180*, 1571–1575. [CrossRef]
38. Cai, L.; Oyeniran, C.; Biswas, D.D.; Allegood, J.; Milstien, S.; Kordula, T.; Maceyka, M.; Spiegel, S. ORMDL proteins regulate ceramide levels during sterile inflammation. *J. Lipid Res.* **2016**, *57*, 1412–1422. [CrossRef]
39. Porez, G.; Gross, B.; Prawitt, J.; Gheeraert, C.; Berrabah, W.; Alexandre, J.; Staels, B.; Lefebvre, P. The hepatic orosomucoid/α1-acid glycoprotein gene cluster is regulated by the nuclear bile acid receptor FXR. *Endocrinology* **2013**, *154*, 3690–3701. [CrossRef]
40. Khovidhunkit, W.; Kim, M.-S.; Memon, R.A.; Shigenaga, J.K.; Moser, A.H.; Feingold, K.R.; Grunfeld, C. Effects of infection and inflammation on lipid and lipoprotein metabolism: Mechanisms and consequences to the host. *J. Lipid Res.* **2004**, *45*, 1169–1196. [CrossRef]
41. Feingold, K.R.; Spady, D.K.; Pollock, A.S.; Moser, A.H.; Grunfeld, C. Endotoxin, TNF, and IL-1 decrease cholesterol 7 α-hydroxylase mRNA levels and activity. *J. Lipid Res.* **1996**, *37*, 223–228. [PubMed]
42. Han, S.; Lone, M.A.; Schneiter, R.; Chang, A. Orm1 and Orm2 are conserved endoplasmic reticulum membrane proteins regulating lipid homeostasis and protein quality control. *Proc. Natl. Acad. Sci. USA* **2010**, *107*, 5851–5856. [CrossRef] [PubMed]
43. Bennett, M.K.; Wallington-Beddoe, C.T.; Pitson, S.M. Sphingolipids and the unfolded protein response. *Biochim. Biophys. Acta Mol. Cell Biol. Lipids* **2019**, *1864*, 1483–1494. [CrossRef] [PubMed]
44. Reverendo, M.; Mendes, A.; Argüello, R.J.; Gatti, E.; Pierre, P. At the crossway of ER-stress and proinflammatory responses. *FEBS J.* **2019**, *286*, 297–310. [CrossRef]
45. Bashiri, A.; Nesan, D.; Tavallaee, G.; Sue-Chue-Lam, I.; Chien, K.; Maguire, G.F.; Naples, M.; Zhang, J.; Magomedova, L.; Adeli, K.; et al. Cellular cholesterol accumulation modulates high fat high sucrose (HFHS) diet-induced ER stress and hepatic inflammasome activation in the development of non-alcoholic steatohepatitis. *Biochim. Biophys. Acta* **2016**, *1861*, 594–605. [CrossRef]

46. Widenmaier, S.B.; Snyder, N.A.; Nguyen, T.B.; Arduini, A.; Lee, G.Y.; Arruda, A.P.; Saksi, J.; Bartelt, A.; Hotamisligil, G.S. NRF1 Is an ER Membrane Sensor that Is Central to Cholesterol Homeostasis. *Cell* **2017**, *171*, 1094–1109.e15. [CrossRef]
47. Silva-Islas, C.A.; Maldonado, P.D. Canonical and non-canonical mechanisms of Nrf2 activation. *Pharmacol. Res.* **2018**, *134*, 92–99. [CrossRef]
48. Lee, M.-H.; Appleton, K.M.; El-Shewy, H.M.; Sorci-Thomas, M.G.; Thomas, M.J.; Lopes-Virella, M.F.; Luttrell, L.M.; Hammad, S.M.; Klein, R.L. S1P in HDL promotes interaction between SR-BI and S1PR1 and activates S1PR1-mediated biological functions: Calcium flux and S1PR1 internalization. *J. Lipid Res.* **2017**, *58*, 325–338. [CrossRef]
49. Motiani, R.K.; Abdullaev, I.F.; Trebak, M. A novel native store-operated calcium channel encoded by Orai3: Selective requirement of Orai3 versus Orai1 in estrogen receptor-positive versus estrogen receptor-negative breast cancer cells. *J. Biol. Chem.* **2010**, *285*, 19173–19183. [CrossRef]

*Article*

# Identification of DNMT3B2 as the Predominant Isoform of DNMT3B in Porcine Alveolar Macrophages and Its Involvement in LPS-Stimulated TNF-α Expression

**Yanbing Zhang, Hui Li, Xiao Xiang, Yan Lu, Mona Sharma, Zongjie Li, Ke Liu, Jianchao Wei, Donghua Shao, Beibei Li, Zhiyong Ma and Yafeng Qiu ***

Shanghai Veterinary Research Institute, Chinese Academy of Agricultural Sciences, Shanghai 200241, China; zhangyanbing129@outlook.com (Y.Z.); lihui022715@outlook.com (H.L.); xiao.xiang@wur.nl (X.X.); yanlu013194@outlook.com (Y.L.); monasharma1990@yahoo.com (M.S.); lizongjie@shvri.ac.cn (Z.L.); liuke@shvri.ac.cn (K.L.); jianchaowei@shvri.ac.cn (J.W.); shaodonghua@shvri.ac.cn (D.S.); lbb@shvri.ac.cn (B.L.); zhiyongma@shvri.ac.cn (Z.M.)

* Correspondence: yafengq@shvri.ac.cn

Received: 20 August 2020; Accepted: 8 September 2020; Published: 10 September 2020

**Abstract:** DNA methyltransferase 3B (DNMT3B) as one member of the DNMT family functions as a de novo methyltransferase, characterized as more than 30 splice variants in humans and mice. However, the expression patterns of DNMT3B in pig as well as the biological function of porcine DNMT3B remain to be determined. In this study, we first examined the expression patterns of DNMT3B in porcine alveolar macrophages (PAM). We demonstrated that only DNMT3B2 and DNMT3B3 were the detectable isoforms in PAM. Furthermore, we revealed that DNTM3B2 was the predominant isoform in PAM. Next, in the model of LPS (lipopolysaccharide)-activated PAM, we showed that in comparison to the unstimulated PAM, (1) expression of DNTM3B is reduced; (2) the methylation level of TNF-α gene promoter is decreased. We further establish that DNMT3B2-mediated methylation of TNF-α gene promoter restricts induction of TNF-α in the LPS-stimulated PAM. In summary, these findings reveal that DNMT3B2 is the predominant isoform in PAM and its downregulation contributes to expression of TNF-α via hypomethylation of TNF-α gene promoter in the LPS-stimulated PAM.

**Keywords:** porcine alveolar macrophages; DNMT3B; DNA methylation; isoform; TNF-α

## 1. Introduction

DNA methyltransferase 3B (DNMT3B) is one member of the DNMT family which comprises DNMT1, DNMT3A, DNMT3B, as well as DNMT3L (DNMT3-like) in mammals. In mammalian systems, DNMT3B similar with DNMT3A serves as de novo methyltransferase for the establishment of DNA methylation [1]; in comparison, DNMT1 acts as maintenance of DNA methylation [2]. Moreover, DNMT3L has no DNA methyltransferase activity, but it can act as an accessory factor of the other DNMTs to regulate DNA methylation [3]. Although the functional characteristics between DNMT3A and DNMT3B are similar, the expression patterns of them are very different, i.e., in comparison to two isoforms of mouse DNMT3A [4], more than 30 isoforms of DNMT3B are identified in humans and mice [5,6].

Although there are more than 30 isoforms of DNMT3B, the expression patterns of DNMT3B appear to be highly conserved, at least in humans and mice. For example, human DNMT3B2 has a 60 bp-deletion (representative of exon 10) in comparison to the canonical isoform DNMT3B1 [5,7]; furthermore, DNMT3B3 of humans and mice has two deletions including a 60-bp deletion and a 189-bp deletion (representative of exon 21 and 22) in comparison to DNMT3B1 [5,8]. Notable, the alternative splicing of DNMT3B could influence DNMT3B functions. Thus, to clarify the expression patterns of DNMT3B is important to understand DNMT3B functions in the other species like pig.

Given the reported data, many spliced variants of DNMT3B are expressed in a tissue, cell, and/or developmental stage-specific manner [6,9], in this study, we focus on the expression pattern of DNMT3B in porcine alveolar macrophages (PAM), which are not only the major immune cells in pig lungs, but also the important resources of the inflammatory cytokines in pneumonia of pigs [10,11]. Though it has been shown that DNA methylation plays role in modulation of lipopolysaccharide (LPS)-induced inflammation in PAM [12], little is known about how DNMT3B is involved in this process. Furthermore, we determined effects of DNA methylation regulated by DNMT3B on TNF-α expression in the LPS-activated PAM. In this study, for the first time we determine the expression pattern of DNMT3B and clarify its effect on TNF-α expression in the LPS-stimulated PAM.

## 2. Materials and Methods

### 2.1. Piglets and Porcine Alveolar Macrophages (PAM)

Clinical healthy 35-day-old piglets (Shanghai great white pig strain) were purchased from the Shanghai Academy of Agricultural Sciences (Shanghai, China). All animal experiments were approved by the Institutional Animal Care and Use Committee of Shanghai Veterinary Research Institute (IACUC No: Shvri-po-201606 0501) and were performed in compliance with the Guidelines on the Humane Treatment of Laboratory Animals (Ministry of Science and Technology of the People's Republic of China, Policy No.2006 398). Porcine alveolar macrophages (PAM) were isolated from piglets as previously described [13] and cultured in RPMI 1640 containing 10% FBS, penicillin, streptomycin, GlutaMAX (all purchased from Thermo Fisher Scientific, Shanghai, China).

### 2.2. Cloning of Porcine DNMT3B Isoforms and Sequence Analysis

Total RNA was extracted from PAM (at least $1.0 \times 10^6$ cells) by using Trizol method [14] (Takara Biotechnology, Dalian, China), and cDNA was prepared by using Super ScriptII Reverse Transcriptase (Thermo Fisher Scientific, Shanghai, China). Three pairs of primers (Table S1) used to amplify full-length porcine DNTM3B were designed based on the porcine DNMT3B1 sequence deposited in GenBank (XM_013985274.2). All PCR products were cloned using pMD19-T Vector Cloning Kit (Takara Biotechnology, Dalian, China), then positive clones were selected and sequenced by using M13 (Bacteriophage M13) forward and reverse primers. The sequences of DNMT3B2 and DNMT3B3 were obtained and deposited in NCBI GenBank with accession numbers MN873575 and MN207312, respectively. Subsequently, porcine DNMT3B2 or DNMT3B3 was inserted into the p3 × Flag-CMV-14 vector (Sigma, St. Louis, MO, USA) by homologous recombination with the ClonExpress MultiS One Step Cloning Kit (Vazyme Biotech, Nanjing, China) (Table S1), named Flag-DNMT3B2 or DNMT3B3, respectively.

Furthermore, the expression of exon 10 was analyzed by RT-PCR (reverse transcription-PCR) with an exon 9 forward primer and an exon 11 reverse primer (Table S1). Moreover, PCR was performed to detect the presence of exon 10 in genomic DNA of PAM with an intron 9 forward primer and an intron 10 reverse primer (Table S1). Furthermore, semi-quantitative RT-PCR was performed with an exon 20 forward primer and an exon 23 reverse prime (Table S1) to analyze the expression profile of DNMT3B2 and DNMT3B3 in PAM. All PCR products were sequenced using gene-specific primers. All of the images for agarose gel electrophoresis were captured by image lab version 5.1 (Bio-Rad Laboratories, Hercules, CA, USA).

The amino acid sequence alignment of human (NP_008823.1), mouse (NP_001003961.2), and three porcine DNMT3B isoforms was performed using the Clustal V method and edited using Genedoc. The phylogenetic tree was constructed using the available DNMT3B proteins by the neighbor-joining method with 1000 bootstrap replicates in MEGA version 6.06 [15,16]

### 2.3. Generation of Polyclonal Antibodies against Porcine DNMT3B

Rabbit experiments were approved by the Institutional Animal Care and Use Committee of Shanghai Veterinary Research Institute (IACUC No: Shvri-po-201606 0501) and were performed in compliance with the Guidelines on the Humane Treatment of Laboratory Animals (Ministry of Science and Technology of the People's Republic of China, Policy No.2006 398). Rabbit against porcine DNMT3B antibodies were generated according to a previous publication [17]. Briefly, a peptide of 15 amino acids (SYTQDLTGDGDGEGE) residues of the porcine DNMT3B was synthesized. Rabbits were immunized five times with the peptide in combination with complete or incomplete Freund's adjuvants every 14 days.After 7 days of the fifth immunization, rabbits were euthanized and serum was generated. The specificity of the generated DNMT3B antibodies was tested by Western blot.

### 2.4. Western Blot

The protein samples were prepared from the cell pellets as previously described [18,19]. Then the samples were separated on SDS-PAGE gel and transferred to NC (nitrocellulose, NC) membrane. After blocking in 5% nonfat milk, primary antibodies were added and incubated with membrane overnight at 4 °C (anti-Flag (1:5000, M2, Sigma), anti-DNMT3B (1:1000), anti-actin (1:10,000, clone C4, Sigma)). Secondary antibody was incubated for 1 h at room temperature (goat anti-mouse HRP (1:5000, Abcam), goat anti-rabbit (1:10,000, Abcam)). The images of Western blot were captured by image lab version 5.1 (Bio-Rad Laboratories, Hercules, California, USA).

### 2.5. Quantitative Real-Time Reverse Transcription-PCR (qRT-PCR)

cDNA was prepared using PrimeScript RT Reagent Kit including gDNA Eraser (Takara, Dalian, China). Then RT-PCR was conducted using a SYBR Premix Ex Taq kit (Takara, Dalian, China). Specific primers are shown in Table S1. The calculation was performed as described in our previous study [20].

### 2.6. Bisulfite Sequencing PCR (BSP)

Genomic DNA was extracted from PAM treated with LPS (1 μg/mL) and vehicle for 6 h. Then genomic DNA (0.8 μg) was subjected to bisulfite treatment using EZ DNA Methylation-Gold Kit (ZymoResearch, Los Angeles, CA, USA). Primers for Bisulfite sequencing PCR (BSP) (Table S1) were designed based on porcine *TNF-α* gene promoter sequence using online software (http://www.urogene.org/cgi-bin/methprimer/methprimer.cgi) [21]. Subsequently, BSP [22] was performed using EpiMark Hot Start Taq DNA Polymerase (New England Biolabs, Ipswich, MA, USA) following the manufacturer's protocol. Briefly, using BSP primers amplified the region of TNF-α promoter, running an agarose gel to recover the PCR products. PCR products were cloned into the pMD19-T vector (Takara, Dalian, China). More than 10 positive clones were randomly selected for DNA sequencing [23,24]. The sequencing data and non-CpG-C-T conversion rates were analyzed using online QUMA software (http://quma.cdb.riken.jp/top/index.html) [25]. The total percentage of methylated CpG was calculated in each group including vehicle-treated, LPS-treated, vector-transfected, and DNMT3B2-transfected groups. Additionally, the difference of methylation level between certain groups was analyzed using Fisher's exact test of the online QUMA software.

### 2.7. Lentivirus Production

HEK293T cells were cultured in Dulbecco's modified Eagle's medium (DMEM) containing 10% FBS, penicillin, streptomycin (Thermo Fisher Scientific, Shanghai, China). The pLenO-DCE-DNMT3B2 or pLenO-DCE-Vector (Invabio, Shanghai, China) was co-transfected with pRsv-REV, pMDlg-pRRE, pMD2G (Addgene) into HEK293T cells using Lipofectamine 2000 reagent (Invitrogen, Carlsbad, CA, USA). The supernatants were collected at 72 h post-transfection and concentrated through ultra-centrifugation (25,000 rpm, 4 °C, 2 h, L7 Ultracentrifuge, Beckman, Duarete, CA, USA) after filtering through a 0.45 μm syringe filter [26,27].

### 2.8. Statistical Analysis

All data shown are arithmetic means ± standard deviations. Statistical significance was assessed using unpaired Student's *t*-test by GraphPad Prism software version 5.01 (GraphPad Software, Inc., La Jolla, CA, USA). 'n' refers to the sample size.

## 3. Results

### 3.1. Identification of DNMT3B2 and DNMT3B3 as the Detectable Isoforms in Porcine Alveolar Macrophages (PAM)

According to the predicted porcine *DNMT3B1* gene sequence (GenBank accession number: XM_013985274.2), we designed the primers to amplify the DNMT3B ORF in PAM cDNA. Interestingly, the full-length sequencing results showed that only DNMT3B2 (GenBank accession number, MN873575) and DNMT3B3 (GenBank accession, MN207312) were identified in PAM (Figure 1A,B). Given that alternative splicing of DNMT3B exon 10 [5,7] distinguished DNMT3B1 (exon 10-included isoforms) with DNMT3B2 and DNMT3B3 (the exon 10-excluded isoforms), we further investigated the expression of DNMT3B exon 10 in PAM. In comparison to the expected fragment (about 160 bp) containing exon 10, a short fragment (about 100 bp) was obtained (Figure 1C). Moreover, the sequencing analysis confirmed that DNMT3B exon 10 was absent in the PCR product. We also investigated the presence of exon 10 in *DNMT3B* gene in PAM by PCR. Successfully, we obtained the fragment containing exon 10 (Figure 1D) which was further confirmed by the sequencing analysis. Taken together, these data reveal that expression of DNMT3B exon 10 is lost in PAM. Consistently, DNMT3B2 and DNMT3B3, the exon 10-excluded isoforms, are detectable in PAM.

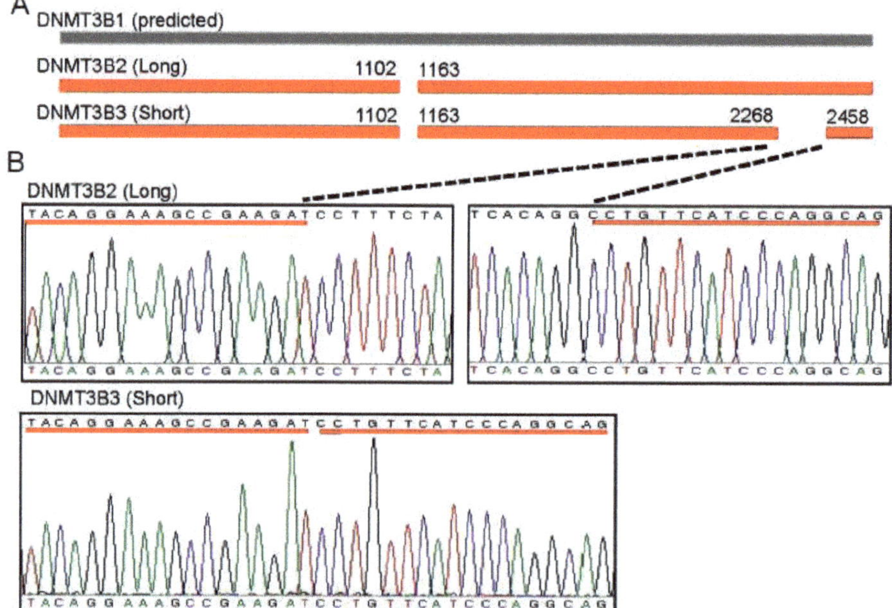


**Figure 1.** *Cont.*

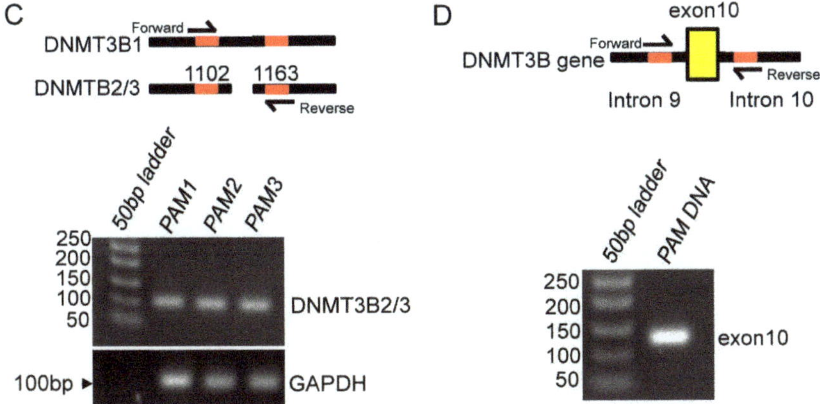


**Figure 1.** Identification of two splice variants of DNA methyltransferase 3B (DNMT3B) in porcine alveolar macrophages (PAM). (**A**) Schematic representation of the predicted DNMT3B1 and two splice variants identified in PAM. (**B**) Sequencing chromatogram showing the deleted region in the short splice variant named DNMT3B3 in comparison with the long splice variant named DNMT3B2. (**C**) Expression analysis of exon 10 of DNTM3B and GAPDH (about 90 bp) in PAM cDNA. Schematic representation of the exon 10-including fragment (about 160 bp) and the exon 10-excluded fragment (about 100 bp) from DNMT3B1 and DNMT3B2/3, respectively. Note that only a short fragment was obtained, indicating that expression of exon 10 of DNMT3B was absent in PAM. (**D**) As shown in the schematic, PCR was performed with an intron 9-forward primer and an intron 10-reverse primer to identify the presence of exon 10 (the positive fragment with 121 bp) in the genomic DNA from PAM. Note that a unique fragment was obtained, which was confirmed by sequencing analysis to show the presence of exon 10 in the *DNMT3B* gene of PAM.

### *3.2. Identification of DNMT3B2 as the Predominant Isoform in Porcine Alveolar Macrophages*

In comparison to DNMT3B2, we observed a 189-bp deletion in DNTM3B3 (Figure 1A,B), which is attributed to lack expression of exon 21 and exon 22 according to the previous reports [5,8]. Based on this expression pattern between DNMT3B2 and DNMT3B3, we set up the RT-PCR with an exon 20 forward primer and an exon 23 reverse prime to analyze the expression profile of DNMT3B2 and DNMT3B3 in PAM. As expected, we totally obtained two fragments by analysis of the RNA samples extracted from PAM: the long fragment (268 bp) is representative of expression level of DNMT3B2; the short fragment (79 bp) is representative of expression level of DNMT3B3 (Figure 2A). Notable, the expression abundance of DNMT3B2 looked greater than that of DNMT3B3 according to the results by agarose gel electrophoresis. The density calculation further confirmed that expression of DNMT3B2 was much higher than that of DNMT3B3 in PAM (Figure 2B). Furthermore, in order to determine the protein level of DNMT3B2 and DNMT3B3, we used an antigen peptide of DNMT3B to generate polyclonal antibodies against DNMT3B. Western blot showed that the polyclonal antibodies were able to specifically detect expression of porcine DNMT3B2 and DNMT3B3 (Figure 2C). Then, we used the polyclonal antibodies to detect the expression profile of DNMT3B in PAM. Our result showed that only DNMT3B2 was detected in PAM (Figure 2D). Collectively, our data reveal that DNMT3B2 is the predominant isoform in PAM.

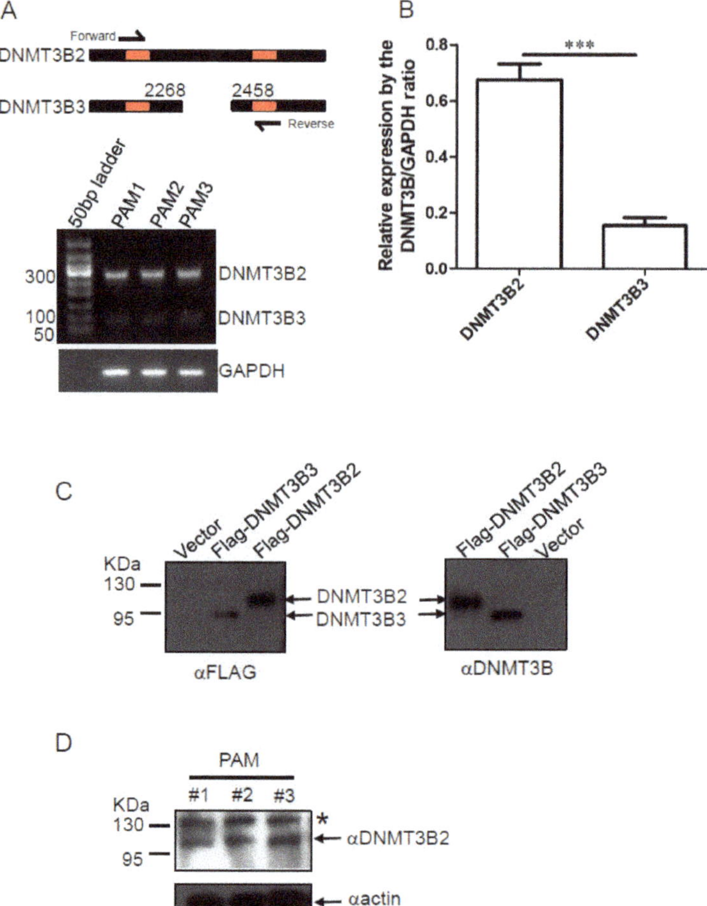


**Figure 2.** (**A**) Total RNA was extracted from PAM and analyzed by RT-PCR with exon 20 forward primer and an exon 23 reverse primer for the expression profile of DNMT3B2 and DNMT3B3 in PAM. Note that two fragments were obtained: one is more than 250 bp, representative expression of DNMT3B2; another is less than 100 bp, representative of DNMT3B3. (**B**) Furthermore, the bar graph represents relative expression of DNMT3B2 or DNMT3B3 by calculating the DNMT3B2/GAPDH ratio. The data shown are mean (SD) ($n$ = 3), ***, $p < 0.001$. (**C**) HEK 293T cells were transiently transfected with recombinant plasmids expressing Flag-pDNMT3B2 and Flag-pDNMT3B3, respectively. Cell lysates were analyzed by Western blot with anti-*p*DNMT3B or anti-Flag antibodies. (**D**) The expression profile of DNMT3B in PAM was analyzed by Western blot. Note that only DNMT3B2 was detected in PAM. * is representative of nonspecific reaction.

### 3.3. *Sequence Analysis of Porcine DNMT3B2 and DNMT3B3*

Having demonstrated that porcine DNMT3B2 and DNMT3B3 share the same expression pattern with that in humans and mice, we next performed sequence analysis to determine the sequence similarity and evolutional relationship of porcine DNMT3B with the other species. Multiple sequence alignment illustrated that the amino acid sequence and function domain of DNMT3B were conserved among human, mouse, and pig (Figure 3A). Notable, both PWWP domain (named as the well-conserved residues, Pro-Trp-Trp-Pro) and PHD domain (the plant homeodomain) in porcine DNMT3B are more than 94% identical to that of human and mouse (data not shown). Moreover, two deletion regions in the porcine DNMT3B3 were founded in comparison to DNMT3B1 of human, mouse and pig, which were attributed to alternative splicing of exon 10, exon 21, and exon 22. Furthermore, phylogenetical analysis of 18 protein sequences of DNMT3B classified them into three branches: mammals, birds, and fishes (Figure 3B), indicating that porcine DNMT3B is clustered with the other mammals.

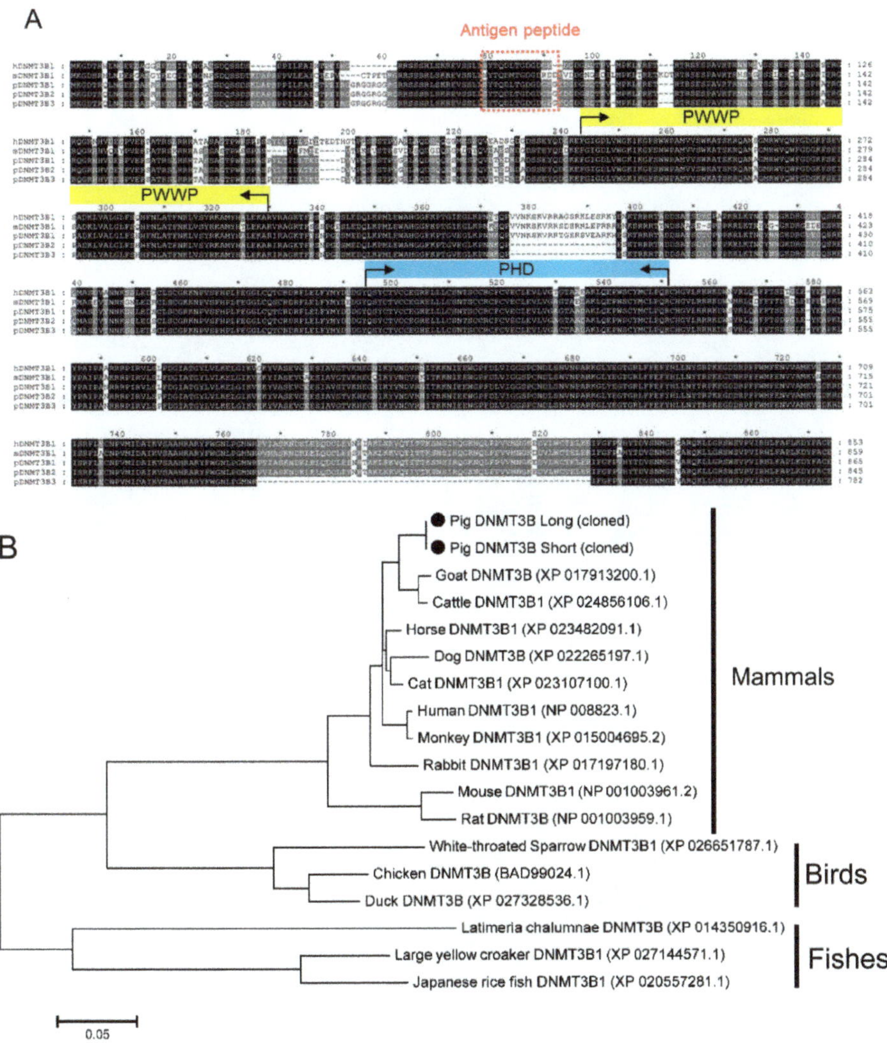


**Figure 3.** (**A**) Comparative sequence analysis of DNMT3B protein from different hosts. The black and gray shading highlight identity in all of the selected sequences and similarity, respectively. The PWWP (yellow line) and PHD (blue line) domains are labeled in reference to the human DNMT3B1 (NP_008823.1), respectively. Moreover, the antigen peptide is shown in the red box. (**B**) Phylogeny analysis of DNMT3B protein from different hosts. The scale bar indicates the genetic distance.

### *3.4. Downregulation of DNMT3B Associates with the Demethylation of TNF-α Gene Promoter in the LPS-Activated PAM*

Furthermore, these results by sequence analysis motivated us to determine the biological function of porcine DNMT3B. Given the reported data that expression of DNMT3B is downregulated in LPS-activated mouse macrophages, which is associated with the global demethylation of DNA [28], thus we chose the LPS-activated PAM as our model for the functional analysis of porcine DNMT3B. Since the alteration of DNMT3B in the LPS-stimulated porcine alveolar macrophages remains unknown, we first investigated the expression profile of DNMT3B in the LPS-activated PAM. In comparison to the untreated PAM, the mRNA level and protein level of DNMT3B was significantly downregulated in LPS-stimulation PAM. Besides, the expression profile of DNMT1 and DNMT3A was also determined in the untreated and LPS-activated PAM (Figure 4A,D). Interestingly, expression of DNMT1 and DNMT3A were not affected by LPS stimulation compared with those in the unstimulated PAM (Figure 4B,C). Thus, these data demonstrated that expression of DNMT3B is reduced in LPS-activated PAM, which might involve in the demethylation of DNA during LPS stimulation.

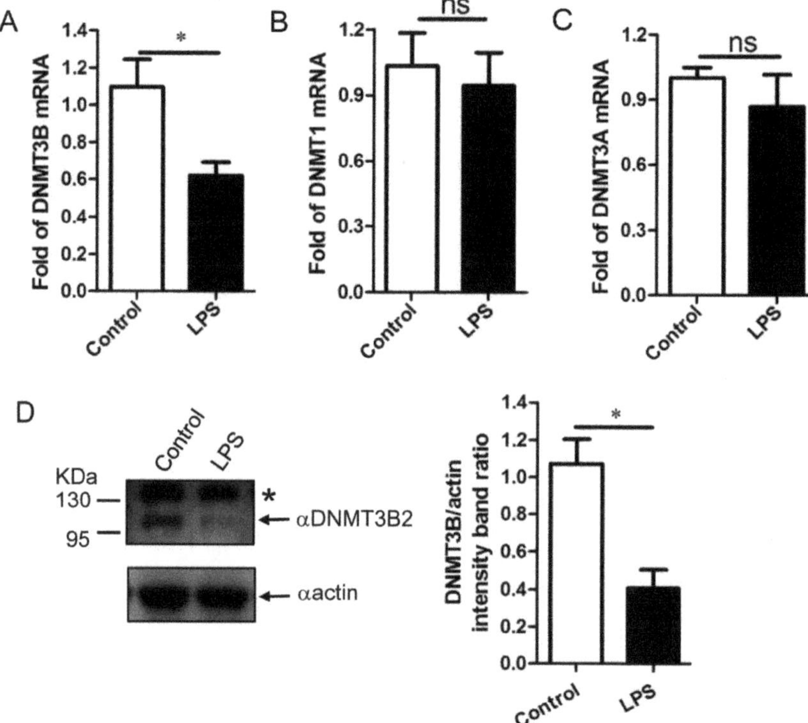


**Figure 4.** Total RNA was extracted from the treated and untreated PAM by LPS and expression of DNMT3B, DNMT1, and DNMT3A was analyzed by qRT-PCR analysis. The relative expression of DNMT3B (**A**), DNMT1 (**B**), and DNMT3A (**C**) was calculated as per material and method. $n = 3$, the data shown are mean (SD), ns is representative of no significant difference between control and LPS stimulated groups, *, $p < 0.05$. (**D**) PAM with or without LPS treatment was harvested and analyzed by Western blot with anti-pDNMT3B antibody. Note that LPS stimulation decreased expression of DNMT3B2 in comparison to control group. The representative bar graph was calculated by DNMT3B2/actin ration. The data shown are mean (SD) ($n = 3$), *, $p < 0.05$.

Given that the methylation status of *TNF-α* gene promoter plays a role in modulation of TNF-α expression [29–31], we next used the model of LPS-activated PAM to determine whether reduction of DNMT3B contributes to the demethylation of the *TNF-α* gene promoter. We performed Bisulfite sequencing PCR to examine the methylation profile of the *TNF-α* gene promoter region (−397 to −151) in the untreated and LPS-activated PAM, respectively. In comparison to the untreated PAM, the methylation level of *TNF-α* gene promoters in the LPS-activated PAM was significantly decreased (Figure 5A), which is accompanied with the robust induction of TNF-α in the stimulated PAM (Figure 5B). Thus, our data reveal that the demethylation of *TNF-α* gene promoter is associated with induction of TNF-α in the LPS-activated PAM.

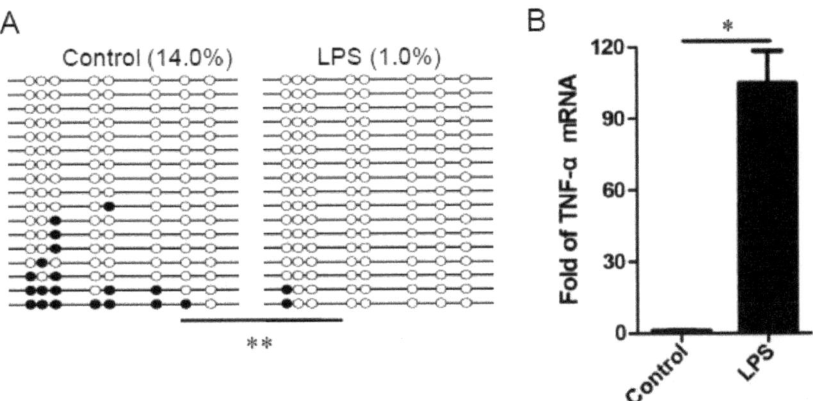


**Figure 5.** (**A**) The DNA methylation level of *TNF-α* promoter region was analyzed by Bisulfite sequencing PCR (BSP). Each line represents one individual sequenced clone. One circle represents one CpG site: the open circle shows unmethylated CpG; the black circle shows methylated CpG. Note that LPS stimulation significantly reduced the DNA methylation level of *TNF-α* promoter region in comparison to control groups. The difference between LPS group ($n$ = 3) and control group ($n$ = 3) was calculated by Fisher's exact test, **, $p < 0.01$. (**B**) The relative expression of TNF-α in PAM with or without LPS treatment was analyzed by qRT-PCR. Note that LPS stimulation promoted TNF-α expression in comparison to control group. $n = 3$, the data shown are mean (SD), *, $p < 0.05$.

### *3.5. DNMT3B2-Mediated Methylation of TNF-α Gene Promoter Restricts Induction of TNF-α in the LPS-Stimulated PAM*

Given that reduction of DNMT3B is associated with the demethylation of *TNF-α* gene promoter and that DNMT3B2 is the predominant isoform in PAM, therefore, our final goal was to determine the biological effect of DNMT3B2 on methylation of *TNF-α* gene promoter as well as expression of TNF-α in PAM. In order to do so, PAM were treated with lentivirus expressing porcine DNTMT3B2 and lentivirus vector (negative control, NC), respectively. After 24 h, the cells were stimulated with LPS and then harvested for the certain analysis. As shown in Figure 6A, in comparison to NC group, PAM treated with lentivirus expressing porcine DNMT3B2 showed a high level of DNMT3B even under the LPS-stimulated conditions. In contrast, the induced TNF-α was restricted by the increased DNMT3B2 in comparison to NC-treated PAM (Figure 6B). Furthermore, the methylation level of *TNF-α* gene promoter (−397 to −151) was analyzed in these two groups. Notable, PAM treated with lentivirus expressing porcine DNMT3B2 showed the higher methylation level of the *TNF-α* gene promoter than that in NC-treated PAM (Figure 6C). Thus, these results reveal that DNMT3B2-mediated methylation of *TNF-α* gene promoter modulates expression of TNF-α in the LPS-stimulated PAM.

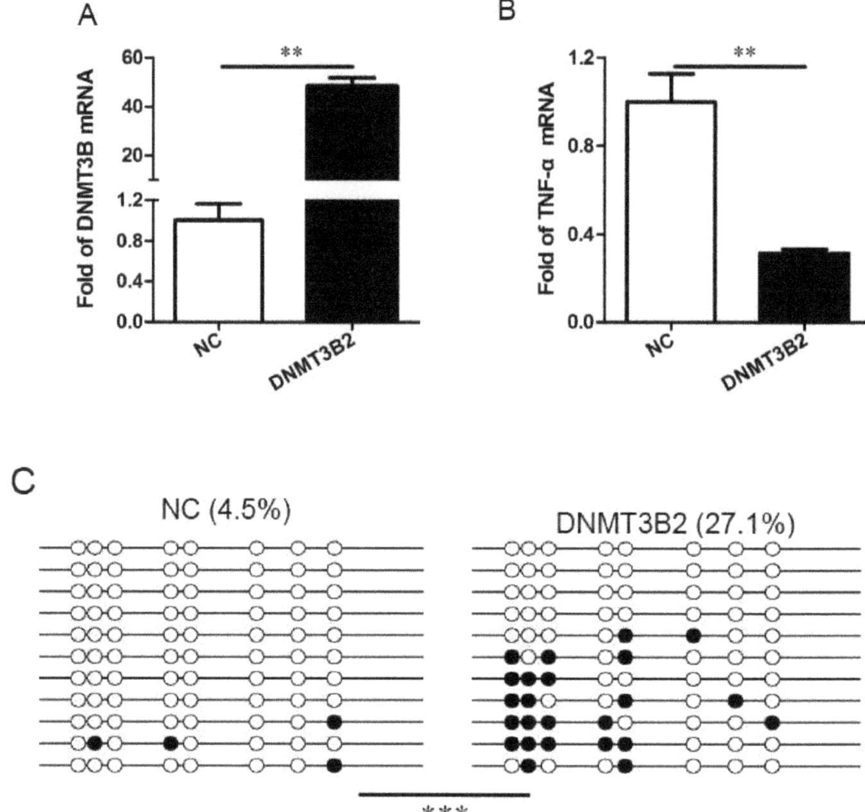


**Figure 6.** Effect of overexpression of DNMT3B2 by lentivirus on TNF-α transcription and methylation profile of *TNF-α* promoter. PAM were infected with lentivirus overexpressing DNMT3B2 or containing control vector. After 48 h of infection, cells were stimulated with LPS (1 μg/mL) for 6 h. Subsequently, cells were harvested, mRNA was extracted and analyzed for expression of DNMT3B2 (**A**) and TNF-α (**B**) by qRT-PCR, respectively. Data are shown as mean (SD) ($n = 3$). *, $p < 0.05$; **, $p < 0.01$. In the meanwhile, cells were treated as shown above, and the total DNA was collected and analyzed for the DNA methylation level of the *TNF-α* promoter region by BSP (**C**). Each line represents one individual sequenced clone. One circle represents one CpG site: the open circle shows unmethylated CpG; the black circle shows methylated CpG. Note that overexpression of DNMT3B2 significantly increased the DNA methylation level of the *TNF-α* promoter region in comparison to control group. The difference between DNMT3B2 group ($n = 3$) and control group ($n = 3$) was calculated by Fisher's exact test, ***, $p < 0.001$.

## 4. Discussion

This study provides novel information regarding the expression pattern of DNMT3B in porcine alveolar macrophages and its effect on TNF-α expression in the LPS-activated PAM. Although there are more than 30 isoforms of DNMT3B identified in humans and mice, yet the expression pattern of DNMT3B in pigs remains to be determined. According to the predicted porcine DNMT3B1 gene sequence (GenBank accession number: XM_013985274.2), we sought to clone the different isoforms of DNMT3B in porcine alveolar macrophages (PAM). In this study, we demonstrate that DNMT3B2 and DNMT3B3 are the detectable isoforms in PAM. In contrast, the canonical isoform, DNMT3B1 and the other splice variants were not founded in the analysis. Since expression of DNMT3B exon 10 was not detected in PAM, alternative splicing of DNMT3B exon 10 [5,7] is important to result in the exon

10-included isoforms (DNMT3B1) and the exon 10-excluded isoforms (DNMT3B2 and DNMT3B3), respectively. Furthermore, we identified DNTM3B2 as the predominant isoform in PAM.

In the model of LPS-activated PAM, we show that in comparison to the unstimulated PAM, (1) expression of DNTM3B is reduced; (2) the methylation level of *TNF-α* gene promoter is decreased. We further establish that DNMT3B2-mediated methylation of *TNF-α* gene promoter restricts induction of TNF-α in the LPS-stimulated PAM. Collectively, our data identify DNMT3B2 as the predominant isoform to have a role in modulation of expression of TNF-α via DNA methylation in the LPS-stimulated PAM.

It has been reported that more than 30 DNMT3B isoforms are identified in humans and mice [5,6]. However, the expression patterns of porcine DNMT3B remains unknown. Here, we demonstrate that DNMT3B2 and DNMT3B3 are detectable in PAM; in comparison, DNMT3B2 is the predominant isoform. Furthermore, we demonstrate that the expression pattern of porcine DNMT3B2 and DNMT3B3 is same as what is shown in humans and mice [5,7,8], i.e., DNMT3B2 is lacking exon10; DNMT3B3 is lacking exon 10, exon 21, and exon 22. Thus, these results provide the evidence that the expression pattern of DNMT3B is conserved in different species [32].

Interestingly, we could not detect expression of DNMT3B1 in PAM despite the presence of exon 10 in the genome. In fact, many spliced variants of DNMT3B are expressed in a tissue, cell, and/or developmental stage-specific manner [6,9]. For example, DNMT3B1 is highly expressed in human ES cells; in contrast, its expression is decreased in the somatic cells which is accompanied with the predominant expression of DNMT3B3 [33]. Therefore, it is possible that porcine DNMT3B1 might be highly expressed in certain cells but not PAM during the development of pigs.

DNMT3B as one member of DNMTs catalytically regulates de novo DNA methylation. However, not all of DNMT3B isoforms have active catalytic domains. Notable, DNM3B1 and DNMT3B2 have the active catalytic domains; in comparison, DNMT3B3 is considered as an inactive catalytic isoform with some deletions in the catalytic domain. Although the role of DNMT3B3 in catalytic activity is controversial, the related studies have shown that DNMT3B3 could regulate DNA methylation by acting as the accessory protein. In this study, we demonstrated porcine DNTM3B2 regulates methylation of the *TNF-α* gene promoter, which has a role in modulation of expression of TNF-α. Of course, it is not excluded that porcine DNMT3B3 still functions as a positive regulator of DNA methylation. Thus, future studies are needed to answer this question.

## 5. Conclusions

In summary, this study for the first time identified DNMT3B2 as the predominant isoform in PAM. We demonstrated an important role of DNMT3B2-mediated DNA methylation in expression of TNF-α in the LPS-stimulated PAM. In this context, this study could provide evidence that DNMT3B2-mediated DNA methylation is the important mechanism responsible for understanding the inflammatory response in the LPS-stimulated PAM, even possible during some bacterial infections in the lungs of pigs.

**Supplementary Materials:** The following are available online at http://www.mdpi.com/2073-4425/11/9/1065/s1, Table S1: Sequence of primers.

**Author Contributions:** Y.Z.: Conceptualization, Methodology, Writing—original daft, Visualization. H.L., X.X., and Y.L.: Conceptualization, Visualization, Methodology. M.S., Z.L., K.L., J.W., D.S., and B.L.: Methodology, Visualization. Z.M.: Supervision. Y.Q.: Conceptualization, Supervision, Writing—review and editing. All authors have read and agreed to the published version of the manuscript.

**Funding:** This study was in part supported by the national key R&D program of China (2018YFD0500101), the National Natural Science Foundation of China (31,972,693), Central Public-interest Scientific Institution Basal Research Fund (No. 2020JB04), and Elite program of CAAS (to Y.Q.).

**Conflicts of Interest:** The authors declare no conflict of interest.

## References


1. Okano, M.; Bell, D.W.; Haber, D.A.; Li, E. DNA methyltransferases Dnmt3a and Dnmt3b are essential for de novo methylation and mammalian development. *Cell* **1999**, *99*, 247–257. [CrossRef]
2. Leonhardt, H.; Page, A.W.; Weier, H.U.; Bestor, T.H. A targeting sequence directs DNA methyltransferase to sites of DNA replication in mammalian nuclei. *Cell* **1992**, *71*, 865–873. [CrossRef]
3. Veland, N.; Lu, Y.; Hardikar, S.; Gaddis, S.; Zeng, Y.; Liu, B.; Estecio, M.R.; Takata, Y.; Lin, K.; Tomida, M.W.; et al. DNMT3L facilitates DNA methylation partly by maintaining DNMT3A stability in mouse embryonic stem cells. *Nucleic Acids Res.* **2019**, *47*, 152–167. [CrossRef]
4. Manzo, M.; Wirz, J.; Ambrosi, C.; Villaseñor, R.; Roschitzki, B.; Baubec, T. Isoform-specific localization of DNMT3A regulates DNA methylation fidelity at bivalent CpG islands. *EMBO J.* **2017**, *36*, 3421–3434. [CrossRef]
5. Ostler, K.R.; Davis, E.M.; Payne, S.L.; Gosalia, B.B.; Expósito-Céspedes, J.; Le Beau, M.M.; Godley, L.A. Cancer cells express aberrant DNMT3B transcripts encoding truncated proteins. *Oncogene* **2007**, *26*, 5553–5563. [CrossRef]
6. Gopalakrishnan, S.; Van Emburgh, B.O.; Shan, J.; Su, Z.; Fields, C.R.; Vieweg, J.; Hamazaki, T.; Schwartz, P.H.; Terada, N.; Robertson, K.D. A novel DNMT3B splice variant expressed in tumor and pluripotent cells modulates genomic DNA methylation patterns and displays altered DNA binding. *Mol. Cancer Res. MCR* **2009**, *7*, 1622–1634. [CrossRef]
7. Gopalakrishna-Pillai, S.; Iverson, L.E. A DNMT3B alternatively spliced exon and encoded peptide are novel biomarkers of human pluripotent stem cells. *PLoS ONE* **2011**, *6*, e20663. [CrossRef]
8. Weisenberger, D.J.; Velicescu, M.; Cheng, J.C.; Gonzales, F.A.; Liang, G.; Jones, P.A. Role of the DNA methyltransferase variant DNMT3b3 in DNA methylation. *Mol. Cancer Res. MCR* **2004**, *2*, 62–72.
9. Robertson, K.D.; Uzvolgyi, E.; Liang, G.; Talmadge, C.; Sumegi, J.; Gonzales, F.A.; Jones, P.A. The human DNA methyltransferases (DNMTs) 1, 3a and 3b: Coordinate mRNA expression in normal tissues and overexpression in tumors. *Nucleic Acids Res.* **1999**, *27*, 2291–2298. [CrossRef]
10. Qiao, S.; Feng, L.; Bao, D.; Guo, J.; Wan, B.; Xiao, Z.; Yang, S.; Zhang, G. Porcine reproductive and respiratory syndrome virus and bacterial endotoxin act in synergy to amplify the inflammatory response of infected macrophages. *Vet. Microbiol.* **2011**, *149*, 213–220. [CrossRef]
11. Li, J.; Wang, S.; Li, C.; Wang, C.; Liu, Y.; Wang, G.; He, X.; Hu, L.; Liu, Y.; Cui, M.; et al. Secondary Haemophilus parasuis infection enhances highly pathogenic porcine reproductive and respiratory syndrome virus (HP-PRRSV) infection-mediated inflammatory responses. *Vet. Microbiol.* **2017**, *204*, 35–42. [CrossRef]
12. Yang, Q.; Pröll, M.J.; Salilew-Wondim, D.; Zhang, R.; Tesfaye, D.; Fan, H.; Cinar, M.U.; Große-Brinkhaus, C.; Tholen, E.; Islam, M.A.; et al. LPS-induced expression of CD14 in the TRIF pathway is epigenetically regulated by sulforaphane in porcine pulmonary alveolar macrophages. *Innate Immun.* **2016**, *22*, 682–695. [CrossRef]
13. Lu, Y.; Zhang, Y.; Xiang, X.; Sharma, M.; Liu, K.; Wei, J.; Shao, D.; Li, B.; Tong, G.; Olszewski, M.A.; et al. Notch signaling contributes to the expression of inflammatory cytokines induced by highly pathogenic porcine reproductive and respiratory syndrome virus (HP-PRRSV) infection in porcine alveolar macrophages. *Dev. Comp. Immunol.* **2020**, *108*, 103690. [CrossRef]
14. Brown, R.A.M.; Epis, M.R.; Horsham, J.L.; Kabir, T.D.; Richardson, K.L.; Leedman, P.J. Total RNA extraction from tissues for microRNA and target gene expression analysis: Not all kits are created equal. *BMC Biotechnol.* **2018**, *18*, 16. [CrossRef]
15. Saitou, N.; Nei, M. The neighbor-joining method: A new method for reconstructing phylogenetic trees. *Mol. Biol. Evol.* **1987**, *4*, 406–425. [CrossRef]
16. Felsenstein, J. Confidence limits on phylogenies: An approach using the bootstrap. *Evolution* **1985**, *39*, 783–791. [CrossRef]
17. Lateef, S.S.; Gupta, S.; Jayathilaka, L.P.; Krishnanchettiar, S.; Huang, J.S.; Lee, B.S. An improved protocol for coupling synthetic peptides to carrier proteins for antibody production using DMF to solubilize peptides. *J. Biomol. Tech. JBT* **2007**, *18*, 173–176.
18. Qiu, Y.; Shen, Y.; Li, X.; Ding, C.; Ma, Z. Molecular cloning and functional characterization of a novel isoform of chicken myeloid differentiation factor 88 (MyD88). *Dev. Comp. Immunol.* **2008**, *32*, 1522–1530. [CrossRef]

19. Xiang, X.; Zhang, Y.; Li, Q.; Wei, J.; Liu, K.; Shao, D.; Li, B.; Olszewski, M.A.; Ma, Z.; Qiu, Y. Expression profile of porcine scavenger receptor A and its role in bacterial phagocytosis by macrophages. *Dev. Comp. Immunol.* **2020**, *104*, 103534. [CrossRef]
20. Neal, L.M.; Qiu, Y.; Chung, J.; Xing, E.; Cho, W.; Malachowski, A.N.; Sandy-Sloat, A.R.; Osterholzer, J.J.; Maillard, I.; Olszewski, M.A. T Cell-Restricted Notch Signaling Contributes to Pulmonary Th1 and Th2 Immunity during Cryptococcus neoformans Infection. *J. Immunol.* **2017**, *199*, 643–655. [CrossRef]
21. Li, L.C.; Dahiya, R. MethPrimer: Designing primers for methylation PCRs. *Bioinformatics* **2002**, *18*, 1427–1431. [CrossRef]
22. Li, L.C. Designing PCR primer for DNA methylation mapping. *Methods Mol. Biol.* **2007**, *402*, 371–384. [CrossRef]
23. von Känel, T.; Huber, A.R. DNA methylation analysis. *Swiss. Med. Wkly.* **2013**, *143*, w13799. [CrossRef]
24. Liu, B.; Pilarsky, C. Analysis of DNA hypermethylation in pancreatic cancer using methylation-specific PCR and bisulfite sequencing. *Methods Mol. Biol.* **2018**, *1856*, 269–282. [CrossRef]
25. Kumaki, Y.; Oda, M.; Okano, M. QUMA: Quantification tool for methylation analysis. *Nucleic Acids Res.* **2008**, *36*, W170–W175. [CrossRef]
26. Kennedy, A.; Cribbs, A.P. Production and concentration of lentivirus for transduction of primary human T cells. *Methods Mol. Biol.* **2016**, *1448*, 85–93. [CrossRef]
27. Benskey, M.J.; Manfredsson, F.P. Lentivirus production and purification. *Methods Mol. Biol.* **2016**, *1382*, 107–114. [CrossRef]
28. Jain, N.; Shahal, T.; Gabrieli, T.; Gilat, N.; Torchinsky, D.; Michaeli, Y.; Vogel, V.; Ebenstein, Y. Global modulation in DNA epigenetics during pro-inflammatory macrophage activation. *Epigenetics* **2019**, *14*, 1183–1193. [CrossRef]
29. Wang, H.; Feng, H.; Sun, J.; Zhou, Y.; Zhu, G.; Wu, S.; Bao, W. Age-associated changes in DNA methylation and expression of the TNF-α gene in pigs. *Genes Genet. Syst.* **2018**, *93*, 191–198. [CrossRef]
30. Zhang, S.; Barros, S.P.; Moretti, A.J.; Yu, N.; Zhou, J.; Preisser, J.S.; Niculescu, M.D.; Offenbacher, S. Epigenetic regulation of TNFA expression in periodontal disease. *J. Periodontol.* **2013**, *84*, 1606–1616. [CrossRef]
31. Kojima, A.; Kobayashi, T.; Ito, S.; Murasawa, A.; Nakazono, K.; Yoshie, H. Tumor necrosis factor-alpha gene promoter methylation in Japanese adults with chronic periodontitis and rheumatoid arthritis. *J. Periodontal Res.* **2016**, *51*, 350–358. [CrossRef] [PubMed]
32. Okano, M.; Xie, S.; Li, E. Cloning and characterization of a family of novel mammalian DNA (cytosine-5) methyltransferases. *Nat. Genet.* **1998**, *19*, 219–220. [CrossRef] [PubMed]
33. Liao, J.; Karnik, R.; Gu, H.; Ziller, M.J.; Clement, K.; Tsankov, A.M.; Akopian, V.; Gifford, C.A.; Donaghey, J.; Galonska, C.; et al. Targeted disruption of DNMT1, DNMT3A and DNMT3B in human embryonic stem cells. *Nat. Genet.* **2015**, *47*, 469–478. [CrossRef] [PubMed]

*Article*

# Generation of Marker-Free *pbd-2* Knock-in Pigs Using the CRISPR/Cas9 and Cre/loxP Systems

**Jing Huang [1,2], Antian Wang [2,3], Chao Huang [1,2], Yufan Sun [2,3], Bingxiao Song [2,3], Rui Zhou [1,4,*] and Lu Li [2,3,*]**

1 State Key Laboratory of Agricultural Microbiology, College of Veterinary Medicine, Huazhong Agricultural University, Wuhan 430070, China; isaac286@live.com (J.H.); hc394511335@163.com (C.H.)
2 Cooperative Innovation Center for Sustainable Pig Production, Huazhong Agricultural University, Wuhan 430070, China; wangantian@webmail.hzau.edu.cn (A.W.); leoandff@webmail.hzau.edu.cn (Y.S.); songbingxiao@webmail.hzau.edu.cn (B.S.)
3 Key Laboratory of Development of Veterinary Diagnostic Products, Ministry of Agriculture and Rural Affairs of China, Wuhan 430070, China
4 International Research Center for Animal Disease, Ministry of Science and Technology of China, Wuhan 430070, China
* Correspondence: rzhou@mail.hzau.edu.cn (R.Z.); lilu@mail.hzau.edu.cn (L.L.)

Received: 17 July 2020; Accepted: 17 August 2020; Published: 18 August 2020

**Abstract:** Porcine β-defensin 2 (PBD-2), expressed by different tissues of pigs, is a multifunctional cationic peptide with antimicrobial, immunomodulatory and growth-promoting abilities. As the latest generation of genome-editing tool, CRISPR/Cas9 system makes it possible to enhance the expression of PBD-2 in pigs by site-specific knock-in of *pbd-2* gene into the pig genome. In this study, we aimed to generate marker-free *pbd-2* knock-in pigs using the CRISPR/Cas9 and Cre/loxP systems. Two copies of *pbd-2* gene linked by a T2A sequence were inserted into the porcine *Rosa26* locus through CRISPR/Cas9-mediated homology-directed repair. The floxed selectable marker gene *neoR*, used for G418 screening of positive cell clones, was removed by cell-penetrating Cre recombinase with a recombination efficiency of 48.3%. Cloned piglets were produced via somatic cell nuclear transfer and correct insertion of *pbd-2* genes was confirmed by PCR and Southern blot. Immunohistochemistry and immunofluorescence analyses indicated that expression levels of PBD-2 in different tissues of transgenic (TG) piglets were significantly higher than those of their wild-type (WT) littermates. Bactericidal assays demonstrated that there was a significant increase in the antimicrobial properties of the cell culture supernatants of porcine ear fibroblasts from the TG pigs in comparison to those from the WT pigs. Altogether, our study improved the protein expression level of PBD-2 in pigs by site-specific integration of *pbd-2* into the pig genome, which not only provided an effective pig model to study the anti-infection mechanisms of PBD-2 but also a promising genetic material for the breeding of disease-resistant pigs.

**Keywords:** porcine β-defensin 2; CRISPR/Cas9; transgenic pigs; antimicrobial peptide; disease-resistant animals

---

## 1. Introduction

Different breeding techniques have been utilized to improve animal production traits including growth rate, milk yield and disease-resistance. With the rapid development of biotechnologies, transgenic (TG) techniques have been gradually applied in animal breeding. Notably, there are three major gene-editing tools used, named as zinc finger nuclease, transcription activator-like effector nuclease and CRISPR/Cas9 [1–3]. Events of using gene-editing tools to improve quantitative traits and welfare of animals, and to eliminate allergens in livestock products, have been well described. Previous studies showed that knock-out of myostatin gene in cattle, goats, pigs and rabbits greatly increased their muscle mass [4–6]. Hornless cattle were produced through integrating *POLLED* gene into the genome, which prevented the suffering of cattle caused by dehorning [7]. Eggs with low allergenicity were produced by gene disruptions of *OVA* and *OVM* genes in hens [8], while β-lactoglobulin-free goat milk was obtained by removing *BLG* gene in goats [9].

The gene-editing technologies have also been widely used to improve disease-resistance in pigs [10]. For instance, pigs expressing porcine reproductive and respiratory syndrome virus (PRRSV)-specific small interfering RNA (siRNA) displayed enhanced resistance to PRRSV infection [11]. Similarly, pigs resistant to foot-and-mouth disease and classical swine fever (CSF) were obtained by inserting a corresponding siRNA expression cassette into the pig genome, respectively [12,13]. The deletion of CD163 SRCR5 domain conferred resilience to PRRSV in pigs [14], while pigs lacking aminopeptidase N acquired insusceptibility to transmissible gastroenteritis virus infection [15]. In addition, it has been identified that overexpression of porcine β-defensin 2 (PBD-2) and histone deacetylase 6 can protect pigs from *Actinobacillus pleuropneumoniae* and PRRSV infection, respectively [16,17]. Fibroblasts isolated from pigs overexpressing Mx1 were less vulnerable to influenza A virus and CSF virus infection [18], and cells from MxA TG pigs could suppress CSF virus replication [19].

Mammal defensins are the major group of cationic host defense peptides which are classified as three subfamilies, α-, β- and θ-defensins, according to their intramolecular disulfide bond pattern [20]. In terms of defensins in pigs, β-defensin is the only subgroup that has been found, with 27 functional β-defensins identified [21,22]. PBD-2 was first discovered by sequence similarity analysis with the well characterized porcine β-defensin 1 [23] and was proved to exist in different organs in pigs [16]. It has been shown that PBD-2 possesses antimicrobial abilities against a broad range of bacteria, both Gram-positive and Gram-negative [24]. The antibacterial mechanism of PBD-2 was elucidated in *Escherichia coli*, being disruption of the membrane integrity and affecting the DNA transcription and translation when PBD-2 entered into the cytoplasm [25]. Previous studies described that PBD-2 could hamper PRRSV replication in vitro [26] and suppress proliferation of pseudorabies virus both in vitro and in TG mice [27]. Besides, PBD-2 has been identified to alleviated inflammation by binding toll-like receptor 4 and inhibiting the subsequent NF-κB activation [28]. Additionally, PBD-2 has been used as a feed additive to improve growth performance, and to prevent post-weaning diarrhea in piglets [29]. Taken together, the *pbd-2* gene could be a promising candidate to generate disease-resistant TG animals.

Although pigs overexpressing PBD-2 have been produced by random gene integration in our previous study, a neomycin-resistance (*neoR*) gene has also been introduced into the pig genome [16]. Given that the CRISPR/Cas9 system is recognized as one of the most efficient tools for precise gene modifications in mammals [30,31], this study aimed to increase the expression level of PBD-2 in pigs by marker-free knock-in of *pbd-2* gene into the porcine *Rosa26* (*pRosa26*) locus, a safe harbor for ubiquitous expression of exogenous genes [32]. The resulting TG pigs would greatly improve resilience of pigs to infectious diseases, providing a potent genetic material for the breeding of disease-resistant pigs.

## 2. Materials and Methods

### 2.1. Cells, Bacterial Strains and Animals

Porcine fetal fibroblasts (PFFs) and porcine ear fibroblasts (PEFs) were grown in Dulbecco's Modified Eagle Medium (DMEM; Thermo Fisher Scientific, Waltham, MA, USA) supplemented with 20% fetal bovine serum (FBS; Thermo Fisher Scientific), while PK-15 cells (ATCC Number: CCL-33) were maintained in DMEM supplemented with 10% FBS. *Streptococcus suis* strain SC19 was cultured in tryptic soy broth (TSB; BD, Franklin Lakes, NJ, USA) with 5% newborn calf serum (NBCS; TIANHANG, Huzhou, China) and on tryptic soy agar (TSA; BD) with 5% NBCS. *A. pleuropneumoniae* strain 4074 was grown in TSB with 5% NBCS and 10 μg/mL of nicotinamide adenine dinucleotide (NAD; Sigma-Aldrich, St. Louis, MO, USA) and on TSA with 5% NBCS and NAD. Experiments involving pigs were performed in accordance with the Hubei Regulations for the Administration of Affairs Concerning Experimental

Animals and were approved by the Scientific Ethical Committee for Experimental Animals of Huazhong Agricultural University, Wuhan, China (HZAUSW-2019-010).

*2.2. Plasmids*

An efficient Kozak sequence (5′-GCCACC-3′) was added upstream of the ATG start codon of the *pbd-2* gene in *pcCAG-PBD2* vector constructed in our previous study [16], with the resulting vector named as *pcCAG-nPBD2* (Figure S1A). To evaluate whether the 2A peptide system could be efficient in linking two copies of *pbd-2* gene, *pcCAG-PBD2-T2A-PBD2* was produced, containing two copies of *pbd-2* gene linked by a T2A peptide sequence (Figure S1B). A 5′ homology arm (HA) of approximately 0.9 kb and a 3′ HA of approximately 0.76 kb was cloned into the *Spe*I and *Bst*Z17I site of *pcCAG-PBD2-T2A-PBD2*, respectively. Subsequently, two loxP sites with the same orientation were inserted, flanking the selectable marker gene (SMG) *neoR*. The resulting vector was designated as *pcCAG-R26-PBD2-T2A-PBD2* (Figure 1A). The *pX330-pRosa26* vector (Figure S2) provided by Professor Bo Zuo (Huazhong Agricultural University) was used to express Cas9 and sgRNA targeting the *pRosa26* locus (5′-GCTCCTTCTCGATTATGGGC-3′).

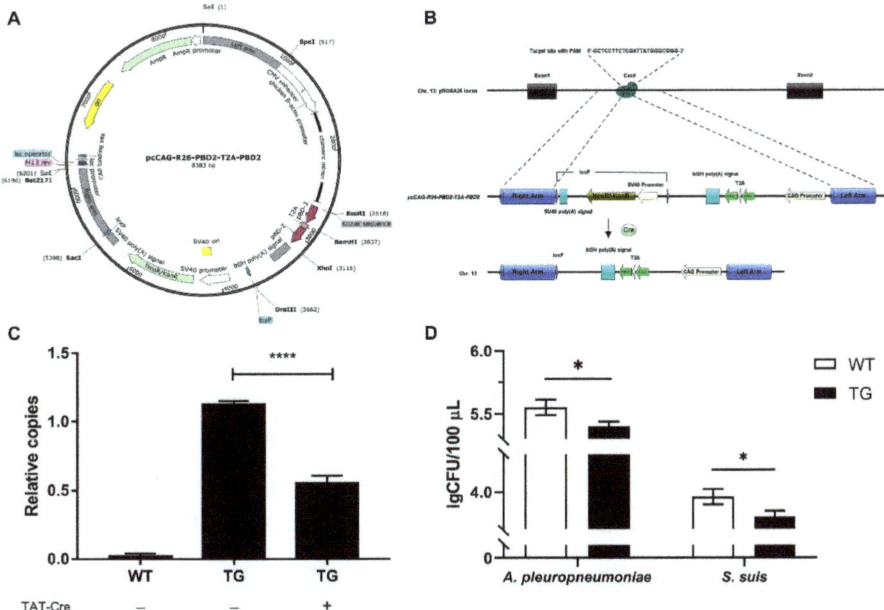


**Figure 1.** Site-specific *pbd-2* knock-in at the porcine *Rosa26* (*pRosa26*) locus. (**A**) Map of the donor plasmid *pcCAG-R26-PBD2-T2A-PBD2* for the site-specific *pbd-2* knock-in. (**B**) Scheme for marker-free targeted *pbd-2* integration in porcine fetal fibroblasts (PFFs) via CRISPR/Cas9-mediated homology-directed repair. The floxed selectable marker gene (SMG) was removed after treatment of Cre recombinase. (**C**) Relative copy number of the SMG *neoR* in wild-type (WT) PFFs, transgenic (TG) PFFs and Cre-recombinase-treated TG PFFs. (**D**) The bactericidal activities of cell culture supernatants of WT PFFs and TG PFFs on *Actinobacillus pleuropneumoniae* and *Streptococcus suis* quantified by bacterial counting. Data are presented as mean ± SD and are plotted from three independent experiments. * $p < 0.05$, **** $p < 0.0001$, unpaired one tailed Student' s *t*-test.

*2.3. Transfection of PK-15 Cells*

At 80% confluency, PK-15 cells were respectively transfected with *pcCAG-nPBD2* and *pcCAG-PBD2-T2A-PBD2* using the Lipofectamine 3000 (Thermo Fisher Scientific) according to the manufacturer's instruction. After 24 h the medium was replaced with fresh DMEM containing 1% FBS. The cell culture supernatant was collected 48 h later for the subsequent bactericidal assay against *A. pleuoropneumoniae*.

## 2.4. Bactericidal Assay

A 10 μL aliquot of *S. suis* (5000 CFU) or *A. pleuropneumoniae* (5000 CFU) was incubated with 90 μL of cell culture supernatants at 37 °C for 1 h, respectively. The mixture was then serially diluted (1:100 and 1:1000 dilution) in DPBS (Thermo Fisher Scientific) and spread onto TBA plates. The plates were left at 37 °C overnight and colonies were counted and analyzed. Bacteria incubated with DPBS was used as a blank control, while those incubated with 60 μg/mL of the synthetic mature PBD-2 (DHYICAKKGGTCNFSPCPLFNRIEGTCYSGKAKCCIR) (ChinaPeptides, Shanghai, China) diluted in DPBS served as a positive control.

## 2.5. Isolation of Genomic DNA

Cells or pig ear tissues were incubated in 500 μL of lysis buffer (10 mM Tris, pH 8.0; 10 mM NaCl; 10 mM EDTA, pH 8.0; 10% SDS; 400 μg/mL Proteinase K) at 56 °C for 8 h. After 200 μL of saturated NaCl was added, tubes were gently shaken for 5 min before centrifugation 13,000× *g* for 10 min. The supernatant was then transferred to a new tube and equal volume of isopropanol (chilled at −20 °C) was added. The mixture was incubated at −20 °C for another 2 h and was subjected to centrifugation 13,000× *g* for 10 min. The supernatant was discarded and the precipitate was washed with 70% ethanol. The tube was then centrifugated 13,000× *g* for another 10 min and the supernatant was removed. After air dry at room temperature (RT) for 10 min, the resulting DNA was resuspended in pre-heated $ddH_2O$ and the concentration was measured using a spectrophotometer (Thermo Fisher Scientific).

## 2.6. Electroporation and Selection of PFFs

PFFs were cultured in a 100-mm dish for two days to reach 80% confluency. After washing with DPBS twice, PFFs were detached from the dish using 0.25% Trypsin-EDTA (Thermo Fisher Scientific) and were centrifugated at 1000× *g* for 5 min. The cell pellet was suspended in 800 μL of Gene Pulser Electroporation Buffer (Bio-Rad, Hercules, CA, USA) and 32 μg of *pcCAG-R26-PBD2-T2A-PBD2* and *pX330-pRosa26* each were added and gently mixed. Cell suspension was added into a 0.2 cm-gap Gene Pulser Electroporation Cuvettes (200 μL) for the subsequent electroporation using the Gene Pulser Xcell system (Bio-Rad) and was left recovery for 24 h. Following selection in 400 μg/mL of G418 for two weeks, single colonies were picked and transferred onto a 48-well plate. As the cells reached confluency, half of them were subjected to genomic DNA isolation which was later used as the PCR template for genotyping. The site-specific knock-in events were identified by PCR using three primer pairs (Table S1) and the positive cell clone was then chosen for the removal of a SMG.

## 2.7. Removal of Selectable Marker

When cells reached about 60% confluency, cells were treated with 3 μM TAT-Cre recombinase (Merck Millipore, Burlington, MA, USA) in DMEM without the presence of antibiotics and FBS. After 3 h of incubation at 37 °C, the medium was removed and the cells were washed twice with PBS. The treated cells were grown in fresh medium and 50% of them were subjected to somatic cell nuclear transfer (SCNT). The remaining cells were used to determine the recombination efficiency by real-time quantitative PCR (RT-qPCR). Briefly, genomic DNA from wild-type (WT) cells, treated and untreated TG cells was extracted and subjected to RT-qPCR using SYBR® Green Realtime PCR Master Mix (TOYOBO, Osaka, Japan). The relative copy number of the SMG *neoR* was analyzed using the $2^{-\Delta\Delta Ct}$ method [33], with *β-actin* as the reference gene. Specific primers for *neoR* are NeoR-F (5′-GCCCCATGGCTGACTAATTTTTTT-3′) and NeoR-R (5′-CGATTGTCTGTTGTGCCCAGTC-3′), while primers for *β-actin* are ACTB-F (5′-GCCTCTCGTCTTGCTTGTTTTAAA-3′) and ACTB-R

(5′-AGCAAGTGAGGGCGTATCCAG-3′). Recombination efficiency = (mean value of treated TG group − mean value of WT group)/(mean value of untreated TG group − mean value of WT group) × 100%. Meanwhile, the cell culture supernatants of WT cells and treated TG cells were collected to measure the bactericidal activity against *A. pleuoropneumoniae* and *S. suis* as described above.

### 2.8. Generation and Genotype Analysis of Cloned Piglets

After removal of the SMG, positive PFFs were subjected to the procedure of SCNT by ViaGen Animal Breeding Resources Development Company (Wuhan, China) in accordance with the protocol described elsewhere [34]. There were eight surrogate sows used and each was transferred with 250 reconstructed embryos.

The resulting piglets were identified by PCR. Genomic DNA was obtained from pig ear tissues of the newborn piglets as described above. Four pairs of primers (Table 1) were used to identify marker-free TG pigs and the PCR products were then sequenced for further confirmation. The Southern blot analysis was performed to confirm a site-specific transgene integration into the pig genome. Briefly, genomic DNA of high-quality was isolated from the PEFs using the QIAamp DNA Mini Kit (QIAGEN, Hilden, Germany) and digested with *Xho*I. The digested DNA was electrophoretically separated on an agarose gel and then transferred onto a nylon membrane. The probe labeled with digoxigenin was hybridized to a single DNA fragment on the membrane, indicating a site-specific insertion of duplicate *pbd-2* genes.

**Table 1.** Information of primer pairs for genotype analysis of cloned pigs.

| Fragment | Primer | Sequence (5′-3′) | Length (bp) |
|---|---|---|---|
| PBD-2 | vTG-F | GCTGGTTGTTGTGCTGTCTCATCA | 547 |
| | vTG-R | CCCTCTAGACTCGAGTCAGGGTCAGC | |
| On-target | OT-F | CTTCCTTTCTCGCCACGTTC | 1235 |
| | OT-R | TCGGTAAATAGCAATCAACTCAG | |
| NeoR | NeoR-F | GCCCCATGGCTGACTAATTTTTTT | 266 |
| | NeoR-R | CGATTGTCTGTTGTGCCCAGTC | |
| β-actin | Control-F | GCCTCTCGTCTTGCTTGTTTTAAA | 169 |
| | Control-R | AGCAAGTGAGGGCGTATCCAG | |

### 2.9. Immunofluorescent and Immunohistochemical Analysis for PBD-2

PEFs were isolated from the ear tissue of cloned pigs and were subjected to the subsequent immunofluorescent analysis. Same amounts of WT and TG PEFs (10,000 cells each) were firstly grown on coverslips for 24 h before being rinsed in PBS, then cells were incubated in chilled acetone (−20 °C) for 10 min. The coverslips were washed three times with ice-cold PBS before being blocked with 1% BSA, 22.52 mg/mL glycine in PBST (PBS+ 0.1% Tween 20) for 30 min. The PEFs were then incubated in self-made PBD-2 mouse monoclonal antibody [16] diluted in 1% BSA in PBST at RT for 1 h. Subsequently, cells were washed five times with PBST for 5 min each and were then incubated with FITC-labeled Goat Anti-Mouse IgG (H+L) (Abclonal, Wuhan, China) diluted at 1:100 in 1% BSA in PBST for 1 h without exposure to light. After that, cells were washed five times with PBST for 5 min each in the dark and then were stained with 4′,6-diamidino-2-phenylindole (Beyotime, Shanghai, China). The results were observed using the EVOS FL Auto Imaging System (Thermo Fisher Scientific).

Besides, a TG founder pig and a WT littermate were both sacrificed to harvest their heart, liver, spleen, lung, kidney, and brain tissues. Tissues were fixed in PBS-buffered 4% formaldehyde followed by paraffin embedding and sectioning. The sections were then subjected to immunohistochemical analysis to detect PBD-2 in different organs. Mouse anti-2A peptide monoclonal antibodies (Novus Biologicals, Littleton, CO, USA) and HRP-conjugated goat anti-mouse IgG were used as primary and secondary antibodies, respectively. PBD-2 was detected after diaminobenzidine staining as a brown coloration.

### 2.10. Transcriptional Analysis of pbd-2 Gene in TG Pigs Using RT-qPCR

Total RNA from heart, liver, spleen, lung, kidney and brain of the TG and WT pigs was extracted using RNAiso Plus (TAKARA, Dalian, China). The obtained RNA of 500 ng from different organs was then reverse-transcribed into cDNA for the subsequent RT-qPCR to measure the relative mRNA levels of *pbd-2*, with *GAPDH* as the reference gene. Specific primers for *pbd-2* are tPBD2-F (5′-AGAGGGCA GAGGAAGTCTGCTAA-3′) and tPBD2-R (5′-TTTAAACGGGCCCTCTAGACTCGA-3′), while primers for *GAPDH* are GAPDH-F (5′-ACCCAGAAGACTGTGGATGGC-3′) and GAPDH-R (5′-AGCCAGAGGCAAAGTGATAGATA-3′).

### 2.11. Off-Target Analysis

Potential off-target sites in pig genome were predicted using an online tool Cas-OFFinder [35]. The maximal mismatch number was set as three, with PAM type being 5′-NGG-3′. After that, fragments containing predicted off-target sites were PCR amplified using primer pairs in Table S2 and subjected to Sanger sequencing for sequence alignment.

### 2.12. Statistical Analysis

Data were analyzed with GraphPad Prism 8 (GraphPad Software, La Jolla, CA, USA) using unpaired one-tailed Student's *t*-test and shown as mean ± SD. * $p < 0.05$, ** $p < 0.01$, *** $p < 0.001$, **** $p < 0.0001$.

## 3. Results

### 3.1. Enhanced Bactericidal Activity by T2A-Linked Dual pbd-2 Expression

To achieve simultaneous expression of two copies of *pbd-2* gene, a T2A self-cleaving peptide sequence was put between two *pbd-2* genes in *pcCAG- PBD2-T2A-PBD2* (Figure S1B). Cell culture supernatants from PK-15 cells transfected with *pcCAG-PBD2-T2A-PBD2* and the control vector *pcCAG-nPBD2* carrying a single copy of *pbd-2* (Figure S1A) were collected and incubated with the same number of *A. pleuropneumoniae* for 1 h, respectively. As shown in Figure S1C, the number of surviving bacteria after incubating with cell culture supernatant of cells transfected with *pcCAG-PBD2-T2A-PBD2* was significantly less than that with *pcCAG-nPBD2*, indicating that T2A-mediated dual *pbd-2* expression could significantly increase the bactericidal activity of cell culture supernatants.

### 3.2. CRISPR/Cas9 and Cre/loxP-Mediated Marker-Free Site-Specific Insertion of pbd-2 in PFFs

The *pcCAG-R26-PBD2-T2A-PBD2* (Figure 1A) along with *pX330-pRosa26* (Figure S2) were co-electroporated into PFFs to achieve the integration of dual *pbd-2* genes, as well as a SMG *neoR*, into the *pRosa26* site. After G418 screening, positive cell colonies was identified by PCR (Figure S3) and the treatment of TAT-Cre recombinase significantly decreased the copy number of *neoR* in TG cells (Figure 1B,C), reaching a recombination efficiency of 48.3% as calculated in accordance with the method described above. In the meantime, the cell culture supernatant of treated TG cells displayed significant increase in the bactericidal activity against both *S. suis* and *A. pleuropneumoniae* (Figure 1D). These indicated that 48.3% of the TG cells were marker-free and exerted enhanced antibacterial activity, suggesting an increase in PBD-2 expression in the treated TG cells, which could be used for the subsequent SCNT.

### 3.3. Genotyping of Cloned Piglets

After transferring 250 reconstructed embryos into the uterus of each recipient, there were three surrogate sows successfully giving birth to seven piglets in total. All the cloned piglets had normal physical appearance when compared with WT pigs (Figure 2A). Regarding identification of TG pigs based on PCR results, the observed bands for the PBD-2 and On-target fragments represented a site-specific *pbd-2* knock-in at the *pRosa26* locus, while no band for the NeoR fragment indicated that the SMG *neoR* was not introduced into the pig genome. As shown in Figure 2B, there were five pigs harboring a site-specific *pbd-2* at the *pRosa26* locus, with no SMG detected. The on-target integration of *pbd-2* in TG pigs was validated by Sanger sequencing analysis on the resulting PBD-2 and On-target fragments (Figure S4). The Southern blot analysis was carried out to further confirm a targeted *pbd-2* insertion into the pig genome. Consistent with the PCR results, there were intended bands for pigs numbered as 220, 222, 224, 226 and 230 (Figure 2C), indicating a site-specific integration of duplicate *pbd-2* genes in these pigs. Together, we obtained five marker-free pigs with the *pbd-2* gene site-specifically incorporated at the *pRosa26* locus.

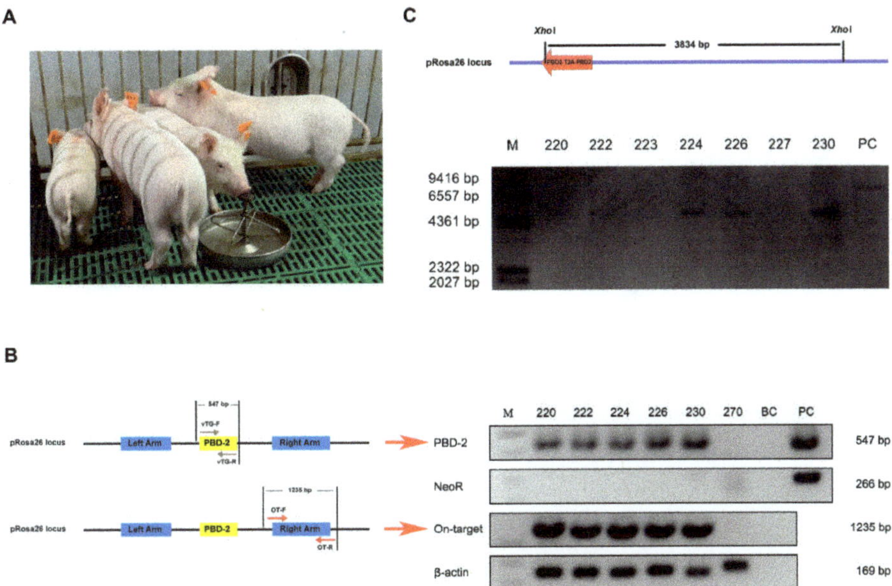


**Figure 2.** Genotyping of cloned piglets. (**A**) Physical appearance of cloned piglets. (**B**) PCR analysis to identify marker-free site-specific *pbd-2* knock-in pigs. PBD-2: Amplification for the dual *pbd-2* gene using primers vTG-F and vTG-R; NeoR: Amplification for the SMG *neoR*; Off-target: Amplification for the fragment which represents on-target insertion of the transgene using primers OT-F and OT-R; β-actin: Amplification for *β-actin*; Lane 220–227: Numbers for pigs; BC: Blank control; PC: Positive control. (**C**) Southern blot analysis for the identification of TG pigs. Genomic DNA from porcine ear fibroblasts (PEFs) was extracted and digested with *Xho*I, followed by Southern blot analysis using a digoxigenin-labeled *pbd-2*-specific probe. M: Molecular mass marker; Lane 220–230: Numbers for pigs; PC: Positive control.

### 3.4. Off-Target Analysis

The potential off-target sites were predicted according to the target sequence (5′-GCTCCTTCTCGATTATGGGCGGG-3′) using Cas-OFFinder. The results showed that there were five potential off-target sites on chromosome 06, 07, 11, 14, 18, respectively (Figure 3A). The fragments containing all these potential off-targets loci were PCR amplified and sequenced. The Sanger sequencing results revealed that no insertions/deletions and site mutations were found within all these potential off-target sites (Figure 3B). These findings suggested no potential off-target effects were observed in TG pigs.

**A**

| | Off-target sequence | Mismatch | Chromosome |
|---|---|---|---|
| OT-1: | GCTCCTcCTCcATTcTGGGCTGG | 3 | 06 |
| OT-2: | GCTCCTTCTCtATgtTGGGCAGG | 3 | 07 |
| OT-3: | GCTCCcTCTCaATTcTGGGCAGG | 3 | 11 |
| OT-4: | GCTCCTTCTCcATgtTGGGCAGG | 3 | 14 |
| OT-5: | GCTCCTTCagGATgATGGGCTGG | 3 | 18 |
| On-target: | 5 '-GCTCCTTCTCGATTATGGGCGGG-3 ' | | |

**B**

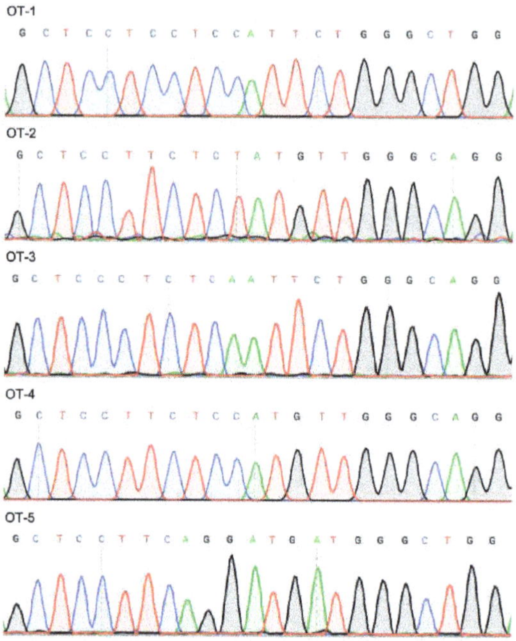

OT-1
G C T C C T C C T C C A T T C T G G G C T G G

OT-2
G C T C C T T C T C T A T G T T G G G C A G G

OT-3
G C T C C C T C T C A A T T C T G G G C A G G

OT-4
G C T C C T T C T C C A T G T T G G G C A G G

OT-5
G C T C C T T C A G G A T G A T G G G C T G G

**Figure 3.** Off-target analysis. (**A**) Predicted off-target sites. Lower case letters represent mismatched bases, while underlined letters represent the 3′ PAM of the target sequence. (**B**) Sanger sequencing results of the PCR products of potential off-target sites.

### *3.5. Characterization of PBD-2 Transcription and Translation*

The transcriptional level of *pbd-2* expression in different organs of TG pigs was quantified by RT-qPCR, with *GAPDH* serving as the reference gene. As shown in Figure 4A, the mRNA levels of *pbd-2* in heart, liver, spleen, lung, kidney and brain tissues of TG pigs were significantly higher than that of their WT littermates. Besides, the kidney showed the highest mRNA level of *pbd-2*, while the transcriptional level of *pbd-2* of lung and spleen ranked the second and the third, respectively.

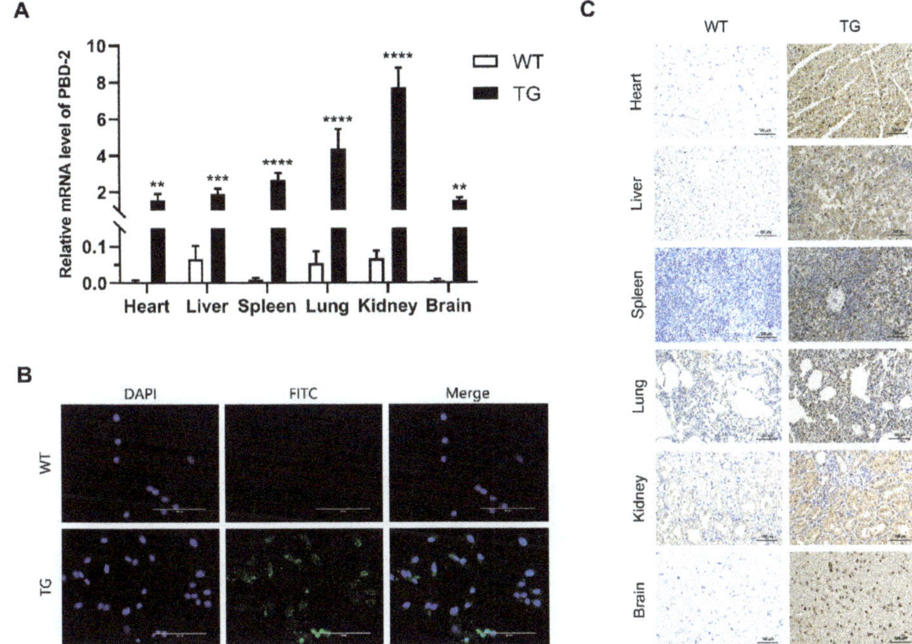


**Figure 4.** Transcriptional and translational analysis of *pbd-2* in cloned pigs. (**A**) The mRNA expression level of *pbd-2* in different organs. Total RNA in different organs of TG and WT pigs was extracted and then subjected to reverse transcription, followed by RT-qPCR to determine the mRNA level of *pbd-2*, *GAPDH* was used as an internal reference gene. Data are presented as mean ± SD and are plotted from three independent experiments. ** $p < 0.01$, *** $p < 0.001$, **** $p < 0.0001$, $n = 3$, unpaired one tailed Student' s *t*-test. (**B**) Immunofluorescent analysis of the expression of porcine β-defensin 2 (PBD-2) using a 40× objective and a 100 ms exposure. The expression of PBD-2 in PEFs from TG pigs was detected by immunofluorescent analysis using mouse monoclonal anti-PBD2 antibody followed by FITC-labeled goat anti-mouse IgG. Cell nuclei were stained with 4′,6-diamidino-2-phenylindole, the PEFs from WT pigs served as a negative control. Scale bar = 100 µm. (**C**) Immunohistochemical analysis of the expression of PBD-2 at a magnification of 200×, with an exposure time of 50 ms. Expression of PBD-2 in heart, liver, spleen, lung, kidney and brain tissues of WT and TG pigs was determined by immunohistochemistry using mouse monoclonal 2A peptide antibody. Brown color represents the detected 2A peptide, which is co-expressed with PBD-2. Scale bars = 100 µm.

To further determine whether the CRISPR/Cas9-mediated knock-in of *pbd-2* could increase the expression of PBD-2, the immunofluorescent analysis was conducted to detect PBD-2 in PEFs isolated from WT and TG pigs, using the self-made mouse anti-PBD-2 monoclonal antibody. The results showed high expression of PBD-2 in TG PEFs, with strong green fluorescence. In contrast, little green signals could be seen in WT PEFs, indicating low expression of PBD-2 in WT pigs (Figure 4B). The immunohistochemical analysis was further performed to detect PBD-2 in different organs of both WT and TG pigs. To exclude the influence of endogenous expression of PBD-2, an anti-2A peptide antibody was used as the primary antibody for that 2A peptide and PBD-2 were co-translated. As was shown in Figure 4C, PBD-2 in different organs of TG pigs was highly expressed, appearing as brown signals, while no obvious brown coloration was observed in tissues of WT pigs, which was in good agreement with those results identified by immunofluorescence and RT-qPCR.

### 3.6. Verification of the Antibacterial Ability of Isolated TG PEFs

To investigate whether overexpression of PBD-2 could endow hosts with enhanced resistance to bacterial infections, PEFs were isolated from TG and WT pigs. Then cell culture supernatants of WT and TG PEFs were harvested to incubate with *A. pleuoropneumoniae* and *S. suis*, and the bactericidal activity of the supernatants was analyzed by bacterial counting. The surviving bacteria of the TG group were significantly less than that of the WT group (Figure 5), suggesting that the site-specific knock-in of *pbd-2* successfully conferred an improved resistance against bacterial infections on TG pigs.

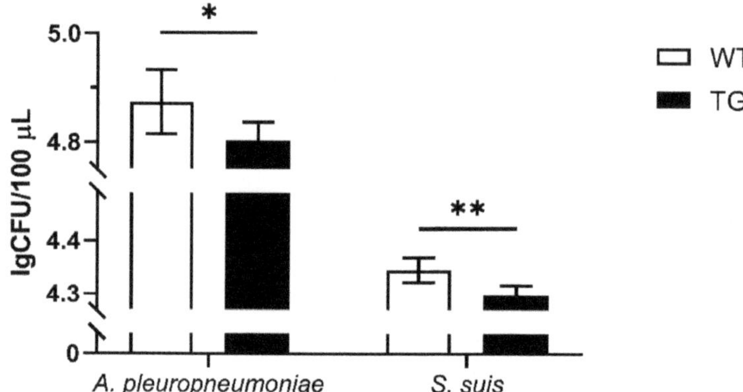


**Figure 5.** Bactericidal activity of cell culture supernatant. The cell culture supernatant of PEFs from TG pigs was incubated with *A. pleuropneumoniae* and *S. suis* for 1 h and the surviving bacteria were then plated on agar for counting. Cell culture supernatant of PEFs from WT pigs was used as a negative control. Data are presented as mean ± SD and are plotted from three independent experiments. * $p < 0.05$, ** $p < 0.01$, unpaired one tailed Student' s *t*-test.

## 4. Discussion

As a major group of cationic peptides widely existing in animals, defensins have been proven to possess broad antimicrobial activities against different pathogens including bacteria, fungi and viruses. Our previous in vivo studies showed that pigs overexpressing PBD-2 displayed increased resistance to *A. pleuoropneumoniae* infection [16], while mice expressing PBD-2 became more resistant to pseudorabies virus and *Salmonella* Typhimurium infections [27,28]. Given that the previous *pbd-2* TG pigs were produced using random gene insertion, which also introduced a SMG, these would raise public concern on the biosafety of TG animals and hamper the commercialization of TG animals derived products. Therefore, this study achieved site-specific integration of *pbd-2* into the *pRosa26* locus through CRISPR/Cas9-mediated homology-directed repair and eliminated the SMG *neoR* using a cell-penetrating TAT-Cre recombinase. Five marker-free *pbd-2* knock-in pigs were generated via SCNT, making it to the overexpression of PBD-2 in different organs of pigs with no off-target effects detected. Additionally, the cell culture supernatant of PEFs of TG pigs exhibited greater antibacterial activities against both Gram-negative (*A. pleuoropneumoniae*) and Gram-positive bacterium (*S. suis*).

The copy number of genes is positively correlated with corresponding protein levels [36]. Fellermann et al. claimed that people carried low copy number of human β-defensin 2 gene were more likely to develop Crohn's disease [37], while a high copy number of β-defensin genes was considered the main reason why East Asians were more resistant to influenza infection [38]. As the 2A self-cleaving peptide has been widely used to generate multi-transgenic pigs [39], we applied T2A peptides, a 2A self-cleaving peptide found in Thosea asigna virus 2A [40], in expression of two copies of the *pbd-2* gene in this study. Similar with other defensins, PBD-2 is firstly produced in the form of pro-peptide and then processed into active form when the N-terminal signal peptide is cleaved, followed by being transported out of cells [16]. Though de Felipe and Ryan claimed that proteins downstream of the 2A peptide could also occur on cell membrane or be secreted despite lack of a signal sequence when a protein upstream of the 2A peptide contained an N-terminal signal sequence [41], while Yan et al. argued that genes following the 2A peptide still demanded a signal sequence for successful secretion expression [42]. Thus, the signal sequence in the second *pbd-2* gene was kept and the bactericidal assay showed that this design could efficiently improve the expression of PBD-2 without hampering its biological activity.

The random integration of exogenous genes has been found to lead to multiple undesired outcomes including gene silencing, unpredictable expression and activation of oncogene expression [43]. In addition to the *pRosa26* locus described above, porcine *H11* and *GAPDH* loci have been well characterized as safe harbors for targeted transgene integrations [44,45]. When introducing a foreign gene into the *Rosa26* locus, higher expression could be achieved by positioning the transgene promoter in an opposite direction to the *Rosa26* promoter [46]. Therefore, this study set *pbd-2* gene in a reverse orientation to the *Rosa26* promoter and achieved overexpression of PBD-2.

The unwanted effects of introducing a SMG *neoR* have been well investigated. Wang et al. argued that the *neoR* gene could inhibit the growth of *Lactobacillus* and *Escherichia-Shigella-Hafnia* in guts of TG pigs [47]. Meanwhile it was worth noting that the expression levels of metabolic genes varied between cells expressing exogenous *neoR* gene and control cells [48]. Also, the integration of *neoR* gene in target loci has been shown to exert a long-distance effect on the expression of downstream or upstream genes [49]. Another study revealed that removing the selectable marker increased the expression level of transgene when promoters of the *Rosa26* and the transgene occurred in opposite orientations [46]. Bi et al. successfully excised the SMG in cells using Cre mRNA [4], while using cell-penetrating TAT-Cre recombinase to delete a SMG could reach a recombination efficiency of up to 55% in pig primary cells [50]. Alike, the use of commercialized TAT-Cre recombinase in this study achieved successful excision of *neoR* in 48.3% PFFs.

It has been quite controversial that whether CRISPR/Cas9-mediated gene-editing can cause genome-wide off-target mutations, while Zuo et al. created a large scale of screening system called GOTI and used to prove that classic CRISPR/Cas9 tool did not induce obvious off-target effects [51]. This study predicted five putative off-target loci using an online tool and the Sanger sequencing of the five sites in TG pigs revealed that no mutations were found among these sites. However, in-depth whole-genome sequencing of the TG pigs is still needed to exclude any off-target mutagenesis in the future.

Chen et al. found that, when compared with crossbred pigs, higher expression of β-defensins in Meishan pigs was one of the main reasons that Meishan pigs exhibited effective disease-resistance traits [52]. The RT-qPCR analysis combined with immunohistochemical and immunofluorescence assays had confirmed that PBD-2 was overexpressed in different organs of the marker-free *pbd-2* knock-in pigs (Figure 4), while the bactericidal assay indicated that PEFs of TG pigs were more resistant to bacterial infections (Figure 5). In addition to the described functions of PBD-2 above, the in vitro studies have shown that PBD-2 could inhibit proliferations of PRRSV, *Staphylococcus aureus* and other pathogens [24,26]. Besides, PBD-2 could be used for growth-promotion among piglets [29,53]. Altogether these suggest excellent prospect for the application of marker-free site-specific *pbd-2* knock-in pigs produced in this study.

In summary, this study generated marker-free site-specific *pbd-2* knock-in pigs, achieving overexpression of PBD-2 in pigs, which provides a novel genetic material for the breeding of disease-resistant pigs. Given that PBD-2 displays multi-functions including antimicrobial and immunomodulatory activities, these pigs are supposed to be reliable animal models to study biofunctions of PBD-2 as well.

**Supplementary Materials:** The following are available online at http://www.mdpi.com/2073-4425/11/8/951/s1, Figure S1: Analysis for the T2A-mediated dual *pbd-2* expression, Figure S2: Map of *pX330-pRosa26*, Figure S3: PCR analysis for the identification of site-specific *pbd-2* knock-in PFFs, Figure S4: Sanger sequencing analyses to confirm on-target integration of *pbd-2* in cloned pigs, Table S1: Information of primer pairs for identification of transgenic PFFs, Table S2: Information of primer pairs for amplification of predicted off-target loci.

**Author Contributions:** Conceptualization: J.H., R.Z. and L.L.; Methodology: J.H. and A.W.; Software: C.H., Y.S. and B.S.; Formal analysis: J.H., A.W., C.H., Y.S. and B.S.; Writing—original draft preparation: J.H., R.Z. and L.L.; supervision, R.Z. and L.L.; funding acquisition, R.Z. and L.L. All authors have read and agreed to the published version of the manuscript.

**Funding:** This research was founded by the National Transgenic Project of China (Grant No. 2016ZX08006003-004), National Key R & D Program of China (Grant No. 2017YFD0500201), and Hubei Province Natural Science Foundation for Innovative Research Groups (Grant No. 2016CFA015).

**Acknowledgments:** We thank Bo Zuo and Shuhong Zhao (Huazhong Agricultural University) for providing *pX330-pRosa26* and PFFs, respectively.

**Conflicts of Interest:** The authors declare no conflict of interest.

## References

1. Cermak, T.; Doyle, E.L.; Christian, M.; Wang, L.; Zhang, Y.; Schmidt, C.; Baller, J.A.; Somia, N.V.; Bogdanove, A.J.; Voytas, D.F. Efficient design and assembly of custom TALEN and other TAL effector-based constructs for DNA targeting. *Nucleic Acids Res.* **2011**, *39*, e82. [CrossRef] [PubMed]
2. Smith, J.; Bibikova, M.; Whitby, F.G.; Reddy, A.R.; Chandrasegaran, S.; Carroll, D. Requirements for double-strand cleavage by chimeric restriction enzymes with zinc finger DNA-recognition domains. *Nucleic Acids Res.* **2000**, *28*, 3361–3369. [CrossRef] [PubMed]
3. Jinek, M.; Chylinski, K.; Fonfara, I.; Hauer, M.; Doudna, J.A.; Charpentier, E. A programmable dual-RNA-guided DNA endonuclease in adaptive bacterial immunity. *Science* **2012**, *337*, 816–821. [CrossRef] [PubMed]
4. Bi, Y.; Hua, Z.; Liu, X.; Hua, W.; Ren, H.; Xiao, H.; Zhang, L.; Li, L.; Wang, Z.; Laible, G.; et al. Isozygous and selectable marker-free MSTN knockout cloned pigs generated by the combined use of CRISPR/Cas9 and Cre/LoxP. *Sci. Rep.* **2016**, *6*, 31729. [CrossRef]
5. Guo, R.; Wan, Y.; Xu, D.; Cui, L.; Deng, M.; Zhang, G.; Jia, R.; Zhou, W.; Wang, Z.; Deng, K.; et al. Generation and evaluation of Myostatin knock-out rabbits and goats using CRISPR/Cas9 system. *Sci. Rep.* **2016**, *6*, 29855. [CrossRef]
6. Luo, J.; Song, Z.; Yu, S.; Cui, D.; Wang, B.; Ding, F.; Li, S.; Dai, Y.; Li, N. Efficient generation of myostatin (MSTN) biallelic mutations in cattle using zinc finger nucleases. *PLoS ONE* **2014**, *9*, e95225. [CrossRef]
7. Carlson, D.F.; Lancto, C.A.; Zang, B.; Kim, E.S.; Walton, M.; Oldeschulte, D.; Seabury, C.; Sonstegard, T.S.; Fahrenkrug, S.C. Production of hornless dairy cattle from genome-edited cell lines. *Nat. Biotechnol.* **2016**, *34*, 479–481. [CrossRef]
8. Oishi, I.; Yoshii, K.; Miyahara, D.; Kagami, H.; Tagami, T. Targeted mutagenesis in chicken using CRISPR/Cas9 system. *Sci. Rep.* **2016**, *6*, 23980. [CrossRef]
9. Cui, C.; Song, Y.; Liu, J.; Ge, H.; Li, Q.; Huang, H.; Hu, L.; Zhu, H.; Jin, Y.; Zhang, Y. Gene targeting by TALEN-induced homologous recombination in goats directs production of β-lactoglobulin-free, high-human lactoferrin milk. *Sci. Rep.* **2015**, *5*, 10482. [CrossRef]
10. Proudfoot, C.; Lillico, S.; Tait-Burkard, C. Genome editing for disease resistance in pigs and chickens. *Anim. Front.* **2019**, *9*, 6–12. [CrossRef] [PubMed]
11. Li, L.; Li, Q.; Bao, Y.; Li, J.; Chen, Z.; Yu, X.; Zhao, Y.; Tian, K.; Li, N. RNAi-based inhibition of porcine reproductive and respiratory syndrome virus replication in transgenic pigs. *J. Biotechnol.* **2014**, *171*, 17–24. [CrossRef]
12. Hu, S.; Qiao, J.; Fu, Q.; Chen, C.; Ni, W.; Wujiafu, S.; Ma, S.; Zhang, H.; Sheng, J.; Wang, P.; et al. Transgenic shRNA pigs reduce susceptibility to foot and mouth disease virus infection. *Elife* **2015**, *4*, e06951. [CrossRef] [PubMed]
13. Xie, Z.; Pang, D.; Yuan, H.; Jiao, H.; Lu, C.; Wang, K.; Yang, Q.; Li, M.; Chen, X.; Yu, T.; et al. Genetically modified pigs are protected from classical swine fever virus. *PLoS Pathog.* **2018**, *14*, e1007193. [CrossRef] [PubMed]
14. Burkard, C.; Lillico, S.G.; Reid, E.; Jackson, B.; Mileham, A.J.; Ait-Ali, T.; Whitelaw, C.B.; Archibald, A.L. Precision engineering for PRRSV resistance in pigs: Macrophages from genome edited pigs lacking CD163

SRCR5 domain are fully resistant to both PRRSV genotypes while maintaining biological function. *PLoS Pathog.* **2017**, *13*, e1006206. [CrossRef] [PubMed]

15. Whitworth, K.M.; Rowland, R.R.R.; Petrovan, V.; Sheahan, M.; Cino-Ozuna, A.G.; Fang, Y.; Hesse, R.; Mileham, A.; Samuel, M.S.; Wells, K.D.; et al. Resistance to coronavirus infection in amino peptidase N-deficient pigs. *Transgenic Res.* **2019**, *28*, 21–32. [CrossRef] [PubMed]
16. Yang, X.; Cheng, Y.T.; Tan, M.F.; Zhang, H.W.; Liu, W.Q.; Zou, G.; Zhang, L.S.; Zhang, C.Y.; Deng, S.M.; Yu, L.; et al. Overexpression of Porcine B-Defensin 2 Enhances Resistance to *Actinobacillus pleuropneumoniae* Infection in Pigs. *Infect. Immun.* **2015**, *83*, 2836–2843. [CrossRef]
17. Lu, T.; Song, Z.; Li, Q.; Li, Z.; Wang, M.; Liu, L.; Tian, K.; Li, N. Overexpression of Histone Deacetylase 6 Enhances Resistance to Porcine Reproductive and Respiratory Syndrome Virus in Pigs. *PLoS ONE* **2017**, *12*, e0169317. [CrossRef]
18. Yan, Q.; Yang, H.; Yang, D.; Zhao, B.; Ouyang, Z.; Liu, Z.; Fan, N.; Ouyang, H.; Gu, W.; Lai, L. Production of transgenic pigs over-expressing the antiviral gene Mx1. *Cell Regen. (Lond.)* **2014**, *3*, 11. [CrossRef]
19. Zhao, Y.; Wang, T.; Yao, L.; Liu, B.; Teng, C.; Ouyang, H. Classical swine fever virus replicated poorly in cells from MxA transgenic pigs. *BMC Vet. Res.* **2016**, *12*, 169. [CrossRef]
20. Mattar, E.H.; Almehdar, H.A.; Yacoub, H.A.; Uversky, V.N.; Redwan, E.M. Antimicrobial potentials and structural disorder of human and animal defensins. *Cytokine Growth Factor Rev.* **2016**, *28*, 95–111. [CrossRef]
21. Sang, Y.; Blecha, F. Porcine host defense peptides: Expanding repertoire and functions. *Dev. Comp. Immunol.* **2009**, *33*, 334–343. [CrossRef] [PubMed]
22. Choi, M.K.; Le, M.T.; Nguyen, D.T.; Choi, H.; Kim, W.; Kim, J.H.; Chun, J.; Hyeon, J.; Seo, K.; Park, C. Genome-level identification, gene expression, and comparative analysis of porcine ss-defensin genes. *BMC Genet.* **2012**, *13*, 98. [CrossRef] [PubMed]
23. Sang, Y.; Patil, A.A.; Zhang, G.; Ross, C.R.; Blecha, F. Bioinformatic and expression analysis of novel porcine β-defensins. *Mamm. Genome* **2006**, *17*, 332–339. [CrossRef] [PubMed]
24. Peng, Z.; Wang, A.; Feng, Q.; Wang, Z.; Ivanova, I.V.; He, X.; Zhang, B.; Song, W. High-level expression, purification and characterisation of porcine β-defensin 2 in *Pichia pastoris* and its potential as a cost-efficient growth promoter in porcine feed. *Appl. Microbiol. Biotechnol.* **2014**, *98*, 5487–5497. [CrossRef] [PubMed]
25. Chen, R.B.; Zhang, K.; Zhang, H.; Gao, C.Y.; Li, C.L. Analysis of the antimicrobial mechanism of porcine β defensin 2 against *E. coli* by electron microscopy and differentially expressed genes. *Sci. Rep.* **2018**, *8*, 14711. [CrossRef]
26. Veldhuizen, E.J.; Rijnders, M.; Claassen, E.A.; van Dijk, A.; Haagsman, H.P. Porcine β-defensin 2 displays broad antimicrobial activity against pathogenic intestinal bacteria. *Mol. Immunol.* **2008**, *45*, 386–394. [CrossRef]
27. Huang, J.; Qi, Y.; Wang, A.; Huang, C.; Liu, X.; Yang, X.; Li, L.; Zhou, R. Porcine β-defensin 2 inhibits proliferation of pseudorabies virus in vitro and in transgenic mice. *Virol. J.* **2020**, *17*, 18. [CrossRef]
28. Huang, C.; Yang, X.; Huang, J.; Liu, X.; Yang, X.; Jin, H.; Huang, Q.; Li, L.; Zhou, R. Porcine B-Defensin 2 Provides Protection Against Bacterial Infection by a Direct Bactericidal Activity and Alleviates Inflammation via Interference With the TLR4/NF-kappaB Pathway. *Front. Immunol.* **2019**, *10*, 1673. [CrossRef]
29. Peng, Z.; Wang, A.; Xie, L.; Song, W.; Wang, J.; Yin, Z.; Zhou, D.; Li, F. Use of recombinant porcine β-defensin 2 as a medicated feed additive for weaned piglets. *Sci. Rep.* **2016**, *6*, 26790. [CrossRef]
30. Ran, F.A.; Hsu, P.D.; Wright, J.; Agarwala, V.; Scott, D.A.; Zhang, F. Genome engineering using the CRISPR-Cas9 system. *Nat. Protoc.* **2013**, *8*, 2281–2308. [CrossRef]
31. Li, M.; Ouyang, H.; Yuan, H.; Li, J.; Xie, Z.; Wang, K.; Yu, T.; Liu, M.; Chen, X.; Tang, X.; et al. Site-Specific Fat-1 Knock-In Enables Significant Decrease of n-6PUFAs/n-3PUFAs Ratio in Pigs. *G3 (Bethesda)* **2018**, *8*, 1747–1754. [CrossRef] [PubMed]
32. Li, X.; Yang, Y.; Bu, L.; Guo, X.; Tang, C.; Song, J.; Fan, N.; Zhao, B.; Ouyang, Z.; Liu, Z.; et al. Rosa26-targeted swine models for stable gene over-expression and Cre-mediated lineage tracing. *Cell Res.* **2014**, *24*, 501–504. [CrossRef] [PubMed]
33. Livak, K.J.; Schmittgen, T.D. Analysis of relative gene expression data using real-time quantitative PCR and the 2(-Delta Delta C(T)) Method. *Methods* **2001**, *25*, 402–408. [CrossRef] [PubMed]
34. Zhu, X.X.; Zhong, Y.Z.; Ge, Y.W.; Lu, K.H.; Lu, S.S. CRISPR/Cas9-Mediated Generation of Guangxi Bama Minipigs Harboring Three Mutations in alpha-Synuclein Causing Parkinson's Disease. *Sci. Rep.* **2018**, *8*, 12420. [CrossRef] [PubMed]

35. Bae, S.; Park, J.; Kim, J.S. Cas-OFFinder: A fast and versatile algorithm that searches for potential off-target sites of Cas9 RNA-guided endonucleases. *Bioinformatics* **2014**, *30*, 1473–1475. [CrossRef]
36. Xu, J.; Sun, Y.; Carretero, O.A.; Zhu, L.; Harding, P.; Shesely, E.G.; Dai, X.; Rhaleb, N.E.; Peterson, E.; Yang, X.P. Effects of cardiac overexpression of the angiotensin II type 2 receptor on remodeling and dysfunction in mice post-myocardial infarction. *Hypertension* **2014**, *63*, 1251–1259. [CrossRef]
37. Fellermann, K.; Stange, D.E.; Schaeffeler, E.; Schmalzl, H.; Wehkamp, J.; Bevins, C.L.; Reinisch, W.; Teml, A.; Schwab, M.; Lichter, P.; et al. A chromosome 8 gene-cluster polymorphism with low human β-defensin 2 gene copy number predisposes to Crohn disease of the colon. *Am. J. Hum. Genet.* **2006**, *79*, 439–448. [CrossRef]
38. Hardwick, R.J.; Machado, L.R.; Zuccherato, L.W.; Antolinos, S.; Xue, Y.; Shawa, N.; Gilman, R.H.; Cabrera, L.; Berg, D.E.; Tyler-Smith, C.; et al. A worldwide analysis of β-defensin copy number variation suggests recent selection of a high-expressing DEFB103 gene copy in East Asia. *Hum. Mutat.* **2011**, *32*, 743–750. [CrossRef]
39. Deng, W.; Yang, D.; Zhao, B.; Ouyang, Z.; Song, J.; Fan, N.; Liu, Z.; Zhao, Y.; Wu, Q.; Nashun, B.; et al. Use of the 2A peptide for generation of multi-transgenic pigs through a single round of nuclear transfer. *PLoS ONE* **2011**, *6*, e19986. [CrossRef]
40. Liu, Z.; Chen, O.; Wall, J.B.J.; Zheng, M.; Zhou, Y.; Wang, L.; Ruth Vaseghi, H.; Qian, L.; Liu, J. Systematic comparison of 2A peptides for cloning multi-genes in a polycistronic vector. *Sci. Rep.* **2017**, *7*, 2193. [CrossRef]
41. de Felipe, P.; Ryan, M.D. Targeting of proteins derived from self-processing polyproteins containing multiple signal sequences. *Traffic* **2004**, *5*, 616–626. [CrossRef] [PubMed]
42. Yan, J.; Wang, H.; Xu, Q.; Jain, N.; Toxavidis, V.; Tigges, J.; Yang, H.; Yue, G.; Gao, W. Signal sequence is still required in genes downstream of "autocleaving" 2A peptide for secretary or membrane-anchored expression. *Anal. Biochem.* **2010**, *399*, 144–146. [CrossRef] [PubMed]
43. Ma, L.; Wang, Y.; Wang, H.; Hu, Y.; Chen, J.; Tan, T.; Hu, M.; Liu, X.; Zhang, R.; Xing, Y.; et al. Screen and Verification for Transgene Integration Sites in Pigs. *Sci. Rep.* **2018**, *8*, 7433. [CrossRef] [PubMed]
44. Han, X.; Xiong, Y.; Zhao, C.; Xie, S.; Li, C.; Li, X.; Liu, X.; Li, K.; Zhao, S.; Ruan, J. Identification of Glyceraldehyde-3-Phosphate Dehydrogenase Gene as an Alternative Safe Harbor Locus in Pig Genome. *Genes* **2019**, *10*, 660. [CrossRef]
45. Ruan, J.; Li, H.; Xu, K.; Wu, T.; Wei, J.; Zhou, R.; Liu, Z.; Mu, Y.; Yang, S.; Ouyang, H.; et al. Highly efficient CRISPR/Cas9-mediated transgene knockin at the H11 locus in pigs. *Sci. Rep.* **2015**, *5*, 14253. [CrossRef]
46. Strathdee, D.; Ibbotson, H.; Grant, S.G. Expression of transgenes targeted to the *Gt(ROSA)26Sor* locus is orientation dependent. *PLoS ONE* **2006**, *1*, e4. [CrossRef]
47. Wang, Q.; Qian, L.; Jiang, S.; Cai, C.; Ma, D.; Gao, P.; Li, H.; Jiang, K.; Tang, M.; Hou, J.; et al. Safety Evaluation of Neo Transgenic Pigs by Studying Changes in Gut Microbiota Using High-Throughput Sequencing Technology. *PLoS ONE* **2016**, *11*, e0150937. [CrossRef]
48. Valera, A.; Perales, J.C.; Hatzoglou, M.; Bosch, F. Expression of the neomycin-resistance (neo) gene induces alterations in gene expression and metabolism. *Hum. Gene Ther.* **1994**, *5*, 449–456. [CrossRef]
49. Pham, C.T.; MacIvor, D.M.; Hug, B.A.; Heusel, J.W.; Ley, T.J. Long-range disruption of gene expression by a selectable marker cassette. *Proc. Natl. Acad. Sci. USA* **1996**, *93*, 13090–13095. [CrossRef]
50. Kang, Q.; Sun, Z.; Zou, Z.; Wang, M.; Li, Q.; Hu, X.; Li, N. Cell-penetrating peptide-driven Cre recombination in porcine primary cells and generation of marker-free pigs. *PLoS ONE* **2018**, *13*, e0190690. [CrossRef]
51. Zuo, E.; Sun, Y.; Wei, W.; Yuan, T.; Ying, W.; Sun, H.; Yuan, L.; Steinmetz, L.M.; Li, Y.; Yang, H. Cytosine base editor generates substantial off-target single-nucleotide variants in mouse embryos. *Science* **2019**, *364*, 289–292. [CrossRef] [PubMed]
52. Chen, J.; Qi, S.; Guo, R.; Yu, B.; Chen, D. Different messenger RNA expression for the antimicrobial peptides β-defensins between Meishan and crossbred pigs. *Mol. Biol. Rep.* **2010**, *37*, 1633–1639. [CrossRef] [PubMed]
53. Tang, Z.; Xu, L.; Shi, B.; Deng, H.; Lai, X.; Liu, J.; Sun, Z. Oral administration of synthetic porcine β-defensin-2 improves growth performance and cecal microbial flora and down-regulates the expression of intestinal toll-like receptor-4 and inflammatory cytokines in weaned piglets challenged with enterotoxigenic *Escherichia coli*. *Anim. Sci. J.* **2016**, *87*, 1258–1266. [CrossRef] [PubMed]

*Article*

# Exploring the lncRNAs Related to Skeletal Muscle Fiber Types and Meat Quality Traits in Pigs

**Rongyang Li [1], Bojiang Li [2], Aiwen Jiang [1], Yan Cao [1], Liming Hou [1], Zengkai Zhang [1], Xiying Zhang [1], Honglin Liu [1], Kee-Hong Kim [3] and Wangjun Wu [1,*]**

1 Department of Animal Genetics, Breeding and Reproduction, College of Animal Science and Technology, Nanjing Agricultural University, Nanjing 210095, China; 2018205007@njau.edu.cn (R.L.); 2019205006@njau.edu.cn (A.J.); 2018205001@njau.edu.cn (Y.C.); liminghou@njau.edu.cn (L.H.); 2017105031@njau.edu.cn (Z.Z.); 2018105030@njau.edu.cn (X.Z.); wangjunwu4062@163.com (H.L.)

2 College of Animal Science and Veterinary Medicine, Shenyang Agricultural University, Shenyang 110866, China; libojiang12@126.com

3 Department of Food Science, Purdue University, West Lafayette, IN 47897, USA; keehong@purdue.edu

* Correspondence: wuwangjun2012@njau.edu.cn; Tel./Fax: +86-025-8439-9762

Received: 2 July 2020; Accepted: 31 July 2020; Published: 4 August 2020

**Abstract:** The alteration in skeletal muscle fiber is a critical factor affecting livestock meat quality traits and human metabolic diseases. Long non-coding RNAs (lncRNAs) are a diverse class of non-coding RNAs with a length of more than 200 nucleotides. However, the mechanisms underlying the regulation of lncRNAs in skeletal muscle fibers remain elusive. To understand the genetic basis of lncRNA-regulated skeletal muscle fiber development, we performed a transcriptome analysis to identify the key lncRNAs affecting skeletal muscle fiber and meat quality traits on a pig model. We generated the lncRNA expression profiles of fast-twitch *Biceps femoris* (Bf) and slow-twitch *Soleus* (Sol) muscles and identified the differentially expressed (DE) lncRNAs using RNA-seq and performed bioinformatics analyses. This allowed us to identify 4581 lncRNA genes among six RNA libraries and 92 DE lncRNAs between Bf and Sol which are the key candidates for the conversion of skeletal muscle fiber types. Moreover, we detected the expression patterns of lncRNA *MSTRG.42019* in different tissues and skeletal muscles of various development stages. In addition, we performed a correlation analyses between the expression of DE lncRNA *MSTRG.42019* and meat quality traits. Notably, we found that DE lncRNA *MSTRG.42019* was highly expressed in skeletal muscle and its expression was significantly higher in Sol than in Bf, with a positive correlation with the expression of *Myosin heavy chain 7* (*MYH7*) ($r = 0.6597, p = 0.0016$) and a negative correlation with meat quality traits glycolytic potential ($r = -0.5447, p = 0.0130$), as well as drip loss ($r = -0.5085, p = 0.0221$). Moreover, we constructed the lncRNA *MSTRG.42019*–mRNAs regulatory network for a better understanding of a possible mechanism regulating skeletal muscle fiber formation. Our data provide the groundwork for studying the lncRNA regulatory mechanisms of skeletal muscle fiber conversion, and given the importance of skeletal muscle fiber types in muscle-related diseases, our data may provide insight into the treatment of muscular diseases in humans.

**Keywords:** pig; skeletal muscle fiber; meat quality; metabolic diseases; lncRNA; RNA-seq

## 1. Introduction

Skeletal muscle is the major component of body mass accounting for approximately 50% of body mass in a mammal, and its growth and development are critical for maintaining skeletal muscle function. Skeletal muscle is composed of various muscle fibers that exhibit different physiological and metabolic properties, such as glycolysis, oxidative metabolism, and contraction [1]. Notably, dysfunctions of skeletal muscle fiber types are known to many human diseases, such as muscular dystrophy, cardiomyopathic disease, and type 2 diabetes [2–5]. Moreover, the differences in skeletal muscle fiber types directly affect meat quality postmortem in livestock, such as pH, meat color, and drip loss [6]. Therefore, elucidating the underlying mechanisms of skeletal muscle growth and development and skeletal muscle fibers formation will be useful for the treatment of human muscle diseases and the improvement of production traits of livestock.

Myogenesis is a complex process that is controlled by a series of factors, such as myogenic regulatory factors (MRFs) [7,8] and non-coding small molecule RNA microRNAs (miRNAs) [9,10]. In recent years, emerging researches found a key role of long non-coding RNAs (lncRNAs) in skeletal muscle growth and development and found that they are closely related to muscle diseases [11,12]. LncRNAs are a class of non-coding RNAs with a length of more than 200 nt and mainly transcribed by RNA polymerase II in eukaryotic organisms [13]. For example, Cesana et al. found that lncRNA linc-MD1 is expressed explicitly in the differentiating myoblasts and exerts its function through miR-133 and miR-135-regulated expression of muscle-specific transcriptional regulators, MAML1 and MEF2C [14]. Dey et al. found that lncRNA lncRNA-H19 controls skeletal muscle differentiation and regeneration by two miRNAs generated from its own exon 1 [15]. Moreover, it has been reported that lncRNAs affect skeletal muscle growth and development by regulating DNA methylation [16]. Intriguingly, researchers recently found that a skeletal muscle-specific RNA, annotated as a putative the lncRNA RNA, exerts its role in skeletal muscle physiology via encoding a conserved micropeptide [17].

Notably, the types of skeletal muscle fiber are critical factors related to various metabolic diseases in humans and affecting meat quality traits in livestock, but the researches on the genetic basis of skeletal muscle fiber formation involved in lncRNAs are still limited. In this study, we performed a comparative analysis of the whole transcriptomes between *Biceps femoris* (Bf) (fast muscle or white muscle) and *Soleus* (Sol) (slow muscle or red muscle) muscles characterized by noticeable muscle fiber type differences using RNA-seq. We further identified a series of differentially expressed (DE) lncRNAs between Bf and Sol muscles, which represent potential candidate lncRNAs affecting the conversion of skeletal muscle fiber types. Furthermore, we found several important Gene Ontology (GO) terms and metabolic pathways associated with skeletal muscle fiber formation. We proposed that lncRNA *MSTRG.42019* may be a key lncRNAs affecting skeletal muscle fiber types based on its expression profiles and relationship with porcine meat quality traits and its network interaction map. Overall, our data provide the groundwork for an understanding of the lncRNA-regulated skeletal muscle growth and development and muscle fiber conversion.

## 2. Materials and Methods

### 2.1. Ethics Statement

All experimental procedures on the pigs were conducted according to the Guidelines of the Institutional Animal Care and Use Committee of Nanjing Agricultural University, Nanjing, China (SYXK2011-0036).

### 2.2. Animals and Samples Collection

Three full-sibling Duroc × Meishan female pigs derived from the offspring of a Duroc boar crossing with eight Meishan sows were used in this study and the experimental pig population was constructed by Li et al. [18]. Two types of skeletal muscles with different meat color, Bf (Bf28, Bf35, and Bf36) and Sol (Sol28, Sol35, and Sol36) muscles, were dissociated from these three full-sibling Duroc × Meishan pigs. Previously, we have only identified the DE coding genes due to the limitation of the library construction method and sequencing depth [18]. In this study, we reconstructed the RNA sequencing library to identify the DE coding and non-coding genes using the Bf (Bf28, Bf35, and Bf36) and the Sol (Sol28, Sol35, and Sol36) muscles. Moreover, twenty *Longissimus dorsi* muscles from a 279 commercial hybrid pig population [Pietrain (P) × Duroc (D)] × [Landrace (L) × Yorkshire

(Y)] were selected randomly to detect the correlation between the expression of DE lncRNAs and meat quality traits, and the experimental pigs and the production traits were described previously by Dong et al. [19].

*2.3. RNA Isolation and LncRNA Library Construction*

Total RNA was extracted using the Trizol® reagent (Invitrogen, Carlsbad, CA, USA), and the quality of the isolated RNA was evaluated by 1% agarose gels. The purity of total RNA was initially detected using the Kaiao K5500® Spectrophotometer (Kaiao, Beijing, China). Next, RNA integrity and concentration were precisely assessed using the RNA Nano 6000 Assay Kit of the Agilent Bioanalyzer 2100 (Agilent Technologies, Santa Clara, CA, USA). LncRNA libraries were constructed using 3 μg total RNA per sample. Ribo-Zero™ Gold Kits (Ribo, Guangzhou, China) was used to remove the ribosomal RNA (rRNA) from the sample, and the different indexes were used to build the libraries according to the NEBNext® Ultra™ Directional RNA Library Prep Kit for Illumina® (NEB, Ipswich, MA, USA). The specific steps for the library construction were as follows: firstly, rRNA was removed from total RNA and the first-strand cDNA was synthesized with fragmented RNA using a six-base random primer. Next, the second-strand cDNA was synthesized by adding buffer, dNTPs, RNase H, and DNA Polymerase I. Subsequently, the library fragments were recovered by agarose gel electrophoresis after purification using the QIAQuick PCR kit (Qiagen, New York, NY, USA), followed by the end-repair, adding base A and adapter ligation. Finally, the double cDNA strand was digested with uracil-N-glycosylase (UNG) enzyme and subjected to PCR amplification, and the 300 bp target fragments were recovered by gel electrophoresis to obtain the sequencing libraries. The RNA concentration of libraries was measured using the Qubit® RNA Assay Kit (Thermo Fisher, New York, NY, USA) in Qubit® 2.0 (Thermo Fisher, New York, NY, USA), and the size of the inserted fragment was assessed using Agilent Bioanalyzer 2100 (Agilent Technologies, Santa Clara, CA, USA). After cluster generation, sequencing was performed on an Illumina Hiseq 4000 sequencing platform with 150 bp paired-end reads. The raw data of the transcriptome sequence data have been deposited in NCBI SRA (accession codes PRJNA597666).

*2.4. Quality Control and Genomic Alignment*

Raw data were processed with Perl scripts to obtain high-quality clean data for further analyses. The filtering criteria were as follows: (1) remove the adaptor-polluted reads (reads containing more than 5 adapter-polluted bases were regarded as adaptor-polluted reads and were filtered out); (2) remove the low-quality reads (reads with a Phred quality value of less than 19 accounting for more than 15% of total bases are regarded as low-quality reads); (3) remove reads whose number of N bases account for more than 5%. As for paired-end sequencing data, both reads were filtered out if any reads of the paired-end reads were adaptor-polluted. The pig reference genome (*Sscrofa11.1*) and the annotation file were downloaded from the Ensembl database (http://www.ensembl.org/index.html). Bowtie2 v2.2.3 was used for building the genome index [20], and clean data were mapped to the reference genome using HISAT2 v2.0.5 [21].

*2.5. Identification of LncRNAs*

Novel transcripts were reconstructed using the StringTie software (v1.3.3b) with the default parameters using the mapped clean reads [22], and GffCompare was used to screen out known mRNAs and other non-coding RNAs (rRNA, tRNA, snoRNA, snRNA, known lncRNA, etc.). Furthermore, the known lncRNAs were also figured out through comparative analysis. Subsequently, the novel potential lncRNAs transcripts were identified according to the transcript length (≥200 bp), the exon number (≥2), and the read coverage of each transcript (≥5). Then, we used four kinds of tools to predict the protein-coding potential of novel transcripts to confirm novel lncRNAs, including the Coding-Non-Coding Index (CNCI) [23], Coding Potential Calculator (CPC) [24], PFAM protein domain analysis [25,26], and the Coding Potential Assessment Tool (CPAT) [27]. Finally, the co-existing transcripts were designated as novel lncRNAs. Moreover, we performed structure and expression characteristic analyses of the novel lncRNAs and known lncRNAs.

### 2.6. Differentially Expressed Gene Analyses

HTseq v0.6.0 [28] was utilized to calculate the read count for each gene, and RPKM (Reads Per Kilobase Millon Mapped Reads) was used to estimate the expression level of genes in each sample. The formula is shown as

$$\text{RPKM} = \frac{10^6 \times R}{\text{NL}/10^3}. \tag{1}$$

*R* represents the number of reads assigned to a specific gene in each sample, *N* represents the total number of mapped reads in each sample, and L represents the length of a specific gene. Furthermore, DEseq2 [29] was used to identify the differentially expressed genes (DEGs), including protein-coding and non-coding genes, between Bf and Sol groups. DEGSeq (v1.16) [30] was used to determine the DEGs between two paired samples (Bf28 vs. Sol28; Bf35 vs. Sol35; Bf36 vs. Sol36). The *p*-value was adjusted by the Benjamini and Hochberg's approach to control the false discovery rate, and $q \leq 0.05$ and $|\text{log2\_ratio}| \geq 1$ were identified as DEGs.

Furthermore, nine lncRNAs were selected for validating the RNA-seq results using qRT-PCR. cDNA was synthesized with the total RNA derived from the same muscle samples for RNA-seq with the Takara PrimeScript$^{\text{TM}}$ II 1st Strand cDNA Synthesis Kit (TaKaRa, Dalian, China), and qRT-PCR was performed on a Step-One Plus Real-Time PCR System (Applied Biosystems, Carlsbad, CA, USA) using the AceQ qPCR SYBR Green Master Mix (Vazyme, Nanjing, China). The primers for quantitative analysis were shown in Additional File 5: Table S1. Relative expression levels were calculated using the $2^{-\Delta\Delta \text{ct}}$ value method [31] and porcine *GAPDH* was used for normalization of lncRNAs expression as an endogenous reference gene.

### 2.7. Target Prediction of DE LncRNAs and GO and KEGG Enrichment Analyses

The Spearman correlation coefficient between the expression of mRNAs and DE lncRNAs was calculated and the mRNAs with high a Spearman correlation coefficient ($r \geq 0.90$ or $\leq -0.90$) were selected as the trans-targets of DE lncRNAs, while the mRNAs with a distance of less than 50 kb to DE lncRNAs were selected as the cis-targets based on the possible region of target genes of lncRNAs [32,33]. To understand the biological functions of the target genes of DE lncRNAs, we performed the GO (Gene Ontology, http://geneontology.org/) and KEGG (Kyoto Encyclopedia of Genes and Genomes, http://www.kegg.jp/) enrichment analyses. The GO and KEGG enrichment of target genes of DE lncRNAs was implemented by the hypergeometric test in which the *p*-value was adjusted by multiple comparisons as *q*-value. The GO and KEGG terms with $p < 0.05$ were considered to be significantly enriched.

### 2.8. Expression Patterns of LncRNA MSTRG.42019

Among these DE lncRNAs, we found that lncRNA *MSTRG.42019* was a promising DE lncRNA affecting skeletal muscle growth and development and muscle fiber types. To validate its expression pattern, we detected its expression levels in Bf and Sol and analyzed its expression patterns in different tissues of adult pigs and 70-day-old fetuses (P70) and *Longissimus dorsi* muscles derived from different developmental stages, including 110-day-old fetuses (P110), the day of birth (D0), and 70 days after birth (D70). Relative expression levels of lncRNA *MSTRG.42019* were calculated using the $2^{-\Delta\Delta \text{ct}}$ value method [31] and normalized using porcine *GAPDH* as an endogenous reference gene.

### 2.9. Correlation Analyses of LncRNAs MSTRG.42019 and Meat Quality Traits

The skeletal muscle mass and muscle fiber types are closely related to porcine meat quality. To confirm the correlation between lncRNA *MSTRG.42019* and meat quality traits, we randomly selected 20 samples from 279 [(P) × (D)] × [(L) × (Y)] commercial pig populations and detected the expression levels of *Myosin heavy chain 7* (*MYH7*), *Myosin heavy chain 4* (*MYH4*), and *Myoglobin* (*MyoB*) using qRT-PCR and performed the correlation analyses between the expression of lncRNA *MSTRG.42019* and *MYH7*, *MYH4*, and *MyoB*. Furthermore, we performed the correlation analyses between the expression levels of lncRNA *MSTRG.42019* and meat quality traits, including carcass weight, back fat, intramuscular fat, $pH_{45min}$, $L_{24h}$, $a_{24h}$, $b_{24h}$, glycolytic potential, and drip loss.

### 2.10. Construction of LncRNA MSTRG.42019–mRNA Network

DEGs and the *trans-* or *cis*-targets of DE lncRNAs were derived from the RNA-seq data in this study, and the network, including DEGs and *trans-* or *cis*-targets of DE lncRNAs *MSTRG.42019*, was constructed using the Cytoscape software. The visualized network was edited based on the attribution of each gene [34]. Furthermore, we performed a pathway search for the DE target genes of lncRNA *MSTRG.42019* using the KEGG Mapper tool (https://www.kegg.jp/kegg/tool/map_pathway2.html).

### 2.11. Statistical Analyses

Statistical analyses were performed using the Prism 7 software (GraphPad Software, San Diego, CA, USA), and data were expressed as mean ± SE unless otherwise noted. Differences were tested using a two-tailed unpaired Student's *t*-test or a one-way analysis of variance (ANOVA) with Duncan's test. A *p*-value of <0.05 was considered significantly different.

## 3. Results

### 3.1. Generation of Transcriptome Data

In this study, we generated six RNA-seq libraries, namely, Bf28, Bf35, Bf36, Sol28, Sol35, and Sol36. The alignment results showed that the average ratio of rRNA mapping was less than 0.20%, and the ratio of clean Q30 bases (those with a base quality of >30 and an error rate of <0.001) in each library was higher than 94%, indicating high-quality raw reads from each library. After strict filtering, more than 15 Gb clean bases data corresponding to more than 100 million reads were obtained from each library after strick filtering, and the ratio of clean reads in each library reached more than 93%. The transcriptome sequence data generated from this study were deposited in NCBI SRA (accession code: PRJNA597666).

### 3.2. Novel LncRNAs Prediction and Characteristics Analyses

In this study, 5864 overlapped novel lncRNA transcripts without coding potential were identified using four kinds of predicted tools (Additional File 1: Figure S1). Detailed information for the coding potential of novel predicted lncRNAs are listed in Additional File 6: Table S2. Additional File 6: Table S2-1, Table S2-2, Table S2-3, and Table S2-4 represent the predicted results from CNCI, PFAM protein domain analysis, CPAT, and CPC, respectively. Moreover, our results showed that the exon number of lncRNAs was gathered at 2–4, while the exon number of mRNA was distributed in 1–29 (Additional File 2: Figure S2A). The length distribution showed that the length of lncRNAs was shorter than mRNA and mainly distributed in 200 bp and 2900 bp (Additional File 2: Figure S2B). In addition, our results indicated that the relative expression level of lncRNAs was lower than that of mRNA (Additional File 2: Figure S2C).

### 3.3. Identification of DE LncRNAs and Target Genes Prediction

Genomic expression analysis revealed that 4362 novel and 219 known lncRNA genes were detectable from Bf vs. Sol. In these, 3786 novel and 125 known lncRNA genes were detectable from Bf28 vs. Sol28, and 3741 novel and 126 known lncRNA genes were detectable from Bf35 and Sol35, and 3787 novel and 120 known lncRNA genes were detectable from Bf36 and Sol36 (Additional File 7: Table S3). Furthermore, 92 DE lncRNA genes were identified between Bf and Sol, including 35 up-regulated and 57 down-regulated lncRNAs (Figure 1A, Additional File 8: Table S4). Hierarchical clustering showed that lncRNA expression patterns between Bf and Sol were distinguishable (Figure 1B), indicating a substantial difference exists between the Bf and Sol groups, while a small variation exists among three biological replicates. Moreover, 600 up-regulated and 775 down-regulated lncRNA genes were identified in Bf28 vs. Sol28; 1077 up-regulated and 349 down-regulated lncRNA genes were identified in Bf35 vs. Sol35; and 618 up-regulated and 602 down-regulated lncRNA genes were identified in Bf36 vs. Sol36 (Additional File 8: Table S4). In addition, 246 DE coding genes were identified between Bf and Sol (Additional File 9: Table S5).

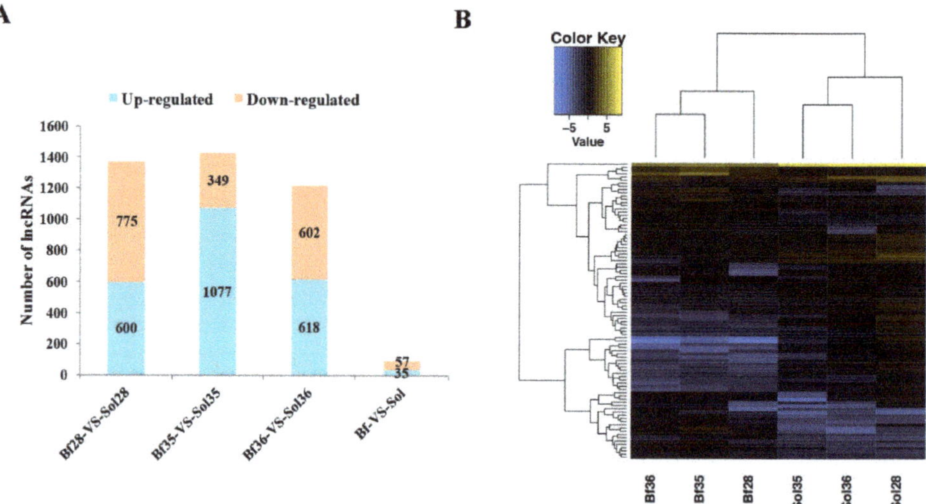


**Figure 1.** Statistics and heat map analyses of differentially expressed (DE) long non-coding RNAs (lncRNAs). (**A**) Statistics of DE lncRNAs. The X-axis represents the different compared groups, *Biceps femoris* (Bf) vs. *Soleus* (Sol) indicates the DE lncRNAs from the DEseq2 method; Bf28 vs. Sol28, Bf35 vs. Sol35, and Bf36 vs. Sol36 indicate DE lncRNAs from the DEGseq method. The Y-axis shows the number of DE lncRNAs. (**B**) Heat map analysis of DE lncRNAs between Bf and Sol muscles. Heat map analysis was conducted with 92 overlapped DE lncRNAs among three different comparative groups (Bf28 vs. Sol28, Bf35 vs. Sol35, and Bf36 vs. Sol36). Each column represents a sample and each row represents a DE lncRNA. Yellow and blue gradients indicate an increase and decrease in gene expression abundance, respectively.

To validate the RNA-Seq results, nine DE lncRNAs, including two known lncRNAs and seven novel lncRNAs, were selected to confirm their expression differences between Bf and Sol by qRT-PCR assay. We found that the expression patterns of these lncRNAs were in agreement with that in RNA-Seq (Figure 2A) and the correlation coefficient of two methods was 0.8957 ($p$ = 0.0011) (Figure 2B). These results indicate that DE lncRNAs identified from RNA-Seq were reliable. In addition, since lncRNAs exert their functions mainly via interaction with protein-coding genes through *cis* or *trans*, *cis* and *trans* target genes prediction of DE lncRNAs from Bf vs. Sol was performed. A total of 111 *cis* target genes and 9,050 *trans* target genes were predicted (Additional File 10: Table S6).

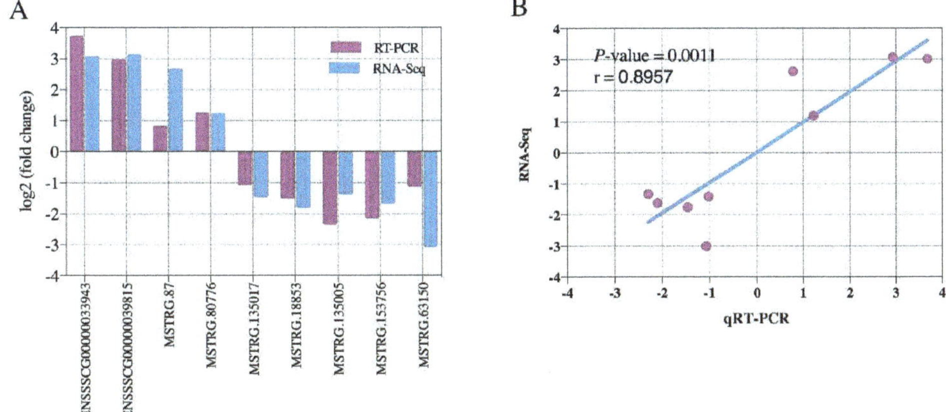


**Figure 2.** Validation of DE lncRNAs and correlation analysis. (**A**) Validation of DE lncRNAs by qRT-PCR ($n = 3$). Relative expression levels were calculated using the $2^{-\Delta\Delta ct}$ value method and porcine *GAPDH* was used for normalization of lncRNAs expression levels as an endogenous reference gene. The X-axis represents DE lncRNAs and the Y-axis represents the $\log_2$ (fold change) for qRT-PCR and RNA-Seq. (**B**) Correlation analysis of the expression of DE lncRNAs between qRT-PCR and RNA-Seq. The X and Y-axis represent the $\log_2$ (fold change) measured by qRT-PCR and RNA-Seq, respectively.

### *3.4. GO and KEGG Pathway Enrichment Analyses of Target Genes of DE LncRNAs*

To understand the function of these DE lncRNAs, we performed the functional enrichment analyses for their target genes. GO analysis results showed that some interesting GO terms related to skeletal muscle fiber properties, such as lipid metabolic process, muscle system process, muscle contraction, muscle structure development, and ATP metabolic process, were significantly enriched in Biological Process (BP); and the mitochondrial part, mitochondrion, mitochondrial membrane part, respiratory chain complex, and mitochondrial membrane were significantly enriched in Cellular Component (CC); the catalytic activity, protein serine/threonine kinase activity, kinase activity, protein kinase activity, and calcium ion binding were significantly enriched in Molecular Function (MF) (Additional File 11: Table S7). KEGG pathway analysis showed that Alzheimer's disease, Parkinson's disease, Huntington's disease, non-alcoholic fatty liver disease, and thermogenesis are the five most significantly enriched pathways; moreover, glycolysis/gluconeogenesis, cardiac muscle contraction, fatty acid metabolism, fatty acid degradation, and pyruvate metabolism pathways were also significantly enriched (Additional file 11: Table S7).

### *3.5. DE LncRNAs Affecting Muscle Fiber Types*

In this study, 53 overlapped DE lncRNAs were filtered out from Bf28 vs. Sol28, Bf35 vs. Sol35, Bf36 vs. Sol36, and Bf and Sol (Figure 3A and Additional File 12: Table S8), and the heat map of 53 DE lncRNAs with distinguishable expression patterns is shown in Figure 3B. Among these DE lncRNAs, we noted that the lncRNA *MSTRG.42019* expression level was significantly higher in Sol muscle than that in Bf muscle (Figures 3B and 4A) and was relatively highly expressed in skeletal muscle than that in other tissues in adult pig (Figure 4B and Figure S3A) and fetuses (Figure 4C). In addition, the expression patterns of lncRNA *MSTRG.42019* were shown to up-regulate before birth from 110 days of pregnancy (P110) to the day of birth (D0) and then down-regulated after birth (Figure 4D and Figure S3B).

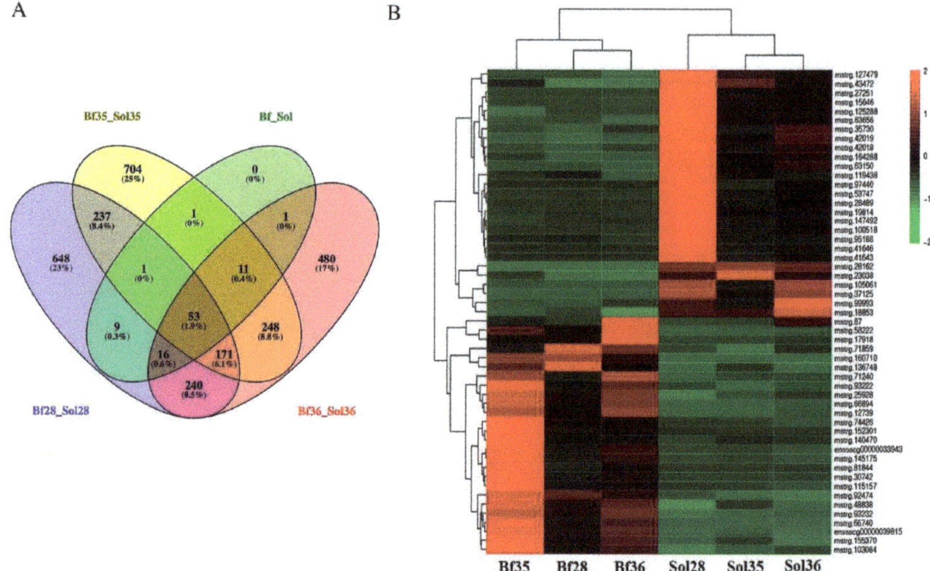


**Figure 3.** Overlapped and heat map analyses of DE lncRNAs. (**A**) Venn diagram of DE lncRNAs. A total of 53 overlapped DE lncRNAs were obtained from DEseq2 and DEGseq methods. Different color represents a different combination, and the number in the overlapped region represents the overlapped DE lncRNAs number. (**B**) Heat map analysis of 53 DE lncRNAs. Heat map analysis was conducted with 53 overlapped DE lncRNAs from four different comparative groups (Bf28 vs. Sol28, Bf35 vs. Sol35, Bf36 vs. Sol36, and Bf vs. Sol). Each column represents a sample and each row represents a DE lncRNA. Red and green gradients indicate an increase and decrease in gene expression abundance, respectively.

### *3.6. Correlation between LncRNA MSTRG.42019 Expression and Meat Quality Traits*

To elucidate the relationship between lncRNA *MSTRG.42019* and meat quality traits, we detected the expression levels of lncRNA *MSTRG.42019* and *MYH7*, *MYH4*, and *MyoB* in 20 *Longissimus dorsi* muscles derived from a 279 [(P) × (D)] × [(L) × (Y)] commercial hybrid pig population and performed correlation analyses between the expression of lncRNA *MSTRG.42019* and *MYH7*, *MYH4*, *MyoB*, and meat quality traits. Our results showed that expression levels of lncRNA *MSTRG.42019* were positively correlated with the mRNA expression level of *MYH7* ($r = 0.659$, $p = 0.0016$) (Figure 5A), with no correlation with the expression levels of *MyoB* and *MYH4* (Additional File 4: Figure S4A,B). Additionally, the expression level of lncRNA *MSTRG.42019* was negatively correlated with glycolytic potential ($r = -0.5447$, $p = 0.013$) (Figure 5B) and drip loss ($r = -0.5085$, $p = 0.0221$) (Figure 5C), respectively. Furthermore, no correlation was observed between the expression of lncRNA *MSTRG.42019* and the carcass weight (Additional File 4: Figure S4C), back fat (Additional File 4: Figure S4D), $pH_{45min}$ (Additional File 4: Figure S4E), $L_{24h}$ (Additional File 4: Figure S4F), $a_{24h}$ (Additional File 4: Figure S4G), $b_{24h}$ (Additional File 4: Figure S4H), and intramuscular fat (Additional File 4: Figure S4I).

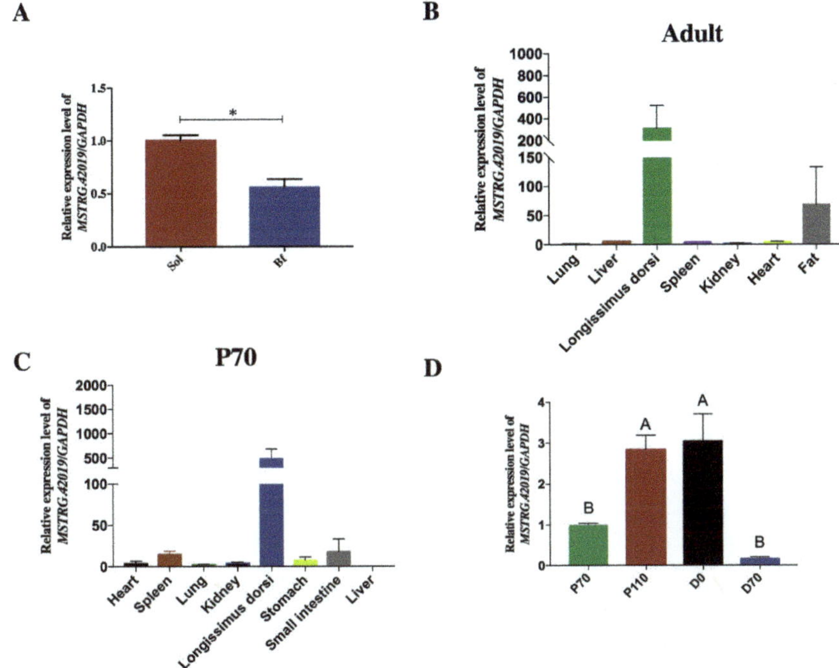


**Figure 4.** Expression patterns of lncRNA *MSTRG.42019*. (**A**) Expression of lncRNA *MSTRG.42019* in Bf and Sol muscles. (**B**) Expression profile of lncRNA *MSTRG.42019* in different tissues of adult pigs, $n = 3$. (**C**) Expression profile of lncRNA *MSTRG.42019* in different tissues of 70-day-old fetuses, $n = 3$. (**D**) Expression patterns of lncRNA *MSTRG.42019* in *Longissimus dorsi* muscles derived from different developmental stages, $n = 3$. Expression levels were determined by qRT-PCR and relative expression levels were calculated using the $2^{-\Delta\Delta ct}$ method and normalized to the expression of *GAPDH*. The expression of lncRNA *MSTRG.42019* in the sample of the first column was determined as control and normalized to 1. All data are presented as mean ± SE, and an unpaired Student's $t$-test in the Prism 7 software was performed to evaluate significant differences between Bf and Sol. ANOVA with Duncan's test was used to evaluate significant differences between groups P70 (70 days of pregnancy), P110 (110 days of pregnancy), D0 (the day of birth), and D70 (70 days after birth). * $p \leq 0.05$, different letters above the bars indicate significant differences; $p < 0.01$.

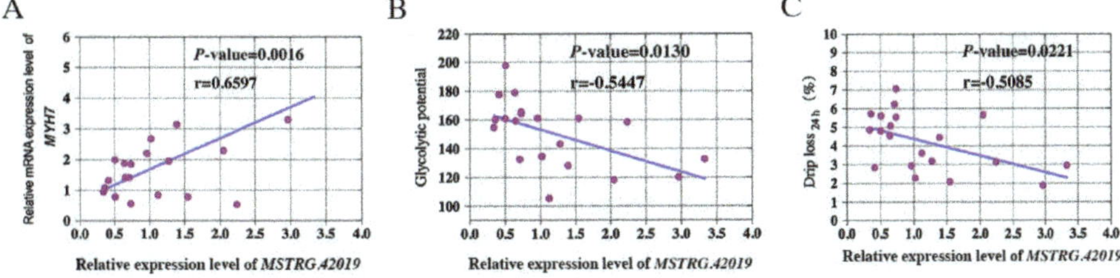


**Figure 5.** Correlation analyses between the expression of lncRNA *MSTRG.42019* and muscle fiber-related genes and meat quality traits. (**A**) Correlation between the expression of lncRNA *MSTRG.42019* and *Myosin heavy chain 7* (*MYH7*); (**B**) correlation between the expression of lncRNA *MSTRG.42019* and glycolytic potential of *Longissimus dorsi* muscles; (**C**) correlation between the expression of lncRNA *MSTRG.42019* and drip loss of *Longissimus dorsi* muscles. Twenty *Longissimus dorsi* muscles were derived from a 279 [(P) × (D)] × [(L) × (Y)] commercial hybrid pig population.

### 3.7. LncRNA *MSTRG.42019*–mRNA Interaction Network

To further explore the underlying function of lncRNA *MSTRG.42019*, we constructed a network map of lncRNA *MSTRG.42019* and its target genes and performed the KEGG pathway search for the DE target genes of lncRNA *MSTRG.42019* (Figure 6). The results showed that the *parvalbumin* (*PVALB*) was the down-regulated gene targeted by lncRNA *MSTRG.42019*. Moreover, the up-regulated expressed genes, including *Glycerol-3-phosphate acyltransferase* (*GPAT*), *Tropomyosin-3* (*TPM3*), *R-spondin 3* (*RSPO3*), *Heat shock protein family B member 6* (*HSPB6*), *Leucine-rich glioma inactivated 1* (*LGI1*), *Cholinergic receptor nicotinic alpha 6 subunit* (*CHRNA6*), *Solute carrier family 35 member F1* (*SLC35F1*), *Acyl-CoA dehydrogenase very long chain* (*ACADVL*), and *Microtubule-associated serine/threonine kinase 2* (*MAST2*) were also predicted as the targets of lncRNA *MSTRG.42019*. In addition, many non-DE genes were targeted by lncRNA *MSTRG.42019*. Furthermore, the KEGG pathway search showed that the DEGs *TPM3*, *RSPO3*, *ACADVL*, and *CHRNA6* were mapped to specific pathways.

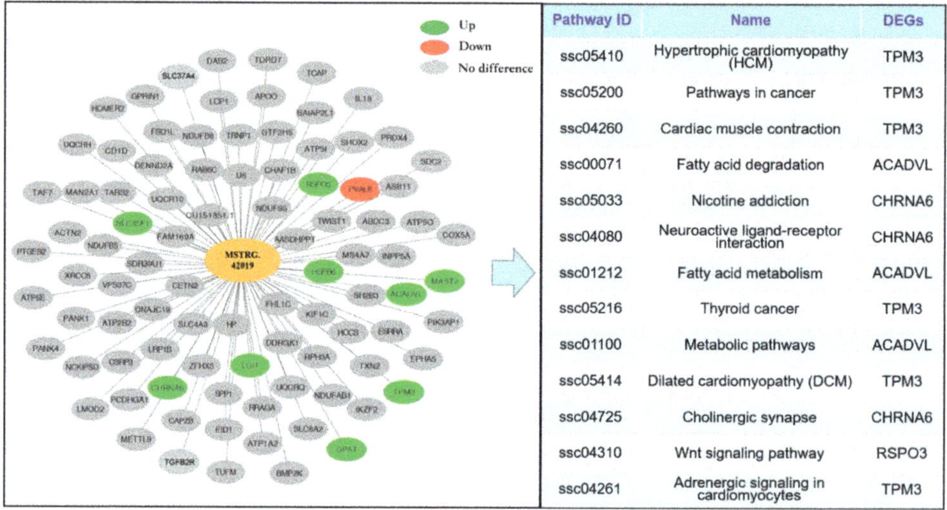


| Pathway ID | Name | DEGs |
|---|---|---|
| ssc05410 | Hypertrophic cardiomyopathy (HCM) | TPM3 |
| ssc05200 | Pathways in cancer | TPM3 |
| ssc04260 | Cardiac muscle contraction | TPM3 |
| ssc00071 | Fatty acid degradation | ACADVL |
| ssc05033 | Nicotine addiction | CHRNA6 |
| ssc04080 | Neuroactive ligand-receptor interaction | CHRNA6 |
| ssc01212 | Fatty acid metabolism | ACADVL |
| ssc05216 | Thyroid cancer | TPM3 |
| ssc01100 | Metabolic pathways | ACADVL |
| ssc05414 | Dilated cardiomyopathy (DCM) | TPM3 |
| ssc04725 | Cholinergic synapse | CHRNA6 |
| ssc04310 | Wnt signaling pathway | RSPO3 |
| ssc04261 | Adrenergic signaling in cardiomyocytes | TPM3 |

**Figure 6.** LncRNA *MSTRG.42019*–target mRNA interaction network and KEGG pathway search. Cytoscape was used to construct the interaction network between lncRNA *MSTRG.42019* and target mRNAs. Red and green represent down-regulated and up-regulated target genes from Bf and Sol, respectively, and gray indicates non-DE target genes. The KEGG pathway search for DE target genes was conducted using the website tool KEGG Mapper (https://www.kegg.jp/kegg/tool/map_pathway2.html).

## 4. Discussion

In this study, we predicted 5864 novel lncRNA transcripts from the six skeletal muscle libraries, and our data showed that the characteristics of exon number, length distribution, and genomic expression level of these lncRNAs are similar to the previous report [35] and distinguished with the mRNAs. These predicted novel lncRNAs may play an important role in skeletal muscle growth and development, but the specific function still needs to be further explored. Moreover, 92 DE lncRNAs were identified between Bf and Sol from the RNA-seq analyses. Bf mainly composed of type IIB muscle fiber is a type of fast-twitch muscle, whereas Sol enriched in type I muscle fiber is a type of slow-twitch muscle [18]. Thus, these DE lncRNAs are the key candidates for the regulation of skeletal muscle fiber types. Certain types of skeletal muscle fiber are known to play a critical role in determining the meat quality postmortem in livestock, such as pH, meat color, and drip loss [6]. Thus, the DE lncRNAs identified in this study are the promising candidates influencing meat quality traits. Moreover, it is known that the alteration of muscle fiber types leads to the disorder of muscle physiological and metabolic properties, and thereby may cause many human diseases, such as muscular dystrophy, cardiomyopathic disease, and type 2 diabetes [2–5]. Thus, our data might provide some insights into human muscle diseases.

To understand the potential function of the identified lncRNAs, the target genes of DE lncRNAs were subjected to functional enrichment analyses. In the BP category, we found that the ATP metabolic process is one of the significantly enriched GO terms. ATP is the energy currency of living cells. Many biological processes are depended on the energy of ATP hydrolysis, for instance, the chromatin-structure dynamics critical for the regulation of gene expression and the chromosome function relies on ATP-dependent chromatin-remodeling complexes [36,37]. Skeletal muscle growth and development are also ATP-dependent, where the ATP induces muscle hypertrophy in slow muscle Sol via activating mTOR signaling pathway [38]. Furthermore, ATP is involved in the regulation of muscle fiber types and glucose metabolism by activation of P2Y receptors, phosphatidylinositol 3-kinase, Akt, and AS160 [39]. However, the regulatory mechanisms of lncRNAs on ATP metabolism and the effect of the lncRNA-ATP signaling pathway on skeletal muscle growth and muscle fiber conversion are still unknown.

Our KEGG pathway analysis showed that the cardiac muscle contraction pathway is significantly enriched. This result suggests that the DE lncRNAs identified in the skeletal muscle also play an important role in myocardial development. In addition, the glycolysis/gluconeogenesis, closely related to skeletal muscle fiber types, is significantly enriched. Moreover, fatty acid metabolism and fatty acid degradation pathways are also significantly enriched, while recent studies found that the change of fatty acid composition was accompanied by the transformation of skeletal muscle fiber types [40,41]. Thus, it will be interesting to investigate the specific regulatory mechanisms of the skeletal muscle fiber types and fatty acid composition mediated by the DE lncRNAs identified in this study. Different skeletal muscle fibers exhibit different glucose metabolic properties, while the pyruvate metabolism pathway critical for glucose metabolism is significantly enriched. The significantly enriched oxidative phosphorylation is the major source of ATP, while ATP is essential for many biological processes, including the growth and development and the fiber types of skeletal muscle [38,39]. Therefore, it is intriguing to investigate the potential mechanisms of the oxidative phosphorylation pathway regulating muscle fiber types which are mediated by the DE lncRNAs identified in this study. Furthermore, Alzheimer's disease, Huntington's disease, and Parkinson's disease are three of the most significantly enriched pathways, thus the relationship between the skeletal muscle fiber types and these diseases is worth exploring.

To screen the key candidate lncRNAs affecting skeletal muscle fiber types, we performed the overlapped analysis of DE lncRNAs from Bf28 vs. Sol28, Bf35 vs. Sol35, Bf36 vs. Sol36, and Bf vs. Sol and identified 53 overlapped DE lncRNAs. Sol muscle (slow muscle) and Bf muscles (fast muscle) are two different skeletal muscles with various muscle fibers and have different physiological and metabolic properties, such as glycolytic potential. Therefore, these overlapped DE lncRNAs are most likely to regulate the formation of different muscle fiber types and participate in the regulation of glucose metabolism in skeletal muscle. Intriguingly, among these DE lncRNAs, we found that lncRNA *MSTRG.42019* is highly expressed in skeletal muscle and its expression is significantly higher in Sol muscle than in Bf muscle, suggesting that lncRNA *MSTRG.42019* may contribute to skeletal muscle fiber type conversion. Moreover, we found that the expression patterns of lncRNA *MSTRG.42019* were shown to up-regulate before birth from 110 days of pregnancy (P110) to the day of birth (D0) and then down-regulate after birth. However, the complete expression pattern of lncRNA *MSTRG.42019* at different developmental stages needs to be verified using more stages. Overall, our data suggest that lncRNA *MSTRG.42019* may participate in the regulation of skeletal muscle growth and development and are involved in skeletal muscle fibers conversion. However, understanding of the specific regulatory mechanisms warrants further study.

The above studies suggest that lncRNA *MSTRG.42019* is a promising lncRNAs affecting skeletal muscle fiber types, while skeletal muscle fibers are closely related to meat quality traits [42]. Therefore, we subsequently performed a correlation analysis between the lncRNA *MSTRG.42019* expression and meat quality traits and skeletal muscle fiber-related marker genes, *MYH7*, *MYH4*, and *MyoB*. Our results showed that lncRNA *MSTRG.42019* expression is significantly positively correlated with the *MYH7* expression, which is consistent with the expression of lncRNA *MSTRG.42019* higher in the Sol muscle (slow muscle) than that in the Bf muscle (fast muscle). These results suggest that lncRNA *MSTRG.42019* may regulate the formation of slow muscle fiber. Furthermore, we observed that the lncRNA *MSTRG.42019* expression is significantly correlated with glycolytic potential and drip loss. The slow muscle fibers (type I) display stronger oxidative metabolism capacity. On the other hand, fast muscle fibers (type IIb) exhibit higher glycolytic capacity, which is in accordance with the result of the lncRNA *MSTRG.42019* expression level significantly negatively correlated with glycolytic potential. The glycolytic potential is critical for pH decline postmortem, in turn, the change of pH is closely related to drip loss [43,44]. The change of pH in slow muscle with lower glycolytic potential postmortem is relatively moderate when compared with fast muscle with higher glycolytic potential, and the drip loss in slow muscle is relatively lower. The lncRNA *MSTRG.42019* expression level is higher in slow muscle than fast muscle, which is consistent with the result of its expression negatively correlated with the drip loss.

The predicted interaction network is very helpful for further study of the potential regulatory mechanism. In this study, we constructed a network map using the lncRNA *MSTRG.42019* and its target genes. *Parvalbumin* (*PVALB*) is the only down-regulated gene targeted by lncRNA *MSTRG.42019* in the interaction network. PVALB, a $Ca^{2+}$ binding protein, is known to play a key role in muscle relaxation and it acts as an intracellular $Ca^{2+}$ buffer to determine the duration and magnitude of the activating $Ca^{2+}$ signal and the force and duration of contraction [45,46]. *PVALB* is positively associated with relaxation speed in skeletal muscle, but less PVALB expressed in the slow-twitch skeletal muscles, cardiac, or smooth muscle [47]. Moreover, ectopic expression of *PVALB* by injection of its cDNA into the slow-twitch muscle of rats leads to an increased relaxation rate [48]. Conversely, the fast-twitch muscle of the *PVALB*-knockout mouse showed a slower $Ca^{2+}$ transient and relaxation rate than the wild-type mouse [49]. Since our result showed that lncRNA *MSTRG.42019* is highly expressed in Sol, we speculated that lncRNA *MSTRG.42019* could promote the formation of slow muscle fiber via $Ca^{2+}$ signaling pathway mediated by *PVLAB*. Moreover, Glycerol-3-phosphate acyltransferase (GPAT) is a limited enzyme that can acylate the glycerol 3-phosphate in the first committed step in triacylglycerol (TAG) synthesis phosphate pathway [50]. Therefore, lncRNA *MSTRG.42019* may affect the intramuscular fat content and fatty acid composition by up-regulating *GPAT* expression. *Tropomyosin-3* (*TPM3*) is required for the early steps of myofibril formation in zebrafish [51] and our result showed that porcine lncRNA *MSTRG.42019* is highly expressed in P110, which suggests that the effects of lncRNA *MSTRG.42019* on embryonic or fetal skeletal muscle development might be associated with the *TPM3*. In addition, lncRNA *MSTRG.42019* also may up-regulate the expression of *RSPO3*, *HSPB6*, *LGI1*, *CHRNA6*, *SL35F1*, *ACADVL*, and *MAST2*. Furthermore, the KEGG pathway search showed that the DEGs *TPM3* involves in cardiomyopathy, *RSPO3* participates in Wnt signaling pathway, and *ACADVL* is critical for fatty acid metabolism. However, the relationship between lncRNA *MSTRG.42019* and its target genes, the exact mechanisms of lncRNA *MSTRG.42019*-related growth and development, and the conversion of the fiber types of skeletal muscle still need to be further explored.

In addition, many non-DE genes are targeted by lncRNA *MSTRG.42019*, such as *Cytochrome C Oxidase Subunit 5A* (*COX5A*). *COX5A* encodes a subunit of the respiratory chain complex IV (cytochrome c oxidase) [52] and its expression is associated with niacin-induced changes in fiber type distribution and expression of MHC isoforms. Moreover, we found that lncRNA *MSTRG.42019* targets several ATPase and ATP synthetase genes, including *ATP1A2*, *ATP2B2*, *ATP5E*, *ATP5I*, and *ATP5O*. These results were consistent with the results that the ATP metabolic process is significantly enriched in GO analysis, suggesting that lncRNA *MSTRG.42019* may play a pivotal role in energy metabolism. Collectively, our data provide preliminary evidence that lncRNA *MSTRG.42019* may play an important role in muscle fiber type specification, mitochondrial respiration activity, and energy metabolism.

## 5. Conclusions

We identified 92 DE lncRNAs between Bf and Sol which are promising candidates regulating skeletal muscle fibers and muscle fiber-related metabolic processes and found some potential signaling pathways related to skeletal muscle fiber formation and metabolic properties mediated by these DE lncRNAs, such as ATP metabolic process pathway, fatty acid metabolic pathway, and pyruvate metabolism. Furthermore, we identified that a skeletal muscle highly expressed DE lncRNA *MSTRG.42019* and documented that it is closely associated with meat quality traits. Overall, our data provide a solid foundation for an in-depth investigation of the lncRNA regulatory mechanisms of skeletal muscle growth and development and skeletal muscle fiber conversion.

**Supplementary Materials:** The following are available online at http://www.mdpi.com/2073-4425/11/8/883/s1, Figure S1: Venn diagram of non-coding potential novel lncRNAs. The coding potential of novel lncRNAs was predicted using CNCI, CPC, PFAM, CPAT analysis. Figure S2: Characteristics of lncRNAs and mRNAs. (A) Length distribution of lncRNAs and mRNAs. (B) The number of lncRNAs and mRNAs exon. (C) Distribution of lncRNAs and mRNAs expression. Table S1: Primers used for qRT-PCR analysis. Figure S3: Expression patterns of lncRNA *MSTRG.42019* in Bf and Sol muscles, different tissues of adult pigs, and *Longissimus dorsi* muscles derived from different developmental stages, $n = 3$. Expression levels were determined by qRT-PCR and relative expression levels were calculated using the $2^{-\Delta\Delta ct}$ method and normalized to the expression of *HPRT*. The expression of lncRNA *MSTRG.42019* in the sample of the first column was determined as control and normalized to 1. All data are presented as mean ± SE. Figure S4: Correlation analyses between the expression of lncRNA *MSTRG.42019* and genes and meat quality traits. Correlation between the expression of lncRNA *MSTRG.42019* and *MyoB* (A), *MYH4* (B), carcass weight (C), back fat (D), $pH_{45min}$ (E), $L_{24h}$ (F), $a_{24h}$ (G), $b_{24h}$ (H), and intramuscular fat (I), respectively. Twenty *Longissimus dorsi* muscles were derived from a 279 [(P) × (D)] × [(L) × (Y)] commercial hybrid pig population. Table S2: Novel predicted lncRNA transcripts; Table S3: Genomic expression analysis of all lncRNA genes; Table S4: DE lncRNAs from Bf and Sol groups; Table S5: DE mRNAs from Bf and Sol groups; Table S6: Target gene prediction of lncRNAs; Table S7: GO and KEGG enrichment analyses; Table S8: 53 DE lncRNAs.

**Author Contributions:** W.W. conceived the experiments; R.L., B.L., and A.J. performed the experiments and analyzed the data. Y.C., Z.Z., X.Z., R.L., and W.W. wrote and revised the paper. L.H. and K.-H.K. revised the paper. H.L. reviewed the paper. All authors have read and agreed to the published version of the manuscript.

**Funding:** This research was supported by the National Natural Science Foundation (31591920) and Nanjing Agricultural University Research Start-up Funding (060804008).

**Conflicts of Interest:** The authors declare that they have no competing interests.

## Abbreviations

Bf: *Biceps femoris*; Sol: *Soleus*; DE: differentially expressed; GO: Gene Ontology; KEGG: Kyoto Encyclopedia of Genes and Genomes; lncRNAs: long non-coding RNAs. *MYH4*: *Myosin heavy chain 4*; *MYH7*: *Myosin heavy chain 7*.

## References

1. Gundersen, K. Excitation-transcription coupling in skeletal muscle: The molecular pathways of exercise. *Biol. Rev. Camb. Philos. Soc.* **2011**, *86*, 564–600. [CrossRef] [PubMed]
2. Bassel-Duby, R.; Olson, E.N. Signaling pathways in skeletal muscle remodeling. *Annu. Rev. Biochem.* **2006**, *75*, 19–37. [CrossRef] [PubMed]
3. Reyes, N.L.; Banks, G.B.; Tsang, M.; Margineantu, D.; Gu, H.; Djukovic, D.; Chan, J.; Torres, M.; Liggitt, H.D.; Hirenallur, S.D.; et al. Fnip1 regulates skeletal muscle fiber type specification, fatigue resistance, and susceptibility to muscular dystrophy. *Proc. Natl. Acad. Sci. USA* **2015**, *112*, 424–429. [CrossRef] [PubMed]
4. Petchey, L.K.; Risebro, C.A.; Vieira, J.M.; Roberts, T.; Bryson, J.B.; Greensmith, L.; Lythgoe, M.F.; Riley, P.R. Loss of Prox1 in striated muscle causes slow to fast skeletal muscle fiber conversion and dilated cardiomyopathy. *Proc. Natl. Acad. Sci. USA* **2014**, *111*, 9515–9520. [CrossRef]
5. Boyer, J.G.; Prasad, V.; Song, T.; Lee, D.; Fu, X.; Grimes, K.M.; Sargent, M.A.; Sadayappan, S.; Molkentin, J.D. ERK1/2 signaling induces skeletal muscle slow fiber-type switching and reduces muscular dystrophy disease severity. *JCI Insight* **2019**, 5. [CrossRef] [PubMed]
6. Ryu, Y.C.; Kim, B.C. The relationship between muscle fiber characteristics, postmortem metabolic rate, and meat quality of pig longissimus dorsi muscle. *Meat Sci.* **2005**, *71*, 351–357. [CrossRef]

7. Zammit, P.S. Function of the myogenic regulatory factors Myf5, MyoD, Myogenin and MRF4 in skeletal muscle, satellite cells and regenerative myogenesis. *Semin. Cell Dev. Biol.* **2017**, *72*, 19–32. [CrossRef]
8. Blais, A.; Tsikitis, M.; Acosta-Alvear, D.; Sharan, R.; Kluger, Y.; Dynlacht, B.D. An initial blueprint for myogenic differentiation. *Genes Dev.* **2005**, *19*, 553–569. [CrossRef]
9. Williams, A.H.; Liu, N.; van Rooij, E.; Olson, E.N. MicroRNA control of muscle development and disease. *Curr. Opin. Cell Biol.* **2009**, *21*, 461–469. [CrossRef]
10. Mok, G.F.; Lozano-Velasco, E.; Munsterberg, A. microRNAs in skeletal muscle development. *Semin. Cell Dev. Biol.* **2017**, *72*, 67–76. [CrossRef]
11. Li, Y.; Chen, X.; Sun, H.; Wang, H. Long non-coding RNAs in the regulation of skeletal myogenesis and muscle diseases. *Cancer Lett.* **2018**, *417*, 58–64. [CrossRef] [PubMed]
12. Simionescu-Bankston, A.; Kumar, A. Noncoding RNAs in the regulation of skeletal muscle biology in health and disease. *J. Mol. Med.* **2016**, *94*, 853–866. [CrossRef] [PubMed]
13. Wilusz, J.E.; Sunwoo, H.; Spector, D.L. Long noncoding RNAs: Functional surprises from the RNA world. *Genes Dev.* **2009**, *23*, 1494–1504. [CrossRef]
14. Cesana, M.; Cacchiarelli, D.; Legnini, I.; Santini, T.; Sthandier, O.; Chinappi, M.; Tramontano, A.; Bozzoni, I. A long noncoding RNA controls muscle differentiation by functioning as a competing endogenous RNA. *Cell* **2011**, *147*, 358–369. [CrossRef] [PubMed]
15. Dey, B.K.; Pfeifer, K.; Dutta, A. The H19 long noncoding RNA gives rise to microRNAs miR-675-3p and miR-675-5p to promote skeletal muscle differentiation and regeneration. *Genes Dev.* **2014**, *28*, 491–501. [CrossRef]
16. Wang, L.; Zhao, Y.; Bao, X.; Zhu, X.; Kwok, Y.K.; Sun, K.; Chen, X.; Huang, Y.; Jauch, R.; Esteban, M.A.; et al. LncRNA Dum interacts with Dnmts to regulate Dppa2 expression during myogenic differentiation and muscle regeneration. *Cell Res.* **2015**, *25*, 335–350. [CrossRef]
17. Anderson, D.M.; Anderson, K.M.; Chang, C.L.; Makarewich, C.A.; Nelson, B.R.; McAnally, J.R.; Kasaragod, P.; Shelton, J.M.; Liou, J.; Bassel-Duby, R.; et al. A micropeptide encoded by a putative long noncoding RNA regulates muscle performance. *Cell* **2015**, *160*, 595–606. [CrossRef]
18. Li, B.; Dong, C.; Li, P.; Ren, Z.; Wang, H.; Yu, F.; Ning, C.; Liu, K.; Wei, W.; Huang, R.; et al. Identification of candidate genes associated with porcine meat color traits by genome-wide transcriptome analysis. *Sci. Rep.* **2016**, *6*, 35224. [CrossRef]
19. Dong, C.; Zhang, X.; Liu, K.; Li, B.; Chao, Z.; Jiang, A.; Li, R.; Li, P.; Liu, H.; Wu, W. Comprehensive Analysis of Porcine Prox1 Gene and Its Relationship with Meat Quality Traits. *Animals* **2019**, *9*, 744. [CrossRef]
20. Langmead, B.; Salzberg, S.L. Fast gapped-read alignment with Bowtie 2. *Nat. Methods* **2012**, *9*, 357–359. [CrossRef]
21. Kim, D.; Langmead, B.; Salzberg, S.L. HISAT: A fast spliced aligner with low memory requirements. *Nat. Methods* **2015**, *12*, 357–360. [CrossRef] [PubMed]
22. Pertea, M.; Pertea, G.M.; Antonescu, C.M.; Chang, T.C.; Mendell, J.T.; Salzberg, S.L. StringTie enables improved reconstruction of a transcriptome from RNA-seq reads. *Nat. Biotechnol.* **2015**, *33*, 290–295. [CrossRef] [PubMed]
23. Sun, L.; Luo, H.; Bu, D.; Zhao, G.; Yu, K.; Zhang, C.; Liu, Y.; Chen, R.; Zhao, Y. Utilizing sequence intrinsic composition to classify protein-coding and long non-coding transcripts. *Nucleic Acids Res.* **2013**, *41*, e166. [CrossRef] [PubMed]
24. Kong, L.; Zhang, Y.; Ye, Z.Q.; Liu, X.Q.; Zhao, S.Q.; Wei, L.; Gao, G. CPC: Assess the protein-coding potential of transcripts using sequence features and support vector machine. *Nucleic Acids Res.* **2007**, *35*, W345–W349. [CrossRef] [PubMed]
25. Bateman, A.; Birney, E.; Cerruti, L.; Durbin, R.; Etwiller, L.; Eddy, S.R.; Griffiths-Jones, S.; Howe, K.L.; Marshall, M.; Sonnhammer, E.L. The Pfam protein families database. *Nucleic Acids Res.* **2002**, *30*, 276–280. [CrossRef] [PubMed]
26. Mistry, J.; Bateman, A.; Finn, R.D. Predicting active site residue annotations in the Pfam database. *BMC Bioinform.* **2007**, *8*, 298. [CrossRef]
27. Wang, L.; Park, H.J.; Dasari, S.; Wang, S.; Kocher, J.P.; Li, W. CPAT: Coding-Potential Assessment Tool using an alignment-free logistic regression model. *Nucleic Acids Res.* **2013**, *41*, e74. [CrossRef]
28. Anders, S.; Pyl, P.T.; Huber, W. HTSeq—A Python framework to work with high-throughput sequencing data. *Bioinformatics* **2015**, *31*, 166–169. [CrossRef]

29. Love, M.I.; Huber, W.; Anders, S. Moderated estimation of fold change and dispersion for RNA-seq data with DESeq2. *Genome Biol.* **2014**, *15*, 550. [CrossRef]
30. Wang, L.; Feng, Z.; Wang, X.; Wang, X.; Zhang, X. DEGseq: An R package for identifying differentially expressed genes from RNA-seq data. *Bioinformatics* **2010**, *26*, 136–138. [CrossRef]
31. Livak, K.J.; Schmittgen, T.D. Analysis of relative gene expression data using real-time quantitative PCR and the 2(-Delta Delta C(T)) Method. *Methods* **2001**, *25*, 402–408. [CrossRef] [PubMed]
32. Guttman, M.; Amit, I.; Garber, M.; French, C.; Lin, M.F.; Feldser, D.; Huarte, M.; Zuk, O.; Carey, B.W.; Cassady, J.P.; et al. Chromatin signature reveals over a thousand highly conserved large non-coding RNAs in mammals. *Nature* **2009**, *458*, 223–227. [CrossRef] [PubMed]
33. Wang, L.; Yang, X.; Zhu, Y.; Zhan, S.; Chao, Z.; Zhong, T.; Guo, J.; Wang, Y.; Li, L.; Zhang, H. Genome-Wide Identification and Characterization of Long Noncoding RNAs of Brown to White Adipose Tissue Transformation in Goats. *Cells* **2019**, *8*, 904. [CrossRef] [PubMed]
34. Shannon, P.; Markiel, A.; Ozier, O.; Baliga, N.S.; Wang, J.T.; Ramage, D.; Amin, N.; Schwikowski, B.; Ideker, T. Cytoscape: A software environment for integrated models of biomolecular interaction networks. *Genome Res.* **2003**, *13*, 2498–2504. [CrossRef]
35. Yotsukura, S.; duVerle, D.; Hancock, T.; Natsume-Kitatani, Y.; Mamitsuka, H. Computational recognition for long non-coding RNA (lncRNA): Software and databases. *Brief. Bioinform.* **2017**, *18*, 9–27. [CrossRef]
36. Zhou, C.Y.; Johnson, S.L.; Gamarra, N.I.; Narlikar, G.J. Mechanisms of ATP-Dependent Chromatin Remodeling Motors. *Annu. Rev. Biophys.* **2016**, *45*, 153–181. [CrossRef]
37. Clapier, C.R.; Iwasa, J.; Cairns, B.R.; Peterson, C.L. Mechanisms of action and regulation of ATP-dependent chromatin-remodelling complexes. *Nat. Rev. Mol. Cell Biol.* **2017**, *18*, 407–422. [CrossRef]
38. Ito, N.; Ruegg, U.T.; Takeda, S. ATP-Induced Increase in Intracellular Calcium Levels and Subsequent Activation of mTOR as Regulators of Skeletal Muscle Hypertrophy. *Int. J. Mol. Sci.* **2018**, *19*, 2804. [CrossRef]
39. Casas, M.; Buvinic, S.; Jaimovich, E. ATP signaling in skeletal muscle: From fiber plasticity to regulation of metabolism. *Exerc. Sport Sci. Rev.* **2014**, *42*, 110–116. [CrossRef]
40. Tucci, S.; Mingirulli, N.; Wehbe, Z.; Dumit, V.I.; Kirschner, J.; Spiekerkoetter, U. Mitochondrial fatty acid biosynthesis and muscle fiber plasticity in very long-chain acyl-CoA dehydrogenase-deficient mice. *FEBS Lett.* **2018**, *592*, 219–232. [CrossRef]
41. Joo, S.T.; Joo, S.H.; Hwang, Y.H. The Relationships between Muscle Fiber Characteristics, Intramuscular Fat Content, and Fatty Acid Compositions in M. longissimus lumborum of Hanwoo Steers. *Korean J. Food Sci. Anim. Resour.* **2017**, *37*, 780–786. [CrossRef] [PubMed]
42. Lee, S.H.; Joo, S.T.; Ryu, Y.C. Skeletal muscle fiber type and myofibrillar proteins in relation to meat quality. *Meat Sci.* **2010**, *86*, 166–170. [CrossRef] [PubMed]
43. Bowker, B.; Zhuang, H. Relationship between water-holding capacity and protein denaturation in broiler breast meat. *Poult. Sci.* **2015**, *94*, 1657–1664. [CrossRef] [PubMed]
44. Bee, G.; Anderson, A.L.; Lonergan, S.M.; Huff-Lonergan, E. Rate and extent of pH decline affect proteolysis of cytoskeletal proteins and water-holding capacity in pork. *Meat Sci.* **2007**, *76*, 359–365. [CrossRef]
45. Brownridge, P.; de Mello, L.V.; Peters, M.; McLean, L.; Claydon, A.; Cossins, A.R.; Whitfield, P.D.; Young, I.S. Regional variation in parvalbumin isoform expression correlates with muscle performance in common carp (*Cyprinus carpio*). *J. Exp. Biol.* **2009**, *212*, 184–193. [CrossRef]
46. Pauls, T.L.; Cox, J.A.; Berchtold, M.W. The $Ca^{2+}$-binding proteins parvalbumin and oncomodulin and their genes: New structural and functional findings. *Biochim. Biophys. Acta (BBA) Gene Struct. Expr.* **1996**, *1306*, 39–54. [CrossRef]
47. Heizmann, C.W. Parvalbumin, an intracellular calcium-binding protein; distribution, properties and possible roles in mammalian cells. *Experientia* **1984**, *40*, 910–921. [CrossRef]
48. Muntener, M.; Kaser, L.; Weber, J.; Berchtold, M.W. Increase of skeletal muscle relaxation speed by direct injection of parvalbumin cDNA. *Proc. Natl. Acad. Sci. USA* **1995**, *92*, 6504–6508. [CrossRef]
49. Schwaller, B.; Dick, J.; Dhoot, G.; Carroll, S.; Vrbova, G.; Nicotera, P.; Pette, D.; Wyss, A.; Bluethmann, H.; Hunziker, W.; et al. Prolonged contraction-relaxation cycle of fast-twitch muscles in parvalbumin knockout mice. *Am. J. Physiol.* **1999**, *276*, C395–C403. [CrossRef]
50. Takeuchi, K.; Reue, K. Biochemistry, physiology, and genetics of GPAT, AGPAT, and lipin enzymes in triglyceride synthesis. *Am. J. Physiol. Endocrinol. Metab.* **2009**, *296*, E1195–E1209. [CrossRef]

51. Bonnet, A.; Lambert, G.; Ernest, S.; Dutrieux, F.X.; Coulpier, F.; Lemoine, S.; Lobbardi, R.; Rosa, F.M. Quaking RNA-Binding Proteins Control Early Myofibril Formation by Modulating Tropomyosin. *Dev. Cell* **2017**, *42*, 527–541.e4. [CrossRef] [PubMed]
52. Khan, M.; Couturier, A.; Kubens, J.F.; Most, E.; Mooren, F.C.; Kruger, K.; Ringseis, R.; Eder, K. Niacin supplementation induces type II to type I muscle fiber transition in skeletal muscle of sheep. *Acta Vet. Scand.* **2013**, *55*, 85. [CrossRef] [PubMed]

*Article*

# Evaluation of the CRISPR/Cas9 Genetic Constructs in Efficient Disruption of Porcine Genes for Xenotransplantation Purposes Along with an Assessment of the Off-Target Mutation Formation

**Natalia Ryczek [1,*], Magdalena Hryhorowicz [1], Daniel Lipiński [1], Joanna Zeyland [1] and Ryszard Słomski [2]**

1 Department of Biochemistry and Biotechnology, Poznan University of Life Sciences, Dojazd 11, 60-632 Poznań, Poland; magdalena.hryhorowicz@gmail.com (M.H.); lipinskidaniel71@gmail.com (D.L.); jzeyland@gmail.com (J.Z.)
2 Institute of Human Genetics, Polish Academy of Sciences, Strzeszyńska 32, 60-479 Poznań, Poland; ryszard.slomski@up.poznan.pl
* Correspondence: nataliaryczek.nr@gmail.com

Received: 21 May 2020; Accepted: 24 June 2020; Published: 26 June 2020

**Abstract:** The increasing life expectancy of humans has led to an increase in the number of patients with chronic diseases and organ failure. However, the imbalance between the supply and the demand for human organs is a serious problem in modern transplantology. One of many solutions to overcome this problem is the use of xenotransplantation. The domestic pig (*Sus scrofa domestica*) is currently considered as the most suitable for human organ procurement. However, there are discrepancies between pigs and humans that lead to the creation of immunological barriers preventing the direct xenograft. The introduction of appropriate modifications to the pig genome to prevent xenograft rejection is crucial in xenotransplantation studies. In this study, porcine *GGTA1*, *CMAH*, *β4GalNT2*, *vWF*, *ASGR1* genes were selected to introduce genetic modifications. The evaluation of three selected gRNAs within each gene was obtained, which enabled the selection of the best site for efficient introduction of changes. Modifications were examined after nucleofection of porcine primary kidney fibroblasts with CRISPR/Cas9 system genetic constructs, followed by the tracking of indels by decomposition (TIDE) analysis. In addition, off-target analysis was carried out for selected best gRNAs using the TIDE tool, which is new in the research conducted so far and shows the utility of this tool in these studies.

**Keywords:** xenoantigen; coagulation system dysregulation; CRISPR/Cas9 system; genome modifications; non-homologous DNA ends joining (NHEJ); TIDE analysis; off-target

---

## 1. Introduction

The increasing life expectancy of humans has led to an increase in the number of patients with chronic diseases and organ failure. Organ transplantation is an effective approach in the treatment of the end-stage organ failure. However, the imbalance between the supply and the demand for human organs is a serious problem in modern transplantology. In view of the above, it is assumed that alternative approaches would reduce or eliminate this problem. One of many solutions is the use of xenotransplantation [1,2].

Xenotransplantation is any procedure that involves the transplantation, implantation, or infusion of a recipient (in this case a human) with zoonotic cells, tissues, or organs. In addition, therapies using human body fluids, tissues, organs, or cells that have had ex vivo contact with animal organs, tissues, or cells are also subject to this term [3]. The domestic pig (*S. s. domestica*) is currently considered as the most suitable species for human organ procurement. The reasons for choosing the pig as a donor animal include its relatively large litter size and short puberty, the size and physiological similarity of its organs to human organs, and the low risk of xenozoonosis transmission [4]. However, there are discrepancies between pigs and humans that lead to the creation of immunological barriers preventing the direct xenograft. These differences cause xenograft rejection [5]. The introduction of appropriate modifications into the porcine genome to prevent xenograft rejection is crucial in xenotransplantation studies. There are three main types of xenograft rejection consecutively—hyperacute rejection (HAR), acute humoral xenograft rejection (AHXR), and acute cellular rejection (ACR) [6–8]. In addition, dysregulation of the recipient's coagulation system is a barrier in xenotransplantation, which appears in parallel with HAR, AXHR, and ACR [9,10]. Accordingly, this study focuses on preventing HAR and coagulation dysregulation in xenotransplantation.

HAR is a process that occurs within a few minutes to several hours after xenotransplantation. HAR is a type of humoral rejection mediated by IgM antibodies naturally occurring in the recipient. The association of recipient antibodies with epitopes present on the porcine endothelial cells activates the complement system [11]. To date, three epitopes have been described that constitute a barrier to xenotransplantation and are responsible for HAR. Galactose-α1,3-galactose (α-Gal) is the major xenoantigen involved in xenograft hyperacute rejection. This epitope is synthesized by α-1,3-galactosyltransferase encoded by the porcine *GGTA1* gene [11,12]. The second significant epitope is Neu5Gc, which is formed in an enzymatic reaction involving cytidine monophospho-N-acetylneuraminic acid hydroxylase encoded by the porcine *CMAH* gene [13,14]. β-1,4 N-acetylgalactosaminyltransferase 2 encoded by the porcine *β4GalNT2* gene is involved in the synthesis of Sd(a) antigen [15–17].

Dysregulation of the recipient's coagulation system is one of the main barriers in xenotransplantation. It causes the development of thrombotic microangiopathy in xenograft. Features of thrombotic microangiopathy include fibrin deposition and platelet aggregation, which causes thrombosis within the transplant blood vessels and ultimately ischemic damage [18]. With the development of disorders of the coagulation system, systemic consumption coagulopathy is often observed in the recipient, which can lead to his death [19]. There are also factors expressed in specific organs that pose a problem in xenotransplantation. One of them is the von Willebrand factor encoded by the porcine *vWF* gene, which is involved in the pathogenesis of transplant failure in lung xenotransplantation [20,21]. Fatal thrombocytopenia accompanying liver xenotransplantation is another barrier resulting from differences in the coagulation system. Human platelets are bound by asialoglycoprotein receptor (ASGR) encoded by porcine *ASGR1* and *ASGR2* genes. They are expressed in liver sinusoidal endothelial cells (LSEC) [22,23].

The described research focuses on introducing changes within genes that are involved in the immune response that is the biggest barrier in xenotransplantation. Porcine *GGTA1*, *CMAH*, and *β4GalNT2* genes involved in the synthesis of epitopes responsible for the xenograft hyperacute rejection. Additionally, two genes responsible for the synthesis of porcine proteins causing dysregulation of the recipient's coagulation system have been chosen—the porcine *vWF* gene and porcine *ASGR1* gene.

The use of the CRISPR/Cas9 system for genome modification has led to enormous progress in the field of animal transgenesis [24]. Through projectable short gRNA, it is possible to precisely generate double-strand DNA breaks (DSBs). DSBs of DNA generated by the CRISPR/Cas9 system are repaired by NHEJ or by homologous directed repair (HDR) after delivery of properly designed donor DNA [25,26]. There are, however, some restrictions related to the use of the CRISPR/Cas9 system. One of them is off-target effect [27]. In this study, the usefulness of the TIDE online tool has been confirmed not only to analyze indel-type mutations arising as a result of DSBs repair via NHEJ but also to analyze additional Cas9 hydrolysis sites linked to a specific gRNA—off-target sites.

## 2. Materials and Methods

### 2.1. Selection of Short Oligonucleotides

Short oligonucleotides (gRNA) and PCR primers were selected using the Benchling platform, San Francisco, CA, USA (www.benchling.com). Nucleotides have been added to the selected gRNA sequences at the 5′ and 3′ ends for molecular cloning. The designed oligonucleotides were ordered from the Institute of Biochemistry and Biophysics Polish Academy of Sciences.

### 2.2. Preparation of Genetic Constructions in the CRISPR/Cas9 System

The constructions were prepared using the vector *pSpCas9(BB)-2A-Puro (PX459) V2.0*, which was a gift from Feng Zhang (Addgene plasmid #62988; http://n2t.net/addgene:62988; RRID: Addgene_62988). For each pair of oligonucleotides, a hybridization mixture containing 100 μM F and R gRNA sequences was prepared. The mixtures were incubated 5 min at 95 °C and then incubated at room temperature for 10 min. Plasmid DNA was hydrolyzed with the restriction enzyme *Bbs*I-HF® (New England Biolabs, Ipswich, MA, USA) for 1.5 h at 37 °C. To combine oligonucleotides after hybridization with plasmids in linear form, ligation was performed using T4 DNA Ligase (New England Biolabs, Ipswich, MA, USA). The mixtures were incubated for 18 h at 16 °C, followed by additional hydrolysis with the use of enzyme *Bbs*I-HF® for 30 min at 37 °C. Then *Escherichia coli* bacterial cells were transformed with purified constructions and positive clones were selected. Plasmid DNA isolation was performed using the Plasmid Maxi Kit (Qiagen, Germantown, MD, USA).

### 2.3. Isolation and Culture of Porcine Primary Kidney Cells In Vitro

For the isolation of porcine primary kidney fibroblasts, a 2 × 2 × 2 cm kidney cortical tissue was excised. Then, after tissue fragmentation, enzymatic-mechanical disintegration was carried out using collagenase II (Sigma-Aldrich, Darmstadt, Germany) and a magnetic stirrer. Then the incubation was carried out for 60 min while heating the mixture to 37 °C. The mixture was then filtered through a filter (0.5 mm pore diameter) and washed several times in the culture medium (DMEM (Sigma-Aldrich, Darmstadt, Germany), 20% FBS (Sigma-Aldrich, Darmstadt, Germany), 1% MEM Non-Essential Amino Acid Solution (100×) (Sigma-Aldrich, Darmstadt, Germany)), 1% Antibiotic Antimycotic Solution (100×) (Sigma-Aldrich, Darmstadt, Germany), 1% Sodium pyruvate solution (100 mM) (Sigma-Aldrich, Darmstadt, Germany), 1% L-Glutamine solution (200 mM) (Sigma-Aldrich, Darmstadt, Germany)). In vitro cell culture was carried out under standard aseptic conditions (5% $CO_2$, 37 °C).

### 2.4. Nucleofection

Primary porcine kidney fibroblasts were detached from the surface of T-flasks by Accutase® solution (Sigma-Aldrich, Darmstadt, Germany) and counted using a Scepter™ Handheld Automated Cell Counter, version 2.0 (Merck, Darmstadt, Germany). One million cells were used for the transfection. Nucleofection was carried out using the Mouse Embryonic Fibroblast Nucleofector™ Kit 1 (Lonza, Basel, Switzerland) and program T-007 on Amaxa™ Nucleofector™ II device (Lonza, Basel, Switzerland).

### 2.5. Antibiotic Selection

After 24 h since performing nucleofection, the culture medium has been changed to selective culture medium with the addition of puromycin at a concentration of 1 μg/mL. Incubation was then carried out for 48 h, after which the selection medium was removed, and standard medium was added again. The culture was carried out until the expected cell confluency was obtained.

### 2.6. Analysis of Introduced Genetic Modifications

Prior to analysis of the introduced genetic modifications, DNA isolation from modified primary porcine kidney fibroblasts was performed using the DNeasy Blood & Tissue Kit (Qiagen, Germantown,

MD, USA). Then PCR reactions were performed using the StartWarm HS-PCR Mix (A&A Biotechnology, Gdynia, Poland) to amplify DNA fragments that include modified *loci* and DNA fragments containing potential off-target sites. The purified PCR products were sequenced in the Sequencing Laboratory of the Faculty of Biology at the Adam Mickiewicz University in Poznan.

### 2.7. Sequencing Analysis Using the TIDE Tool

The presence of modifications at the target locus and off-target sites after nucleofection of primary porcine kidney fibroblasts was identified using the TIDE online tool, version 2.0.1 by The Netherlands Cancer Institute, Amsterdam, Netherlands (https://tide.deskgen.com/).

## 3. Results

### 3.1. Selection of the Potential Modifications Location in Porcine Genome

The first stage of research was to find the best bioinformatically selected gRNA using the Benchling platform. The choice was made based on several guidelines: the localization of the introduced modifications, the values of the on-target score (from 0 to 100—the higher score, the better) and the off-targets score (from 0 to 100—the higher the score, the lower the chance of additional genomic cleavage sites). The localization of the three selected gRNAs for each of the tested *loci* in the porcine genome is shown in Table 1.

**Table 1.** Localization of the potential modification *loci* in porcine genome.

| Porcine Genome Locus | gRNA | Exon | Chromosome Localization [1] |
|---|---|---|---|
| *GGTA1* | gGGTA1 F1/R1 | Exon 8 | Chromosome 1, c261513705-261513686 (NC_010443.5) |
| | gGGTA1 F2/R2 | Exon 8 | Chromosome 1, c261513541-261513522 (NC_010443.5) |
| | gGGTA1 F3/R3 | Exon 8 | Chromosome 1, c261513764-261513745 (NC_010443.5) |
| *CMAH* | gCMAH F1/R1 | Exon 6 | Chromosome 7, c19902027-19902008 (NC_010449.5) |
| | gCMAH F2/R2 | Exon 3 | Chromosome 7, c19917616-19917597 (NC_010449.5) |
| | gCMAH F3/R3 | Exon 5 | Chromosome 7, c19903792-19903773 (NC_010449.5) |
| *β4GalNT2* | gβ4GalNT2 F1/R1 | Exon 2 | Chromosome 12, c25388178-25388159 (NC_010454.4) |
| | gβ4GalNT2 F2/R2 | Exon 3 | Chromosome 12, c25386323-25386304 (NC_010454.4) |
| | gβ4GalNT2 F3/R3 | Exon 6 | Chromosome 12, c25381330-25381311 (NC_010454.4) |
| *vWR* | gvWR F1/R1 | Exon 2 | Chromosome 5, 64553818-64553837 (NC_010447.5) |
| | gvWR F2/R2 | Exon 3 | Chromosome 5, 64556041-64556060 (NC_010447.5) |
| | gvWR F3/R3 | Exon 4 | Chromosome 5, 64557621-64557640 (NC_010447.5) |
| *ASGR1* | gASGR1 F1/R1 | Exon 3 | Chromosome 12, c52538530-52538511 (NC_010454.4) |
| | gASGR1 F2/R2 | Exon 7 | Chromosome 12, c52537633-52537614 (NC_010454.4) |
| | gASGR1 F3/R3 | Exon 9 | Chromosome 12, c52537146-52537127 (NC_010454.4) |

[1] Based on NCBI: Sus scrofa isolate TJ Tabasco breed Duroc chromosome 12, Sscrofa11.1, whole genome shotgun sequence, GCF_000003025.6.

### 3.2. Analysis of On-Target Modification Sites

#### 3.2.1. Nucleofection Efficiency

To obtain genetically modified primary porcine renal fibroblasts, nucleofections were performed. Transfection efficiency for isolated cells was checked using the *pmaxGFP™ Vector* from the nucleofection kit and visualized using a ZOE Fluorescent Cell Imager (Figure 1). After counting 10,000 cells from multiple views, transfection efficiency was estimated to be around 65–70%.

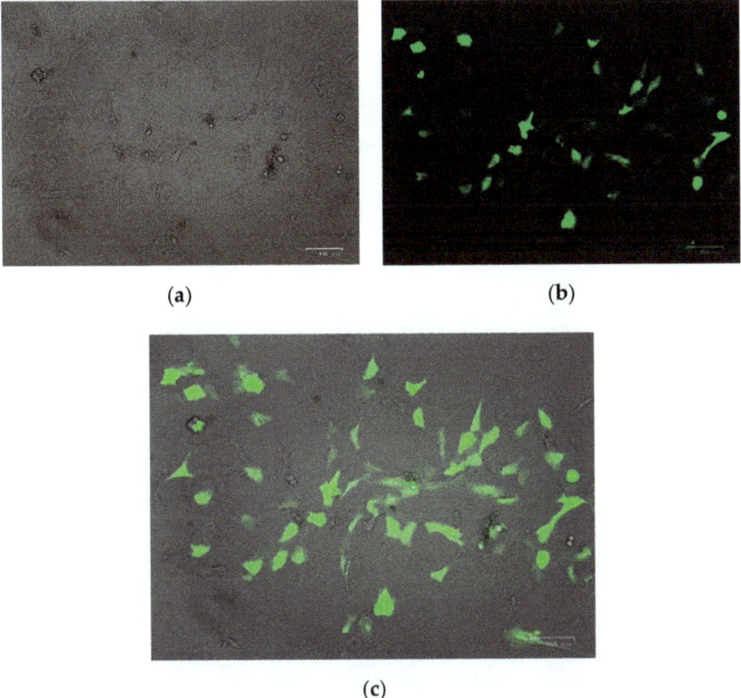

Figure 1. Nucleofection efficiency obtained on porcine primary kidney fibroblasts. Cells visualization was performed using a ZOE Fluorescent Cell Imager 24 h after nucleofection—(a) brightfield; (b) fluorescence detection lamp; (c) merge of view (a,b). The scale is 100 μm.

#### 3.2.2. The Efficiency of Introducing Modifications within the Examined Porcine Genes Using Genetic Constructions Selected as the Best

PCR products obtained on DNA template isolated from modified and unmodified porcine primary kidney fibroblasts were purified using a CleanUp kit (A&A Biotechnology, Gdynia, Poland). Then, sequencing was commissioned, the results of which were analyzed using the TIDE online tool, version 2.0.1. One, best gRNA from the three directed at each of the examined genes was selected.

The gGGTA1 F1/R1 was chosen as the best for the disruption of the porcine *GGTA1* gene. The gGGTA1 F1/R1, when combined with Cas9, enabled on modification of porcine primary kidney fibroblasts with total efficiency of 70.1%. A statistically significant mutation occurring with the highest frequency was a single nucleotide deletion in 51.1% sequences. It was determined that the insertion of one nucleotide occurred at 11.6% (adenosine nucleotide was incorporated in 47.6% sequences).

From the genetic constructs of the CRISPR/Cas9 system directed to the porcine *CMAH* gene, the one containing gCMAH F3/R3 was chosen as the best, as it enabled obtaining 61.2% total efficiency of modification. The most common mutation was the insertion of a single nucleotide, occurring in 57% of the tested sequences. In 93.6% of sequences, the inserted nucleotide was the cytidine.

The next gene tested was porcine *β4GalNT2*. For genetic construction containing gβ4GalNT2 F3/R3, a total modification efficiency of 45.2% was obtained. The most common mutation was a deletion of one nucleotide, which occurred in 28.8% sequences. The second most common mutation was the insertion of a single nucleotide, occurring in 8.4% of cases. Adenosine nucleotide was inserted in 79.8% of sequences.

Another analysis was performed for genetic constructions directed at the porcine *vWF* gene. A total efficiency of 85.1% was obtained for the plasmid containing gvWF F2/R2. The most common deletion of one nucleotide occurred with a frequency of 39.9%. The second most common mutation was the insertion of one nucleotide, which occurred in 34.3% of sequences. An adenosine nucleotide was inserted in 83.6% of sequences.

The last gene tested was porcine *ASGR1*. A total efficiency of 80.5% was obtained for the plasmid containing gASGR1 F3/R3. The most common deletion of one nucleotide occurred in 58.6% of sequences. The second most common mutation was the deletion of two nucleotides.

Detailed results of indel spectrum and inserted nucleotide probability for all chosen as the best gRNAs are presented in Figure 2.

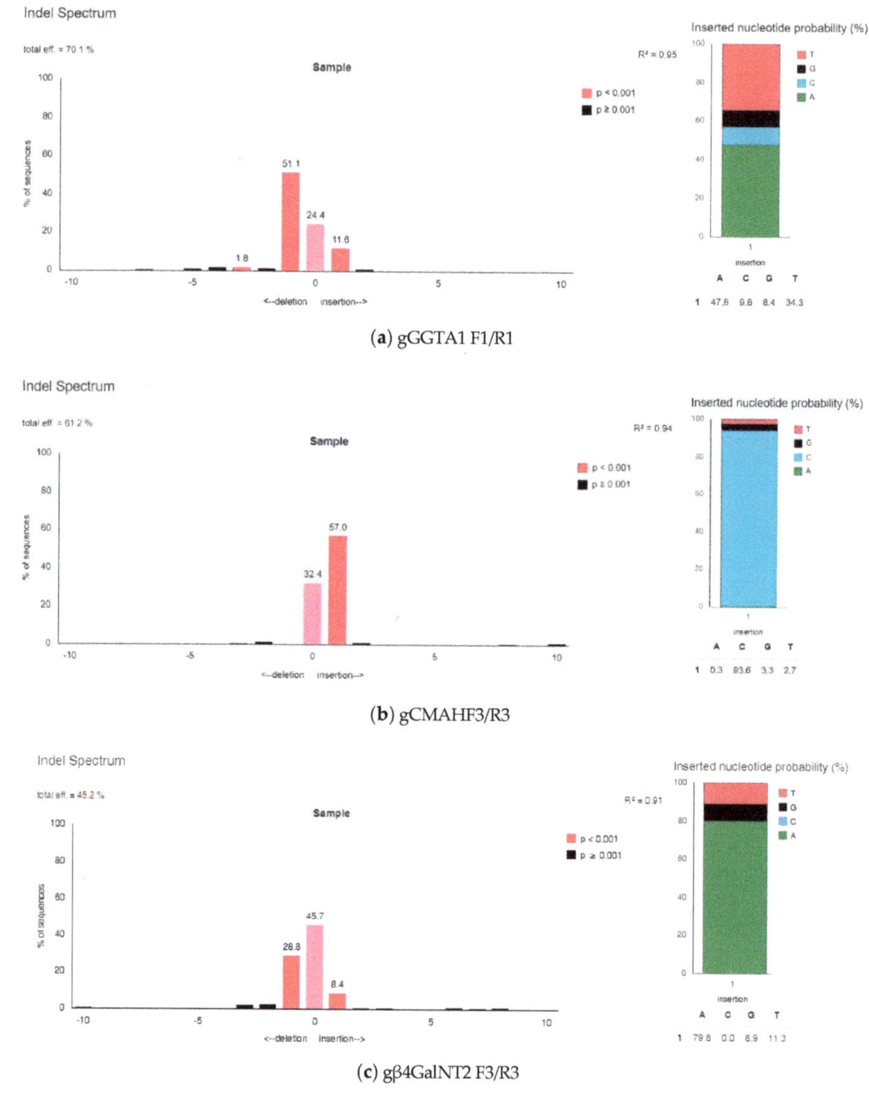


(**a**) gGGTA1 F1/R1



(**b**) gCMAHF3/R3



(**c**) gβ4GalNT2 F3/R3

**Figure 2.** *Cont.*

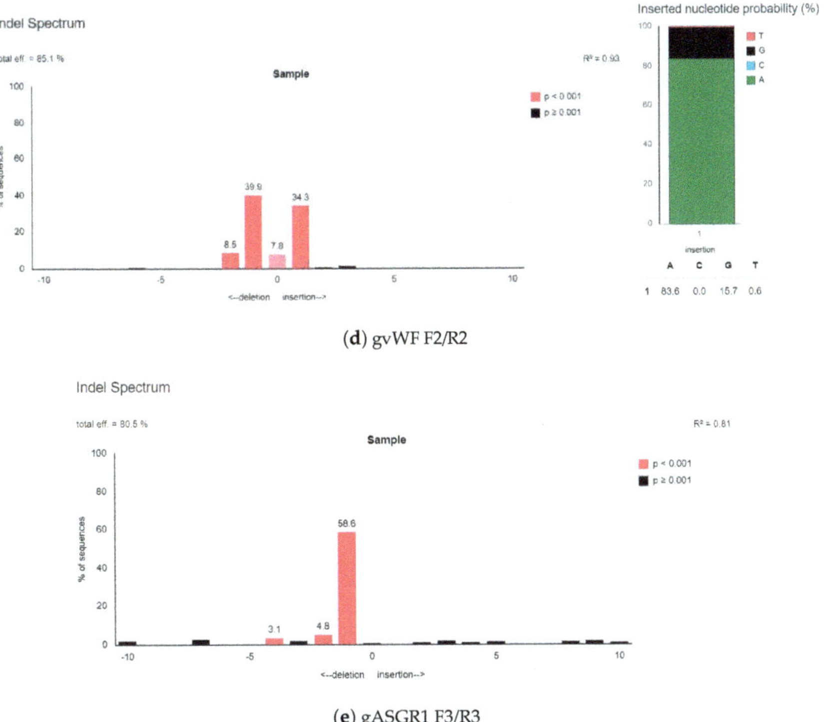


(**d**) gvWF F2/R2

(**e**) gASGR1 F3/R3

**Figure 2.** The indel spectrum and inserted nucleotide probability results for the CRISPR/Cas9 genetic constructs containing gRNA chosen as the best for disruption of tested porcine genes. Results obtained after use the plasmids with (**a**) gGGTA1 F1/R1; (**b**) gCMAH F3/R3; (**c**) gβ4GalNT2 F3/R3; (**d**) gvWF F2/R2; (**e**) gASGR1 F3/R3.

The indel spectrum and inserted nucleotide probability results, alignment, and decomposition quality controls for the other CRISPR/Cas9 genetic constructs containing gRNA tested for disruption of the porcine genes are present in the Supplement (Figure S1). The alignment and decomposition quality controls for the CRISPR/Cas9 genetic constructs containing gRNA chosen as the best for the disruption of the porcine genes are present in the Supplement (Figure S2).

### 3.2.3. Comparison of Bioinformatically Predicted DNA Diruption Efficiency Results with Those Obtained in the In Vitro Cultured Cells

The total efficiency results of modifications obtained using the CRISPR/Cas9 system predicted during bioinformatics analysis in silico were compared with those obtained in in vitro cultured porcine primary kidney fibroblasts. Comparison of the results for all tested construction variants is shown in Table 2.

**Table 2.** Comparison of the results of total efficiency predicted in silico with the results obtained in the in vitro cultured cells.

| gRNA | In Silico Analysis Predicted Total Efficiency | In Vitro Analysis Total Efficiency |
|---|---|---|
| gGGTA1 F1/R1 | 78.9% | 70.1% * |
| gGGTA1 F2/R2 | 72.2% | 27.1% |
| gGGTA1 F3/R3 | 70.6% | 18.2% |
| gCMAH F1/R1 | 61.5% | 7.8% |

**Table 2.** *Cont.*

| gRNA | In Silico Analysis Predicted Total Efficiency | In Vitro Analysis Total Efficiency |
|---|---|---|
| gCMAH F2/R2 | 73.5% | 12.9% |
| gCMAH F3/R3 | 70.1% | 61.2% * |
| gβ4GalNT2 F1/R1 | 53.3% | 8.1% |
| gβ4GalNT2 F2/R2 | 66% | 17.6% |
| gβ4GalNT2 F3/R3 | 75.8% | 45.2% * |
| gvWF F1/R1 | 48.9% | 5.2% |
| gvWF F2/R2 | 73.9% | 85.1% * |
| gvWF F3/R3 | 76% | 39.6% |
| gASGR1 F1/R1 | 72.7% | 13.9% |
| gASGR1 F2/R2 | 67.3% | 33.5% |
| gASGR1 F3/R3 | 68.7% | 80.5% * |

* the gRNA together with total efficiency was marked, which was chosen as the best for disruption of the studied genes.

## 3.3. *Analysis of the Potential Off-Target Sites in Porcine Genome*

### 3.3.1. Selection of the Potential Off-Target Sites Location

After selecting the best genetic constructs containing gRNAs that mediate in the efficient disruption of studied porcine loci, off-target sites were predicted. For each targeted modification site, a minimum of three loci have been selected that can lead to mutations outside the target locus using the Benchling internet platform. The main criterion considered when choosing the tested off-target sites was the score showing the probability of a DNA break at a given off-target site by the testes construct is equal to or higher than 1.0. In addition, all off-target loci predicted in the coding sequences for selected constructs were checked. The list of tested potential off-target sites is presented in Table 3.

**Table 3.** A summary of bioinformatically predicted potential off-target sites for selected genetic constructs.

| *Chosen* gRNA | * | Sequence | Porcine Genome Localization |
|---|---|---|---|
| gGGTA1 F1/R1 | 1. | GCTGC**ACT**TGAAGACCATCG | chr7: +33592796 |
| | 2. | GAT**AGT**CATG**G**AGACCATCG | chr7: +1925303 (*RIPK1* gene: ENSSSCG00000001009) |
| | 3. | **C**CTGCGC**G**TGAAGACCA**A**CG | chr2: -44588168 (*OTOG* gene: ENSSSCG00000013376) |
| | 4. | GA**G**GTGCATGAAGA**A**CATC**T** | chr2: +13174290 |
| gCMAH F3/R3 | 5. | AT**TCG**ATCCTCCTAACCCC**T** | chr15: +40937188 |
| | 6. | **TCTT**AACCCTC**A**TAACCCGT | chr4: -97128319 |
| | 7. | AATAAATC**A**CCCTAACC**A**GT | chr4: +116709576 (*HIPK1* gene: ENSSSCG00000006760) |
| gβ4Gal NT2 F3/R3 | 8. | A**A**ACTACCA**G**CTCCACAGAG | chr16: -5677136 |
| | 9. | AT**T**GTACCACCTCCACAGA**C** | chr10: -13115375 |
| | 10. | **T**CAGTA**T**CACCTCCACAGAG | chr7: -109695811 |
| gvWF F2/R2 | 11. | CC**T**TCT**G**C**T**TCATGCCCGCG | chr6: +157052374 |
| | 12. | **G**C**A**CGTACTCC**T**TGCCCGCG | chr4: -347951 (*ARHGAP39* gene: ENSSSCG00000005894) |
| | 13. | CC**G**TGTC**G**TCCA**G**GCCCGCG | chr6: -9684887 (*WWOX* gene: ENSSSCG00000027415) |
| | 14. | CCCTGTCC**TG**CA**G**GCC**T**GCG | chr14: -55036169 (*COMT* gene: ENSSSCG00000010132) |

**Table 3.** *Cont.*

| *Chosen* gRNA | * | Sequence | Porcine Genome Localization |
|---|---|---|---|
| gASGR1 F3/R3 | 15. | **G**C**AT**ATG**T**CTGGTACGGGCA | chr6: +4310372 |
| | 16. | **G**C**AT**ATG**T**CTGGTACGGGCA | chr6: -4126764 |
| | 17. | C**ACA**ATGAC**A**GGTACGGGCA | chr5: +67768944 (*KCNA1* gene: ENSSSCG00000000716) |
| | 18. | CC**A**GA**C**GACTGG**C**ACGGGCA | chr12: -54751930 |
| | 19. | CCCG**C**TG**T**CTGG**G**ACGGGCA | chr6: +155398812 (*C1orf210* gene: ENSSSCG00000003951) |

Nucleotides not complementary to the target template for a given gRNA were determined with the bold font. * The number of the selected off-target site.

### 3.3.2. TIDE Analysis of the Chosen Potential Off-Target Sites

The presence of unwanted changes—the indel mutations, in eight off-target sites from the 19 tested loci was confirmed using the TIDE tool. The off-target sites were confirmed in *loci* numbers 1, 4, 5, 10, 11, 14, 18, and 19. Only in one locus the occurred modification was statistically significant at the level $p < 0.001$ and it was site number 1. It was determined that the off-target site number 1 for the genetic construct of gGGTA1 F1/R1 was cut with a total efficiency of 3.9%. A mutation arising with the statistical significance was the insertion of one nucleotide, which occurred in 1.7% sequences. Guanosine nucleotide was inserted in the 50.8% sequences. The indel spectrum and inserted nucleotide probability for the number 1 off-target locus is presented in Figure 3.

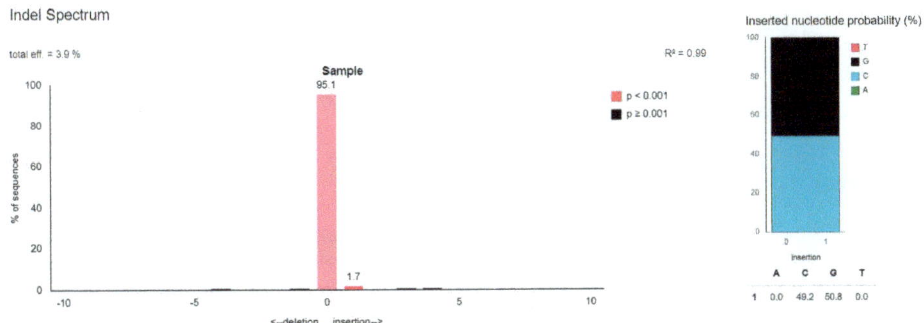


**Figure 3.** The indel spectrum and inserted nucleotide probability result for the number 1 off-target locus after the use of CRISPR/Cas9 genetic construct containing gGGTA1 F1/R1 chosen as the best for disruption of porcine *GGTA1* gene.

### 3.3.3. Comparison of Bioinformatically Predicted DNA Diruption Efficiency Results in the Off-Target Sites with Those Obtained in the In Vitro Cultured Cells

The efficiency results of off-target mutation using the CRISPR/Cas9 system obtained during bioinformatics analysis were compared with those obtained in in vitro cultures. The results are summarized in Table 4.

**Table 4.** Comparison of bioinformatic analysis with the results of laboratory analyses—analysis of the off-target sites presence.

| gRNA | * | DNA Hydrolysis Efficiency at a Potential Off-Target Site | |
|---|---|---|---|
| | | **Bioinformatic Analysis** | **In Vitro Cultured Cells** |
| **gGGTA1 F1/R1** | 1. | 1.6% | 3.9% |
| | 2. | 0.5% | 0% |
| | 3. | 0.4% | 0% |
| | 4. | 0.2% | 0.8% |
| **gCMAH F3/R3** | 5. | 0.6% | 1.1% |
| | 6. | 0.6% | 0% |
| | 7. | 0.2% | 0% |
| **gB4GalNT2 F3/R3** | 8. | 2.6% | 0% |
| | 9. | 1.8% | 0% |
| | 10. | 1.8% | 2.2% |
| **gvWF F2/R2** | 11. | 0.9% | 1.8% |
| | 12. | 0.8% | 0% |
| | 13. | 0.4% | 0% |
| | 14. | 0.2% | 3.4% |
| **gASGR1 F3/R3** | 15. | 1.5% | 0% |
| | 16. | 1.5% | 0% |
| | 17. | 1.4% | 0% |
| | 18. | 0.7% | 1.7% |
| | 19. | 0.6% | 1.6% |

* The number of the selected off-target site.

## 4. Discussion

The research presented in this paper shows the importance of the verification of the bioinformatically selected gRNAs in the context of the modifications obtaining efficiency. To compare the efficiency of individual gRNAs in modifications obtaining in cells cultured in vitro, it was necessary to use the third-generation CRISPR/Cas9 system, which enables to carry out antibiotic selection using puromycin after transfection. This process eliminates the impact of the transfection efficiency associated with the chosen method on the efficiency of obtaining modifications via individual gRNAs [28–30]. After antibiotic selection, a population of positively transfected cells after nucleofection was examined. In this way, it was possible to determine the efficiency of indel mutation formation at target sites for specific gRNAs. It was shown that the results obtained in the cells prepared in this way slightly coincided with bioinformatics predictions.

Therefore, the selection of gRNA itself, which would mediate the disruption of specific genomic site should be given the great importance. There are many different online tools that enable gRNA design. The limiting factor are the available databases related to the genomes of various organisms. Thus, the selection of the appropriate tool for prediction of Cas9 DNA hydrolysis target sites is based on the chosen research model [31,32].

The results of sequencing of PCR products that were amplified within the target sites of designed gRNAs were analyzed using the TIDE tool. This method makes it easy to determine, based on the results of sequencing, the indel mutation profile that occurs in a population of genetically modified cells using the CRISPR/Cas9 system. The indel detection by amplicon analysis (IDAA) method is also used for this purpose. It is based on three-primer amplicon labeling and detection by capillary electrophoresis [33]. The sensitivity and precision of both methods to determine the insertion-deletion profile of the modification is similar. The choice of method used depends on the researcher's preferences and available infrastructure [34,35].

Based on the results obtained in this study, the best targeted sites were selected within the genes tested, thanks to which it is possible to obtain modifications efficiently. The assessment of individual gRNAs is very important in the context of obtaining simultaneous multimodification of the porcine genome for xenotransplantation purposes. This is the latest goal of researchers in this field. Research related to this work is part of this trend. Modifications used in the research are aimed at counteracting the two immune barriers existing in xenotransplantation. The knock-out of the porcine *GGTA1*, *CMAH* and *β4GalNT2* genes aims at the elimination of the hyperacute xenograft rejection. Prevention of the coagulation dysregulation can be achieved by the disruption of the porcine *vWF* gene for lung xenotransplantation and knock-out of the porcine *ASGR1* gene for liver xenotransplantation.

An important aspect related to these results is the ethical effect. Preliminary evaluation of bioinformatic data allows to limit the participation of animals in research. This is in line with the Polish recommendations of the National Ethics Committee for Animal Experiments described in Resolution No. 14/2017, as well as with international guidelines and trends.

This study also analyzes the off-target sites generated by Cas9 nuclease in combination with the best gRNAs selected. The formation of additional DNA hydrolysis sites by Cas9 is a common problem observed in CRISPR/Cas9-related studies. Off-target mutations are the biggest threat associated with the use of the new genome editing technology. In total, 19 potential off-target sites were tested for genetic constructs selected during experiments (Figure 3, Figures S3–S7). Off-target mutations were confirmed at eight loci. Only in one of them the changes occurred with statistical significance at the level of $p < 0.001$. This was site No. 1 (for genetic construction containing gGGTA1 F1/R1). Eight potential off-target sites were in the coding sequences of the porcine genome. Experimentally, in genetically modified porcine primary kidney fibroblast cultures, the presence of the off-target mutations in two gene coding sequences was confirmed. The first locus in which off-target mutation has been reviled after the use of the genetic construction with gvWF F2/R2 lies within the porcine *COMT* gene. Another one confirmed off-target site for gASGR1 F3/R3 containing constructs is within the porcine *C1orf210* gene. These loci must be checked after receiving genetically modified animals using obtained using these two genetic constructs.

The obtained results correlate with the research that describe the influence of the non-complementary nucleotides (mismatch nucleotides) position between gRNA and the sequence of the potential off-target site. The efficiency of the off-target mutation formation is higher when incomplementarity occurs in the eighth nucleotide and in the first three nucleotides of the gRNA sequence (5′→3′). For nucleotide in the eighth position (5′→3′) such a relationship was confirmed for two out of eight off-target sites (numbers 1 and 19) for the examined gRNAs in this study. In addition, as many as seven out of eight off-target sites (numbers 1, 4, 5, 10, 11, 18, and 19) had a mismatch of at least one nucleotide in the first, second, or third position of the gRNA sequence (5′→3′) [36]. Interestingly, this study also revealed that in three out of eight confirmed off-target sites, complementarity was found for the seventh position nucleotide. Studies show that the CRISPR/Cas9 system tolerates three to five mismatches in the distal gRNA region from the protospacer adjacent motif (PAM) sequence. In addition, it has been reported that the complementary alignment of the ten nucleotides of the proximal gRNA region from the PAM sequence is sufficient to mediate the Cas9 nuclease DNA hydrolysis [37]. It has also been shown that the occurrence of nucleotide incomplementarity at positions 15, 16, and 17 of the gRNA sequence (5′→3′) to a potential off-target site abolishes CRISPR/Cas9 system activity [38].

The research presented above proved the usefulness of the TIDE online tool for off-target sites checking. There are other, more accurate methods used for this purpose. One of them is DISCOVER-Seq (discovery of in situ Cas off-targets and verification by sequencing), which uses recruitment of the factors involved in DNA repair. To this end, the binding of double-strand break DNA repair protein (MRE11) is tracked. Thanks to this it is possible to check the double-strand DNA breaks in the whole genome [39]. Another method is CIRCLE-seq (circularization for in vitro reporting of cleavage effects by sequencing), it consists of next generation sequencing and leads to the analysis of the presence of off-target sites throughout the genome [40]. Both above methods have many advantages, so it would be worth to additionally perform these analyzes for the best genetic constructions selected in this paper. However, their use is only possible in in vitro cell culture studies and it is not possible to evaluate off-target sites in genetically modified animals. In contrast, the TIDE tool enables quick and cheap recognition of basic off-target sites, and its use can apply to both cellular and model animals research.

New approaches have emerged to reduce the risk of off-target mutations after using the CRISPR/Cas9 system. One of them is the use of new algorithms in bioinformatics tools, thanks to which it is possible to more accurately assess the efficiency of modifications within individual off-target sites [41–44]. Another approach is to use modified Cas9 with nickase activity. It has been proven that introducing modifications with the CRISPR/Cas9 system modified in such a way reduces the number of unwanted mutations [45,46]. The second approach is to provide Cas9 protein and gRNA in the form of a ribonucleoprotein complex (RNP). It has been shown that this method of delivery of the CRISPR/Cas9 system enabled obtaining the modifications at the target site with high efficiency. The number of the off-target mutations that have arisen has significantly decreased. However, the use of the RNP complex has a major drawback. There is a problem with the number of complexes delivered. Very high concentrations are used because only such concentrations can guarantee that some of them will function in the cell nuclei of genetically modified cell cultures in vitro. High concentration of the RNP complexes may have a cytotoxic effect on some cell lines [47–51]. Accordingly, ribonucleoprotein complexes can only be used in some experiments.

## 5. Conclusions

The research presented above enabled the usefulness assessment of the CRISPR/Cas9 system genetic constructions containing gRNAs to obtain modifications for xenotransplantation purposes. Selected gRNAs, which in combination with Cas9 nuclease enable for the efficient disruption of the studied genes, can be used to obtain genetically modified pigs. To obtain an effect associated with counteracting the immune response in the recipient—the primate animal models or human, all the modifications tested should be present in the porcine genome simultaneously. The genetically modified porcine primary kidney fibroblasts cells nuclei for the SCNT procedure or verified genetic constructs for microinjection of the porcine zygotes can be used to obtain the desired porcine genome modifications. After receiving genetically modified animals, it is necessary to check off-target sites, which presence has been confirmed by the above analyzes.

**Supplementary Materials:** The indel spectrum results, insertion nucleotides probability results, alignment and decomposition quality controls for studied genetic constructs are available online at http://www.mdpi.com/2073-4425/11/6/713/s1, Figure S1: The indel spectrum and inserted nucleotide probability results, alignment, and decomposition quality controls for the other CRISPR/Cas9 genetic constructs containing gRNA tested for disruption of the porcine genes.; Figure S2: The alignment and decomposition quality controls for the CRISPR/Cas9 genetic constructs containing gRNA chosen as the best for the disruption of the porcine genes; Figure S3: The indel spectrum for the numbers 2–4 off-target loci after the use of CRISPR/Cas9 genetic construct containing gGGTA1 F1/R1 chosen as the best for disruption of porcine *GGTA1* gene; Figure S4: The indel spectrum for the numbers 5–7 off-target loci after the use of CRISPR/Cas9 genetic construct containing gCMAH F3/R3 chosen as the best for disruption of porcine *CMAH* gene; Figure S5: The indel spectrum for the number 8–10 off-target loci after the use of CRISPR/Cas9 genetic construct containing gβ4GalNT2 F3/R3 chosen as the best for disruption of porcine *β4GalNT2* gene; Figure S6: The indel spectrum for the numbers 11–14 off-target loci after the use of CRISPR/Cas9 genetic construct containing gvWF F2/R2 chosen as the best for disruption of porcine *vWF* gene; Figure S7: The indel spectrum for the numbers 15–19 off-target loci after the use of CRISPR/Cas9 genetic construct containing gASGR1 F3/R3 chosen as the best for disruption of porcine *ASGR1* gene.

**Author Contributions:** Investigation and writing—original draft, review and editing, N.R.; supervision and conceptualization, M.H., D.L.; supervision, J.Z.; funding acquisition and supervision, R.S. All authors have read and agreed to the published version of the manuscript.

**Funding:** This research was partially funded by the National Centre for Research and Development (Grant No. INNOMED/I/17/NCBR/2014) within the framework of the INNOMED programme entitled "Development of an Innovative Technology Using Transgenic Porcine Tissues for Bio-Medical Purposes," acronym: MEDPIG. The publication was co-financed within the framework of a Ministry of Science and Higher Education program as "Regional Initiative Excellence" in the years 2019-2022, Project No. 005/RID/2018/19.

**Conflicts of Interest:** The authors declare no conflict of interest.

## References

1. Ekser, B.; Li, P.; Cooper, D.K.C. Xenotransplantation: Past, present, and future. *Curr. Opin. Organ. Transplant.* **2017**, *22*, 513–521. [CrossRef] [PubMed]
2. Cooper, D.K.C.; Ekser, B.; Tector, J. A brief history of clinical xenotransplantation. *Int. J. Surg.* **2015**, *23*, 205–210. [CrossRef] [PubMed]
3. Aristizabal, A.M.; Caicedo, L.A.; Martínez, J.M.; Moreno, M.; Echeverri, G.J. Clinical xenotransplantation, a closer reality: Literature review. *Cirugía Española (English Ed.)* **2017**, *95*, 62–72. [CrossRef]
4. Cooper, D.K.C.; Iwase, H.; Wang, L.; Yamamoto, T.; Li, Q.; Li, J.; Zhou, H.; Hara, H. Bringing home the bacon: Update on the state of kidney xenotransplantation. *Blood Purif.* **2018**, *45*, 254–259. [CrossRef] [PubMed]
5. Vadori, M.; Cozzi, E. The immunological barriers to xenotransplantation. *Tissue Antigens* **2015**, *86*, 239–253. [CrossRef] [PubMed]
6. Cooper, D.K.C.; Ekser, B.; Tector, J. Immunobiological barriers to xenotransplantation. *Int. J. Surg.* **2015**, *23*, 211–216. [CrossRef]
7. Fischer, K.; Rieblinger, B.; Hein, R.; Sfriso, R.; Zuber, J.; Fischer, A.; Klinger, B.; Liang, W.; Flisikowski, K.; Kurome, M.; et al. Viable pigs after simultaneous inactivation of porcine MHC class I and three xenoreactive antigen genes GGTA1, CMAH and B4GALNT2. *Xenotransplantation* **2020**, *27*, e12560. [CrossRef]
8. Ganji, M.-R.; Broumand, B. Acute cellular rejection. *Iran. J. Kidney Dis.* **2007**, *1*, 54–56.
9. Lin, C.C.; Cooper, D.K.C.; Dorling, A. Coagulation dysregulation as a barrier to xenotransplantation in the primate. *Transpl. Immunol.* **2009**, *21*, 75–80. [CrossRef]
10. Cooper, D.K.C.; Ezzelarab, M.B.; Hara, H.; Iwase, H.; Lee, W.; Wijkstrom, M.; Bottino, R. The pathobiology of pig-to-primate xenotransplantation: A historical review. *Xenotransplantation* **2013**, *23*, 83–105. [CrossRef]
11. Chuang, C.K.; Chen, C.H.; Huang, C.L.; Su, Y.H.; Peng, S.H.; Lin, T.Y.; Tai, H.C.; Yang, T.S.; Tu, C.F. Generation of GGTA1 mutant pigs by direct pronuclear microinjection of CRISPR/Cas9 plasmid vectors. *Anim. Biotechnol.* **2017**, *28*, 174–181. [CrossRef] [PubMed]
12. Lin, S.S.; Hanaway, M.J.; Gonzalez-Stawinski, G.V.; Lau, C.L.; Parker, W.; Davis, R.D.; Byrne, G.W.; Diamond, L.E.; Logan, J.S.; Platt, J.L. The role of anti-Galalpha1-3Gal antibodies in acute vascular rejection and accommodation of xenografts. *Transplantation* **2000**, *70*, 1667–1674. [CrossRef] [PubMed]
13. Scobie, L.; Padler-Karavani, V.; Le Bas-Bernardet, S.; Crossan, C.; Blaha, J.; Matouskova, M.; Hector, R.D.; Cozzi, E.; Vanhove, B.; Charreau, B.; et al. Long-term IgG response to porcine Neu5Gc antigens without transmission of PERV in burn patients treated with porcine skin xenografts. *J. Immunol.* **2013**, *191*, 2907–2915. [CrossRef] [PubMed]
14. Varki, A. Loss of N-glycolylneuraminic acid in humans: Mechanisms, consequences, and implications for hominid evolution. *Am. J. Phys. Anthropol.* **2001**, *116*, 54–69. [CrossRef] [PubMed]
15. Byrne, G.; Ahmad-Villiers, S.; Du, Z.; McGregor, C. B4GalNT2 and xenotransplantation: A newly appreciated xenogeneic antigen. *Xenotransplantation* **2018**, *25*, e12394. [CrossRef] [PubMed]
16. Byrne, G.W.; Du, Z.; Stalboerger, P.; Kogelberg, H.; McGregor, C.G.A. Cloning and expression of porcine β1,4 N-acetylgalactosaminyl transferase encoding a new xenoreactive antigen. *Xenotransplantation* **2014**, *21*, 543–554. [CrossRef] [PubMed]
17. Wang, Z.Y.; Burlak, C.; Estrada, J.L.; Li, P.; Tector, M.F.; Tector, A.J. Erythrocytes from GGTA1/CMAH knockout pigs: Implications for xenotransfusion and testing in non-human primates. *Xenotransplantation* **2014**, *21*, 376–384. [CrossRef] [PubMed]
18. Cooper, D.K.C.; Bottino, R. Recent advances in understanding xenotransplantation: Implications for the clinic. *Expert Rev. Clin. Immunol.* **2015**, *11*, 1379–1390. [CrossRef]
19. Bühler, L.; Basker, M.; Alwayn, I.P.; Goepfert, C.; Kitamura, H.; Kawai, T.; Gojo, S.; Kozlowski, T.; Ierino, F.L.; Awwad, M.; et al. Coagulation and thrombotic disorders associated with pig organ and hematopoietic cell transplantation in nonhuman primates. *Transplantation* **2000**, *70*, 1323–1331. [CrossRef]

20. Lau, C.L.; Cantu, E.; Gonzalez-Stawinski, G.V.; Holzknecht, Z.E.; Nichols, T.C.; Posther, K.E.; Rayborn, C.A.; Platt, J.L.; Parker, W.; Davis, R.D. The role of antibodies and von Willebrand factor in discordant pulmonary xenotransplantation. *Am. J. Transplant.* **2003**, *3*, 1065–1075. [CrossRef]
21. Cantu, E.; Balsara, K.R.; Li, B.; Lau, C.; Gibson, S.; Wyse, A.; Baig, K.; Gaca, J.; Gonzalez-Stawinski, G.V.; Nichols, T.; et al. Prolonged function of macrophage, von Willebrand factor-deficient porcine pulmonary xenografts. *Am. J. Transplant.* **2007**, *7*, 66–75. [CrossRef] [PubMed]
22. Paris, L.L.; Estrada, J.L.; Li, P.; Blankenship, R.L.; Sidner, R.A.; Reyes, L.M.; Montgomery, J.B.; Burlak, C.; Butler, J.R.; Downey, S.M.; et al. Reduced human platelet uptake by pig livers deficient in the asialoglycoprotein receptor 1 protein. *Xenotransplantation* **2015**, *22*, 203–210. [CrossRef] [PubMed]
23. Li, P.; Estrada, J.L.; Burlak, C.; Montgomery, J.; Butler, J.R.; Santos, R.M.; Wang, Z.Y.; Paris, L.L.; Blankenship, R.L.; Downey, S.M.; et al. Efficient generation of genetically distinct pigs in a single pregnancy using multiplexed single-guide RNA and carbohydrate selection. *Xenotransplantation* **2015**, *22*, 20–31. [CrossRef] [PubMed]
24. Jinek, M.; Chylinski, K.; Fonfara, I.; Hauer, M.; Doudna, J.A.; Charpentier, E. A programmable dual-RNA-guided DNA endonuclease in adaptive bacterial immunity. *Science* **2012**, *337*, 816–821. [CrossRef]
25. Du, J.; Yin, N.; Xie, T.; Zheng, Y.; Xia, N.; Shang, J.; Chen, F.; Zhang, H.; Yu, J.; Liu, F. Quantitative assessment of HR and NHEJ activities via CRISPR/Cas9-induced oligodeoxynucleotide-mediated DSB repair. *DNA Repair* **2018**, *70*, 67–71. [CrossRef]
26. Miyaoka, Y.; Mayerl, S.J.; Chan, A.H.; Conklin, B.R. Detection and quantification of HDR and NHEJ induced by genome editing at endogenous gene loci using droplet digital PCR. *Methods Mol. Biol.* **2018**, *1768*, 349–362. [CrossRef]
27. Fu, Y.; Foden, J.A.; Khayter, C.; Maeder, M.L.; Reyon, D.; Joung, J.K.; Sander, J.D. High-frequency off-target mutagenesis induced by CRISPR-Cas nucleases in human cells. *Nat. Biotechnol.* **2013**, *31*, 822–826. [CrossRef]
28. Lentsch, E.; Li, L.; Pfeffer, S.; Ekici, A.B.; Taher, L.; Pilarsky, C.; Grützmann, R. CRISPR/Cas9-mediated knock-out of krasG12D mutated pancreatic cancer cell lines. *Int. J. Mol. Sci.* **2019**, *20*, 5706. [CrossRef]
29. Yang, H.; Wang, J.; Zhao, M.; Zhu, J.; Zhang, M.; Wang, Z.; Gao, Y.; Zhu, W.; Lu, H. Feasible development of stable HEK293 clones by CRISPR/Cas9-mediated site-specific integration for biopharmaceuticals production. *Biotechnol. Lett.* **2019**, *41*, 941–950. [CrossRef]
30. Steyer, B.; Bu, Q.; Cory, E.; Jiang, K.; Duong, S.; Sinha, D.; Steltzer, S.; Gamm, D.; Chang, Q.; Saha, K. Scarless Genome Editing of Human Pluripotent Stem Cells via Transient Puromycin Selection. *Stem Cell Rep.* **2018**, *10*, 642–654. [CrossRef]
31. Doench, J.G.; Fusi, N.; Sullender, M.; Hegde, M.; Vaimberg, E.W.; Donovan, K.F.; Smith, I.; Tothova, Z.; Wilen, C.; Orchard, R.; et al. Optimized sgRNA design to maximize activity and minimize off-target effects of CRISPR-Cas9. *Nat. Biotechnol.* **2016**, *34*, 184–191. [CrossRef]
32. Cui, Y.; Xu, J.; Cheng, M.; Liao, X.; Peng, S. Review of CRISPR/Cas9 sgRNA Design Tools. *Interdiscip. Sci.* **2018**, *10*, 455–465. [CrossRef] [PubMed]
33. Yang, Z.; Steentoft, C.; Hauge, C.; Hansen, L.; Thomsen, A.L.; Niola, F.; Vester-Christensen, M.B.; Frödin, M.; Clausen, H.; Wandall, H.H.; et al. Fast and sensitive detection of indels induced by precise gene targeting. *Nucleic Acids Res.* **2015**, *43*, e59. [CrossRef] [PubMed]
34. Sentmanat, M.F.; Peters, S.T.; Florian, C.P.; Connelly, J.P.; Pruett-Miller, S.M. A survey of validation strategies for CRISPR-Cas9 editing. *Sci. Rep.* **2018**, *8*, 888. [CrossRef] [PubMed]
35. Kosicki, M.; Rajan, S.S.; Lorenzetti, F.C.; Wandall, H.H.; Narimatsu, Y.; Metzakopian, E.; Bennett, E.P. Dynamics of indel profiles induced by various CRISPR/Cas9 delivery methods. *Prog. Mol. Biol. Transl. Sci.* **2017**, *152*, 49–67. [CrossRef]
36. Peng, H.; Zheng, Y.; Zhao, Z.; Liu, T.; Li, J. Recognition of CRISPR/Cas9 off-target sites through ensemble learning of uneven mismatch distributions. *Bioinformatics* **2018**, *34*, i757–i765. [CrossRef]
37. Kuscu, C.; Arslan, S.; Singh, R.; Thorpe, J.; Adli, M. Genome-wide analysis reveals characteristics of off-target sites bound by the Cas9 endonuclease. *Nat. Biotechnol.* **2014**, *32*, 677–683. [CrossRef]
38. Zheng, T.; Hou, Y.; Zhang, P.; Zhang, Z.; Xu, Y.; Zhang, L.; Niu, L.; Yang, Y.; Liang, D.; Yi, F.; et al. Profiling single-guide RNA specificity reveals a mismatch sensitive core sequence. *Sci. Rep.* **2017**, *7*, 40638. [CrossRef]
39. Wienert, B.; Wyman, S.K.; Richardson, C.D.; Yeh, C.D.; Akcakaya, P.; Porritt, M.J.; Morlock, M.; Vu, J.T.; Kazane, K.R.; Watry, H.L.; et al. Unbiased detection of CRISPR off-targets in vivo using DISCOVER-Seq. *Science* **2019**, *364*, 286–289. [CrossRef]

40. Tsai, S.Q.; Nguyen, N.T.; Malagon-Lopez, J.; Topkar, V.V.; Aryee, M.J.; Joung, J.K. CIRCLE-seq: A highly sensitive in vitro screen for genome-wide CRISPR-Cas9 nuclease off-targets. *Nat. Methods* **2017**, *14*, 607–614. [CrossRef]
41. Pliatsika, V.; Rigoutsos, I. Off-Spotter': Very fast and exhaustive enumeration of genomic lookalikes for designing CRISPR/Cas guide RNAs. *Biol. Direct* **2015**, *10*, 4. [CrossRef] [PubMed]
42. Xiao, A.; Cheng, Z.; Kong, L.; Zhu, Z.; Lin, S.; Gao, G.; Zhang, B. CasOT: A genome-wide Cas9/gRNA off-target searching tool. *Bioinformatics* **2014**, *30*, 1180–1182. [CrossRef] [PubMed]
43. Xiong, Y.; Xie, X.; Wang, Y.; Ma, W.; Liang, P.; Songyang, Z.; Dai, Z. pgRNAFinder: A web-based tool to design distance independent paired-gRNA. *Bioinformatics* **2017**, *33*, 3642–3644. [CrossRef] [PubMed]
44. Zhu, L.J.; Holmes, B.R.; Aronin, N.; Brodsky, M.H. CRISPRseek: A Bioconductor package to identify target-specific guide RNAs for CRISPR-Cas9 genome-editing systems. *PLoS ONE* **2014**, *9*, e108424. [CrossRef]
45. Cong, L.; Ran, F.A.; Cox, D.; Lin, S.; Barretto, R.; Habib, N.; Hsu, P.D.; Wu, X.; Jiang, W.; Marraffini, L.A.; et al. Multiplex Genome Engineering Using CRISPR/Cas Systems. *Science* **2013**, *339*, 819–823. [CrossRef]
46. Fu, B.; Smith, J.D.; Fuchs, R.T.; Mabuchi, M.; Curcuru, J.; Robb, G.B.; Fire, A.Z. Target-dependent nickase activities of the CRISPR–Cas nucleases Cpf1 and Cas9. *Nat. Microbiol.* **2019**, *4*, 888–897. [CrossRef]
47. Lyu, P.; Javidi-Parsijani, P.; Atala, A.; Lu, B. Delivering Cas9/sgRNA ribonucleoprotein (RNP) by lentiviral capsid-based bionanoparticles for efficient 'hit-and-run' genome editing. *Nucleic Acids Res.* **2019**, *47*, e99. [CrossRef]
48. Martin, R.M.; Ikeda, K.; Cromer, M.K.; Uchida, N.; Nishimura, T.; Romano, R.; Tong, A.J.; Lemgart, V.T.; Camarena, J.; Pavel-Dinu, M.; et al. Highly Efficient and Marker-free Genome Editing of Human Pluripotent Stem Cells by CRISPR-Cas9 RNP and AAV6 Donor-Mediated Homologous Recombination. *Cell Stem Cell* **2019**, *24*, 821–828.e5. [CrossRef]
49. Oh, S.A.; Seki, A.; Rutz, S. Ribonucleoprotein Transfection for CRISPR/Cas9-Mediated Gene Knockout in Primary T Cells. *Curr. Protoc. Immunol.* **2019**, *124*, e69. [CrossRef]
50. Seki, A.; Rutz, S. Optimized RNP transfection for highly efficient CRI SPR/Cas9-mediated gene knockout in primary T cells. *J. Exp. Med.* **2018**, *215*, 985–997. [CrossRef]
51. Vakulskas, C.A.; Behlke, M.A. Evaluation and reduction of crispr off-target cleavage events. *Nucleic Acid Ther.* **2019**, *29*, 167–174. [CrossRef] [PubMed]

*Review*

# Application of Genetically Engineered Pigs in Biomedical Research

**Magdalena Hryhorowicz [1,*], Daniel Lipiński [1], Szymon Hryhorowicz [2], Agnieszka Nowak-Terpiłowska [1], Natalia Ryczek [1] and Joanna Zeyland [1]**

[1] Department of Biochemistry and Biotechnology, Poznan University of Life Sciences, Dojazd 11, 60-632 Poznań, Poland; lipinskidaniel71@gmail.com (D.L.); nwk.agnieszka@gmail.com (A.N.-T.); nataliaryczek.nr@gmail.com (N.R.); joanna.zeyland@up.poznan.pl (J.Z.)

[2] Institute of Human Genetics, Polish Academy of Sciences, Strzeszyńska 32, 60-479 Poznań, Poland; szymon.hryhorowicz@igcz.poznan.pl

* Correspondence: magdalena.hryhorowicz@gmail.com

Received: 4 May 2020; Accepted: 17 June 2020; Published: 19 June 2020

**Abstract:** Progress in genetic engineering over the past few decades has made it possible to develop methods that have led to the production of transgenic animals. The development of transgenesis has created new directions in research and possibilities for its practical application. Generating transgenic animal species is not only aimed towards accelerating traditional breeding programs and improving animal health and the quality of animal products for consumption but can also be used in biomedicine. Animal studies are conducted to develop models used in gene function and regulation research and the genetic determinants of certain human diseases. Another direction of research, described in this review, focuses on the use of transgenic animals as a source of high-quality biopharmaceuticals, such as recombinant proteins. The further aspect discussed is the use of genetically modified animals as a source of cells, tissues, and organs for transplantation into human recipients, i.e., xenotransplantation. Numerous studies have shown that the pig (*Sus scrofa domestica*) is the most suitable species both as a research model for human diseases and as an optimal organ donor for xenotransplantation. Short pregnancy, short generation interval, and high litter size make the production of transgenic pigs less time-consuming in comparison with other livestock species This review describes genetically modified pigs used for biomedical research and the future challenges and perspectives for the use of the swine animal models.

**Keywords:** genetically modified pigs; genome modifications; transgenic pigs; genetic engineering; disease models; recombinant proteins; xenotransplantation

## 1. Introduction

Pigs have been extensively used in biomedical research due to anatomical and physiological similarities to humans. Moreover, progress in gene editing platforms and construct delivery methods allow efficiently, targeted modifications of the porcine genome and significantly broadened the application of pig models in biopharming and biomedicine.

Target editing is possible through site-specific nucleases, of which the following are most commonly used: zinc finger nucleases (ZFNs), transcription activator-like effectors (TALEs), and nucleases from the CRISPR/Cas (clustered regularly interspaced short palindromic repeats/CRISPR associated) system. Introducing modifications in a specific site of the genome is possible due to cellular processes of repairing double-strand breaks induced by site-specific nucleases. Double-strand breaks may be repaired in two ways: by non-homologous end joining (NHEJ) or by homologous recombination (HR). Repair provided by NHEJ may lead to the formation of indel (insertion/deletion) mutations in the target site that may be used to inactivate selected genes. In turn, repairing double-strand breaks by HR using a donor template allows for the introduction or elimination of specific point mutations and the introduction of any foreign genes. Pigs can be modified to create porcine models of human disease [1], produce recombinant proteins [2], or provide tissue and organs for xenotransplantation [3].

## 2. Techniques for Producing Genetically Engineered Pigs

Currently, microinjection or somatic cell nuclear transfer (SCNT) is most commonly used to obtain genetically modified animals. The microinjection is the oldest method for producing transgenic animals [4,5]. This well-developed technology involves the injection of the DNA material into the male pronucleus, the RNA material into the cytoplasm, or proteins into cytoplasm or pronucleus. Because of the difficulty in recognizing the male pronucleus, often, microinjection is simply made into one pronucleus. The efficiency of microinjection depends, among others, on the solution purity, its concentration, material form (DNA/RNA/protein), the length/size of the introduced structure (with increasing length/size the efficiency decreases), the skills of the operator, procedure, or embryo development stage. Due to the low efficiency of this procedure of about 10% in mice, 2%–3% in pig, 4% in rabbit, and less than 1% in cattle, researchers are introducing modifications to increase the number of successful microinjections and further transgenesis [6–8]. One of the introduced changes was the simultaneous injection of the transgene into both pronuclei. Increased transgene integration efficiency has been observed with the high invasiveness of the process resulting in high embryo mortality [9]. Subsequent tests were performed using DNA microinjection directly into the embryo cytoplasm. In this method, the problem of high invasiveness of injections was eliminated, while a problem arose with the integration of exogenous DNA and a further drastic decrease in the procedure efficiency [10]. This problem was eliminated by providing genetic material into the cytoplasm in the form of RNA. In this case, the stability of the material is a critical element. The third modification of DNA microinjection carried out by scientists was the delivery of the transgene into the cell nuclei of a two-blastomeric embryo. It was aimed at increasing the survival rate of the embryos. However, due to the higher probability of obtaining mosaic than transgenic animals, it is not commonly used in transgenesis [11]. Another method by which transgenic animals can be obtained is the use of transformed cells for somatic cloning. Interest in somatic cell nuclear transfer has been increasing since 1996 when the Dolly sheep was cloned for the first time [12]. This method, despite very low efficiency (0.5%–1.0% in livestock animals), allows for the use of a wide range of available approaches related to the cells' genetic modification whose cell nuclei are later used for cloning. The choice of delivery method for exogenous DNA depends on its length, used cells, and transfection efficiency. After the identification of cells with appropriate modification, they should be brought to the $G_0$ phase of the cell cycle. The cell nuclei of the modified cells are then transferred into the enucleated oocytes. The quality of oocytes and their age affect the cloning process. The whole method of obtaining transgenic animals by using transformed cells for somatic cloning has many advantages, but its main disadvantage is high fetal mortality resulting primarily from genetic defects [13]. A summary of the advantages and disadvantages of microinjection and SCNT is presented in Table 1.

**Table 1.** Summary of the most important advantages and disadvantages of the methods for obtaining genetically modified animals.

| | Microinjection | SCNT |
|---|---|---|
| **Advantages** | increased efficiency of the transgene integration<br>the number of damaged zygotes do not exceed 10% | precise transformation and selection of modified cells used in cloning<br>the obtained animals do not exhibit mosaicism<br>modification is also revealed in germ cells—transgenic offspring |
| **Disadvantages** | low process efficiency (2%–3% in pigs)<br>the possibility of random integration of the transgene<br>high process invasiveness | very low efficiency<br>early fetal mortality<br>the possibility of genetic defects |

## 3. Pig Models for Human Diseases

Genetically modified animals as research models for human diseases are a very important tool in searching for and developing new methods of therapy. A suitable model organism should be characterized by rapid growth, a high number of offspring, easy and inexpensive breeding, ability to be easily manipulated, and having a sequenced genome. Initially, only rodents were used as a model in biomedical research. Experiments on mice contributed to understanding the genetic background of numerous diseases. However, not every genetic disease induced in mice has the same clinical manifestation as in humans. Furthermore, the short life span, together with a higher metabolic rate, makes the analysis of some hereditary diseases challenging.

Currently, pigs are one of the most important large animal models for biomedical research. Many human diseases, such as cardiovascular diseases, obesity, and diabetes, have their counterparts in this species. The use of model organisms makes it possible to analyze diseases that occur naturally in animals with specific mutations and those deliberately induced. Introduced animal genome changes may reflect mutations occurring in people suffering from specific genetic disorders. Moreover, accurate and efficient genome editing can be used in the treatment of monogenic diseases. A slightly different direction of research is the use of genetically modified animal models in toxicological studies for testing drugs. Genetically modified pigs are used as model organisms in research into various diseases, including cardiovascular and neurodegenerative diseases, neoplasms, and diabetes.

### *3.1. Cystic Fibrosis*

Cystic fibrosis (CF) is an autosomal recessive disorder manifested by bronchopulmonary failure and pancreatic enzyme insufficiency. CF is caused by a mutation in the gene responsible for the synthesis of the CFTR (cystic fibrosis transmembrane conductance regulator) chloride channel, altering the mucosal function in the respiratory epithelium, pancreatic ducts, intestines, and sweat glands. The most common cystic fibrosis-causing mutation is the deletion of a phenylalanine at amino acid position 508 (dF508). CF lung disease is the main cause of morbidity and mortality in CF patients. Porcine lungs share many anatomical and histological similarities with humans. It has been shown that pigs in which the *CFTR* gene was inactivated develop all symptoms of the disease occurring in humans, such as meconium ileus, defective chloride transport, pancreatic destruction, and focal biliary cirrhosis. This makes them a very good model species for this disease [14–16]. Cystic fibrosis is a monogenic disease, and the insertion of the functional *CFTR* gene into CF patient cells should theoretically restore the CFTR channel function. Therefore, pigs have also been used in gene therapy. The treatment with viral vectors successfully improved anion transport and inhibited bacterial growth [17,18]. Currently, the research focuses on improving the CF gene therapy with the use of the CRISPR/Cas9 system. These efforts are focused on increasing the delivery efficiency of CRISPR/Cas9 elements to target locus and obtaining sustained expression of the *CFTR* transgene [19,20]. It was demonstrated that precise integration of the human *CFTR* gene at a porcine safe harbor locus through CRISPR/Cas9-induced HDR-mediated knock-in allowed the achievement of persistent in vitro expression of the transgene in transduced cells. These results can help design effective gene therapy to treat CF patients [20].

### 3.2. Duchenne Muscular Dystrophy

Duchenne muscular dystrophy (DMD) is a progressive, monogenic, X-linked lethal disease characterized by degenerative changes in muscle fibers and the connective tissue. It involves the degeneration of subsequent muscles—skeletal, respiratory, and cardiac—and progressive muscular dystrophy. Muscular dystrophy is caused by a frameshift mutation in the *DMD* gene, which encodes dystrophin, a protein in muscle cells that connects the cytoskeleton with the cell membrane. The dystrophin gene contains 79 exons, with exons 3–7 and 45–55 being the most susceptible to mutations. *DMD* gene mutations are usually large deletions or duplications of one or several exons, as well as point mutations, leading to a change in the reading frame, the appearance of a premature stop codon, and failure to produce a stable protein. Muscular dystrophy is most often diagnosed in early childhood, and patients become wheelchair dependent by 12 years of age. Untreated boys die of cardiorespiratory complications around their 20 years. The rapid progress in gene editing gives hope for effective targeted therapies for DMD. Moreover, the use of an animal model can facilitate the development of personalized treatment approaches. Pigs with a DMD gene mutation (exon 52 deletion) develop human disease symptoms, such as lack of dystrophin in skeletal muscles, increased serum creatine kinase levels, progressive muscle dystrophy, and impaired mobility [21]. However, these animals died prematurely (up to 3 months old at most) what precluded natural breeding. The histological evaluation of skeletal muscles and diaphragm confirmed the presence of excessive fiber size variation, hypercontracted fibers, and segmentally necrotic fibers, resembled that of human DMD patients [21]. Moreover, proteome analysis of biceps femoris muscle was performed. An increased amount of muscle repair-related proteins and reduced amount of respiratory chain proteins was found in tissue from 3-month-old DMD pigs. This indicated severe disturbances in aerobic energy production and a decrease in functional muscle tissue [22]. As the deletion of exon 52 in the human *DMD* gene is a common cause of Duchenne muscular dystrophy, pigs can make an accurate research model for gene therapy. Another porcine DMD model is genetically modified miniature pigs with a mutation in exon 27 in the *DMD* gene obtained by the CRISPR/Cas9 system. In addition, these animals have shown symptoms of skeletal and heart muscle degeneration, characteristic of human patients with Duchenne muscular dystrophy. Reduced thickness of smooth muscle in the stomach and intestine was also observed in the pigs studied. However, founder pigs died of unreported causes [23]. Although mutations in exon 27 are not reported in human DMD patients, pigs with this deletion constitute another useful animal DMD model. Recently, Moretti et al. demonstrated the restoration of dystrophin by intramuscular injection of CRISPR/Cas9 components with the use of adeno-associated viral vectors in a pig model. In this study, pigs with DMD carrying a deletion of DMD exon 52 (d52DMD), resulting in a complete loss of dystrophin expression, were used. The restoration of dystrophin expression was possible due to the excision of exon 51 and the restoration of the DMD reading frame. The internally truncated d51-52DMD sufficed to improve skeletal muscle function, prevent malignant arrhythmias as well as prolonging lifespan of DMD pigs [24]. In the future, this strategy may prove useful in the clinical treatment of patients with d52DMD.

### 3.3. Alzheimer's Disease

Alzheimer's disease (AD) is an age-related, progressive neurodegenerative disorder characterized by memory dysfunction followed by cognitive decline and disorientation. AD accounts for 50%–80% of human dementia cases. Familial forms of AD are caused by autosomal mutations in the genes encoding presenilin 1 (*PSEN1*) and presenilin 2 (*PSEN2*) and amyloid precursor protein (APP). These mutations are associated with the accumulation of amyloid β (Aβ) peptide in senile plaques and phosphorylated tau protein in neurofibrillary tangles (NFTs), which leads to synaptic damage and neuronal dysfunction [25].

The first AD model with the use of transgenic pigs was generated in 2009 by Kragh et al. They produced Göttingen mini pigs that carried a randomly integrated construct containing the cDNA of the human *APP* gene with AD causing a dominant mutation known as the Swedish mutation (APPsw) and a human PDGFβ promoter fragment [26]. Although the transgene was specifically expressed in brain tissue at a high level, no AD phenotype was observed in mutant pigs. The same group also obtained Göttingen minipigs with the human *PSEN1* gene carrying the AD-causing Met146Ile mutation (*PSEN1* M146I) and driven by a cytomegalovirus (CMV)-enhanced human UbiC promoter. Pigs were generated with the use of a site-specific integration system—recombinase-mediated cassette exchange (RMCE) [27]. The PSEN1 M146I protein was expressed and tolerated well in the porcine brain, but also in this case, no symptoms of the AD disease were noticed. Therefore, this group generated double transgenic Göttingen minipigs with both APPsw and *PSEN1* M146I mutations. Such a solution allowed the increase in intraneuronal accumulation of Aβ [28]. In turn, another group obtained AD transgenic pigs using a retroviral multi-cistronic vector containing three AD-related human genes: APP, Tau, and *PSEN1*, with a total of six well-characterized mutations under the control of a fusion promoter: CMVE+ hPDGFβ promoter region. They confirmed that transgenes were expressed at high levels in brain tissue and demonstrated a two-fold increase in Aβ levels in the brains of transgenic pigs compared to wild-type [29].

### 3.4. Porcine Cancer Models

Cancer is a genetic disease involving uncontrolled, abnormal cell growth in the blood or solid organs resulting from acquired or inherited mutations. Pigs represent a useful animal for the development and validation of new medicines and procedures in human tumor models. There are many resemblances in cancer biology between pigs and humans. These animals can correctly mimic human tumors and show similar pharmacokinetic responses to humans. Adam et al. demonstrated that autologous transplantation of primary porcine cells transformed with retroviral oncogenic vectors caused tumorigenesis akin to those found in humans [30]. In turn, Schook et al. induced tumor formation in pigs by introducing random transgenes that encode Cre-dependent KRAS (Kirsten rat sarcoma viral oncogene homolog) G12D and TP53 (tumor protein 53) R167H oncogenic mutations (orthologous to human TP53 R175H) [31]. Moreover, Saalfrank et al. reported that porcine mesenchymal stem cells (MSCs) resemble human MSCs requiring disturbance of p53, KRAS, and MYC signaling pathways to become a fully transformed phenotype [32]. At present, pig models commonly used in cancer research include the TP53 knock-out model of osteosarcoma and APC (adenomatous polyposis coli) mutations model of familial adenomatous polyposis (FAP). *TP53* is a known tumor suppressor gene, and a germline mutation within this gene leads to Li–Fraumeni Syndrome, a rare, autosomal dominant disorder that predisposes carriers to cancer development. The first model of Li–Fraumeni Syndrome using genetically modified pigs has been described by Leuchs et al. They generated pigs carrying a latent *TP53* R167H mutation that can be activated by the Cre-lox recombinase system [33]. After several years of observation, it was noted that both pigs with homozygous *TP53* knock-out and pigs with heterozygous knock-out of *TP53* showed osteosarcoma development. The heterozygous knock-out caused the development of spontaneous osteosarcoma in older animals, while homozygous *TP53* knock-out resulted in multiple large osteosarcomas in 7 to 8-month-old pigs [32]. Moreover, Sieren and colleagues generated genetically modified Yucatan minipigs that carried the TP53 R167H mutation. Animals heterozygous for this mutant allele showed no tumorigenesis process, whereas homozygotes that reached sexual maturity developed lymphomas, osteogenic tumors, and renal tumors at varying rates. The tumor formations were validated by computed tomography, histopathological evaluation, and magnetic resonance imaging [34]. Familial adenomatous polyposis is an inherited disorder characterized by the development of numerous adenomatous polyps in the colon and rectum which greatly increases the risk of colorectal cancer. The mutations in the *APC* tumor-suppressor gene are responsible for FAP and may result in a hereditary predisposition to colorectal cancer. Flisikowska et al. generated gene-targeted cloned pigs with translational stop signals at codon 1311 in porcine

APC (APC 1311), orthologous to common germline *APC* 1309 mutations in human FAP. Evaluation of one-year-old pigs carrying the *APC* 1311 mutation showed aberrant crypt foci and adenomatous polyps with low- to high-grade intraepithelial dysplasia, similar to tumor progression as in human FAP [35]. The APC 1311 pig model resulting in the development of polyposis in the colon and rectum can be useful in the diagnosis and therapy of colorectal cancer.

### 3.5. Cardiovascular Diseases

Cardiovascular diseases (CVDs) are the major cause of morbidity and mortality worldwide. CVD is a group of disorders of the heart and blood vessels that involve coronary heart disease (such as angina and myocardial infarction), deep vein thrombosis and pulmonary embolism, peripheral arterial disease, cerebrovascular disease, and rheumatic heart disease. The dominant cause of CVD is atherosclerosis, which is characterized by the narrowing of arteries due to the accumulation of lipid and plaque formation. The plaque buildup restricts blood flow, and plaque burst can entail blood clots. Similarities in heart anatomy and physiology, vessel size, blood parameters, coronary artery system anatomy, and lipoprotein metabolism make pigs a suitable model for the human cardiovascular system. Atherosclerosis starts with the buildup of serum low-density lipoprotein (LDL), and mutations in the LDL receptor (*LDLR*) gene may cause familial hypercholesterolemia (FH). A porcine FH model has been generated in Yucatan miniature pigs through recombinant adeno-associated virus-mediated targeted disruptions of the endogenous *LDLR* gene. LDLR+/− heterozygous pigs exhibited mild hypercholesterolemia, while LDLR−/− homozygotes animals were born with severe hypercholesterolemia and developed atherosclerotic lesions in the coronary arteries. These phenotypes were accelerated by high fat and high cholesterol diets [36]. The utilization of LDLR-deficient Yucatan minipigs in the preclinical evaluation of therapeutics was also demonstrated. LDLR+/− and LDLR−/− pigs were used to assess the ability of novel drug—bempedoic acid (BemA)—to reduce cholesterol biosynthesis. Long-term treatment with BemA decreased LDL cholesterol and attenuated aortic and coronary atherosclerosis in this FA model [37]. Moreover, a model of FA and atherosclerosis was created by using the Yucatan miniature pigs with liver-specific expression of a human proprotein convertase subtilisin/kexin type 9 (PCSK9) carrying the gain-of-function mutation D374Y. PCSK9 plays important functions in cholesterol homeostasis by reducing LDLR levels on the plasma membrane. Gain-of-function mutations in this protein cause increased levels of plasma LDL cholesterol, which in turn may result in more susceptibility to coronary heart disease. PCSK9 D374Y transgenic pigs exhibited decreased hepatic LDLR levels, severe hypercholesterolemia on high-fat, high-cholesterol diets, and atherosclerotic lesions [38]. It is also considered that hypertriglyceridemia is an independent risk factor for coronary heart disease, in which apolipoprotein (Apo)CIII is associated with plasma triglyceride levels. The hypertriglyceridemic ApoCIII transgenic miniature pig model was generated for the examination of the correlation between hyperlipidemia and atherosclerosis. Transgenic pigs expressing human ApoCIII exhibited increased plasma triglyceride levels with their delayed clearance and reduced lipoprotein lipase activity compared to non-transgenic controls [39].

### 3.6. Diabetes Mellitus

Diabetes mellitus (DM) is a group of metabolic disorders characterized by hyperglycemia (elevated levels of blood sugar over a prolonged period), which results from deficiency or ineffectiveness of insulin. DM may lead over time to cardiovascular disease, chronic kidney disease, damage to the nerves and eyes. There are two main types of diabetes mellitus, called type 1 and type 2. Type 1 DM, also referred to as juvenile diabetes or insulin-dependent diabetes mellitus, is caused by the pancreas's failure to produce enough insulin. The most common is type 2 diabetes, which is characterized by insulin resistance (reduced tissue sensitivity to insulin) that may be combined with relative insulin deficiency. The anatomical and physiological resemblance to the human pancreas and islets makes pigs excellent animals for metabolic diseases modeling. Moreover, the structure of porcine and human insulin is also very similar (differs by only one amino acid). A transgenic pig model for type 2 DM was generated to evaluate the role of impaired glucose-dependent insulinotropic poly-peptide (GIP). The main function of incretin hormones GIP and glucagon-like peptide-1 (GLP1) is stimulated insulin secretion from pancreatic beta cells in a glucose-dependent manner. In type 2 DM, the insulinotropic action of GIP is impaired, which may suggest its association with early disease pathogenesis. The transgenic pigs expressing a dominant negative GIP receptor (GIPRdn) in pancreatic cells were produced by lentiviral vectors. A significant reduction in oral glucose tolerance due to delayed insulin secretion as well as in β-cell mass caused by diminished cell proliferation was observed in GIPRdn animals [40]. These observations resemble the characteristic features of human type 2 diabetes, which makes the porcine GIPRdn model useful for testing incretin-based therapeutic strategies. Further analyses revealed characteristic changes in plasma concentrations of seven amino acids (Phe, Orn, Val, xLeu, His, Arg, and Tyr) and specific lipids (sphingomyelins, diacylglycerols, and ether phospholipids) in the plasma of 5-month-old GIPRdn transgenic pigs that correlate significantly with β-cell mass [41]. These metabolites represent possible biomarkers for the early stages of prediabetes. Moreover, the porcine GIPRdn model has been used to test liraglutide, GLP1 receptor agonist, which improve glycemic control in type 2 diabetic patients. Ninety-day liraglutide treatment of adolescent transgenic pigs resulted in improved glycemic control and insulin sensitivity as well as reduction in body weight gain and food intake compared to placebo-treated animals. However, the use of liraglutide did not stimulate beta-cell proliferation in the endocrine pancreas [42]. Another type of diabetes, type 3 DM, is maturity-onset diabetes of the young (MODY3). MODY3 is a noninsulin-dependent type of diabetes with an autosomal dominant inheritance and is caused by mutations in the human hepatocyte nuclear factor 1 α (*HNF1α*) gene. Mutation in *HNF1α* gene leads to pancreatic β-cell dysfunction and impaired insulin secretion. A pig model for MODY3 was generated by expressing a mutant human *HNF1α* gene (HNF1α P291fsinsC) using intracytoplasmic sperm injection-mediated gene transfer and somatic cell nuclear transfer. The transgenic piglets exhibited the pathophysiological characteristics of diabetes, including high glucose level and reduced insulin secretion from the small and irregularly formed Langerhans Islets [43]. Furthermore, HNF1α P291fsinsC pigs revealed nodular lesions in the renal glomeruli, diabetic retinopathy, and cataract, complications similar to those in patients with DM [44]. Mutations in the insulin (INS) gene may result in permanent neonatal diabetes mellitus (PNDM) in humans. A PNDM large animal model was establish by generated pigs expressing a mutant porcine *INS* gene (INS C94Y), orthologous to human INS C96Y. Transgenic animals showed signs of PNDM, such as lower fasting insulin levels, decreased β-cell mass, reduced body weight, and cataract development. In addition, INS C94Y pigs exhibited significant β-cell impairment, including the reduction in insulin secretory granules and dilation of the endoplasmic reticulum [45]. The porcine INS C94Y model was further used to perform analysis of pathological changes in retinas and evaluation of the liver of transgenic pigs. The studies revealed several features of diabetic retinopathy, such as intraretinal microvascular abnormalities or central retinal edema [46]. Moreover, the multi-omics analysis of the liver demonstrated higher activities in amino acid metabolism, oxidation of fatty acids, gluconeogenesis, and ketogenesis, characteristic of insulin-deficient diabetes mellitus [47]. The genetically modified pig models for human diseases described in this review are summarized in Table 2.

**Table 2.** Selected genetically engineered pig models for human diseases.

| Human Disease | Genetic Modification | Reference |
|---|---|---|
| Cystic fibrosis | targeted disruption of *CFTR* gene | [14–16] |
| Duchenne muscular dystrophy | targeted deletion of *DMD* exon 52 | [21] |
| | targeted knock-out of *DMD* gene | [23] |
| Alzheimer's disease | expression of human *APPsw* and *PSEN1* M146I genes | [28] |
| | expression of human *APP* (K670N/M671L, I716V, V717I), *Tau* (P301L), and *PSEN1* (M146V, L286P) genes | [29] |
| Osteosarcoma | targeted knock-out of *TP53* gene | [32] |
| | targeted homozygous *TP53* R167H mutation | [34] |
| Colorectal cancer | targeted heterozygous *APC* 1311 mutation | [35] |
| Cardiovascular Diseases | targeted disruption of *LDLR* gene | [36] |
| | expression of human *PCSK9* D374Y gene | [38] |
| | expression of human *ApoCIII* gene | [39] |
| Diabetes mellitus | expression of human *GIPRdn* gene | [40] |
| | expression of human *HNF1α* P291fsinsC gene | [43] |
| | expression of porcine *INS* C94Y gene | [45] |

## 4. Pigs as Bioreactors for Pharmaceutical Products

Human-derived proteins have long been used as therapeutics in the treatment of numerous diseases. However, their quantities are limited by the availability of human tissues. Thanks to the development of biotechnology and genetic engineering, modified animals can be used as "bioreactors" to produce recombinant proteins for pharmaceutical use. By using adequate regulatory sequences, promoters, the expression of transgenes can be directed to selected cells and organs. The therapeutic proteins can be obtained from milk, blood, urine, seminal plasma, egg white, or salivary gland that can be collected, purified, and used at an industrial scale. Moreover, it is possible to generate multi-transgenic animals that produce many biopharmaceuticals or vaccines in a single organism. The use of an animal platform allows for the relatively low-cost production of pharmacologically valuable preparations in high quantity and quality. The mammary gland is considered to be an excellent bioreactor system for pharmaceutical protein production. The advantage of milk is that it contains large amounts of foreign proteins that do not affect the animal's health during lactation as well as the ease of product collection and purification. While cows are the best species for obtaining large amounts of pharmaceuticals in milk, the cost and time necessary to carry out successful transgenesis make rabbits, sheep, goats, and pigs more popular species. Although the pig is not a typical dairy animal, a lactating sow can give about 300 L of milk per year. Velander et al. generated transgenic pigs that synthesized human protein C in the mammary gland. Protein C plays an important role in human blood clotting, which makes it a potentially attractive drug. The collected milk contained 1 g/L of this protein [48]. Other recombinant human proteins involved in the coagulation process, such as factor VIII [49], factor IX [50,51], von Willebrand factor [52], were also successfully obtained in the porcine mammary gland. Furthermore, the line of transgenic pigs producing functional recombinant human erythropoietin in their milk was demonstrated. Erythropoietin regulates red blood cell production (erythropoiesis) in the bone marrow by binding to a specific membrane receptor and has been used in the treatment of anemia. This bioreactor system generates active recombinant human erythropoietin at concentrations of approximately 877.9 ± 92.8 IU/1 mL [53]. In turn, Lu et al. generated transgenic cloned pigs expressing large quantities of recombinant human lysozyme in milk. Lysozyme is a natural broad-spectrum antimicrobial enzyme which constitutes part of the innate immune system. The authors demonstrated that the highest concentration of recombinant human lysozyme with in vitro bioactivity was 2759.6 ± 265.0 mg/L [54]. Biopharmaceuticals can also be synthesized in pigs with the use of alternative systems, such as blood, urine, and semen. The blood of transgenic animals can be a source of human blood proteins, such as hemoglobin. Swanson et al. and Sharma et al. obtained transgenic pigs that produced recombinant human hemoglobin in their blood cells at a high level, with the ability to bind oxygen identical to that of human blood hemoglobin [55,56]. There remains, however, the issue of obtaining large amounts of animal-generated therapeutics easily and inexpensively, without killing the animal. Moreover, blood cannot store high levels of recombinant proteins for a long time, which are innately unstable, and bioactive proteins in the blood may affect the metabolism of the animals [57]. For this reason, research is being conducted into the production of recombinant proteins secreted into the urine or semen. The advantage of semen is that it is easily obtained and produced in high amounts in species such as pigs (boars can produce 200–300 mL of semen 2–3 times a week), while the advantage of urine is that proteins can be obtained from animals of both sexes throughout their lives. In addition, urine contains few proteins, which facilitates the purification of the protein product, and the urine-based systems pose a low risk to the animal's health. However, the limitation of protein production in the bladder is low yield [58]. The recombinant pharmaceutical proteins produced from transgenic pigs are listed in Table 3.

**Table 3.** Recombinant proteins produced from transgenic pigs for pharmaceutical use.

| Protein | Production System | Yield | Reference |
|---|---|---|---|
| Human protein C | milk | up to 1 g/L | [48] |
| Human factor VIII | milk | up to 2.7 μg/mL | [49] |
| Human factor IX | milk | up to 0.25 mg/mL | [50] |
| Human von Willebrand factor | milk | mean 280 μg/mL | [52] |
| Human erythropoietin | milk | mean 877.9 IU/1 mL | [53] |
| Human lysozyme | milk | up to 2759.6 mg/L | [54] |
| Human hemoglobin | blood | up to 32 g/L | [56] |

## 5. Pig-to-Human Xenotransplantation

Genetically modified pigs can also be used as a source of cells, tissues, and organs for transplantation into human recipients. Despite the growing knowledge and ability to perform transplants, the shortage of organs means that the number of patients awaiting a transplant is constantly increasing. Xenotransplantation is any procedure involving the transplantation, implantation, or infusion of cells, tissues or animal donor organs, and also body fluids, cells, tissues, and human organs (or their fragments), which had ex vivo contact with animal cells, tissues, or organs into a human recipient. Organ xenotransplantation would give us an unlimited and predictable source of organs and enable careful planning of the surgery and preoperative drug treatment of the donor. The animal that best meets the criteria for xenotransplantation is the domestic pig (*Sus scrofa domestica*). Pig and human organs show great anatomical and physiological similarities. However, the significant phylogenetic distance results in serious immunological problems after transplantation. Despite major difficulties, the pig is currently the focus of all research aimed towards eliminating the problem of organ shortage for human transplantation in the future. Thus, the challenge now is to overcome interspecies differences that cause xenograft rejection by the human immune system. The solution, therefore, is to modify pigs in such a way that their organs are not rejected as belonging to another species. Advances in genetic engineering have brought scientists closer to obtaining modified animals that would be useful for pig to human transplants. A number of studies have reached the preclinical stage, using primates as model organisms.

### 5.1. *Hyperacute Xenograft Rejection*

Pig organs transplanted into human recipients are immediately rejected as a result of the so-called hyperacute immunological reaction. Xenograft rejection is mainly caused by the Gal antigen found on the donor's cell surface, which is synthesized by the GGTA-1 enzyme. Humans lack both the Gal antigen and the GGTA-1 enzyme, but have xenoreactive antibodies directed against the porcine Gal antigen, which leads to the so-called enzymatic complement cascade in the recipient. The sequence of reactions results in a formation membrane attack complex, lysis, and destruction of the graft cells.

The best possible solution to the problem of hyperacute rejection is to inactivate the gene encoding the GGTA-1 enzyme responsible for the formation of the Gal antigen. In 2001, the first heterozygous GGTA1 knock-out pigs were produced [59], and one year later, the first piglets with two knock-out alleles of the GGTA1 gene were born [60]. A series of GGTA1 knock-out pigs has also been generated by using ZFNs [61], TALENs [62], and the CRISPR/Cas9 system [63]. Moreover, other carbohydrate xenoantigens present on pig cells but absent in humans have been identified and include Neu5Gc antigen (N-glycolylneuraminic acid) catalyzed by cytidine monophosphate-N-acetylneuraminic acid hydroxylase (CMAH) and the SDa antigen produced by beta-1,4-N-acetyl-galactosaminyltransferase 2 (β4GALNT2). Pigs with GGTA1/CMAH/β4GalNT2 triple gene knock-out were generated using the CRISPR/Cas9 system. Cells from these genetically modified animals exhibited a reduced level of human IgM and IgG binding resulting in diminished porcine xenoantigenicity [64]. To prevent hyperacute rejection, it is possible to introduce human genes regulating the enzymatic complement cascade into the porcine genome. As the complement system may undergo spontaneous autoactivation and attack the body's own cells, defense mechanisms have developed in the course of evolution. They regulate complement activity through a family of structurally and functionally similar proteins blocking complement activation and preventing the formation of a membrane attack complex (MAC). Introduction of human genes encoding complement inhibitors, such as CD55 (DAF, decay-accelerating factor), CD46 (MCP, membrane cofactor protein), CD59 (membrane inhibitor of reactive lysis), into the porcine genome may overcome xenogeneic hyperacute organ rejection [65]. It was demonstrated that the expression of human complement-regulatory proteins can prevent complement-mediated xenograft injury and prolong the survival time of the xenotransplant [66–68]. Studies have shown that the absence of GGTA1 and additional human CD55, CD59, or CD46 expression has greater survival rates than just GGTA1 knock-out [69,70]. Many genetically modified pigs with human complement inhibitors and other modifications important for xenotransplantation were also generated [71–73]. The modifications of the porcine genome described above largely resolved the problem of hyperacute rejection. However, xenogenic transplant becomes subject to less severe rejection mechanisms resulting from coagulation dysregulation, natural killer (NK) cells-mediated cytotoxicity, macrophage-mediated cytotoxicity as well as T-cell response.

### 5.2. Coagulation Dysregulation

The coagulative disorders result from incompatibilities between pig anticoagulants and human coagulation factors. Overcoming coagulation dysregulation in xenotransplantation will require the introduction of human gene encoding coagulation–regulatory proteins into the porcine genome, for example, thrombomodulin (TBM), endothelial cell protein C receptor (EPCR), tissue factor pathway inhibitor (TFPI), and ectonucleoside triphosphate diphosphohydrolase 1 (CD39). Thrombomodulin binds thrombin and functions as a cofactor for the activation of protein C, which is strongly anticoagulative. Porcine TBM binds human thrombin less strongly and cannot effectively activate protein C. It was demonstrated that expressing human TBM (hTBM) in porcine aortic endothelial cells (PAECs) suppresses prothrombinase activity and delays clotting time [71]. The endothelial protein C receptor enhances the activation of protein C and decreases proinflammatory cytokine synthesis. In vitro studies revealed the correlation between human EPCR (hEPCR) expression in PAECs and reduced human platelet aggregation [74]. A meta-analysis of multiple genetic modifications on pig lung xenotransplant showed that hEPCR was one of the modifications that had a positive effect on xenograft survival prolongation in the ex vivo organ perfusion model with human blood [75]. Further study demonstrated that kidneys from genetically-engineered pigs (carrying six modifications) functioned in baboons for 237 and 260 days. The authors suggested that prolonged survival time was associated, among others, with the expression of the human EPCR gene [76]. Tissue factor pathway inhibitor is the primary physiological regulator of the early stage of coagulation. TFPI binds to factor Xa, and then Xa/TFPI inhibits the procoagulant activity of the tissue factor (TF)/factor VIIa complex. It was demonstrated that the expression of human TFPI in PAECs can inhibit TF activity, suggesting potential for controlling the TF-dependent pathway of blood coagulation in xenotransplantation [77]. More recently, multi-modified pigs carrying human TFPI transgene were produced [76,78]. CD39 is an ectoenzyme that plays a key role in reducing platelet activation. CD39 converts adenosine triphosphate (ATP) and adenosine diphosphate (ADP) to adenosine monophosphate (AMP), which in turn is further degraded by ecto-5′-nucleotidase (CD73) to antithrombotic adenosine. Transgenic pigs with human CD39 (hCD39) gene were generated. The study showed that hCD39 expression protects against myocardial injury in a model of myocardial acute ischemia-reperfusion injury [79].

### 5.3. Inflammatory Response

Another approach to xenograft protection may be introducing a human gene that protects against the inflammatory response into the porcine genome. Transgenic pigs expressing antiapoptotic and anti-inflammatory proteins, such as human heme oxygenase-1 (HO-1) and human tumor necrosis factor-alpha-induced protein 3 (A20), were produced [80,81]. Porcine aortic endothelial cells derived from pigs carrying human A20 transgene were protected against TNF-α (tumor necrosis factor alpha)-mediated apoptosis and less susceptible to cell death induced by CD95 (Fas) ligands [81]. Similarly, overexpression of human HO-1 ensured prolonged porcine kidney survival in an ex vivo perfusion model with human blood and PAECs protection from TNF-α-mediated apoptosis [80]. Furthermore, pigs with a combined expression of human A20 and HO-1 on a GGTA1 knock-out background were generated. That transgenic approach alleviated rejection and ischemia-reperfusion damage during ex vivo kidney perfusion [82].

### 5.4. Cellular Xenograft Rejection

The cellular immune response is another barrier to xenotransplantation. Human NK cells can activate the endothelium and lyse porcine cells through direct NK cytotoxicity and by antibody-dependent cellular mechanisms. Direct NK cytotoxicity is regulated by activating and inhibitory receptor-ligand interactions. To prevent NK-mediated lysis through the inhibitory CD94/NKG2A receptor, pigs with human leukocyte antigens- E (HLA-E) were obtained [83,84]. The study showed that the expression of HLA-E in endothelial cells from transgenic pigs markedly reduces xenogeneic human NK responses. In addition, it was demonstrated that the introduction of the HLA-E gene into the porcine genome may also protect pig cells from macrophage-mediated cytotoxicity [85]. More recently, GGTA1 knock-out pigs with hCD46 and HLA-E/human β2-microglobulin transgenes were produced. The study showed that multiple genetically modified porcine hearts were protected from complement activation and myocardial natural killer cell infiltration in an ex vivo perfusion model with human blood [86]. Another approach to inhibit direct xenogeneic NK cytotoxicity is the elimination of porcine UL-16-binding protein 1 (ULBP1), which binds to NKG2D activating NK receptors. CRISPR technology was adapted to create genetically modified pigs with a disrupted UL16-binding protein 1 gene. In vitro studies confirmed that porcine aortic endothelial cells derived from ULBP1 knock-out pigs were less susceptible to NK-cells' cytotoxic effects [87]. Macrophages also play an important role in xenograft rejection and can be activated by direct interactions between receptors present on their surface and donor endothelial antigens as well as by xenoreactive T lymphocytes. The binding CD47 antigen to macrophage surface signaling regulatory protein (SIRP-α) delivers a signal to prevent phagocytosis. However, the interaction between porcine CD47 and human SIRP-α does not supply the inhibitory effect on macrophages [88]. Therefore, the introduction of human CD47 (hCD47) into the porcine genome can overcome macrophage-mediated responses in xenotransplantation. The overexpression of hCD47 in porcine endothelial cells suppressed the phagocytic and cytotoxic activity of macrophages, decreased inflammatory cytokine (TNF-α, IL-6, IL-1β) secretion and inhibited the infiltration of human T cells [89]. The pigs with GGTA1 knock-out and hCD47 were obtained [90]. It was demonstrated that the expression of human CD47 markedly prolonged survival of donor porcine skin xenografts on baboons in the absence of immunosuppression [91]. Another challenge in xenotransplantation is the prevention of T cell-mediated rejection. T cells can be induced directly by swine leukocyte antigen (SLA) class I and class II on porcine antigen-presenting cells (APCs) or by swine donor peptides presented on recipient APCs. The main co-stimulatory signals regulating T cell function include CD40–CD154 and CD28–CD80/86 pathways. The cytotoxic T-lymphocyte antigen 4-immunoglobulin (CTLA4) can inhibit the CD28–CD80/86 co-stimulatory pathway. Therefore, the introduction of human CTLA4-Ig (hCTLA4-Ig) into the porcine genome may alleviate T cell response in xenografts. It was shown that neuronal expression of hCTLA4-Ig in pigs reduced human T lymphocyte proliferation [92]. Moreover, transgenic hCTLA4-Ig protein in pigs extended the survival time of porcine skin grafts in a xenogeneic rat transplantation model [93]. Another approach to inhibit T-cell immune response may be the deletion of swine leukocyte antigen class I. Reyes et al. created SLA class I knock-out pigs using gRNA and the Cas9 endonuclease. The obtained animals revealed decreased levels of CD4−CD8+ T cells in peripheral blood [94]. Recently, pigs carrying functional knock-outs of GGTA1, CMAH, B4GALNT2, and SLA class I with multi-transgenic background (hCD46, hCD55, hCD59, hHO1, hA20) were produced. In vitro study presented that the four-fold knock-out reduced the binding of human IgG and IgM to porcine kidney cells [95].

### 5.5. Porcine Endogenous Retroviruses

Beyond immune barriers in xenotransplantation, there is also concern about the risk of cross-species pathogens infection. The main problem constitutes porcine endogenous retroviruses (PERV), which are integrated into multiple locations in the pig genome. Utilizing the CRISPR/Cas9 technology gives great hopes for the complete elimination of the risk of PERV transmission. Niu et al. using the CRISPR/Cas9 system inactivated all 25 copies of functional PERVs in a porcine primary cell line and successfully generated healthy PERV-inactivated pigs via somatic cell nuclear transfer. What is more, no reinfection was observed in the obtained pigs [96].

### 5.6. Preclinical Studies in Xenotransplantation

Advances in genetic engineering and immunosuppressive therapies prolong organ survival time in preclinical pig-to-non-human primate (NHP) xenotransplantation models. The first xenotransplantation using pig hearts with eliminated Gal antigen into immunosuppressed baboons was performed in 2005. The longest surviving heterotopic graft functioned in the recipient for 179 days [97], in comparison to 4-6 hours of survival time with the use of wild-type pig hearts [98]. Introducing additional modifications extended the xenograft survival time even more. The longest survival was obtained for heterotopic cardiac xenotransplantation–up to 945 days. The authors used hearts derived from genetically multimodified pigs (GGTA1 knock-out, hCD46, hTBM) and chimeric 2C10R4 anti-CD40 antibody therapy [99]. Additional expression of hTBM in GGTA1 knock-out, hCD46 genetically modified pigs prevented early dysregulation of coagulation and prolonged the cardiac xenografts survival time [99,100]. Using the same genetic background, orthotopic heart xenotransplantation was performed, resulting in a maximum survival of 195 days [101]. However, xenograft survival time depends on the types of transplanted organs. In the case of kidneys in pig-to-NHP transplantation models, the longest survival of a life-sustaining xenograft was 499 days. GGTA1 knock-out pigs carrying hCD55 gene as well as immunosuppression with transient pan–T cell depletion and an anti-CD154–based regimen were used in the experiments. Moreover, the selection of recipients with low-titer anti-pig antibodies improved the long-term survival of pig-to-rhesus macaque renal xenotransplants [102]. The success of porcine liver and lung xenotransplantation remains limited, which is mainly associated with the occurrence of coagulation disorders [103]. The longest survival time for orthotopic liver xenografts (29 days) was achieved using GGTA1 knock-out pigs, exogenous human coagulation factors, and immunosuppression, including co-stimulation blockade [104]. In turn, Watanabe et al. demonstrated prolonged survival time of lung xenotransplants (14 days) from GGTA1 knock-out, hCD47/hCD55 donor pigs in immunosuppressed baboons [105]. The authors indicated the important role of hCD47 expression in reducing immunologic damages and extending lung graft survival in the pig-to-NHP model. However, additional genetic modifications of the porcine genome and immunosuppressive regimen strategy are necessary for the clinical application of xenotransplantation. Table 4 summarizes the most important genetic modifications of the porcine genome for xenotransplantation purposes.

**Table 4.** Selected genetically engineered pigs for xenotransplantation.

| Genetic Modification | Function | Reference |
|---|---|---|
| GGTA1 knock-out | deletion of Gal xenoantigen | [59,60] |
| CMAH knock-out | deletion of Neu5Gc xenoantigen | [106] |
| β4GALNT2 knock-out | deletion of SDa xenoantigen | [64] |
| expression of human CD55 gene | complement regulation | [107] |
| expression of human CD46 gene | complement regulation | [108] |
| expression of human CD59 gene | complement regulation | [65] |
| expression of human TBM gene | coagulation regulation | [109] |
| expression of human EPCR gene | coagulation regulation | [74] |
| expression of human TFPI gene | coagulation regulation | [77] |
| expression of human CD39 gene | coagulation regulation | [79] |
| expression of human HO-1 gene | anti-inflammatory/antiapoptotic | [80] |
| expression of human A20 gene | anti-inflammatory/antiapoptotic | [81] |
| expression of HLA-E | regulation of NK-cells-mediated responses | [83,84] |
| ULBP1 knock-out | regulation of NK-cells-mediated responses | [87] |
| expression of human CD47 gene | regulation of macrophage-mediated responses | [90] |
| expression of human CTLA4-Ig | regulation of T cells-mediated responses | [92] |
| SLA class I knock-out | regulation of T cells-mediated responses | [94] |
| PERV inactivation | xenozoonosis | [96] |

## 6. Conclusions and Future Perspectives

The anatomical and physiological similarity between pigs and humans makes this species very interesting for biomedical research [110]. The rapid development of genetic engineering in recent years has allowed for precise and efficient modification of the animal genome using site-specific nucleases. The nuclease-mediated editing of the porcine genome, as well as potential applications of genetically modified pigs in biomedicine, are shown in Figure 1. Certainly, the driving force for development is the human mind and ideas that arise in it. One of the factors limiting the possibilities of using the potential of our ideas is the technical aspect. The transfer of new technologies, tested on smaller animal models, for example, is often limited and requires optimization for a large animal model. In the case of CRISPR/Cas9 technology, a lot of emphasis should be placed on the possibility of reducing the risk of so-called off-targets by improving this system. Paired nicking has the potential to reduce off-target activity in mice from 50–1000 times without compromising on-target performance [111]. Another strategy to limit the number of undesirable off-targets is to increase the specificity of the system. In this situation, one can focus on enhancing or improving Cas9 protein or sgRNA modifications. Protein Cas9 properties can be modified, or their lifespan can be changed [112,113]. For the future clinical success, it is also important to improve the efficiency of HR-mediated gene correction, especially in the situation of treating disease in which a template sequence is delivered to replace the mutated variant. Another important goal to achieve is the possibility of applying HR not only for dividing cells but also for cells in the post-mitotic stage. Hopes are placed in the fusion of the CRISPR/Cas9 technique and AAV (adeno-associated virus) as a donor template provider [114]. Considering the low immunogenicity of the AAV virus, the ability to transduce a wide spectrum of cells in terms of both type and developmental stage and strong limiting factor-capacity, it should be considered to minimize the CRISPR/Cas9 system or use more than one separate virus to simultaneously exploit the potential of both technologies. The use of other delivery systems, e.g., nanoparticles, is also worth considering [115].

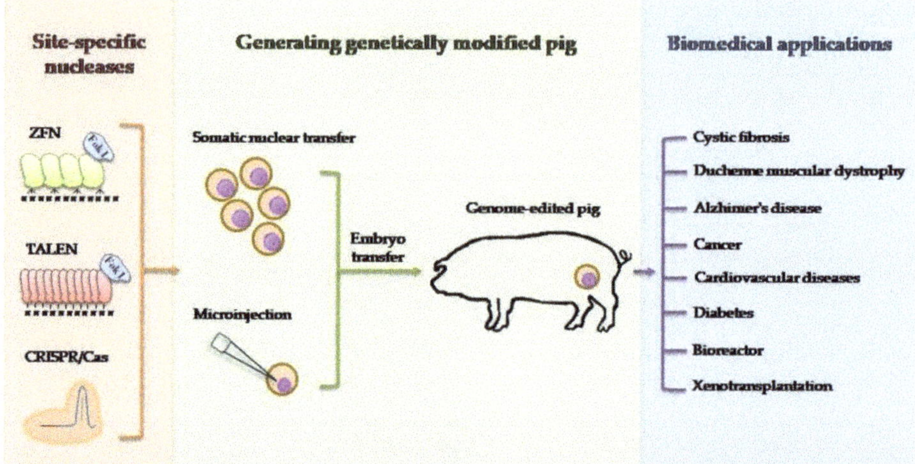


**Figure 1.** Schematic diagram of generation genetically engineered pig for biomedical purposes. The different site-specific nucleases (ZFN, TALEN, CRISPR/Cas9) used for genome editing and two techniques (somatic nuclear transfer and microinjection) to produce genetically modified pigs are shown. Biomedical applications for which genetically engineered pigs are generated include modeling human diseases, production of pharmaceutical proteins, and xenotransplantation.

Genetically modified pigs serve as an important large animal model for studying the genetic background of human diseases, testing novel drugs and therapy methods as well as developing models for gene therapy [116–118]. Pigs can be used as anatomical (e.g., endovascular), surgical, behavioral, and cytotoxic models. Ideas for new models of large animals are provided by the reality that shapes current demand. A lot has been done (pig model for influenza A infection), but still there is a need for pig models of other human viral diseases (hepatitis B; human immunodeficiency virus, HIV; severe acute respiratory syndrome coronavirus 2, SARS-CoV-2) [119]. HIV has been modeled in mice, filoviruses (Ebolavirus, Marburgvirus) have been modeled in small animals (i.e., mice, hamsters), but still, we need large models to investigate vaccines and antiviral drugs [120,121]. Transgenic pigs can also be a promising source of recombinant proteins used as pharmacological preparations. Actually, the possibility of using pigs for the production of biopharmaceuticals has been slowed in recent years. Some studies demonstrated that the pig mammary gland can be used as a complex recombinant protein source with appropriate post-translational modifications [122]. Despite the advantages of pig animal platform (natural secretion, correct posttranslational modifications, constant production), some ethical doubts are probably limiting the boost. Finally, the use of genetically engineered pigs for xenotransplantation is becoming an increasingly feasible alternative to standard allogeneic transplants and a potential solution to the problem of organ shortage. The combination of various multi-modified pigs and immunosuppressive therapies is required for overcoming immune rejection and effective xenotransplantation of different solid organs [123–125]. When it comes to treating end-stage organ failure, biomedical research could go a step further and try to create chimeric genetically modified pigs that would be carriers of human organs [126].

**Author Contributions:** Conceptualization, M.H., D.L.; writing—original draft preparation, M.H., S.H., A.N.-T., N.R.; writing—review and editing, M.H., J.Z.; supervision, J.Z., D.L. All authors have read and agreed to the published version of the manuscript.

**Funding:** The publication was co-financed within the framework of a Ministry of Science and Higher Education program as "Regional Initiative Excellence" in the years 2019-2022, Project No. 005/RID/2018/19.

**Conflicts of Interest:** The authors declare no conflict of interest.

## References


1. Perleberg, C.; Kind, A.; Schnieke, A. Genetically engineered pigs as models for human disease. *Dis. Model. Mech.* **2018**, *11*, dmm030783. [CrossRef]
2. Gifre, L.; Arís, A.; Bach, À.; Garcia-Fruitós, E. Trends in recombinant protein use in animal production. *Microb. Cell Fact.* **2017**, *16*, 40. [CrossRef]
3. Hryhorowicz, M.; Zeyland, J.; Słomski, R.; Lipiński, D. Genetically modified pigs as organ donors for xenotransplantation. *Mol. Biotechnol.* **2017**, *59*, 435–444. [CrossRef] [PubMed]
4. Gordon, J.; Scangost, G.A.; Plotkin, D.J.; Barbosaf, J.A. Genetic transformation of mouse embryos by microinjection of purified DNA. *Proc. Natl. Acad. Sci. USA* **1980**, *77*, 7380–7384. [CrossRef]
5. Palmiter, R.D.; Brinster, R.L.; Hammer, R.E.; Trumbauer, M.E.; Rosenfeld, M.G.; Birnberg, N.C.; Evans, R.M. Dramatic growth of mice that develop from eggs microinjected with metallothionein-growth hormone fusion genes. *Nature* **1982**, *300*, 611–615. [CrossRef] [PubMed]
6. Wall, R.J. Pronuclear microinjection. *Cloning Stem Cells* **2001**, *3*, 209–220. [CrossRef] [PubMed]
7. Wolf, E.; Schernthaner, W.; Zakhartchenko, V.; Prelle, K.; Stojkovic, M.; Brem, G. Transgenic technology in farm animals-progress and perspectives. *Exp. Physiol.* **2000**, *85*, 615–625. [CrossRef]
8. Maga, E.A.; Sargent, R.G.; Zeng, H.; Pati, S.; Zarling, D.A.; Oppenheim, S.M.; Collette, N.M.; Moyer, A.L.; Conrad-Brink, J.S.; Rowe, J.D.; et al. Increased efficiency of transgenic livestock production. *Transgenic Res.* **2003**, *12*, 485–496. [CrossRef] [PubMed]
9. Chrenek, P.; Vasicek, D.; Makarevich, A.V.; Jurcik, R.; Suvegova, K.; Parkanyi, V.; Bauer, M.; Rafay, J.; Batorova, A.; Paleyanda, R.K. Increased transgene integration efficiency upon microinjection of DNA into both pronuclei of rabbit embryos. *Transgenic Res.* **2005**, *14*, 417–428. [CrossRef]
10. Sumiyama, K.; Kawakami, K.; Yagita, K. A simple and highly efficient transgenesis method in mice with the Tol2 transposon system and cytoplasmic microinjection. *Genomics* **2010**, *95*, 306–311. [CrossRef]
11. Wang, Y.; Du, Y.; Shen, B.; Zhou, X.; Li, J.; Liu, Y.; Wang, J.; Zhou, J.; Hu, B.; Kang, N.; et al. Efficient generation of gene-modified pigs via injection of zygote with Cas9/sgRNA. *Sci. Rep.* **2015**, *5*, 8256. [CrossRef] [PubMed]
12. Campbell, K.H.S.; McWhir, J.; Ritchie, W.A.; Wilmut, I. Sheep cloned by nuclear transfer from a cultured cell line. *Nature* **1996**, *380*, 64–66. [CrossRef] [PubMed]
13. Campbell, K.H.S. A background to nuclear transfer and its applications in agriculture and human therapeutic medicine. *J. Anat.* **2002**, *200*, 267–275. [CrossRef] [PubMed]
14. Rogers, C.S.; Stoltz, D.A.; Meyerholz, D.K.; Ostedgaard, L.S.; Rokhlina, T.; Taft, P.J.; Rogan, M.P.; Pezzulo, A.A.; Karp, P.H.; Itani, O.A.; et al. Disruption of the CFTR gene produces a model of cystic fibrosis in newborn pigs. *Science* **2008**, *321*, 1837–1841. [CrossRef]
15. Ostedgaard, L.S.; Meyerholz, D.K.; Chen, J.H.; Pezzulo, A.A.; Karp, P.H.; Rokhlina, T.; Ernst, S.E.; Hanfland, R.A.; Reznikov, L.R.; Ludwig, P.S.; et al. The DeltaF508 mutation causes CFTR misprocessing and cystic fibrosis-like disease in pigs. *Sci. Transl. Med.* **2011**, *3*, 74ra24. [CrossRef]
16. Klymiuk, N.; Mundhenk, L.; Kraehe, K.; Wuensch, A.; Plog, S.; Emrich, D.; Langenmayer, M.C.; Stehr, M.; Holzinger, A.; Kröner, C.; et al. Sequential targeting of CFTR by BAC vectors generates a novel pig model of cystic fibrosis. *J. Mol. Med.* **2012**, *90*, 597–608. [CrossRef]
17. Cooney, A.L.; Abou Alaiwa, M.H.; Shah, V.S.; Bouzek, D.C.; Stroik, M.R.; Powers, L.S.; Gansemer, N.D.; Meyerholz, D.K.; Welsh, M.J.; Stoltz, D.A.; et al. Lentiviral-mediated phenotypic correction of cystic fibrosis pigs. *JCI Insight* **2016**, *1*, e88730. [CrossRef]
18. Steines, B.; Dickey, D.D.; Bergen, J.; Excoffon, K.J.; Weinstein, J.R.; Li, X.; Yan, Z.; Abou Alaiwa, M.H.; Shah, V.S.; Bouzek, D.C.; et al. CFTR gene transfer with AAV improves early cystic fibrosis pig phenotypes. *JCI Insight* **2016**, *1*, e88728. [CrossRef]
19. Ruan, J.; Hirai, H.; Yang, D.; Ma, L.; Hou, X.; Jiang, H.; Wei, H.; Rajagopalan, C.; Mou, H.; Wang, G.; et al. Efficient gene editing at major CFTR mutation loci. *Mol. Ther. Nucleic Acids* **2019**, *16*, 73–81. [CrossRef]
20. Zhou, Z.P.; Yang, L.L.; Cao, H.; Chen, Z.R.; Zhang, Y.; Wen, X.Y.; Hu, J. In vitro validation of a CRISPR-mediated CFTR correction strategy for preclinical translation in pigs. *Hum. Gene Ther.* **2019**, *30*, 1101–1116. [CrossRef]
21. Klymiuk, N.; Blutke, A.; Graf, A.; Krause, S.; Burkhardt, K.; Wuensch, A.; Krebs, S.; Kessler, B.; Zakhartchenko, V.; Kurome, M.; et al. Dystrophin-deficient pigs provide new insights into the hierarchy of physiological derangements of dystrophic muscle. *Hum. Mol. Genet.* **2013**, *22*, 4368–4382. [CrossRef]

22. Fröhlich, T.; Kemter, E.; Flenkenthaler, F.; Klymiuk, N.; Otte, K.A.; Blutke, A.; Krause, S.; Walter, M.C.; Wanke, R.; Wolf, E.; et al. Progressive muscle proteome changes in a clinically relevant pig model of Duchenne muscular dystrophy. *Sci. Rep.* **2016**, *6*, 33362. [CrossRef] [PubMed]
23. Yu, H.H.; Zhao, H.; Qing, Y.B.; Pan, W.R.; Jia, B.Y.; Zhao, H.Y.; Huang, X.X.; Wei, H.J. Porcine zygote injection with Cas9/sgRNA results in DMD-modified pig with muscle dystrophy. *Int. J. Mol. Sci.* **2016**, *17*, 1668. [CrossRef] [PubMed]
24. Moretti, A.; Fonteyne, L.; Giesert, F.; Hoppmann, P.; Meier, A.B.; Bozoglu, T.; Baehr, A.; Schneider, C.M.; Sinnecker, D.; Klett, K.; et al. Somatic gene editing ameliorates skeletal and cardiac muscle failure in pig and human models of Duchenne muscular dystrophy. *Nat. Med.* **2020**, *26*, 207–214. [CrossRef]
25. Walsh, D.M.; Klyubin, I.; Shankar, G.M.; Townsend, M.; Fadeeva, J.V.; Betts, V.; Podlisny, M.B.; Cleary, J.P.; Ashe, K.H.; Rowan, M.J.; et al. The role of cell-derived oligomers of Abeta in Alzheimer's disease and avenues for therapeutic intervention. *Biochem. Soc. Trans.* **2005**, *33*, 1087–1090. [CrossRef]
26. Kragh, P.M.; Nielsen, A.L.; Li, J.; Du, Y.; Lin, L.; Schmidt, M.; Bøgh, I.B.; Holm, I.E.; Jakobsen, J.E.; Johansen, M.G.; et al. Hemizygous minipigs produced by random gene insertion and handmade cloning express the Alzheimer's disease-causing dominant mutation APPsw. *Transgenic Res.* **2009**, *18*, 545–558. [CrossRef] [PubMed]
27. Jakobsen, J.E.; Johansen, M.G.; Schmidt, M.; Dagnæs-Hansen, F.; Dam, K.; Gunnarsson, A.; Liu, Y.; Kragh, P.M.; Li, R.; Holm, I.E.; et al. Generation of minipigs with targeted transgene insertion by recombinase-mediated cassette exchange (RMCE) and somatic cell nuclear transfer (SCNT). *Transgenic Res.* **2013**, *22*, 709–723. [CrossRef] [PubMed]
28. Jakobsen, J.E.; Johansen, M.G.; Schmidt, M.; Liu, Y.; Li, R.; Callesen, H.; Melnikova, M.; Habekost, M.; Matrone, C.; Bouter, Y.; et al. Expression of the Alzheimer's disease mutations AβPP695sw and PSEN1M146I in double-transgenic Göttingen minipigs. *J. Alzheimer's Dis.* **2016**, *53*, 1617–1630. [CrossRef]
29. Lee, S.E.; Hyun, H.; Park, M.R.; Choi, Y.; Son, Y.J.; Park, Y.G.; Jeong, S.G.; Shin, M.Y.; Ha, H.J.; Hong, H.S.; et al. Production of transgenic pig as an Alzheimer's disease model using a multi-cistronic vector system. *PLoS ONE* **2017**, *12*, e0177933. [CrossRef]
30. Adam, S.J.; Rund, L.A.; Kuzmuk, K.N.; Zachary, J.F.; Schook, L.B.; Counter, C.M. Genetic induction of tumorigenesis in swine. *Oncogene* **2007**, *26*, 1038–1045. [CrossRef]
31. Schook, L.B.; Collares, T.V.; Hu, W.; Liang, Y.; Rodrigues, F.M.; Rund, L.A.; Schachtschneider, K.M.; Seixas, F.K.; Singh, K.; Wells, K.D.; et al. A genetic porcine model of cancer. *PLoS ONE* **2015**, *10*, e0128864. [CrossRef] [PubMed]
32. Saalfrank, A.; Janssen, K.P.; Ravon, M.; Flisikowski, K.; Eser, S.; Steiger, K.; Flisikowska, T.; Müller-Fliedner, P.; Schulze, É.; Brönner, C.; et al. A porcine model of osteosarcoma. *Oncogenesis* **2016**, *5*, e210. [CrossRef]
33. Leuchs, S.; Saalfrank, A.; Merkl, C.; Flisikowska, T.; Edlinger, M.; Durkovic, M.; Rezaei, N.; Kurome, M.; Zakhartchenko, V.; Kessler, B.; et al. Inactivation and inducible oncogenic mutation of p53 in gene targeted pigs. *PLoS ONE* **2012**, *7*, e43323. [CrossRef]
34. Sieren, J.C.; Meyerholz, D.K.; Wang, X.J.; Davis, B.T.; Newell, J.D., Jr.; Hammond, E.; Rohret, J.A.; Rohret, F.A.; Struzynski, J.T.; Goeken, J.A.; et al. Development and translational imaging of a TP53 porcine tumorigenesis model. *J. Clin. Invest.* **2014**, *124*, 4052–4066. [CrossRef]
35. Flisikowska, T.; Merkl, C.; Landmann, M.; Eser, S.; Rezaei, N.; Cui, X.; Kurome, M.; Zakhartchenko, V.; Kessler, B.; Wieland, H.; et al. A porcine model of familial adenomatous polyposis. *Gastroenterology* **2012**, *143*, 1173–1175. [CrossRef]
36. Davis, B.T.; Wang, X.J.; Rohret, J.A.; Struzynski, J.T.; Merricks, E.P.; Bellinger, D.A.; Rohret, F.A.; Nichols, T.C.; Rogers, C.S. Targeted disruption of LDLR causes hypercholesterolemia and atherosclerosis in Yucatan miniature pigs. *PLoS ONE* **2014**, *9*, e93457. [CrossRef]
37. Burke, A.C.; Telford, D.E.; Sutherland, B.G.; Edwards, J.Y.; Sawyez, C.G.; Barrett, P.H.R.; Newton, R.S.; Pickering, J.G.; Huff, M.W. Bempedoic acid lowers low-density lipoprotein cholesterol and attenuates atherosclerosis in low-density lipoprotein receptor-deficient (LDLR+/− and LDLR−/−) Yucatan miniature pigs. *Arterioscler. Thromb. Vasc. Biol.* **2018**, *38*, 1178–1190. [CrossRef] [PubMed]
38. Al-Mashhadi, R.H.; Sørensen, C.B.; Kragh, P.M.; Christoffersen, C.; Mortensen, M.B.; Tolbod, L.P.; Thim, T.; Du, Y.; Li, J.; Liu, Y.; et al. Familial hypercholesterolemia and atherosclerosis in cloned minipigs created by DNA transposition of a human PCSK9 gain-of-function mutant. *Sci. Transl. Med.* **2013**, *5*, 166ra1. [CrossRef]

39. Wei, J.; Ouyang, H.; Wang, Y.; Pang, D.; Cong, N.X.; Wang, T.; Leng, B.; Li, D.; Li, X.; Wu, R.; et al. Characterization of a hypertriglyceridemic transgenic miniature pig model expressing human apolipoprotein CIII. *FEBS J.* **2012**, *279*, 91–99. [CrossRef]
40. Renner, S.; Fehlings, C.; Herbach, N.; Hofmann, A.; von Waldthausen, D.C.; Kessler, B.; Ulrichs, K.; Chodnevskaja, I.; Moskalenko, V.; Amselgruber, W.; et al. Glucose intolerance and reduced proliferation of pancreatic beta-cells in transgenic pigs with impaired glucose-dependent insulinotropic polypeptide function. *Diabetes* **2010**, *59*, 1228–1238. [CrossRef] [PubMed]
41. Renner, S.; Römisch-Margl, W.; Prehn, C.; Krebs, S.; Adamski, J.; Göke, B.; Blum, H.; Suhre, K.; Roscher, A.A.; Wolf, E. Changing metabolic signatures of amino acids and lipids during the prediabetic period in a pig model with impaired incretin function and reduced β-cell mass. *Diabetes* **2012**, *61*, 2166–2175. [CrossRef] [PubMed]
42. Streckel, E.; Braun-Reichhart, C.; Herbach, N.; Dahlhoff, M.; Kessler, B.; Blutke, A.; Bähr, A.; Übel, N.; Eddicks, M.; Ritzmann, M.; et al. Effects of the glucagon-like peptide-1 receptor agonist liraglutide in juvenile transgenic pigs modeling a pre-diabetic condition. *J. Transl. Med.* **2015**, *13*, 73. [CrossRef] [PubMed]
43. Umeyama, K.; Watanabe, M.; Saito, H.; Kurome, M.; Tohi, S.; Matsunari, H.; Miki, K.; Nagashima, H. Dominant-negative mutant hepatocyte nuclear factor 1α induces diabetes in transgenic-cloned pigs. *Transgenic Res.* **2009**, *18*, 697–706. [CrossRef] [PubMed]
44. Umeyama, K.; Nakajima, M.; Yokoo, T.; Nagaya, M.; Nagashima, H. Diabetic phenotype of transgenic pigs introduced by dominant-negative mutant hepatocyte nuclear factor 1α. *J. Diabetes Complicat.* **2017**, *31*, 796–803. [CrossRef]
45. Renner, S.; Braun-Reichhart, C.; Blutke, A.; Herbach, N.; Emrich, D.; Streckel, E.; Wünsch, A.; Kessler, B.; Kurome, M.; Bähr, A.; et al. Permanent neonatal diabetes in INS(C94Y) transgenic pigs. *Diabetes* **2013**, *62*, 1505–1511. [CrossRef]
46. Kleinwort, K.J.H.; Amann, B.; Hauck, S.M.; Hirmer, S.; Blutke, A.; Renner, S.; Uhl, P.B.; Lutterberg, K.; Sekundo, W.; Wolf, E.; et al. Retinopathy with central oedema in an INS C94Y transgenic pig model of long-term diabetes. *Diabetologia* **2017**, *60*, 1541–1549. [CrossRef]
47. Backman, M.; Flenkenthaler, F.; Blutke, A.; Dahlhoff, M.; Ländström, E.; Renner, S.; Philippou-Massier, J.; Krebs, S.; Rathkolb, B.; Prehn, C.; et al. Multi-omics insights into functional alterations of the liver in insulin-deficient diabetes mellitus. *Mol. Metab.* **2019**, *26*, 30–44. [CrossRef]
48. Velander, W.H.; Johnson, J.L.; Page, R.L.; Russell, C.G.; Subramanian, A.; Wilkins, T.D.; Gwazdauskas, F.C.; Pittius, C.; Drohan, W.N. High-level expression of a heterologous protein in the milk of transgenic swine using the cDNA encoding human protein C. *Proc. Natl. Acad. Sci. USA* **1992**, *89*, 12003–12007. [CrossRef]
49. Paleyanda, R.K.; Velander, W.H.; Lee, T.K.; Scandella, D.H.; Gwazdauskas, F.C.; Knight, J.W.; Hoyer, L.W.; Drohan, W.N.; Lubon, H. Transgenic pigs produce functional human factor VIII in milk. *Nat. Biotechnol.* **1997**, *15*, 971–975. [CrossRef]
50. Lee, M.H.; Lin, Y.S.; Tu, C.F.; Yen, C.H. Recombinant human factor IX produced from transgenic porcine milk. *Biomed. Res. Int.* **2014**, *2014*, 315375. [CrossRef]
51. Zhao, J.; Xu, W.; Ross, J.W.; Walters, E.M.; Butler, S.P.; Whyte, J.J.; Kelso, L.; Fatemi, M.; Vanderslice, N.C.; Giroux, K.; et al. Engineering protein processing of the mammary gland to produce abundant hemophilia B therapy in milk. *Sci. Rep.* **2015**, *5*, 14176. [CrossRef] [PubMed]
52. Lee, H.G.; Lee, H.C.; Kim, S.W.; Lee, P.; Chung, H.J.; Lee, Y.K.; Han, J.H.; Hwang, I.S.; Yoo, J.I.; Kim, Y.K.; et al. Production of recombinant human von Willebrand factor in the milk of transgenic pigs. *J. Reprod. Dev.* **2009**, *55*, 484–490. [CrossRef] [PubMed]
53. Park, J.K.; Lee, Y.K.; Lee, P.; Chung, H.J.; Kim, S.; Lee, H.G.; Seo, M.K.; Han, J.H.; Park, C.G.; Kim, H.T.; et al. Recombinant human erythropoietin produced in milk of transgenic pigs. *J. Biotechnol.* **2006**, *122*, 362–371. [CrossRef]
54. Lu, D.; Liu, S.; Shang, S.; Wu, F.; Wen, X.; Li, Z.; Li, Y.; Hu, X.; Zhao, Y.; Li, Q.; et al. Production of transgenic-cloned pigs expressing large quantities of recombinant human lysozyme in milk. *PLoS ONE* **2015**, *10*, e0123551. [CrossRef]
55. Swanson, M.E.; Martin, M.J.; O'Donnell, J.K.; Hoover, K.; Lago, W.; Huntress, V.; Parsons, C.T.; Pinkert, C.A.; Pilder, S.; Logan, J.S. Production of functional human hemoglobin in transgenic swine. *Biotechnology (N. Y.)* **1992**, *10*, 557–559. [CrossRef]

56. Sharma, A.; Martin, M.J.; Okabe, J.F.; Truglio, R.A.; Dhanjal, N.K.; Logan, J.S.; Kumar, R. An isologous porcine promoter permits high level expression of human hemoglobin in transgenic swine. *Biotechnology (N. Y.)* **1994**, *12*, 55–59. [CrossRef] [PubMed]
57. Houdebine, L.M. Production of pharmaceutical proteins by transgenic animals. *Rev. Sci. Tech.* **2018**, *37*, 131–139. [CrossRef] [PubMed]
58. Wang, Y.; Zhao, S.; Bai, L.; Fan, J.; Liu, E. Expression systems and species used for transgenic animal bioreactors. *Biomed. Res. Int.* **2013**, *2013*, 580463. [CrossRef] [PubMed]
59. Lai, L.; Kolber-Simonds, D.; Park, K.W.; Cheong, H.T.; Greenstein, J.L.; Im, G.S.; Samuel, M.; Bonk, A.; Rieke, A.; Day, B.N.; et al. Production of alpha-1,3-galactosyltransferase knockout pigs by nuclear transfer cloning. *Science* **2002**, *295*, 1089–1092. [CrossRef]
60. Phelps, C.J.; Koike, C.; Vaught, T.D.; Boone, J.; Wells, K.D.; Chen, S.H.; Ball, S.; Specht, S.M.; Polejaeva, I.A.; Monahan, J.A.; et al. Production of alpha 1,3-galactosyltransferase-deficient pigs. *Science* **2003**, *299*, 411–414. [CrossRef]
61. Lipiński, D.; Nowak-Terpiłowska, A.; Hryhorowicz, M.; Jura, J.; Korcz, A.; Słomski, R.; Juzwa, W.; Mazurkiewicz, N.; Smorąg, Z.; Zeyland, J. Production of ZFN-mediated GGTA1 knock-out pigs by microinjection of gene constructs into pronuclei of zygotes. *Pol. J. Vet. Sci.* **2019**, *22*, 91–100. [PubMed]
62. Kang, J.T.; Kwon, D.K.; Park, A.R.; Lee, E.J.; Yun, Y.J.; Ji, D.Y.; Lee, K.; Park, K.W. Production of α1,3-galactosyltransferase targeted pigs using transcription activator-like effector nuclease-mediated genome editing technology. *J. Vet. Sci.* **2016**, *17*, 89–96. [CrossRef] [PubMed]
63. Chuang, C.K.; Chen, C.H.; Huang, C.L.; Su, Y.H.; Peng, S.H.; Lin, T.Y.; Tai, H.C.; Yang, T.S.; Tu, C.F. Generation of GGTA1 mutant pigs by direct pronuclear microinjection of CRISPR/Cas9 plasmid vectors. *Animal Biotechnol.* **2017**, *28*, 174–181. [CrossRef]
64. Estrada, J.L.; Martens, G.; Li, P.; Adams, A.; Newell, K.A.; Ford, M.L.; Butler, J.R.; Sidner, R.; Tector, M.; Tector, J. Evaluation of human and non-human primate antibody binding to pig cells lacking GGTA1/CMAH/β4GalNT2 genes. *Xenotransplantation* **2015**, *22*, 194–202. [CrossRef]
65. Fodor, W.L.; Williams, B.L.; Matis, L.A.; Madri, J.A.; Rollins, S.A.; Knight, J.W.; Velander, W.; Squinto, S.P. Expression of a functional human complement inhibitor in a transgenic pig as a model for the prevention of xenogeneic hyperacute organ rejection. *Proc. Natl. Acad. Sci. USA* **1994**, *91*, 11153–11157. [CrossRef] [PubMed]
66. McGregor, C.G.; Davies, W.R.; Oi, K.; Teotia, S.S.; Schirmer, J.M.; Risdahl, J.M.; Tazelaar, H.D.; Kremers, W.K.; Walker, R.C.; Byrne, G.W.; et al. Cardiac xenotransplantation: Recent preclinical progress with 3-month median survival. *J. Thorac. Cardiovasc. Surg.* **2005**, *130*, 844–851. [CrossRef]
67. Niemann, H.; Verhoeyen, E.; Wonigeit, K.; Lorenz, R.; Hecker, J.; Schwinzer, R.; Hauser, H.; Kues, W.A.; Halter, R.; Lemme, E.; et al. Cytomegalovirus early promoter induced expression of hCD59 in porcine organs provides protection against hyperacute rejection. *Transplantation* **2001**, *72*, 1898–1906. [CrossRef]
68. Cozzi, E.; Bhatti, F.; Schmoeckel, M.; Chavez, G.; Smith, K.G.; Zaidi, A.; Bradley, J.R.; Thiru, S.; Goddard, M.; Vial, C.; et al. Long-term survival of nonhuman primates receiving life-supporting transgenic porcine kidney xenografts. *Transplantation* **2000**, *70*, 15–21.
69. Liu, F.; Liu, J.; Yuan, Z.; Qing, Y.; Li, H.; Xu, K.; Zhu, W.; Zhao, H.; Jia, B.; Pan, W.; et al. Generation of GTKO diannan miniature pig expressing human complementary regulator proteins hCD55 and hCD59 via T2A peptide-based bicistronic vectors and SCNT. *Mol. Biotechnol.* **2018**, *60*, 550–562. [CrossRef]
70. Mohiuddin, M.M.; Corcoran, P.C.; Singh, A.K.; Azimzadeh, A.; Hoyt, R.F., Jr.; Thomas, M.L.; Eckhaus, M.A.; Seavey, C.; Ayares, D.; Pierson, R.N., 3rd; et al. B-cell depletion extends the survival of GTKO.hCD46*Tg* pig heart xenografts in baboons for up to 8 months. *Am. J. Transplant.* **2012**, *12*, 763–771. [CrossRef]
71. Miwa, Y.; Yamamoto, K.; Onishi, A.; Iwamoto, M.; Yazaki, S.; Haneda, M.; Iwasaki, K.; Liu, D.; Ogawa, H.; Nagasaka, T.; et al. Potential value of human thrombomodulin and DAF expression for coagulation control in pig-to-human xenotransplantation. *Xenotransplantation* **2010**, *17*, 26–37. [CrossRef]
72. Bongoni, A.K.; Kiermeir, D.; Jenni, H.; Bähr, A.; Ayares, D.; Klymiuk, N.; Wolf, E.; Voegelin, E.; Constantinescu, M.A.; Seebach, J.D.; et al. Complement dependent early immunological responses during ex vivo xenoperfusion of hCD46/HLA-E double transgenic pig forelimbs with human blood. *Xenotransplantation* **2014**, *21*, 230–243. [CrossRef] [PubMed]

73. Nowak-Terpiłowska, A.; Lipiński, D.; Hryhorowicz, M.; Juzwa, W.; Jura, J.; Słomski, R.; Mazurkiewicz, N.; Gawrońska, B.; Zeyland, J. Production of ULBP1-KO pigs with human CD55 expression using CRISPR technology. *J. Appl. Animal Res.* **2020**, *48*, 93–101. [CrossRef]
74. Iwase, H.; Ekser, B.; Hara, H.; Phelps, C.; Ayares, D.; Cooper, D.K.; Ezzelarab, M.B. Regulation of human platelet aggregation by genetically modified pig endothelial cells and thrombin inhibition. *Xenotransplantation* **2014**, *21*, 72–83. [CrossRef] [PubMed]
75. Harris, D.G.; Quinn, K.J.; French, B.M.; Schwartz, E.; Kang, E.; Dahi, S.; Phelps, C.J.; Ayares, D.L.; Burdorf, L.; Azimzadeh, A.M.; et al. Meta-analysis of the independent and cumulative effects of multiple genetic modifications on pig lung xenograft performance during ex vivo perfusion with human blood. *Xenotransplantation* **2015**, *22*, 102–111. [CrossRef]
76. Iwase, H.; Hara, H.; Ezzelarab, M.; Li, T.; Zhang, Z.; Gao, B.; Liu, H.; Long, C.; Wang, Y.; Cassano, A.; et al. Immunological and physiological observations in baboons with life-supporting genetically engineered pig kidney grafts. *Xenotransplantation* **2017**, *24*. [CrossRef]
77. Lin, C.C.; Ezzelarab, M.; Hara, H.; Long, C.; Lin, C.W.; Dorling, A.; Cooper, D.K. Atorvastatin or transgenic expression of TFPI inhibits coagulation initiated by anti-nonGal IgG binding to porcine aortic endothelial cells. *J. Thromb. Haemost.* **2010**, *8*, 2001–2010. [CrossRef]
78. Kwon, D.J.; Kim, D.H.; Hwang, I.S.; Kim, D.E.; Kim, H.J.; Kim, J.S.; Lee, K.; Im, G.S.; Lee, J.W.; Hwang, S. Generation of α-1,3-galactosyltransferase knocked-out transgenic cloned pigs with knocked-in five human genes. *Transgenic Res.* **2017**, *26*, 153–163. [CrossRef]
79. Wheeler, D.G.; Joseph, M.E.; Mahamud, S.D.; Aurand, W.L.; Mohler, P.J.; Pompili, V.J.; Dwyer, K.M.; Nottle, M.B.; Harrison, S.J.; d'Apice, A.J.F.; et al. Transgenic swine: Expression of human CD39 protects against myocardial injury. *J. Mol. Cell. Cardiol.* **2012**, *52*, 958–961. [CrossRef]
80. Petersen, B.; Ramackers, W.; Lucas-Hahn, A.; Lemme, E.; Hassel, P.; Queisser, A.L.; Herrmann, D.; Barg-Kues, B.; Carnwath, J.W.; Klose, J.; et al. Transgenic expression of human heme oxygenase-1 in pigs confers resistance against xenograft rejection during ex vivo perfusion of porcine kidneys. *Xenotransplantation* **2011**, *18*, 355–368. [CrossRef]
81. Oropeza, M.; Petersen, B.; Carnwath, J.W.; Lucas-Hahn, A.; Lemme, E.; Hassel, P.; Herrmann, D.; Barg-Kues, B.; Holler, S.; Queisser, A.L.; et al. Transgenic expression of the human A20 gene in cloned pigs provides protection against apoptotic and inflammatory stimuli. *Xenotransplantation* **2009**, *16*, 522–534. [CrossRef]
82. Ahrens, H.E.; Petersen, B.; Ramackers, W.; Petkov, S.; Herrmann, D.; Hauschild-Quintern, J.; Lucas-Hahn, A.; Hassel, P.; Ziegler, M.; Baars, W.; et al. Kidneys from α1,3-galactosyltransferase knockout/human heme oxygenase-1/human A20 transgenic pigs are protected from rejection during ex vivo perfusion with human blood. *Transplant. Direct* **2015**, *1*, e23. [CrossRef]
83. Weiss, E.H.; Lilienfeld, B.G.; Müller, S.; Müller, E.; Herbach, N.; Kessler, B.; Wanke, R.; Schwinzer, R.; Seebach, J.D.; Wolf, E.; et al. HLA-E/human beta2-microglobulin transgenic pigs: Protection against xenogeneic human anti-pig natural killer cell cytotoxicity. *Transplantation* **2009**, *87*, 35–43. [CrossRef] [PubMed]
84. Hryhorowicz, M.; Zeyland, J.; Nowak-Terpiłowska, A.; Jura, J.; Juzwa, W.; Słomski, R.; Bocianowski, J.; Smorąg, Z.; Woźniak, A.; Lipiński, D. Characterization of three generations of transgenic pigs expressing the HLA-E gene. *Ann. Animal Sci.* **2018**, *18*, 919–935. [CrossRef]
85. Maeda, A.; Kawamura, T.; Ueno, T.; Usui, N.; Eguchi, H.; Miyagawa, S. The suppression of inflammatory macrophage-mediated cytotoxicity and proinflammatory cytokine production by transgenic expression of HLA-E. *Transpl. Immunol.* **2013**, *29*, 76–81. [CrossRef]
86. Abicht, J.M.; Sfriso, R.; Reichart, B.; Längin, M.; Gahle, K.; Puga Yung, G.L.; Seebach, J.D.; Rieben, R.; Ayares, D.; Wolf, E.; et al. Multiple genetically modified GTKO/hCD46/HLA-E/hβ2-mg porcine hearts are protected from complement activation and natural killer cell infiltration during ex vivo perfusion with human blood. *Xenotransplantation* **2018**, *25*, e12390. [CrossRef]
87. Zeyland, J.; Hryhorowicz, M.; Nowak-Terpiłowska, A.; Jura, J.; Słomski, R.; Smorąg, Z.; Gajda, B.; Lipiński, D. The production of UL16-binding protein 1 targeted pigs using CRISPR technology. *3 Biotech* **2018**, *8*, 70. [CrossRef] [PubMed]
88. Ide, K.; Wang, H.; Tahara, H.; Liu, J.; Wang, X.; Asahara, T.; Sykes, M.; Yang, Y.G.; Ohdan, H. Role for CD47-SIRPalpha signaling in xenograft rejection by macrophages. *Proc. Natl. Acad. Sci. USA* **2007**, *104*, 5062–5066. [CrossRef] [PubMed]

89. Yan, J.J.; Koo, T.Y.; Lee, H.S.; Lee, W.B.; Kang, B.; Lee, J.G.; Jang, J.Y.; Fang, T.; Ryu, J.H.; Ahn, C.; et al. Role of human CD200 overexpression in pig-to-human xenogeneic immune response compared with human CD47 overexpression. *Transplantation* **2018**, *102*, 406–416. [CrossRef] [PubMed]
90. Tena, A.; Kurtz, J.; Leonard, D.A.; Dobrinsky, J.R.; Terlouw, S.L.; Mtango, N.; Verstegen, J.; Germana, S.; Mallard, C.; Arn, J.S.; et al. Transgenic expression of human CD47 markedly increases engraftment in a murine model of pig-to-human hematopoietic cell transplantation. *Am. J. Transplant.* **2014**, *14*, 2713–2722. [CrossRef] [PubMed]
91. Tena, A.A.; Sachs, D.H.; Mallard, C.; Yang, Y.G.; Tasaki, M.; Farkash, E.; Rosales, I.A.; Colvin, R.B.; Leonard, D.A.; Hawley, R.J. Prolonged survival of pig skin on baboons after administration of pig cells expressing human CD47. *Transplantation* **2017**, *101*, 316–321. [CrossRef]
92. Martin, C.; Plat, M.; Nerriére-Daguin, V.; Coulon, F.; Uzbekova, S.; Venturi, E.; Condé, F.; Hermel, J.M.; Hantraye, P.; Tesson, L.; et al. Transgenic expression of CTLA4-Ig by fetal pig neurons for xenotransplantation. *Transgenic Res.* **2005**, *14*, 373–384. [CrossRef] [PubMed]
93. Wang, Y.; Yang, H.Q.; Jiang, W.; Fan, N.N.; Zhao, B.T.; Ou-Yang, Z.; Liu, Z.M.; Zhao, Y.; Yang, D.S.; Zhou, X.Y.; et al. Transgenic expression of human cytoxic T-lymphocyte associated antigen4-immunoglobulin (hCTLA4Ig) by porcine skin for xenogeneic skin grafting. *Transgenic Res.* **2015**, *24*, 199–211. [CrossRef] [PubMed]
94. Reyes, L.M.; Estrada, J.L.; Wang, Z.Y.; Blosser, R.J.; Smith, R.F.; Sidner, R.A.; Paris, L.L.; Blankenship, R.L.; Ray, C.N.; Miner, A.C.; et al. Creating class I MHC-null pigs using guide RNA and the Cas9 endonuclease. *J. Immunol.* **2014**, *193*, 5751–5757. [CrossRef]
95. Fischer, K.; Rieblinger, B.; Hein, R.; Sfriso, R.; Zuber, J.; Fischer, A.; Klinger, B.; Liang, W.; Flisikowski, K.; Kurome, M.; et al. Viable pigs after simultaneous inactivation of porcine MHC class I and three xenoreactive antigen genes GGTA1, CMAH and B4GALNT2. *Xenotransplantation* **2020**, *27*, e12560. [CrossRef]
96. Niu, D.; Wei, H.J.; Lin, L.; George, H.; Wang, T.; Lee, I.H.; Zhao, H.Y.; Wang, Y.; Kan, Y.; Shrock, E.; et al. Inactivation of porcine endogenous retrovirus in pigs using CRISPR-Cas9. *Science* **2017**, *357*, 1303–1307. [CrossRef]
97. Tseng, Y.L.; Kuwaki, K.; Dor, F.J.; Shimizu, A.; Houser, S.; Hisashi, Y.; Yamada, K.; Robson, S.C.; Awwad, M.; Schuurman, H.J.; et al. α1,3-Galactosyltransferase gene-knockout pig heart transplantation in baboons with survival approaching 6 months. *Transplantation* **2005**, *80*, 1493–1500. [CrossRef]
98. Simon, P.M.; Neethling, F.A.; Taniguchi, S.; Goode, P.L.; Zopf, D.; Hancock, W.W.; Cooper, D.K. Intravenous infusion of Galα-3Gal oligosaccharides in baboons delays hyperacute rejection of porcine heart xenografts. *Transplantation* **1998**, *65*, 346–353. [CrossRef]
99. Mohiuddin, M.M.; Singh, A.K.; Corcoran, P.C.; Thomas Iii, M.L.; Clark, T.; Lewis, B.G.; Hoyt, R.F.; Eckhaus, M.; Pierson, R.N., III; Belli, A.J.; et al. Chimeric 2C10R4 anti-CD40 antibody therapy is critical for long-term survival of *GTKO.hCD46.hTBM* pig-to-primate cardiac xenograft. *Nat. Commun.* **2016**, *7*, 11138. [CrossRef] [PubMed]
100. Singh, A.K.; Chan, J.L.; DiChiacchio, L.; Hardy, N.L.; Corcoran, P.C.; Lewis, B.G.T.; Thomas, M.L.; Burke, A.P.; Ayares, D.; Horvath, K.A.; et al. Cardiac xenografts show reduced survival in the absence of transgenic human thrombomodulin expression in donor pigs. *Xenotransplantation* **2019**, *26*, e12465. [CrossRef]
101. Längin, M.; Mayr, T.; Reichart, B.; Michel, S.; Buchholz, S.; Guethoff, S.; Dashkevich, A.; Baehr, A.; Egerer, S.; Bauer, A.; et al. Consistent success in life-supporting porcine cardiac xenotransplantation. *Nature* **2018**, *564*, 430–433. [CrossRef]
102. Kim, S.C.; Mathews, D.V.; Breeden, C.P.; Higginbotham, L.B.; Ladowski, J.; Martens, G.; Stephenson, A.; Farris, A.B.; Strobert, E.A.; Jenkins, J.; et al. Long-term survival of pig-to-rhesus macaque renal xenografts is dependent on CD4 T cell depletion. *Am. J. Transplant.* **2019**, *19*, 2174–2185. [CrossRef] [PubMed]
103. Wang, L.; Cooper, D.K.C.; Burdorf, L.; Wang, Y.; Iwase, H. Overcoming coagulation dysregulation in pig solid organ transplantation in nonhuman primates: Recent progress. *Transplantation* **2018**, *102*, 1050–1058. [CrossRef] [PubMed]
104. Shah, J.A.; Patel, M.S.; Elias, N.; Navarro-Alvarez, N.; Rosales, I.; Wilkinson, R.A.; Louras, N.J.; Hertl, M.; Fishman, J.A.; Colvin, R.B.; et al. Prolonged survival following pig-to-primate liver xenotransplantation utilizing exogenous coagulation factors and costimulation blockade. *Am. J. Transplant.* **2017**, *17*, 2178–2185. [CrossRef]
105. Watanabe, H.; Ariyoshi, Y.; Pomposelli, T.; Takeuchi, K.; Ekanayake-Alper, D.K.; Boyd, L.K.; Arn, S.J.; Sahara, H.; Shimizu, A.; Ayares, D.; et al. Intra-bone bone marrow transplantation from hCD47 transgenic pigs to baboons prolongs chimerism to >60 days and promotes increased porcine lung transplant survival. *Xenotransplantation* **2020**, *27*, e12552. [CrossRef] [PubMed]

106. Kwon, D.N.; Lee, K.; Kang, M.J.; Choi, Y.J.; Park, C.; Whyte, J.J.; Brown, A.N.; Kim, J.H.; Samuel, M.; Mao, J.; et al. Production of biallelic CMP-Neu5Ac hydroxylase knock-out pigs. *Sci. Rep.* **2013**, *3*, 1981. [CrossRef]
107. Cozzi, E.; White, D.J. The generation of transgenic pigs as potential organ donors for humans. *Nat. Med.* **1995**, *1*, 964–966. [CrossRef] [PubMed]
108. Diamond, L.E.; Quinn, C.M.; Martin, M.J.; Lawson, J.; Platt, J.L.; Logan, J.S. A human CD46 transgenic pig model system for the study of discordant xenotransplantation. *Transplantation* **2001**, *71*, 132–142. [CrossRef]
109. Petersen, B.; Ramackers, W.; Tiede, A.; Lucas-Hahn, A.; Herrmann, D.; Barg-Kues, B.; Schuettler, W.; Friedrich, L.; Schwinzer, R.; Winkler, M.; et al. Pigs transgenic for human thrombomodulin have elevated production of activated protein C. *Xenotransplantation* **2009**, *16*, 486–495. [CrossRef]
110. Aigner, B.; Renner, S.; Kessler, B.; Klymiuk, N.; Kurome, M.; Wünsch, A.; Wolf, E. Transgenic pigs as models for translational biomedical research. *J. Mol. Med. (Berl.)* **2010**, *88*, 653–664. [CrossRef]
111. Ran, F.A.; Hsu, P.D.; Lin, C.Y.; Gootenberg, J.S.; Konermann, S.; Trevino, A.E.; Scott, D.A.; Inoue, A.; Matoba, S.; Zhang, Y.; et al. Double nicking by RNA-guided CRISPR Cas9 for enhanced genome editing specificity. *Cell* **2013**, *154*, 1380–1389. [CrossRef] [PubMed]
112. Kleinstiver, B.P.; Pattanayak, V.; Prew, M.S.; Tsai, S.Q.; Nguyen, N.T.; Zheng, Z.; Joung, J.K. High-fidelity CRISPR-Cas9 nucleases with no detectable genome-wide off-target effects. *Nature* **2016**, *529*, 490–495. [CrossRef] [PubMed]
113. Lin, S.; Staahl, B.T.; Alla, R.K.; Doudna, J.A. Enhanced homology-directed human genome engineering by controlled timing of CRISPR/Cas9 delivery. *eLife* **2014**, *3*, e04766. [CrossRef] [PubMed]
114. Nishiyama, J.; Mikuni, T.; Yasuda, R. Virus-mediated genome editing via homology-directed repair in mitotic and postmitotic cells in mammalian brain. *Neuron* **2017**, *96*, 755–768. [CrossRef] [PubMed]
115. Hryhorowicz, M.; Grześkowiak, B.; Mazurkiewicz, N.; Śledziński, P.; Lipiński, D.; Słomski, R. Improved delivery of CRISPR/Cas9 system using magnetic nanoparticles into porcine fibroblast. *Mol. Biotechnol.* **2019**, *61*, 173–180. [CrossRef]
116. Gün, G.; Kues, W.A. Current progress of genetically engineered pig models for biomedical research. *Biores. Open Access* **2014**, *3*, 255–264. [CrossRef]
117. Yang, H.; Wu, Z. Genome editing of pigs for agriculture and biomedicine. *Front. Genet.* **2018**, *9*, 360. [CrossRef]
118. Kalla, D.; Kind, A.; Schnieke, A. Genetically engineered pigs to study cancer. *Int. J. Mol. Sci.* **2020**, *21*, 488. [CrossRef]
119. Iwatsuki-Horimoto, K.; Nakajima, N.; Shibata, M.; Takahashi, K.; Sato, Y.; Kiso, M.; Yamayoshi, S.; Ito, M.; Enya, S.; Otake, M.; et al. The microminipig as an animal model for influenza a virus infection. *J. Virol.* **2017**, *91*, e01716-16. [CrossRef]
120. Whitney, J.B.; Brad Jones, R. In vitro and in vivo models of HIV latency. *Adv. Exp. Med. Biol.* **2018**, *1075*, 241–263.
121. Siragam, V.; Wong, G.; Qiu, X.G. Animal models for filovirus infections. *Zool. Res.* **2018**, *39*, 15–24. [CrossRef] [PubMed]
122. Xu, W. Production of Recombinant Human Coagulation Factor IX by Transgenic Pig. Ph.D. Thesis, University of Nebraska, Lincoln, NE, USA, 2014. Available online: https://digitalcommons.unl.edu/dissertations/AAI3632487 (accessed on 10 August 2014).
123. Cooper, D.K.; Ekser, B.; Ramsoondar, J.; Phelps, C.; Ayares, D. The role of genetically engineered pigs in xenotransplantation research. *J. Pathol.* **2016**, *238*, 288–299. [CrossRef] [PubMed]
124. Wolf, E.; Kemter, E.; Klymiuk, N.; Reichart, B. Genetically modified pigs as donors of cells, tissues, and organs for xenotransplantation. *Animal Front.* **2019**, *9*, 13–20. [CrossRef] [PubMed]
125. Lu, T.; Yang, B.; Wang, R.; Qin, C. Xenotransplantation: Current status in preclinical research. *Front. Immunol.* **2020**, *10*, 3060. [CrossRef]
126. Wu, J.; Platero-Luengo, A.; Sakurai, M.; Sugawara, A.; Gil, M.A.; Yamauchi, T.; Suzuki, K.; Bogliotti, Y.S.; Cuello, C.; Morales Valencia, M.; et al. Interspecies chimerism with mammalian pluripotent stem cells. *Cell* **2017**, *168*, 473–486. [CrossRef]

*Article*

# Identification of Molecular Mechanisms Related to Pig Fatness at the Transcriptome and miRNAome Levels

**Katarzyna Ropka-Molik [1,*], Klaudia Pawlina-Tyszko [1], Kacper Żukowski [2], Mirosław Tyra [3], Natalia Derebecka [4], Joanna Wesoły [4], Tomasz Szmatoła [1,5] and Katarzyna Piórkowska [1]**

1 Department of Animal Molecular Biology, National Research Institute of Animal Production, Krakowska 1, 32-083 Balice, Poland; klaudia.pawlina@izoo.krakow.pl (K.P.-T.); tomasz.szmatola@izoo.krakow.pl (T.S.); katarzyna.piorkowska@izoo.krakow.pl (K.P.)

2 Department of Cattle Breeding, National Research Institute of Animal Production, Krakowska 1, 32-083 Balice, Poland; kacper.zukowski@izoo.krakow.pl

3 Department of Pig Breeding, National Research Institute of Animal Production, Krakowska 1, 32-083 Balice, Poland; miroslaw.tyra@izoo.krakow.pl

4 Laboratory of High Throughput Technologies, Institute of Molecular Biology and Biotechnology, Faculty of Biology, Uniwersytetu Poznanskiego street 6, 61-614 Poznań, Poland; nataliad@amu.edu.pl (N.D.); j.wesoly@amu.edu.pl (J.W.)

5 University Centre of Veterinary Medicine, University of Agriculture in Kraków, Al. Mickiewicza 24/28, 30-059 Kraków, Poland

* Correspondence: katarzyna.ropka@izoo.krakow.pl; Tel.: +48-666-081-208; Fax: +48-012-285-6733

Received: 13 April 2020; Accepted: 27 May 2020; Published: 29 May 2020

**Abstract:** Fat deposition and growth rate are closely related to pork quality and fattening efficiency. The next-generation sequencing (NGS) approach for transcriptome and miRNAome massive parallel sequencing of adipocyte tissue was applied to search for a molecular network related to fat deposition in pigs. Pigs were represented by three breeds (Large White, Pietrain, and Hampshire) that varied in fat content within each breed. The obtained results allowed for the detection of significant enrichment of Gene Ontology (GO) terms and pathways associated directly and indirectly with fat deposition via regulation of fatty acid metabolism, fat cell differentiation, inflammatory response, and extracellular matrix (ECM) organization and disassembly. Moreover, the results showed that adipocyte tissue content strongly affected the expression of leptin and other genes related to a response to excessive feed intake. The findings indicated that modification of genes and miRNAs involved in ECM rearrangements can be essential during fat tissue growth and development in pigs. The identified molecular network within genes and miRNAs that were deregulated depending on the subcutaneous fat level are proposed as candidate factors determining adipogenesis, fatness, and selected fattening characteristics in pigs.

**Keywords:** pig; fatness; obesity; extracellular matrix; fat deposition; lipid metabolism; NGS

---

## 1. Introduction

The quantity of fat is a critical factor influencing meat quality and fattening efficiency in pig production. The fat level and mass of adipocyte tissue control food intake and body weight via leptin secretion [1,2]. Leptin is considered an essential hormone regulating adipose tissue metabolism and energy expenditure. Leptin acts as an appetite regulation factor (hunger inhibition and satiety stimulation) through interaction with hypothalamic receptors [3] and controls glucose/lipid metabolism and body weight homeostasis through gluconeogenesis regulation [4]. Moreover, pigs can be a suitable animal model for studies of fat accumulation and obesity [5]. The identification of genetic factors determining fatness levels in pigs can be a valuable resource for medical research in humans.

In humans, genetic predisposition is a key contributing factor in obesity [6]. It has been estimated that the genetic basis of phenotypic variations in obesity can range from 40% to 70% [6,7]. In pigs, the high heritability coefficient ($h^2 = 0.5$) indicates the genetic background of fatness traits and the ability to improve them by selection [5]. The identification of genome regions, genes, or single nucleotide polymorphisms (SNPs) related to fatness in pigs has been conducted using different methodological approaches. In 2011, the usage of expression and single nucleotide polymorphism microarrays allowed for the detection of the panel of porcine genes relevant to fatness-associated traits [8]. In 2015, the genome-wide association study (GWAS) method was applied to identify QTLs (quantitative trait loci) located on Sus scrofa chromosome 7 (SSC7) and SSC4 strongly associated with growth and fatness traits based on Chinese pig breeds [9]. Guo et al. [10] using the GWAS method also indicated the involvement of the SSC7 region in the regulation of fatness and growth features. Another approach that can allow for the identification of molecular networks related to fat metabolism and growth traits is transcriptomic research concentrated on examining the hypothalamus–pituitary axis. The hypothalamus is a key brain structure controlling food intake and fat accumulation [11]. The pituitary is a crucial gland involved in fetal adipose metabolism via controlling the fatty acid synthesis and receptor-mediated lipolytic response [12]. Pérez-Montarelo et al. [13] confirmed that the usage of network interaction based on hypothalamic transcriptome analysis could pinpoint candidate genes for fatness in pigs. The transcriptomic profiling of the swine pituitary showed the strong association of this gland with postnatal growth and development [14], and also with fatness-associated traits [15].

A significant number of studies have confirmed an important involvement of adipocyte tissue in food intake [16] and body-weight regulation [17]. Such findings indicated the necessity of the analysis of adipocyte tissue to obtain the full view of metabolic processes related to fatness in pigs. To date, in pigs, transcriptomic research has been applied to identify the regulation processes determining feed efficiency [18] and to examine the body's response to different nutritional treatments [19]. Processes based on adipocyte tissue gene expression profiling in two Portuguese local pig breeds to determine the fat deposition via controlling of de novo fatty acid synthesis have been proposed, [20]. Furthermore, gene expression profiling in combination with genome resequencing has allowed for the detection of genes potentially related to backfat thickness [21].

The present study aimed to perform comprehensive whole transcriptome and miRNAome analyses [20] of adipocyte tissue to identify the molecular networks strongly related to the most crucial production traits, namely, fatness and feed intake in pigs.

## 2. Materials and Methods

### 2.1. Animals, Phenotype Data Collection, and Tissue Sampling

Animals used for gene and miRNA expression analyses were selected from a large pig population, based on fatness phenotypic traits measured after dissection. All pigs, representing three breeds (Pietrain, Pi; Hampshire, Hp; and Large White, LW) were unrelated (at least three generations back), females, and were maintained under the same housing and feeding conditions according to a procedure previously described [22]. The pigs were kept in individual pens and fed to a weight of 105 kg (±2.5 kg) (based on the Pig Test Station (SKURTCh) of the National Research Institute of Animal Production in Chorzelów). Half-carcasses were dissected 24 h after slaughter and several carcass characteristics, including fatness traits, were evaluated. The traits included average backfat thickness (ABT), calculated from five measurement points (cm) (at the thickest point over the shoulder; on the back above the joint between the last thoracic and first lumbar vertebrae; and at three points over the loin: above the rostral edge (loin I), above the middle (loin II), and the caudal edge (loin III) of gluteal muscle cross-section), and the weight of peritoneal fat (kg), using a procedure described by Tyra and Żak [23].

Pigs for next-generation sequencing (NGS) analysis were selected based on the most important fatness parameter, i.e., ABT. In each analyzed breed (Pietrain, 8, Hampshire, 8, and Large White, 8), pigs with different fatness phenotypes were selected (4 pigs with high and 4 with low fatness). The separation of the two groups with extreme fatness within each breed allowed for avoidance of any potential breed effect on obtained genes or miRNA expression profiles (Table 1). Immediately after slaughter, the samples of fat tissue (subcutaneous fat) were collected into the RNAlater solution (Ambion, Thermo Fisher Scientific, Waltham, MA, USA) and stored at −20 °C.

**Table 1.** The differences in fatness traits detected in all three breeds and obtained pig groups used in genetic analyses.

| | **Backfat Thickness (cm)** | | | | **Weight of Peritoneal Fat (kg)** | | | |
|---|---|---|---|---|---|---|---|---|
| | **L** | | **H** | | **L** | | **H** | |
| Pietrain | 0.71 | ±0.0.1 $^{b}$ | 1.29 | ±0.10 $^{a}$ | 0.15 | ±0.001 $^{b}$ | 0.31 | ±0.051 $^{a}$ |
| Hampshire | 0.99 | ±0.03 $^{b}$ | 1.54 | ±0.20 $^{a}$ | 0.30 | ±0.016 | 0.29 | ±0.067 |
| Large White | 1.04 | ±0.06 | 1.40 | ±0.14 | 0.23 | ±0.070 | 0.31 | ±0.122 |

L, low fatness; H, high fatness. Data are presented as means ± standard error; means with different letters (a,b differ significantly, $p$-value = 0.05).

The research did not require the approval of the Animal Experimentation Committee due to the fact that all biological material was collected from animals which were slaughtered and dissected, and, after carcass evaluation, meat was intended for consumption. The pigs were maintained and slaughtered according to Directive 2010/63/EU of the European Parliament and the Council of 22 September 2010.

*2.2. Whole Transcriptome Sequencing (RNA-seq)*

Extracted total RNA (Direct-zol RNA Mini Prep kit (Zymo Research, Orange, CA, USA)) was subjected to quantity and quality controls using a NanoDrop 2000 spectrophotometer (Thermo Fisher Scientific, Waltham, MA, USA) and a TapeStation 2200 instrument (Agilent, Santa Clara, CA, USA), respectively. Samples with RIN value (RNA integrity number) above 7 were chosen for further analyses. A quantity of 300 ng of total RNA was used for cDNA library preparation according to the TruSeq RNA Kit v2 kit protocol (Illumina, San Diego, CA, USA). The quality and quantity of cDNA libraries were checked using Qubit 2.0 and TapeStation 2200. The individual cDNA libraries were ligated with adaptors with different indexes and pooled. After pooling, each cDNA library was loaded into at least four flow cell lanes as four technical replicates. Sequencing of RNA was performed on the HiScanSQ system (Illumina, San Diego, CA, USA) using the TruSeq SR Cluster Kit v3- CBOT-HS kit and TruSeq SBS Kit v 3 - HS chemistry (Illumina, San Diego, CA, USA) according to the standard protocols and with 81 single-end reads.

The quality of raw data was analyzed using FastQC software (overrepresented sequences, sequence duplication levels, adapter contents, sequence length distribution, sequence quality scores, and GC content). The adapter's sequence reads shorter than 36 bp and/or reads with a quality score lower than 20 were removed using Flexbar software v2.5 [24]. The estimation of transcript abundance was conducted using RSEM software (v1.2.22) [25] supported by STAR aligner software (STAR_2.4.2a) [26]. The reads were aligned to the Sus scrofa reference genome (assembly Sscrofa11.1, Ensemble release 94). The whole procedure was followed by the alignment parameter evaluation using ENCODE3's STAR-RSEM pipeline. Differentially-expressed genes (DEGs) were identified according to the following criteria: fold change ≥ |1.0| and adjusted $p$-value < 0.05 separately for each breed using DESeq2 software (version 1.12.4) [26].

### 2.3. MicroRNA Sequencing

MicroRNA libraries were prepared with the use of the NEBNext Multiplex Small RNA Library Prep Set for Illumina (New England Biolabs, Ipswich, MA, USA) according to the protocol. The libraries were barcoded with different indexed primers to allow multiplexing of the samples during NGS sequencing. The amplified libraries were purified and size-selected on a Novex 6% TBE PAGE gel (Invitrogen, Thermo Fisher Scientific, Waltham, MA, USA). The quality and quantity of obtained libraries were measured with a Qubit 2.0 Fluorometer (Thermo Fisher Scientific, USA) and size-assessed with a 2200 TapeStation instrument (Agilent, Santa Clara, CA, USA). The whole miRNA profile sequencing was performed in 36 cycles on a HiScanSQ (Illumina, San Diego, CA, USA) instrument according to the manufacturer's protocol.

The raw miRNA reads were quality controlled using FastQC software (parameters as for the transcriptome) [27]. The adaptor trimming and filtering were performed with the UEA sRNA Workbench v3.2 helper tool [28], while the miRCat tool v1.0 was used to detect known and novel miRNA sequences. The identification was carried out on the basis of the *Sus scrofa* genome (assembly Sscrofa 10.2) and miRBase v21.0 (Griffiths-Jones lab at the Faculty of Biology, Medicine, and Health, University of Manchester, USA) [29,30], due to the lack of 11.1 assembly in miRbase dataset. The obtained miRNAs genome localization based on 10.2 reference were converted to 11.1 reference using NCBI Genome Remapping tool. The default animal parameters were set except for minimum abundance (6 reads), minimum length (17 nt), and maximum length (25 nt). In the next step, potential other non-coding RNAs were identified and excluded employing the RNAcentral database (the RNAcentral Consortium, 2017). Moreover, isomiR-SEA software (version 1.60) [31] with the default settings was used to identify microRNA length and sequence variants (isomiRs). miRNAs differentially expressed between pigs with low and high fatness in each breed were detected using DESeq2 software (v.1.12.4) [26]. The identification of experimentally-validated target genes and enriched biological processes (KEGG, GO) was performed using the mirPath v.3.0 DIANA Tools web application (DIANA-Lab, Athens, Greece) [32], including TarBase v7.0 (DIANA-Lab, Athens, Greece) as a reference [33]. The analysis was performed based on human homologues deposited in miRBase (21.0) due to the lack of data for pig microRNAs.

### 2.4. Validation of NGS Results

Validation of the RNA-seq results was carried out using a real-time PCR method, for 10 selected DEGs. The primers for selected DEGs were designed based on reference sequences showed in Supplementary Table S1 and using Primer3 Input (version 0.4.0) The detailed information about the chosen genes is presented in Supplementary Table S1. The cDNA was prepared from 300 ng of total RNA using a Maxima First Strand cDNA Synthesis Kit for RT-qPCR (Thermo Fisher Scientific, Waltham, MA, USA) according to the protocol. The exact gene expression levels were measured on a QuantStudio 7 flex instrument (Applied Biosystems, Thermo Fisher Scientific, Waltham, MA, USA) using Sensitive RT HS-PCR EvaGreen Mix (A&A Biotechnology, Gdynia, Poland). The reaction was carried out in three replications for each gene. The expression was calculated using the delta-delta CT method, according to Pfaffl [34] and based on two reference controls *OAZ1* [35] and *RPS29* [36].

Validation of the detected expression levels of 10 selected microRNAs was carried out using the qPCR method. A quantity of 10 ng of total RNA was reverse transcribed to cDNA using a TaqMan Advanced miRNA cDNA Synthesis Kit (Thermo Fisher Scientific, USA) following the protocol. The quantitative estimation of miRNAs was performed on a QuantStudio 7 flex instrument using TaqMan® Advanced miRNA Assays labelled with VIC or FAM fluorescent dye and with TaqMan® Fast Advanced Master Mix (Thermo Fisher Scientific, USA) according to the standard protocol (Supplementary Table S1). The relative expression levels of selected miRNAs were calculated using the delta-delta CT method according to Pfaffl [34] and based on mir-451a as a reference control. The comparison between NGS data (RNA-seq, miRNA-seq) and relative quantity obtained by the real-time PCR method was performed using the Pearson correlation (SAS software, version 8.02).

## 3. Results

### 3.1. RNA Sequencing Results—Differentially-Expressed Genes

The whole transcriptome sequencing of adipocyte tissue allowed for the identification of differentially-expressed genes (DEGs) between pigs varying in backfat thickness. The investigation of three pig breeds enabled the detection of genes related to fatness traits regardless of breed factor. According to the NGS approach, 167 DEGs were identified between Large White pigs with different fatness phenotypes (85 up-regulated and 82 down-regulated genes in pigs with thicker backfat). A total of 32 of these were recognized as novel genes. In turn, in the Pietrain breed, 247 DEGs were detected (53 up-regulated and 194 down-regulated), of which 50 were novel genes or uncharacterized proteins. The highest number of DEGs was identified for the Hampshire breed: 128 up-regulated and 173 down-regulated genes, for a total of 301 genes (46 novel).

The comparison of the detected sets of DEGs across breeds enabled the identification of eight known genes (*SCD, SFRP2, MMRN1, PCK1, TNC, C2, CXCl10,* and *CXCL9*) and one novel gene common to all three breeds (Figure 1). The largest group of common genes identified as differentially-expressed was found for Large White and Pietrain breeds (44 DEGs and 12 novel genes). The lowest similarity in the identified gene panel was observed for Hampshire pigs and the other two breeds.

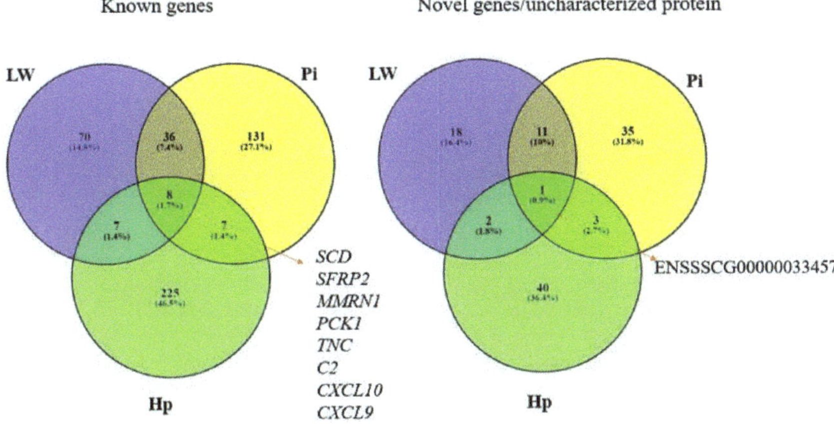


**Figure 1.** Comparison of differentially-expressed genes (DEGs) between subcutaneous fat tissue varying in thickness in each of three breeds (LW, Large White; Pi, Pietrain; and Hp, Hampshire).

The Gene Ontology (GO) analysis performed for the DEGs identified in at least two breeds showed significantly-overrepresented GO terms (Table 2). The highest number of genes was assigned to innate immune response (GO:0045087), 14 genes; inflammatory response (GO:0006954), 11 genes; and positive regulation of apoptotic process (GO:0043065), 10 genes. The DEGs were also included in GO terms related to lipid metabolism: fatty acid biosynthetic process, long-chain fatty acid biosynthetic process, positive regulation of fat cell differentiation, positive regulation of mast cell degranulation, and response to excessive food intake (Table 2; David software). Furthermore, GO and pathways analyses using over-representation analysis (WebGestalt software) confirmed that detected DEGs were involved in the biosynthesis of unsaturated fatty acids, fatty acid metabolism, and regulation of fat cell differentiation and brown fat differentiation (not significant GO) (Figure 2). The DEGs associated with several identified GO terms were as follows: *ACACA* (acetyl-CoA Carboxylase α), *SCD* (stearoyl-CoA desaturase), *SCD5* (stearoyl-CoA desaturase 5), *FASN* (fatty acid synthase), *LEP* (leptin), and *CTGF* (connective tissue growth factor) (Table 3; Figure 3).

**Table 2.** Significant, enriched Gene Ontology terms for differentially-expressed genes (DEGs) in relation to backfat thickness.

| Gene Ontology/Accession Number | FDR * | *N* | Gene Name |
|---|---|---|---|
| Extracellular matrix organization (GO:0030198) | $9.2 \times 10^{-3}$ | 6 | *CYR61; ELN; GFAP; HPSE; POSTN; VTN* |
| Positive regulation of mast cell degranulation (GO:0043306) | $2.4 \times 10^{-3}$ | 4 | *FGR; FCER1A; FCER1G; ZAP70* |
| Cell adhesion (GO:0007155) | $3.5 \times 10^{-3}$ | 12 | *TNFAIP6; WISP1; CTGF; CYR61; HAS1; LYVE1; NOV; POSTN; SELL; TNC; THBS2; THBS3* |
| Innate immune response (GO:0045087) | $5.9 \times 10^{-2}$ | 14 | *FGR; FCER1G; MX1; MX2; S100A8; B2M; BST2; FGB; IFIH1; JCHAIN; LCN2; PML; TLR2; ZAP70;* |
| Fatty acid biosynthetic process (GO:0006633) | $8.4 \times 10^{-3}$ | 4 | *ACACA; SCD; FASN; ACLY* |
| Positive regulation of apoptotic process (GO:0043065) | $5.6 \times 10^{-3}$ | 10 | *BMF; ALDH1A2; CLU; CYP1B1; GADD45G; IGFBP3; SFRP2; TOP2A; TGM2; ZBTB16* |
| Long-chain fatty acid biosynthetic process (GO:0042759) | $6.0 \times 10^{-3}$ | 3 | *PLPP1; SCD; SCD5* |
| Response to dietary excess (GO:0002021) | $8.6 \times 10^{-3}$ | 3 | *PPARGC1A; LEP; PCSK1N* |
| Positive regulation of fat cell differentiation (GO:0045600) | $6.8 \times 10^{-3}$ | 5 | *CCDC3; MEDAG; SFRP2; ZBTB16; ZNF385A* |
| Inflammatory response (GO:0006954) | $8.6 \times 10^{-3}$ | 11 | *CCL5; CCR5; CD180; FAS; CXCL10; C5AR1; PLP1; TSPAN2; TBXA2R; TLR2; ZAP70* |

* FDR (false discovery rate); *p*-values are shown after Benjamini correction and according to David software; *N*, number of identified DEGs.–

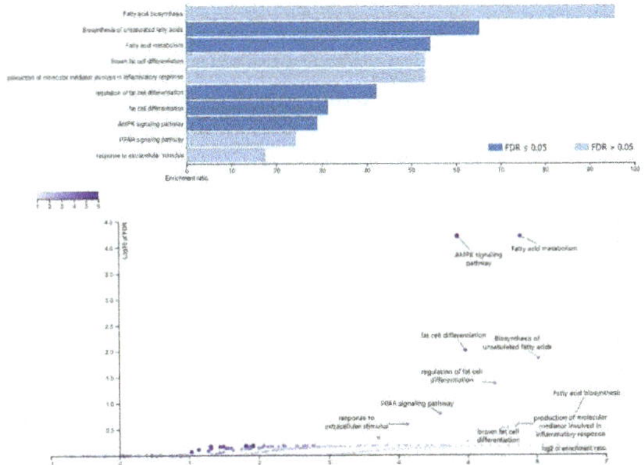

**Figure 2.** The significant enrichment GO terms and pathways identified based on backfat thickness differences and related to lipid metabolism (WebGestalt software).

**Table 3.** The fold-change values of DEGs related to lipid metabolism and detected as significant for at least two breeds.

| Gene | Accession Number | Pietrain | | Large White | | Hampshire | |
|---|---|---|---|---|---|---|---|
| | | FC | FDR | FC | FDR | FC | FDR |
| *LEP* | ENSSSCG00000040464 | 1.41 | 0.03 | 2.14 | 0.001 | 1.10 | ns |
| *ACACA* | ENSSSCG00000017694 | 1.62 | 0.03 | 1.67 | 0.001 | 1.52 | ns |
| *SCD* | ENSSSCG00000010554 | 2.95 | 0.001 | 1.66 | 0.001 | 2.23 | 0.04 |
| *SCD5* | ENSSSCG00000009245 | −1.32 | 0.04 | −1.09 | ns | −2.28 | 0.001 |
| *FASN* | ENSSSCG00000029944 | 1.58 | ns | 1.47 | 0.02 | 3.07 | 0.005 |
| *ACOX3* | ENSSSCG00000008724 | 1.40 | 0.05 | 1.32 | ns | 2.06 | 0.001 |
| *C2* | ENSSSCG00000001422 | −3.18 | 0.001 | −2.00 | 0.001 | 2.31 | 0.01 |
| *ACLY* | ENSSSCG00000017421 | 1.65 | 0.05 | 2.83 | 0.001 | 1.13 | ns |
| *TNC* | ENSSSCG00000005494 | −3.23 | 0.0004 | −2.55 | 0.001 | −4.17 | ns |
| *PPARGC1A* | ENSSSCG00000029275 | 1.16 | ns | 13.45 | 0.0001 | 3.27 | 0.05 |
| *PCSK1N* | ENSSSCG00000021328 | 1.07 | 0.05 | −1.52 | ns | −2.99 | 0.05 |
| *TLR2* | ENSSSCG00000009002 | −2.63 | 0.001 | −1.59 | 0.05 | −1.34 | ns |
| *FGR* | ENSSSCG00000003578 | 1.13 | 0.05 | 1.62 | 0.001 | −1.10 | ns |
| *FCER1A* | ENSSSCG00000006413 | 1.40 | 0.001 | −1.19 | 0.01 | −2.69 | ns |
| *FCER1G* | ENSSSCG00000006357 | −2.40 | 0.001 | −1.40 | 0.05 | −1.22 | ns |

FDR (false discovery rate); FC, fold change; ns, not significant.

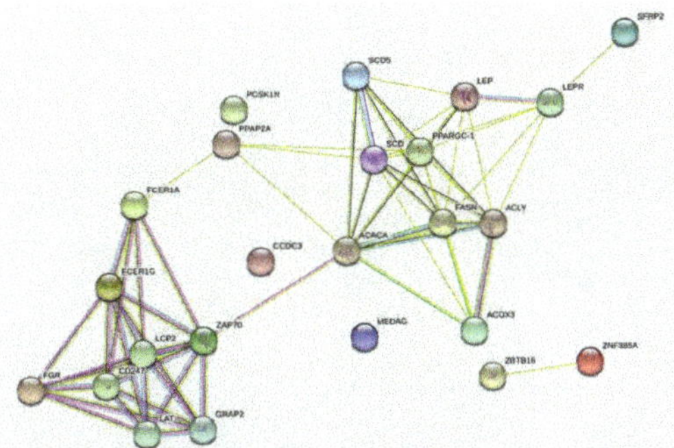

**Figure 3.** The detected differentially-expressed genes involved in GO terms related to lipid metabolism processes (String software based on Sus scrofa reference; detected genes showed no more than five interactions. Line shape indicates the predicted mode of action: red, interactions that were experimentally determined; blue, interactions from curated databases; black, co-expression; green, text mining associations and interactions based on relevant publications mentioning a transfer from other organisms; yellow, transcriptional regulation).

### 3.2. MiRNA Sequencing Results—Differentially-Expressed miRNAs

The comparison of whole miRNAome profiles between groups of pigs with significant-different fatness traits allowed for the identification of differentially-expressed (DE) miRNAs: 46 for LW (21 up-regulated and 25 down-regulated genes in pigs with thicker backfat), 61 for Pietrain (36 up-regulated and 25 down-regulated), and 41 for Hampshire (8 up-regulated and 31 down-regulated). As for DEGs, analogous comparison across breeds was also carried out for DE miRNAs and showed the presence of 14 miRNAs common to all three breeds (Figure 4). The lowest number of common miRNAs was detected for Hampshire and Large White pigs. Detailed data on the chromosomal localization of the identified miRNAs according to Sscrofa10.2 and Sscrofa11.1 assemblies are in the Supplementary Table S2.

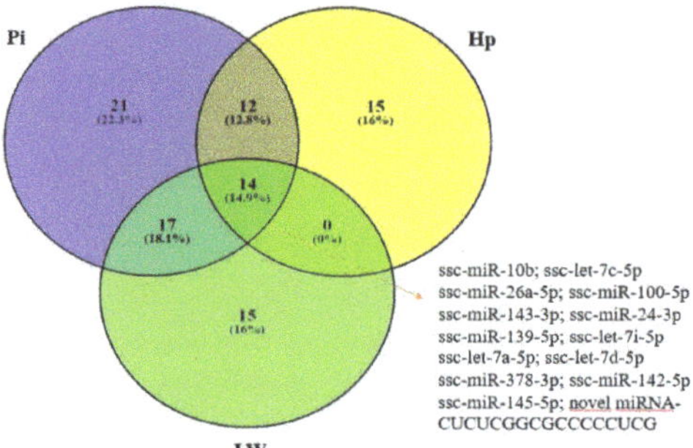


**Figure 4.** Venn diagram of differentially-expressed miRNAs in subcutaneous fat tissue dependent on fat content in three analyzed breeds (LW, Large White; Pi, Pietrain; and Hp, Hampshire).

The GO analysis performed using the mirPath v.3 DIANA tool showed that differentially-expressed miRNAs were involved in extracellular matrix organization (GO:0030198), extracellular matrix disassembly (GO:0022617), and innate immune response (GO:0045087), where the highest number of predicted targeted genes for miRNAs was detected. Eight miRNAs were associated with cellular lipid metabolic process (GO:0044255) (Table 4). Moreover, miR-100-5p, miR-143-3p, miR-10b-5p, and miR-24-3p were involved in fatty acid metabolism and fatty acid biosynthesis (Figure 5). The predicted targeted genes for differentially-expressed miRNAs are summarized in Supplementary Table S3.

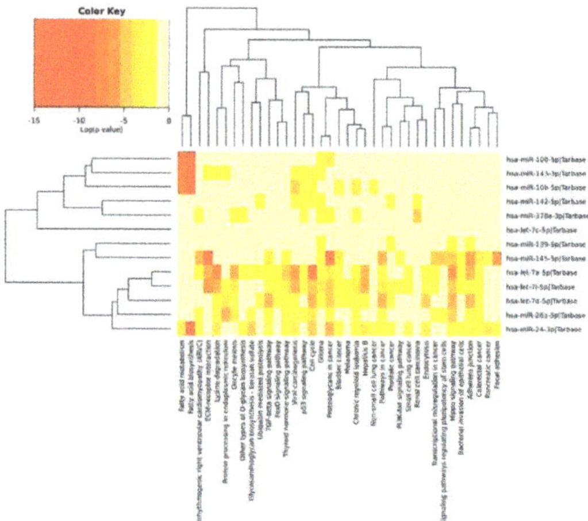


**Figure 5.** Venn diagram of differentially-expressed miRNAs in subcutaneous fat tissue dependent on fat content in three analyzed breeds (LW, Large White; Pi, Pietrain; and Hp, Hampshire).

**Table 4.** Significant, enriched Gene Ontology terms for differentially-expressed miRNAs in relation to backfat thickness.

| Gene Ontology/Accession Number | FDR * | *N* Target Genes | *N* miRNAs | miRNAs |
| --- | --- | --- | --- | --- |
| Extracellular matrix organization (GO:0030198) | $<1.0 \times 10^{-325}$ | 113 | 5 | hsa-let-7a-5p; hsa-let-7i-5p; hsa-miR-24-3p; hsa-miR-145-5p; hsa-let-7d-5p |
| Extracellular matrix disassembly (GO:0022617) | $<1.0 \times 10^{-325}$ | 46 | 6 | hsa-let-7a-5p; hsa-let-7i-5p; hsa-miR-24-3p; has-miR-26a-5p; hsa-miR-145-5p; hsa-let-7d-5p |
| Cellular lipid metabolic process (GO:0044255) | $3.1 \times 10^{-13}$ | 63 | 8 | hsa-let-7a-5p; hsa-let-7i-5p; hsa-miR-24-3p; has-miR-26a-5p; has-miR-143-3p; has-miR-142-5p; has-miR-145-5p; hsa-let-7d-5p |
| Cell junction organization (GO:0034330) | $<1.0 \times 10^{-325}$ | 66 | 6 | hsa-let-7a-5p; hsa-let-7i-5p; hsa-miR-24-3p; has-miR-143-3p; hsa-miR-145-5p; hsa-let-7d-5p |
| MyD88-independent toll-like receptor signaling pathway (GO:0002756) | $<1.0 \times 10^{-325}$ | 44 | 7 | hsa-let-7a-5p; hsa-let-7i-5p; hsa-miR-24-3p; has-miR-26a-5p; hsa-miR-145-5p; hsa-let-7d-5p; has-miR-378a-3p |
| Cellular component disassembly involved in execution phase of apoptosis (GO:0006921) | $<1.0 \times 10^{-325}$ | 28 | 7 | hsa-let-7a-5p; hsa-let-7i-5p; hsa-miR-24-3p; has-miR-26a-5p; hsa-miR-145-5p; hsa-miR-145-5p; hsa-miR-378a-3p |
| Toll-like receptor TLR1:TLR2 signaling pathway (GO:0038123)Toll-like receptor TLR6:TLR2 signaling pathway (GO:0038124) | $<1.0 \times 10^{-325}$ | 39 | 7 | hsa-let-7a-5p; hsa-let-7i-5p; hsa-miR-24-3p; has-miR-26a-5p; hsa-miR-139-5p; has-miR-378a-3p; hsa-let-7d-5p |
| Innate immune response (GO:0045087) | $<1.0 \times 10^{-325}$ | 213 | 7 | hsa-let-7a-5p; hsa-let-7i-5p; hsa-miR-24-3p; has-miR-26a-5p; hsa-miR-139-5p; has-miR-142-5p; has-miR-145-5p; hsa-let-7d-5p |

* *p*-values are presented after FDR correction and according to the mirPath v.3 tool; *N* target genes, the number of predicted target genes of differentially-expressed miRNAs; *N* miRNAs, number of DE miRNAs.

### 3.3. Analysis of Pathways Common for DEGs and DE miRNAs

The pathways analysis was performed for both DEGs and DE miRNAs identified in at least two breeds. Furthermore, the predicted genes regulated by the identified miRNAs, and involved in selected pathways, detected using mirPath v.3 DIANA tool are shown in Table 5.

An enriched pathway significant for both DEGs and DE miRNAs was fatty acid metabolism (KEGG ID: hsa01212 for miRNA/ ssc01212 for DEGs) for which four miRNAs and five genes were detected. Three of all DEGs involved in fatty acid metabolism overlapped with predicted genes regulated by detection miRNAs, namely: *SCD, FASN,* and *ACACA*. A similar situation was observed for the fatty acid biosynthesis pathway ($p$-value for miRNAs $< 1.0 \times 10^{-325}$; not significant for DEGs); *FASN* and *ACACA* differentially-expressed genes overlapped with miRNA-predicted targets. The pathways identified as significant and common for DE miRNAs and DEGs were also the Hippo-signaling pathway, ECM−receptor interaction, cell cycle, and P53-signaling pathways.

The pathway analysis of DE miRNAs also showed significant enrichment for the set of pathways related to toll-like receptors (toll-like receptor 2, 3, 4, 5, 9, and 10 signaling pathways; toll-like receptor TLR1:TLR2 signaling pathways; and toll-like receptor TLR6:TLR2 signaling pathways), as well as TRIF-dependent toll-like receptor signaling pathways (Figure S1).

**Table 5.** Significant pathways detected for both DEGs (significant for at least two breeds) and DE miRNAs (14 miRNAs common for all breeds).

| | DE miRNAs | | | | | DEGs | | |
|---|---|---|---|---|---|---|---|---|
| **Pathways** | **FDR *** | ***N*** | **miRNAs** | ***N*** | ***N* Target Genes** | **FDR*** | ***N*** | **Genes** |
| Fatty acid metabolism (hsa01212/ssc01212) | $6.6 \times 10^{-16}$ | 4 | hsa-miR-100-5p; hsa-miR-10b-5p; hsa-miR-143-3p; hsa-miR-24-3p | 9 | *FASN; ACACA;MCAT; ACAA2; ACAA1; CPT1A; CPT2; SCD; ELOVL5* | $5.4 \times 10^{-3}$ | 5 | *ACACA;ACOX3; FASN; SCD5; SCD* |
| ECM–receptor interaction (hsa04512/ssc04512) | <1.0 × $10^{-325}$ | 4 | hsa-let-7a-5p; hsa-let-7i-5p; hsa-miR-143-3p; hsa-miR-145-5p | 35 | *SPP1; CD44; LAMC1; LAMC3; ITGA3; ITGA5; FN1; TNC; COL1A2; COL4A2; COL5A1; COL5A2; COL6A1; COL27A1; ITGB1; GP5; THBS1* | $2.4 \times 10^{-2}$ | 11 | *CHAD; COL1A1; COL5A3; COL6A6; COL11A1; LAMC2; TNC; THBS2; THBS3; THBS4; VTN* |
| Hippo-signaling pathway (hsa04390/ssc04390) | <1.0 × $10^{-325}$ | 7 | hsa-let-7a-5p; hsa-let-7i-5p; hsa-miR-145-5p; hsa-miR-24-3p; hsa-miR-139-5p; hsa-let-7d-5p; hsa-miR-26a-5p | 87 | *PPP2CA; BMP5; BMP2; BMP7; FGF1; ACTG1; PPP1CB; CTGF;* Figure S2 | 0.01 | 7 | *YWHAH; CTGF; FZD2; FZD4; NKD1; TGFB3; WNT2B* |
| Fatty acid biosynthesis (hsa00061/ssc00061) | <1.0 × $10^{-325}$ | 4 | hsa-miR-100-5p; hsa-miR-10b-5p; hsa-miR-143-3p; hsa-miR-24-3p | 3 | *FASN; ACACA; MCAT* | ns | 2 | *FASN; ACACA* |
| Cell cycle (hsa04110/ssc04110) | <1.0 × $10^{-325}$ | 9 | hsa-let-7a-5p; hsa-let-7i-5p; hsa-miR-10b-5p; hsa-miR-143-3p; hsa-miR-24-3p; hsa-miR-26a-5p; hsa-miR-142-5p; hsa-let-7d-5p | 78 | *ESPL1; CDC6; GSK3B; CCNB2; RBL2; PCNA; E2F1E2F2;YWHAE; MCM6; MCM2* Figure S3 | 0.003 | 5 | *E2F2; GADD45G; MCM2; TGFB3; YWHAH;* |
| P53-signaling pathways (hsa04115/ ssc04151) | $2.5 \times 10^{-9}$ | 7 | hsa-let-7a-5p; hsa-let-7i-5p; hsa-miR-26a-5p; hsa-miR-143-3p; hsa-miR-10b-5p; hsa-miR-142-5p; hsa-miR-378-3p | 60 | *ZMAT3; CCNB2; CCNB1; CDK4; BID; THBS1; CDK2 CCND2; PERP; RRM2B; CDK1; CDKN2A; CDK6 CHEK1; TP53; APAF1; PMAIP1; CD82; CASP3* | 0.0002 | 13 | *CHAD; COL1A1; COL6A6; LAMC2; PCK1; THBS2; THBS3; THBS4; TLR2; TNC; VTN; YWHAH* |

Accession numbers are presented for DE miRNAs/DEGs from the mirPath v.3 and KEGG database, respectively; * *p*-values are presented after FRD correction and according to the mirPath v.3 tool for DE miRNAs and after Benjamini correction and according to David software for DEGs; *N*, number of identified DEGs; ns, not significant; highlighted genes were identified both as differentially-expressed and as predicted genes regulated by miRNAs. The bold and underline were used to highlighted the genes identified in both–DEGs and targeted genes groups.

### 3.4. Validation of Obtained Data

The validation of RNA-seq results was performed for eight DEGs and six miRNAs (Table 6). For the majority of analyzed DEGs, the RNA-seq data and RQ results showed high and significant correlation coefficients (from 0.66 to 0.95), which confirmed the reliability of the NGS results. The highest similarity of results was obtained for *PCK1*, *ACACA*, and *LEP* genes. For four miRNAs, high correlation coefficients were detected, but these were significant only for miR-26a-5p, mir-100-5p, and mir-103a-3p.

**Table 6.** Correlation coefficients for NGS results and qPCR data for both miRNAs and DEGs.

| DEGs | Correlation | miRNAs | Correlation |
|---|---|---|---|
| *ROCK1* | 0.56 | hsa-miR-26a-5p | 0.81 * |
| *LRP12* | 0.66 * | hsa-let-7a-5p | 0.50 |
| *ACACA* | 0.91 ** | hsa-mir-100-5p | 0.75 * |
| *HK2* | 0.87 ** | hsa-mir-378a-3p | 0.40 |
| *LRP6* | 0.75 * | hsa-mir-103a-3p | 0.88 * |
| *LEP* | 0.94 ** | hsa-miR-125b-5p | 0.73 |
| *TNC* | 0.83 * | | |
| *PCK1* | 0.95 ** | | |

* $p$-value < 0.05; ** $p < 0.01$.

## 4. Discussion

In pigs, fatness traits are one of the important production features due to their strong relationship with meat quality and fattening performance. The inverse correlation between lean meat content and fatness traits results in the decrease of the fatness level in carcasses as meatiness increases [37]. Moreover, the proper fat content, including of intramuscular fat (IMF), is critical to achieving the desired meat quality parameters. The adipocyte tissue, as a secretory organ, also determines the regulation of food intake; thus, the body fat content can influence food intake and body weight [2]. In turn, the food intake and feed conversion ratio (FCR) are primary factors which determine the economic efficiency of pig production.

The high phenotypic variability of fatness characteristics, as well as the heritability coefficient (about 0.5), strongly indicate the genetic background of this group of traits and the possibility of their improvement or modification using genetic markers [5]. From a breeding perspective, the most important step would be to establish such a genetic marker that enables the controlling of pig fatness without losing the already-achieved level of meatiness. The latest successes of high throughput genetic methods, such as NGS sequencing technology and the GWAS approach, have become more applicable and create new possibilities in the search for and identification of the molecular background of phenotypic variations.

In the presented report, comprehensive whole mRNA and miRNA profiling of adipocyte tissue was applied to detect the molecular network related to fat deposition in pigs. The use of three pig breeds representing different usage types (maternal line, Large White; sire lines, Pietrain and Hampshire) allowed indication of a universal genetic basis of fatness characteristics that was not associated with any breed. The transcriptome sequencing of fat tissue delivered from pigs diverse in terms of backfat thickness enabled the pinpointing of significant, enriched Gene Ontology terms and pathways possibly related to fat deposition. For both DEGs and DE miRNAs, the fatty acid metabolism pathway was detected as significant. This pathway involved three DEGs, namely, *ACACA*, *FASN*, and *SCD*, and differentially-expressed miRNAs that recognized these genes as their targets. Moreover, the stearoyl-CoA desaturase gene (*SCD*) was identified as significant for all three analyzed breeds, and the *SCD* expression level was higher in pig groups with thicker fat cover. The stearoyl-CoA desaturase plays a vital role during the biosynthesis of monounsaturated fatty acids, within palmitoyl- and oleoyl-CoA, making them readily available to the body [38]. The higher *SCD* expression observed in this study, in pigs with higher fat content was in accord with previously-found human data. Hulver et al. [39] showed that an increased *SCD* level is correlated to obesity. Similarly, decrease of *SCD* transcription leads to a reduction of adiposity [40]. The deficiency of stearoyl-CoA desaturase results in leanness and leads to an increase of metabolic rate as well as insulin sensitivity [38]. Moreover, a study performed on ob/ob mice indicated that *SCD* is involved in metabolic response to leptin and down-regulation of *SCD* can be a key element of the metabolic actions of leptin [41].

Reports of many authors have indicated the association of the *SCD* gene and selected fatness traits in pigs. Using the GWAS approach, Yang et al. [42] showed a possible relationship between the *SCD* locus and C18:0 content. Similarly, results from the expression Genome-Wide Association Study (eGWAS) indicated that the *SCD* gene is related to intramuscular fat composition in pigs [43]. Xing et al. [21] compared transcriptomes of subcutaneous adipose tissue of Landrace pigs depending on variable backfat levels, which showed significantly-higher *SCD* expression in animals with increased fatness. The authors proposed stearoyl-CoA desaturase as a candidate gene for fat deposition. Our results, based on three different pig breeds, confirmed previous observations, thus strongly supporting an essential role of the *SCD* gene in the determination of fat-associated traits.

Moreover, the present miRNAome sequencing identified that expression of miR-24-3p is also dependent on subcutaneous fat level. In turn, the mirPath v.3 and TarBase v7.0 DIANA bioinformatic tools found that this microRNA modulates expression of the *SCD* gene, as well as fatty acid synthase (*FASN*), acetyl-CoA carboxylase α (*ACACA*), acetyl-coA acyltransferase 1 and 2 (*ACAA1* and *ACAA2*), and malonyl coA-acyl carrier protein transacylase (*MCAT*). miR-24-3p has been established to play a role in adipogenesis via activating adipocyte differentiation by targeting genes engaged in MAPK7-signaling pathways [44,45]. To date, only a few reports have been published concerning the possible involvement of miR-24-3p in lipid metabolism processes and obesity development. The present study indicates that this miRNA can be related to the regulation of fat deposition processes by targeting the key genes of fatty acid metabolism.

The analogues gene expression profile—increased transcript level in all fatty pigs independent of breed—was also observed for *ACACA* and *FASN* genes. A recent study applied the RNA-seq method to analyze the whole transcriptome profile of subcutaneous fat in native Puławska pigs that differed in backfat tissue [46]. The authors identified the higher transcript level of all three genes—*ACACA*, *FASN*, and SCD—in pigs with increased backfat thickness. Previous research showed the significant association of the *FASN* gene with meat quality and the *ACACA* gene with IMF content. In Iberian pigs, the *ACACA* gene was proposed as a candidate gene responsible for intramuscular content of saturated fatty acids and monounsaturated fatty acids fatty acids [47]. Furthermore, the up-regulation of *FASN* and *SCD* genes were reported in Alentejano pigs and related to the higher fat deposition observed in this breed [20]. The research performed on Iberian × Landrace pig cross lines indicated a significant association of the *ACACA* gene and IMF palmitic fatty acid percentage [48]. Similarly, the *FASN* gene was previously correlated to meat quality, fatty acid content, and composition [49,50]. Our research, supported by previous studies, strongly confirms that the identified DEGs determine fat deposition in pigs.

*Enriched Metabolic Process and Pathways Associated with Fat Deposition*

Genes whose expression varied between thick and thin backfat groups belonged to several processes and pathways directly related to fatty lipid metabolism, within the fatty acid biosynthetic process, the long-chain fatty acid biosynthetic process, positive regulation of fat cell differentiation, and response to dietary excess. The molecular network of food intake controlled by adipocyte tissues has been widely investigated and described. One of the main factors determining the energy balance by hunger inhibition is leptin [51]. The DEGs set comparison showed that *LEP* gene expression was increased in fat tissue of more obese pigs independent of breed. These findings are in accordance with observations made in humans, where the overexpression of the leptin gene in subcutaneous and omental adipose tissues in obese patients was detected [52]. Adipocyte cells accumulate triglyceride and increase the synthesis of leptin during their growth. Leptin acts via the hypothalamus to control energy balance and feeding behavior [53] Thus, leptin, also called an anti-obesity hormone, reduces food intake and increases energy expenditure [54]. Likewise, the other two DEGs identified in the present study—*PPARGC1A* (peroxisome proliferator-activated receptor-$\gamma$ co-activator-1alpha) and *PCSK1N* (proprotein convertase subtilisin/kexin type 1 inhibitor)—have been associated with obesity in humans. A polymorphism in the *PPARGC1A* gene is related to obesity and type 2 diabetes [55], while the down-regulation of *PCSK1N* gene expression is also associated with obesity [56]. In the present research, *PPARGC1A* expression was up-regulated in thicker backfat tissue in all three investigated pig breeds, whereas expression of the *PCSK1N* gene was lowered in obese pigs represented by Hampshire and Large White breeds. This suggests that all identified DEGs related to mast cell (MC) degranulation can affect the increase of fat tissue mass and lipid metabolism.

Research performed in humans and animals strongly indicates that MCs are involved in activation of adipocytes and recruitment of inflammatory cells [57]. The positive regulation of the MC degranulation-GO term was identified as significantly deregulated in fat tissue dependent on fat cover thickness. The research allowed for the detection of *FGR, FCER1A, FCER1G,* and *ZAP70* as DEGs associated with MC degranulation. The function of MC degranulation is associated with the release of multiple enzymes which can influence adipocyte tissue modification. In vitro studies on human white adipocyte tissue showed that MCs can promote adipose initiation in response to cold [58]. Furthermore, MC cells induced remodeling of adipose tissue extracellular matrix [57].

Interestingly, the obtained data showed the significant enrichment of ECM organization (GO:0030198) and ECM−receptor interaction pathways. The panel of DEGs was identified including genes coding for collagens, thrombospondins, and laminin, which belong to extracellular matrix remodeling. A recent report highlighted that ECM plays a critical role in adipose tissue expansion through controlling cell migration during body growth and development [59]. It has been proven that adipose tissue expansion is strongly related to ECM remodeling at the level of matrix individual components—collagens, thrombospondins, metalloproteinases, and their inhibitors (Tissue Inhibitor of Metalloproteinase - TIMPs) [60]. The present research allowed the identification of DE miR-145-5p and its target gene, tenascin C (*TNC*), both of which are involved in the ECM−receptor interaction pathway. Tenascin C, as a glycoprotein member of ECM, is related to tissue developmental processes, while miR-145 is involved in abdominal obesity, insulin resistance, and lipid metabolism [61]. The exact role of miR-145 has not been well established but the increased expression of miR-145 was observed in adipose and liver tissues in obese patients [62]. It has been proposed that this miRNA plays a role in lipolysis and our results supports this thesis, and also confirm that miR-145 overexpression in fat tissue is associated with obese individuals, which was independently observed in the investigated pig breeds.

Another interesting observation is the down-regulation of the *TNC* gene identified in pigs with thicker fat cover. Tenascin C determines the formation of the collagen network in adipocyte tissue controlling cell migration and proliferation [63]. On the other hand, the excessive cross-linking of adipocytes by fibrotic components can lead to the reduction of their metabolic activity [64]. We hypothesized that the increased expression of the *TNC* gene in thinner backfat can be related to significant modification of extracellular matrix components, which may result in reduced adipocyte activity and a decreased rate of fat tissue development. Tenascin C is also related to activation of the toll-like receptor 4 (TLR4) signaling pathway, which triggers the obesity-induced inflammatory response [65]. Our results also showed the significant enrichment of both toll-like receptor TLR1:TLR2 signaling and toll-like receptor TLR6:TLR2 signaling pathways for DE miRNAs. Gene expression profiling of human peripheral blood mononuclear cells (PBMCs) showed the increased transcript level of *TLR2* and *TLR4* genes in obese patients [66]. The increase in expression of the *TLR4* gene, reported in differentiating adipocytes in db/db mice suggests that this gene is critical during obesity development processes [67]. In the present study, in addition to the detection of GO terms related to *TLR2* and *TLR6*, the set of DE miRNAs targeting the *TLR4* gene was identified, namely, the members of the let-7 family (let-7a-5p, let-7i-5p, and let-7d-5p). To date, many reports have indicated a significant role of let-7 miRNAs in regulation of adipogenesis, lipolysis, and insulin resistance [68]. The differential expression of several let-7 family members may confirm that, beside *TLR* genes, this miRNA family plays a key role in the fat deposition rate and lipid metabolism.

The other proven role of ECM receptor pathways during fat tissue development is the regulation of adipocyte apoptosis and/or necrosis processes [60]. The results obtained in our study also indicated expression changes of genes involved in the positive regulation of apoptotic process GO term and cell cycle pathways. Other members of the EC matrix—thrombospondins—such as those identified in our study (*THBS2, THBS3,* and *THBS4*), are mainly detected in visceral adipose tissues and their increased expression level is associated with obesity. Moreover, thrombospondin 1 has been proposed as a novel marker related to obesity and metabolic syndrome [69]. The identification of the panel of genes involved in ECM receptor pathways and extracellular matrix assembly confirmed that such signaling is essential for fat deposition processes in pigs irrespective of a breed factor.

## 5. Conclusions

The backfat thickness and growth rate are closely related to pork quality and fattening efficiency, as well as reproductive and immune characteristics. Taking into account whole transcriptome profiling of pigs varying in fat content across three pig breeds, the significant enrichment of Gene Ontology terms and pathways associated directly and indirectly with fat deposition was detected (*ACACA, ACOX3, FASN, SCD5, SCD,* and *PLPP1).* The differentially-expressed miRNAs and genes were involved in fatty acid metabolism, positive regulation of fat cell differentiation, and the inflammatory response. Moreover, the results showed that adipocyte tissue content regulated the expression of leptin and other genes related to a response to dietary excess. Our results confirmed previous findings in humans that showed ECM organization and disassembly are fundamental for fat tissue growth and development. The modification of genes and miRNAs involved in ECM rearrangements can also be essential during fat tissue growth and development in pigs. We also suggest that mast cell degranulation can be one of the significant processes associated with adipocyte tissue development. The pinpointed molecular networks within genes and miRNAs deregulated by subcutaneous fat level are proposed as candidate factors determining adipogenesis, fatness, and selected fattening characteristics in pigs. Identified DEGs and DE miRNAs should be investigated in the future in terms of their use as genetic markers associated with pig production traits.

**Supplementary Materials:** The following are available online at http://www.mdpi.com/2073-4425/11/6/600/s1, Table S1. Detailed information about DEGs and miRNAs whose expression was estimated using real-time PCR methods. Table S2. Detailed data on the chromosomal localization of the identified miRNAs according to Sscrofa10.2 and Sscrofa11.1 ("LW" stands for Large White samples, "Pi" denotes Pietrain samples, and "Hp" stands for Hampshire samples; "NA" denotes potentially novel miRNA identified in this study). Table S3. TarBase experimentally supported interactions—the predicted targeted genes (N) for each of the 14 detected miRNAs. Figure S1. The significant enrichment molecular pathways in which 14 microRNAs commonly detected in all three pig breeds were detected (DIANA-miRPath v3.0, DIANA-Lab, Athens, Greece). Figure S2. The significant enrichment Hippo-signaling pathway (KEGG database, hsa04390) involved genes targeted for identified differentially-expressed miRNAs. Targeted genes are marked orange (genes included in more than one pathway) and yellow (genes included only in one pathway). Figure S3. The significant enrichment cell cycle (KEGG database; hsa04110) involved genes targeted for identified differentially-expressed miRNAs. Targeted genes are marked orange (genes included in more than one pathway) and yellow (genes included only in one pathway).

**Author Contributions:** For research articles with several authors, a short paragraph specifying their individual contributions must be provided. The following statements should be used "Conceptualization, K.R.-M., K.P.-T., and K.P.; methodology, K.R.-M. and K.P.-T.; software, K.Ż. and T.S.; validation, K.P. and K.R.-M.; formal analysis, K.R.-M., K.P.-T., N.D., and J.W.; investigation, K.R.-M., K.P.-T., N.D., and J.W.; resources, M.T.; data curation, K.Ż. and T.S.; writing—original draft preparation, K.R.-M.; writing—review and editing, K.P.-T., K.P., K.Ż., T.S., M.T., N.D., and J.W.; visualization, K.R.-M.; supervision, K.R.-M.; project administration, K.R.-M. and K.P.; funding acquisition, K.R.-M. and K.P. All authors have read and agreed to the published version of the manuscript.

**Funding:** This research was funded by statutory activity of the National Research Institute of Animal Production, grant number 04.-0.14.1 (cost of biological material collection and storage) and grant number 01-18-05-21 (cost of genetic analyses).

**Conflicts of Interest:** The authors declare no conflict of interest.

## References


1. Friedman, J.M.; Halaas, J.L. Leptin and the regulation of body weight in mammals. *Nature* **1998**, *395*, 763–770. [CrossRef]
2. Hussain, Z.; Khan, J.A. Food intake regulation by leptin: Mechanisms mediating gluconeogenesis and energy expenditure. *Asian Pac. J. Trop. Med.* **2017**, *10*, 940–944. [CrossRef] [PubMed]
3. Elmquist, J.K.; Elias, C.F.; Saper, C.B. From lesions to leptin: Hypothalamic control of food intake and body weight. *Neuron* **1999**, *22*, 221–232. [CrossRef]
4. Coelho, M.; Oliveira, T.; Fernandes, R. Biochemistry of adipose tissue: An endocrine organ. *Arch. Med. Sci.* **2013**, *9*, 191–200. [CrossRef] [PubMed]
5. Switonski, M.; Stachowiak, M.; Cieslak, J.; Bartz, M.; Grzes, M. Genetics of fat tissue accumulation in pigs: A comparative approach. *J. Appl. Genet.* **2010**, *51*, 153–168. [CrossRef]
6. Allison, D.B.; Neale, M.C.; Kezis, M.I.; Alfonso, V.C.; Heshka, S.; Heymsfield, S.B. Assortative mating for relative weight: Genetic implications. *Behav. Genet.* **1996**, *26*, 103–111. [CrossRef]
7. Ashrafi, K. Obesity and the Regulation of Fat Metabolism. *WormBook*. 2007. Available online: https://www.ncbi.nlm.nih.gov/books/NBK19757/ (accessed on 28 May 2020).
8. Ponsuksili, S.; Murani, E.; Brand, B.; Schwerin, M.; Wimmers, K. Integrating expression profiling and whole-genome association for dissection of fat traits in a porcine model. *J. Lipid Res.* **2011**, *52*, 668–678. [CrossRef]
9. Qiao, R.; Gao, J.; Zhang, Z.; Li, L.; Xie, X.; Fan, Y.; Cui, L.; Ma, J.; Ai, H.; Ren, J.; et al. Genome-wide association analyses reveal significant loci and strong candidate genes for growth and fatness traits in two pig populations. *Genet. Sel. Evol.* **2015**, *47*, 17. [CrossRef]
10. Guo, Y.; Huang, Y.; Hou, L.; Ma, J.; Chen, C.; Ai, H.; Huang, L.; Ren, J. Genome-wide detection of genetic markers associated with growth and fatness in four pig populations using four approaches. *Genet. Sel. Evol.* **2017**, *49*, 21. [CrossRef]
11. Timper, K.; Brüning, J.C. Hypothalamic circuits regulating appetite and energy homeostasis: Pathways to obesity. *DMM Dis. Model. Mech.* **2017**, *10*, 679–689. [CrossRef] [PubMed]
12. Hausman, D.B.; Hausman, G.J.; Martin, R.J. Influence of the pituitary on lipolysis and lipogenesis in fetal pig adipose tissue. *Horm. Metab. Res.* **1993**, *25*, 17–20. [CrossRef] [PubMed]
13. Pérez-Montarelo, D.; Madsen, O.; Alves, E.; Rodríguez, M.C.; Folch, J.M.; Noguera, J.L.; Groenen, M.A.M.; Fernández, A.I. Identification of genes regulating growth and fatness traits in pig through hypothalamic transcriptome analysis. *Physiol. Genom.* **2014**, *46*, 195–206. [CrossRef]
14. Shan, L.; Wu, Q.; Li, Y.; Shang, H.; Guo, K.; Wu, J.; Wei, H.; Zhao, J.; Yu, J.; Li, M.H. Transcriptome profiling identifies differentially expressed genes in postnatal developing pituitary gland of miniature pig. *DNA Res.* **2014**, *21*, 207–216. [CrossRef]
15. Piórkowska, K.; Żukowski, K.; Tyra, M.; Szyndler-Nędza, M.; Szulc, K.; Skrzypczak, E.; Ropka-Molik, K. The pituitary transcriptional response related to feed conversion in pigs. *Genes* **2019**, *10*, 712. [CrossRef]
16. Henry, B.A.; Clarke, I.J. Adipose Tissue Hormones and the Regulation of Food Intake. *J. Neuroendocrinol.* **2008**, *20*, 842–849. [CrossRef] [PubMed]
17. Havel, P.J. Role of adipose tissue in body-weight regulation: Mechanisms regulating leptin production and energy balance. *Proc. Nutr. Soc.* **2000**, *59*, 359–371. [CrossRef]
18. Xu, Y.; Qi, X.; Hu, M.; Lin, R.; Hou, Y.; Wang, Z.; Zhou, H.; Zhao, Y.; Luan, Y.; Zhao, S.; et al. Transcriptome analysis of adipose tissue indicates that the cAMP signaling pathway affects the feed efficiency of pigs. *Genes* **2018**, *9*, 336. [CrossRef]
19. Benítez, R.; Trakooljul, N.; Núñez, Y.; Isabel, B.; Murani, E.; De Mercado, E.; Gómez-Izquierdo, E.; García-Casco, J.; López-Bote, C.; Wimmers, K.; et al. Breed, diet, and interaction effects on adipose tissue transcriptome in iberian and duroc pigs fed different energy sources. *Genes* **2019**, *10*, 589. [CrossRef]

20. Albuquerque, A.; Óvilo, C.; Núñez, Y.; Benítez, R.; López-Garcia, A.; García, F.; Félix, M.D.R.; Laranjo, M.; Charneca, R.; Martins, J.M. Comparative transcriptomic analysis of subcutaneous adipose tissue from local pig breeds. *Genes* **2020**, *11*, 422. [CrossRef]
21. Xing, K.; Zhu, F.; Zhai, L.; Chen, S.; Tan, Z.; Sun, Y.; Hou, Z.; Wang, C. Identification of genes for controlling swine adipose deposition by integrating transcriptome, whole-genome resequencing, and quantitative trait loci data. *Sci. Rep.* **2016**, *6*, 1–10. [CrossRef] [PubMed]
22. Ropka-Molik, K.; Dusik, A.; Piórkowska, K.; Tyra, M.; Oczkowicz, M.; Szmatoła, T. Polymorphisms of the membrane-associated ring finger 4, ubiquitin protein ligase gene (MARCH4) and its relationship with porcine production traits. *Livest. Sci.* **2015**, *178*, 18–26. [CrossRef]
23. Tyra, M.; Żak, G. Analysis of the possibility of improving the indicators of pork quality through selection with particular consideration of intramuscular fat (imf) conntenttent*. *Ann. Anim. Sci.* **2013**, *13*, 33–44. [CrossRef]
24. Dodt, M.; Roehr, J.T.; Ahmed, R.; Dieterich, C. FLEXBAR-flexible barcode and adapter processing for next-generation sequencing platforms. *Biology* **2012**, *1*, 895–905. [CrossRef]
25. Li, B.; Dewey, C.N. RSEM: Accurate transcript quantification from RNA-Seq data with or without a reference genome. *BMC Bioinformatics* **2011**, *12*, 323. [CrossRef]
26. Love, M.I.; Huber, W.; Anders, S. Moderated estimation of fold change and dispersion for RNA-seq data with DESeq2. *Genome Biol.* **2014**, *15*. [CrossRef]
27. Andrews, S. FastQC A Quality Control tool for High Throughput Sequence Data 2010. Available online: https://www.bioinformatics.babraham.ac.uk/projects/fastqc/ (accessed on 28 May 2020).
28. Stocks, M.B.; Moxon, S.; Mapleson, D.; Woolfenden, H.C.; Mohorianu, I.; Folkes, L.; Schwach, F.; Dalmay, T.; Moulton, V. The UEA sRNA workbench: A suite of tools for analysing and visualizing next generation sequencing microRNA and small RNA datasets. *Bioinformatics* **2012**, *28*, 2059–2061. [CrossRef]
29. Griffiths-Jones, S.; Saini, H.; van Dongen, S.; Enright, A. miRBase: Tools for microRNA genomics. *Nucleic Acids Res.* **2008**, *36*, D154–D158. [CrossRef]
30. Griffiths-Jones, S. miRBase: MicroRNA sequences, targets and gene nomenclature. *Nucleic Acids Res.* **2006**, *34*, D140–D144. [CrossRef]
31. Urgese, G.; Paciello, G.; Acquaviva, A.; Ficarra, E. IsomiR-SEA: An RNA-Seq analysis tool for miRNAs/isomiRs expression level profiling and miRNA-mRNA interaction sites evaluation. *BMC Bioinformatics* **2016**, *17*, 148. [CrossRef]
32. Vlachos, I.S.; Paraskevopoulou, M.D.; Karagkouni, D.; Georgakilas, G.; Vergoulis, T.; Kanellos, I.; Anastasopoulos, I.-L.; Maniou, S.; Karathanou, K.; Kalfakakou, D. DIANA-TarBase v7.0: Indexing more than half a million experimentally supported miRNA:mRNA interactions. *Nucleic Acids Res.* **2015**, *43*, D153–D159. [CrossRef] [PubMed]
33. Vlachos, I.S.; Zagganas, K.; Paraskevopoulou, M.D.; Georgakilas, G.; Karagkouni, D.; Vergoulis, T.; Dalamagas, T.; Hatzigeorgiou, A.G. DIANA-miRPath v3.0: Deciphering microRNA function with experimental support. *Nucleic Acids Res.* **2015**, *43*, W460–W466. [CrossRef] [PubMed]
34. Pfaffl, M.W.; Tichopad, A.; Prgomet, C.; Neuvians, T.P. Determination of stable housekeeping genes, differentially regulated target genes and sample integrity: BestKeeper—Excel-based tool using pair-wise correlations. *Biotechnol. Lett.* **2004**, *26*, 509–515. [CrossRef] [PubMed]
35. Piórkowska, K.; Oczkowicz, M.; Rózycki, M.; Ropka-Molik, K.; Kajtoch, A.P.K. Novel porcine housekeeping genes for real-time RT-PCR experiments normalization in adipose tissue: Assessment of leptin mRNA quantity in different pig breeds. *Meat Sci.* **2011**, *87*, 191–195. [CrossRef] [PubMed]
36. Piórkowska, K.; Tyra, M.; Ropka-Molik, K.; Podbielska, A. Evolution of peroxisomal trans-2-enoyl-CoA reductase (PECR) as candidate gene for meat quality. *Livest. Sci.* **2017**, *201*, 85–91. [CrossRef]
37. Jacyno, E.; Pietruszka, A.; Kawecka, M.; Biel, W.; Kołodziej-Skalska, A. Phenotypic correlations of backfat thickness with meatiness traits, intramuscular fat, Longissimus muscle cholesterol and fatty acid composition in pigs. *South Afr. J. Anim. Sci.* **2015**, *45*, 122–128. [CrossRef]
38. Paton, C.M.; Ntambi, J.M. Biochemical and physiological function of stearoyl-CoA desaturase. *Am. J. Physiol. Endocrinol. Metab.* **2009**, *297*, E28. [CrossRef]
39. Hulver, M.W.; Berggren, J.R.; Carper, M.J.; Miyazaki, M.; Ntambi, J.M.; Hoffman, E.P.; Thyfault, J.P.; Stevens, R.; Dohm, G.L.; Houmard, J.A.; et al. Elevated stearoyl-CoA desaturase-1 expression in skeletal

muscle contributes to abnormal fatty acid partitioning in obese humans. *Cell Metab.* **2005**, *2*, 251–261. [CrossRef]
40. Flowers, M.T.; Ntambi, J.M. Role of stearoyl-coenzyme A desaturase in regulating lipid metabolism. *Curr. Opin. Lipidol.* **2008**, *19*, 248–256. [CrossRef]
41. Cohen, P.; Miyazaki, M.; Socci, N.D.; Hagge-Greenberg, A.; Liedtke, W.; Soukas, A.A.; Sharma, R.; Hudgins, L.C.; Ntambi, J.M.; Friedman, J.M. Role for stearoyl-CoA desaturase-1 in leptin-mediated weight loss. *Science* **2002**, *297*, 240–243. [CrossRef]
42. Yang, B.; Zhang, W.; Zhang, Z.; Fan, Y.; Xie, X.; Ai, H.; Ma, J.; Xiao, S.; Huang, L.; Ren, J. Genome-Wide Association Analyses for Fatty Acid Composition in Porcine Muscle and Abdominal Fat Tissues. *PLoS ONE* **2013**, *8*, e65554. [CrossRef] [PubMed]
43. Puig-Oliveras, A.; Revilla, M.; Castelló, A.; Fernández, A.I.; Folch, J.M.; Ballester, M. Expression-based GWAS identifies variants, gene interactions and key regulators affecting intramuscular fatty acid content and composition in porcine meat. *Sci. Rep.* **2016**, *6*, 31803. [CrossRef] [PubMed]
44. Matoušková, P.; Hanousková, B.; Skálová, L. Micrornas as potential regulators of glutathione peroxidases expression and their role in obesity and related pathologies. *Int. J. Mol. Sci.* **2018**, *19*, 1199. [CrossRef] [PubMed]
45. Jin, M.; Wu, Y.; Wang, J.; Chen, J.; Huang, Y.; Rao, J.; Feng, C. MicroRNA-24 promotes 3T3-L1 adipocyte differentiation by directly targeting the MAPK7 signaling. *Biochem. Biophys. Res. Commun.* **2016**, *474*, 76–82. [CrossRef] [PubMed]
46. Piórkowska, K.; Małopolska, M.; Ropka-Molik, K.; Nędza, M.S.; Wiechniak, A.; Żukowski, K.; Lambert, B.; Tyra, M. Evaluation of scd, acaca and fasn mutations: Effects on pork quality and other production traits in pigs selected based on RNA-seq results. *Animals* **2020**, *10*, 123. [CrossRef] [PubMed]
47. Pena, R.N.; Noguera, J.L.; García-Santana, M.J.; González, E.; Tejeda, J.F.; Ros-Freixedes, R.; Ibáñez-Escriche, N. Five genomic regions have a major impact on fat composition in Iberian pigs. *Sci. Rep.* **2019**, *9*, 1–9. [CrossRef]
48. Muñoz, M.; Alves, E.; Corominas, J.; Folch, J.M.; Casellas, J.; Noguera, J.L.; Silió, L.; Fernández, A.I. Survey of SSC12 regions affecting fatty acid composition of intramuscular fat using high-density SNP data. *Front. Genet.* **2012**, *2*, 101. [CrossRef]
49. Kim, S.-W.; Choi, Y.-I.; Choi, J.-S.; Kim, J.-J.; Choi, B.-H.; Kim, T.-H.; Kim, K.-S. Porcine Fatty Acid Synthase Gene Polymorphisms Are Associated with Meat Quality and Fatty Acid Composition. *Korean J. Food Sci. Anim. Resour.* **2011**, *31*, 356–365. [CrossRef]
50. Grzes, M.; Sadkowski, S.; Rzewuska, K.; Szydlowski, M.; Switonski, M. Pig fatness in relation to FASN and INSIG2 genes polymorphism and their transcript level. *Mol. Biol. Rep.* **2016**, *43*, 381–389. [CrossRef]
51. Brennan, A.M.; Mantzoros, C.S. Drug Insight: The role of leptin in human physiology and pathophysiology—Emerging clinical applications. *Nat. Clin. Pract. Endocrinol. Metab.* **2006**, *2*, 318–327. [CrossRef]
52. Lönnqvist, F.; Arner, P.; Nordfors, L.; Schalling, M. Overexpression of the obese (ob) gene in adipose tissue of human obese subjects. *Nat. Med.* **1995**, *1*, 950–953. [CrossRef] [PubMed]
53. Aslam, M. Leptin: Fights against obesity! *Pak. J. Physiol.* **2006**, *2*, 54–60.
54. Hill, J.O.; Wyatt, H.R.; Peters, J.C. Energy balance and obesity. *Circulation* **2012**, *126*, 126–132. [CrossRef] [PubMed]
55. Ridderstråle, M.; Johansson, L.E.; Rastam, L.; Lindblad, U. Increased risk of obesity associated with the variant allele of the PPARGC1A Gly482Ser polymorphism in physically inactive elderly men. *Diabetologia* **2006**, *49*, 496–500. [CrossRef]
56. Liu, T.; Zhao, Y.; Tang, N.; Feng, R.; Yang, X.; Lu, N.; Wen, J.; Li, L. Pax6 Directly Down-Regulates Pcsk1n Expression Thereby Regulating PC1/3 Dependent Proinsulin Processing. *PLoS ONE* **2012**, *7*, e46934. [CrossRef]
57. Shi, M.A.; Shi, G.P. Different roles of mast cells in obesity and diabetes: Lessons from experimental animals and humans. *Front. Immunol.* **2012**, *3*, 7. [CrossRef]
58. Finlin, B.S.; Confides, A.L.; Zhu, B.; Boulanger, M.C.; Memetimin, H.; Taylor, K.W.; Johnson, Z.R.; Westgate, P.M.; Dupont-Versteegden, E.E.; Kern, P.A. Adipose Tissue Mast Cells Promote Human Adipose Beiging in Response to Cold. *Sci. Rep.* **2019**, *9*, 8658. [CrossRef]
59. Ruiz-Ojeda, F.J.; Méndez-Gutiérrez, A.; Aguilera, C.M.; Plaza-Díaz, J. Extracellular matrix remodeling of adipose tissue in obesity and metabolic diseases. *Int. J. Mol. Sci.* **2019**, *20*, 4888. [CrossRef]

60. Lin, D.; Chun, T.H.; Kang, L. Adipose extracellular matrix remodelling in obesity and insulin resistance. *Biochem. Pharmacol.* **2016**, *119*, 8–16. [CrossRef]
61. Huang, Y.; Yan, Y.; Xv, W.; Qian, G.; Li, C.; Zou, H.; Li, Y. A New Insight into the Roles of MiRNAs in Metabolic Syndrome. *BioMed Res. Int.* **2018**. [CrossRef]
62. Heneghan, H.M.; Miller, N.; McAnena, O.J.; O'brien, T.; Kerin, M.J. Differential miRNA expression in omental adipose tissue and in the circulation of obese patients identifies novel metabolic biomarkers. *J. Clin. Endocrinol. Metab.* **2011**, *96*, E846–E850. [CrossRef] [PubMed]
63. Divoux, A.; Tordjman, J.; Le Lacasa, D.; Veyrie, N.; Hugol, D.; Aissat, A.; Basdevant, A.; Le Guerre-Millo, M.; Poitou, C.; Zucker, J.-D.; et al. Fibrosis in Human Adipose Tissue: Composition, Distribution, and Link With Lipid Metabolism and Fat Mass Loss. *Am. Diabetes Assoc.* **2010**, *59*, 2817–2825. [CrossRef] [PubMed]
64. Pellegrinelli, V.; Heuvingh, J.; Du Roure, O.; Rouault, C.; Devulder, A.; Klein, C.; Lacasa, M.; Clément, E.; Lacasa, D.; Clément, K. Human adipocyte function is impacted by mechanical cues. *J. Pathol.* **2014**, *233*, 183–195. [CrossRef] [PubMed]
65. Rogero, M.M.; Calder, P.C. Obesity, Inflammation, Toll-Like Receptor 4 and Fatty Acids. *Nutrients* **2018**, *10*, 432. [CrossRef]
66. Ahmad, R.; Al-Mass, A.; Atizado, V.; Al-Hubail, A.; Al-Ghimlas, F.; Al-Arouj, M.; Bennakhi, A.; Dermime, S.; Behbehani, K. Elevated expression of the toll like receptors 2 and 4 in obese individuals: Its significance for obesity-induced inflammation. *J. Inflamm.* **2012**, *9*, 48. [CrossRef]
67. Song, M.J.; Kim, K.H.; Yoon, J.M.; Kim, J.B. Activation of Toll-like receptor 4 is associated with insulin resistance in adipocytes. *Biochem. Biophys. Res. Commun.* **2006**, *346*, 739–745. [CrossRef]
68. Deiuliis, J. MicroRNAs as regulators of metabolic disease: Pathophysiologic significance and emerging role as biomarkers and therapeutics. *Int. J. Obes. Vol.* **2016**, *40*, 88–101. [CrossRef]
69. Matsuo, Y.; Tanaka, M.; Yamakage, H.; Metabolism, Y.S.; Satoh-Asahara, N. Thrombospondin 1 as a novel biological marker of obesity and metabolic syndrome. *Metabolism* **2015**, *64*, 1490–1499. [CrossRef]

*Article*

# Genome-Wide Transcriptome and Metabolome Analyses Provide Novel Insights and Suggest a Sex-Specific Response to Heat Stress in Pigs

**Krishnamoorthy Srikanth [1,†], Jong-Eun Park [1,†], Sang Yun Ji [2], Ki Hyun Kim [2], Yoo Kyung Lee [2], Himansu Kumar [1], Minji Kim [2], Youl Chang Baek [2], Hana Kim [1], Gul-Won Jang [1], Bong-Hwan Choi [1] and Sung Dae Lee [2,*]**

1 Amimal Genomics and Bioinformatics Division, National Institute of Animal Science, RDA, Wanju 55365, Korea; kris87@korea.kr (K.S.); jepark0105@korea.kr (J.-E.P.); himansu@korea.kr (H.K.); hanakim0307@gmail.com (H.K.); kwchang@korea.kr (G.-W.J.); bhchoi@korea.kr (B.-H.C.)

2 Animal Nutrition and Physiology Team, National Institute of Animal Science, RDA, Wanju 55365, Korea; syjee@korea.kr (S.Y.J.); kihyun@korea.kr (K.H.K.); yoo3930@korea.kr (Y.K.L.); mjkim00@korea.kr (M.K.); chang4747@korea.kr (Y.C.B.)

* Correspondence: leesd@korea.kr; Tel.: +82-63-238-7454; Fax: +82-63-238-7497

† These authors contributed equally.

Received: 5 March 2020; Accepted: 6 May 2020; Published: 11 May 2020

**Abstract:** Heat stress (HS) negatively impacts pig production and swine health. Therefore, to understand the genetic and metabolic responses of pigs to HS, we used RNA-Seq and high resolution magic angle spinning (HR-MAS) NMR analyses to compare the transcriptomes and metabolomes of Duroc pigs ($n = 6$, 3 barrows and 3 gilts) exposed to heat stress (33 °C and 60% RH) with a control group (25 °C and 60% RH). HS resulted in the differential expression of 552 (236 up, 316 down) and 879 (540 up, 339 down) genes and significant enrichment of 30 and 31 plasma metabolites in female and male pigs, respectively. Apoptosis, response to heat, Toll-like receptor signaling and oxidative stress were enriched among the up-regulated genes, while negative regulation of the immune response, ATP synthesis and the ribosomal pathway were enriched among down-regulated genes. Twelve and ten metabolic pathways were found to be enriched (among them, four metabolic pathways, including arginine and proline metabolism, and three metabolic pathways, including pantothenate and CoA biosynthesis), overlapping between the transcriptome and metabolome analyses in the female and male group respectively. The limited overlap between pathways enriched with differentially expressed genes and enriched plasma metabolites between the sexes suggests a sex-specific response to HS in pigs.

**Keywords:** heat stress; pig; Duroc; RNA-Seq; NMR; metabolome

---

## 1. Introduction

Heat stress affects animal husbandry worldwide and is a major environmental factor that affects animal health and production [1,2]. The increasing environmental temperature due to global warming will particularly affect pigs due to their lack of functional sweat glands, which would have helped in endogenous heat dissipation [3,4]. Added to this, the thick subcutaneous adipose tissue in the pig, impedes effective radiant heat loss [5]. Moreover the increase in metabolic rate due to rapid lean tissue accretion increases endogenous heat production, exacerbating the innate inability of porcine animals to tolerate heat [3,6–9]. The estimated annual economic loss due to heat stress for the swine industry in the US alone is nearly US $300 to $450 million [5,10,11], mainly due to increased mortality and morbidity, altered carcass composition, decreased feed efficiency, inconsistent growth, reduced fecundity and poor sow performance [6], so heat stress in a pressing issue for worldwide pig production.

Quiniou and Noblet [12] have shown that the growth rate is normally between 18 and 25 °C and that the thermoregulatory response in pigs is activated at about 25 °C. Animals respond to heat stress by regulating physiological and metabolic changes, such as redistribution of blood flow from the body core to the periphery, and by reducing feed intake [13] to reduce metabolic heat production [14]. This has a significant effect on animal health and productivity [3]. The heat stress mitigation options available at present include expensive heat abatement processes, such as spray or floor cooling and nutritional supplementation strategies. Since the response to heat stress varies within a population, due to genetic variations, selection for thermal tolerance could be a better alternative [1,15]. However traditional breeding programs have limitations due to the difficulties in collecting precise records of thermal states of individual animals. To overcome this, production traits that are correlated with heat stress are used for selecting animals with higher thermal tolerance [16]. They, however, mask the actual effects of heat stress; therefore, a precise and efficient method for identifying genetic tolerance to heat stress is necessary. Identifying genetic and metabolic markers that are sensitive to heat stress will be the most important step for managing heat stress in pigs. These markers could also be putative candidates for identifying genomic variants for improving heat tolerance in pigs.

Previous studies have shown that heat stress triggers a complex array of gene expression and metabolic changes in livestock, resulting in the modulation of a wide variety of pathways involved in the heat stress response, the inflammatory response, DNA damage repair, chaperone use, etc. [5,17–19]. Similarly, a variety of metabolites involved in carbohydrate, amino acid, fatty acid and amine metabolism are modulated, including glucose, lactate, alanine, cysteine, isoleucine, etc. [20–23]. However, these studies did not involve integrated transcriptome and metabolome profiling, which could lead to better understanding of the effect of heat stress on swine physiology, gene expression and metabolism, which will help in designing and implementing innovative strategies to limit the economic losses due to heat stress on pig farm profitability. Therefore, in this study, we exposed 3-month-old pigs to heat stress and performed transcriptome and metabolome analysis using RNA-Seq and high resolution magic angle spinning (HR-MAS) NMR (nuclear magnetic resonance) to understand the gene expression and metabolome perturbations in response to heat stress.

## 2. Materials and Methods

### *2.1. Animals*

The experiments were performed following the ethical guidelines laid down by the Institutional Animal Care and Use Committee (IACUC) of the National Institute of Animal Science (NIAS). The experimental procedures were reviewed and approved by IACUC of NIAS (NO 2017-1070). A total of 6 (3 barrows and 3 gilts) unrelated, 3-month-old Duroc pigs, weighing 56 ± 2.62 Kg, were used in this study. Animals were fed a corn and soybean meal based diet (Table 1), containing 18% crude protein, 4.9% crude fat, 4.6% crude fiber and 4.4% crude ash; the whole diet was formulated to provide 3450 Kcal/Kg. The animals had ad libitum access to feed and water. Water and feed intake were measured (Table S1). One-way analysis of variance (ANOVA) was performed in R to test for significance. Differences were considered to be significant at $p < 0.05$. Other than the typical heat stress responses, no significant differences in behavior were noted.

**Table 1.** Composition of the total mixed ration (TMR) used during the experimental period.

| Raw Material | Percentage |
|---|---|
| Corn | 55.8 |
| Soybean meal product | 24.4 |
| Wheat Bran | 9 |
| Soybean Hull | 3 |
| Molasses | 3 |
| Soybean Oil | 2 |
| Limestone | 1.1 |
| Lysine | 0.4 |
| Salt | 0.4 |
| Globik SW | 0.3 |
| TCP | 0.4 |
| Methionine-50 | 0.2 |
| **Nutrient** | |
| Calcium | 0.63 |
| Total Phosphorus | 0.5 |
| Crude Protein | 18 |
| Crude Fat | 4.9 |
| Crude Fiber | 4.6 |
| Crude Ash | 4.4 |
| DRY MATTER | 87.5 |
| Arginine | 1.16 |
| Lysine | 1.37 |
| Methionine + Cysteine | 0.7 |
| D.Energy | 3450 Kcal/kg |

### 2.2. Heat Stress Experimental Setup

The heat stress (HS) experiment was carried out in an environmental chamber which had controls for temperature and humidity in NIAS. After acclimation to the chamber for 3 days, the experiments were performed. The animals were first kept at 25 °C and 60% relative humidity for 24 h, and at the end, 10 mL of blood was drawn via venipuncture and 5 mL of blood was stored in a Tempus Blood RNA tube (Life Technlogies, Carlsbad, CA, USA) for RNAseq analysis and the rest of the blood was stored in a vacutainer tube containing anticoagulants (EDTAK2), for the metabolomic analysis. The same animals were then exposed to 33 °C temperature and 60% relative humidity for 24 h and blood was drawn and stored in the same way.

### 2.3. RNA Isolation and Sequencing

The RNA was isolated using TRIzol Reagent (Invitrogen, Carlsbad, CA, USA) following the manufacturer's guidelines. The integrity and quality of the RNA was assessed using 2100 Bioanalyzer and RNA 6000 Nano LabChip Kit (Agilent Technologies, Palo Alto, CA, USA). Only samples with RIN values >8 were used for library construction. The concentration of RNA was determined using NanoDrop ND-1000 spectrophotometer (NanoDrop Technologies, Wilmington, DE, USA). The sequencing library was constructed using Illumina TruSeq RNA sample preparation kit (Illumina, San Diego, CA, USA) following processes previously described [16]. The sequencing was performed on an Illumina HiSeq 2000 sequencer. The raw reads are available for download from sequence read archive (SRA), NCBI under the accession number SUB7043048.

### 2.4. Sample Preperation and Metabolite Analysis with $^{1}$H-NMR

36 μL of Plasma was mixed with 4 μL of $D_2O$ (20 mM TSP-d4), and the sample was then transferred to a 4 mm NMR nanotub, and $^{1}$H-NMR spectra were obtained by high-resolution magic-angle-spinning (HR-MAS) NMR spectra using an Agilent 600 MHz NMR spectrometer (Agilent Technologies, Palo Alto,

CA, USA) with a 4 mm gHX Nanoprobe. A Carr–Purcell–Meiboom–Gill (CPMG) pulse sequence was used to reduce signals generated by macromolecules and water. The $^{1}$H-NMR spectra were measured as described previously [24]. The TSP-$d_4$ peak at 0.0 ppm was used as a reference for calibrating chemical shifts.

### *2.5. Data Analysis*

#### 2.5.1. RNA-Seq Data

The processing of the data was previously described [25]. Briefly, after checking the quality of the raw reads using FastQC (version 0.11.5) [26] and trimming the adaptors and low-quality bases using TRIMMOMATIC (version 0.36) [27], the reads were aligned to the Pig reference genome (*Sus scrofa* 11.1) using HiSAT2 (version 2.05) [28]. The aligned reads were then counted using FeatureCounts (version 1.5.0) [29]. After correcting for batch and unknown effects using Svaseq [30], differential expression analysis was performed using DESeq2 [31]. Significant genes (FDR < 0.1) were identified, and functional enrichment analysis based on Gene Ontology (GO) under Biological Process and Molecular Function was performed with DAVID [32]. KEGG (Kyoto Encylopedia of Genes and Genomes (KEGG) pathway enrichment analysis was performed using ClueGO plugin [33] in cytoscape version 3.7.2 [34]. Goseq [35] was used for metabolic pathway enrichment analysis.

#### 2.5.2. Metabolome Data

The metabolite quantification was performed using Chenomx NMR suite 7.1 (Chenomx, Edmonton, Canada). The generated spectra were binned with a binning size of 0.001 ppm, and normalized to the total area, and binned data were aligned with the icoshift algorithm of MATLAB $R^2$013b (MathWorks, Natick, MA, USA). Principle component analysis (PCA) and statistical analyses were performed using MetaboAnalyst 4.0 [36]. Only features that were detected in at least 50% of samples were used.

#### 2.5.3. Pathway Enrichment Analysis of Differentially-Enriched Metabolites

Significantly differentially-enriched metabolites ($p < 0.05$) were subjected to KEGG pathway analyses using the Pathway analysis module in MetaboAnalyst 4.0 [36]. A combination of quantitative enrichment and topology analysis using only curated metabolic pathways from the KEGG database was used in the analyses.

### *2.6. Real-Time PCR Validation*

Real time reverse transcriptase PCR (qRT-PCR) was performed using gene-specific primes (Supplementary File S1). The PCR was performed on an ABI 7500 Real Time PCR system using Fast SYBR green master mix (Applied Biosystems, Foster City, CA, USA). A total of 18 genes (9 each from male and female differentially expressed gene (DEG) set) were analyzed. β-actin and GADPH (Glyceraldehyde-3-phosphate dehydrogenase) were used as endogenous controls. The stability of expression for each of those two genes was checked using GeNORM (https://genorm.cmgg.be/) against the same concentration of RNA from different samples. β-actin was the most stable and was used for normalizing the expression data of the target genes.

## 3. Results

### *3.1. Transcriptome Alignment, Mapping and Principle Component Analysis*

We investigated the impact of heat stress on Duroc pigs using high throughput RNA-Seq analysis. A total of 423.3 million 100 bp Paired-End (PE) reads corresponding to an average of 35.2 million reads per individual was generated. After trimming for adapters and low-quality reads, 410.5 million reads remained. The reads were mapped to the *Sus scrofa* reference genome at an average alignment rate of 96.8% (File S2). The reads were mapped to a total of 19,283 genes. Principal component analysis (PCA)

(Figure S1) suggested that sex had a large effect on transcriptome difference between the groups, so we decided to compare the heat stress effect on male and female pigs separately. PCA showed that 30% and 36% of the expression variation was due to heat stress in female and male pigs respectively (Figure 1a,c); however, considerable within group variation was also observed, confirming previous reports that the heat stress response varies within populations due to underlying genomic variation [1,15].

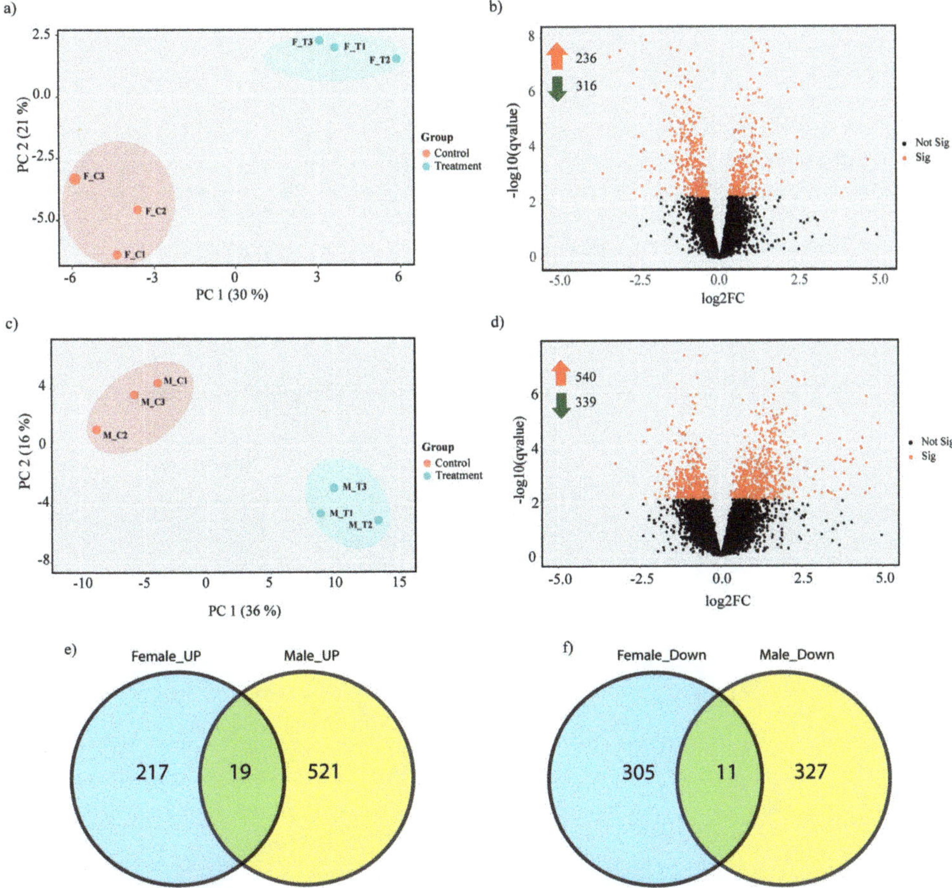


**Figure 1.** Summary of the transcriptome analysis: (**a**) PCA of female group samples; (**b**) volcano plot showing the number and distribution of significantly differentially expressed genes in the female group; (**c**) PCA of male samples; (**d**) volcano plot showing the number and distribution of significantly differentially expressed genes in the male group; Venn diagrams showing differentially expressed genes DEGs common between male and female groups: (**e**) up-regulated and (**f**) down-regulated.

### *3.2. Porcine Transcriptome Response to Heat Stress*

Heat stress resulted in 552 and 879 genes being significantly (FDR < 0.1) differentially expressed in male and female groups respectively. Out of those, 236 and 540 genes were up-regulated and 316 and 339 genes were down-regulated in female and male pigs respectively (Figure 1b,d, File S3). The genes modulated in response to heat stress were considerably different between the male and female groups, with only 19 up-regulated and 11 down-regulated genes being common between the groups (Figure 1e,f). Among the top up-regulated DEGs in the female group were *HIST2H2AA4* (histone cluster 2H2A family member a4), *MYH4* (myosin heavy chain 4), *ABCA6* (ATP binding cassette subfamily A member 6), *HSF4* (heat shock factor 4) and *IL18* (interleukin 18), while top down-regulated DEGs included *CXCL10* (C-X-C motif chemokine 10), *IDO1* (indoleamine 2,3-dioxygenase 1), *HES4* (Hes family BHLH transcription factor 4) and *IFI6* (γ-interferon-inducible protein Ifi-16); in the male group the up-regulated genes included *DES* (Desmin), *RAG1* (recombination activating 1), *LSAMP* (limbic system associated membrane protein), *CD1E* (CD1e molecule), *HSPG2* (heparin sulfate proteoglycan 2) and down-regulated genes included *CEBPE*(CCAAT/enhancer binding protein), *ZNF316* (Zinc Finger Protein 316), *PXDC1*(PX domain containing 1) and *COX4I1* (cytochrome c oxidase subunit 4I 1).

Gene Ontology (GO) enrichment analysis of the significantly DEGs (FDR < 0.1) in the female group (Figure 2a) showed that among the up-regulated DEGs, "angiogenesis," "apoptosis," "response to heat," " heme biosynthetic process" and "ATPase activity" were enriched, whereas "negative regulation of innate immune response," "positive regulation of signal transduction," "response to cytokine," "MAPK cascade," "response to lipopolysaccharide," "threonine-type endopeptidase activity," etc., were enriched amongst the down-regulated genes. KEGG (Kyoto Encyclopedia of Genes and Genomes) pathway enrichment analysis of the female group DEGs (Figure 3a) showed that "apoptosis," "non-small cell lung cancer," "Toll-like receptor signaling pathway," "necroptosis," "NOD-like receptor signaling pathway" and "Influenza A" pathways were enriched.

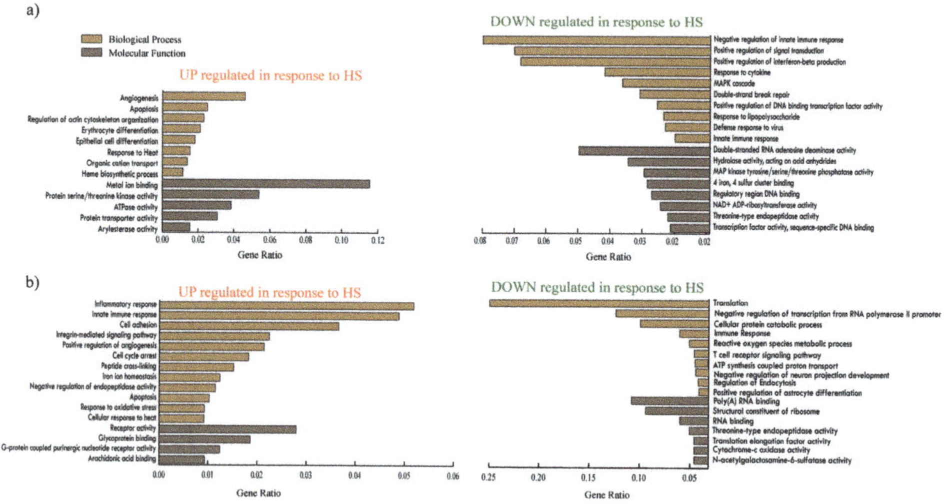


**Figure 2.** Gene Ontology analysis: (**a**) enriched GOs in female group; (**b**) enriched GOs in male group. The GO enrichment was performed under Biological Process and Molecular Functions categories.

In the male group, heat stress resulted in the up-regulation of genes functioning in "inflammatory response," "innate immune response," "cell adhesion," "integrin-mediated signaling pathway," "cell cycle arrest," "positive regulation of angiogenesis," "apoptosis," "response to oxidative stress," "cellular response to heat" and "glycoprotein binding." Genes functioning in the "cellular protein catabolic process," "regulation of endocytosis," "negative regulation of astrocyte differentiation," "cytochrome-c oxidase activity" and "N-acetylgalactosamine-6-sulfatase activity" were down-regulated (Figure 2b). Among KEGG pathways (Figure 3b), genes part of "complement and coagulation cascades," "platelet activation," "fc γ r-mediated phagocytosis," "thyroid hormone signaling," "pathways in cancer," "proteoglycans in cancer," and "Notch signaling" pathways were enriched. The top 10 DEGs from the female and male group were validated using q-RT PR analysis (quantitative Real-time PCR) (Figure 4); the correlation between RNA-Seq and q-RT PCR analyses was 0.861.

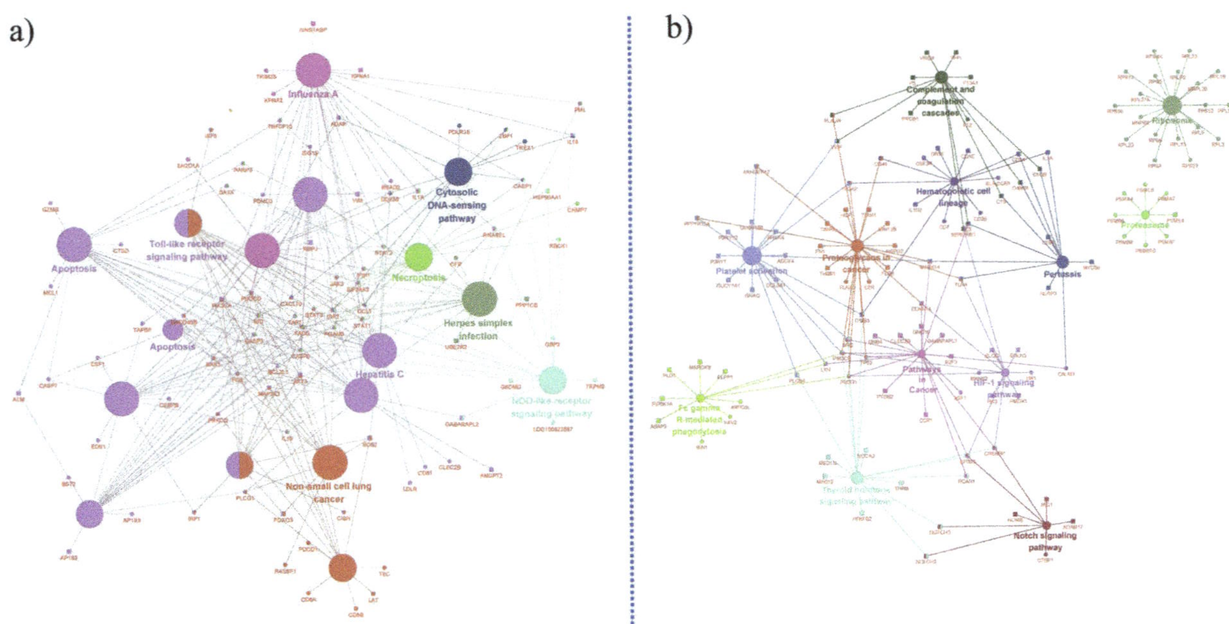


**Figure 3.** KEGG pathway enrichment analysis: (**a**) enriched KEGG pathways in female group; (**b**) enriched KEGG pathways in male group. Nodes are genes and edges represent pathways. Up-regulated genes are represented as rectangles and down-regulated genes are represented as hexagons.

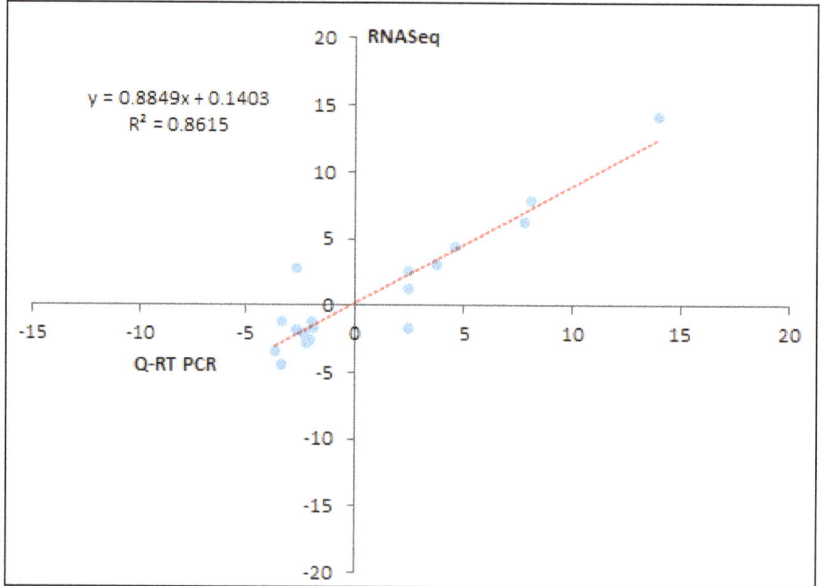


**Figure 4.** qRT-PCR validation of RNA-Seq results, using the top 10 DEGs from male and female groups respectively. The fold change from q-RT PCR was plotted on the *x*-axis, and the $\log_2$ fold change from the RNA-Seq analysis was plotted on the *y*-axis; and the correlation between the two methods is given in the figure.

*3.3. Porcine Metabolome Response to Heat Stress*

We detected and identified a total of 46 metabolites (Figure 5b,d, File S4). After normalizing the sample (quantile normalization) and scaling the data (auto scaling) in MetaboAnalyst 4.0 [36], PCA and a heatmap were generated. The overall metabolic profiles of the groups resolved well in PC1 (Figure 5a,c), and hierarchical clustering based on metabolite concentration showed that the control and heat stress group were well separated. However, PC2 showed that there was considerable within-group variation in both male and female pigs. The fold change was calculated as a ratio between the two groups' means. A Wilcoxon rank-sum test was performed, and 30 and 31 metabolites were found to significantly differ ($p < 0.05$) between the control and heat stressed groups in female and male groups respectively (File S4).

In the female group, out of the 30 metabolites that significantly differed between the heat stress and control groups, the concentrations of 21 metabolites increased while nine metabolites decreased in concentration. The concentrations of tryptophan, arginine, pyruvate, alanine, acetate and lactate were higher, while DL-plus *allo*-δ-hydroxylysine, creatine, urea, 3-aminoisobutric acid, 4-aminobutyric acid and 3-methylhistidine were lower in the heat stress group (File S4). In the male group samples, 18 of the 31 significantly differing metabolites increased, among which where aspartic acid, pyruvate, betaine, creatine and lactate, while 13 metabolites, including tryptophan, carnosine, valine, alanine and creatine, were significantly reduced in the heat stress group compared to the control group (File S4).

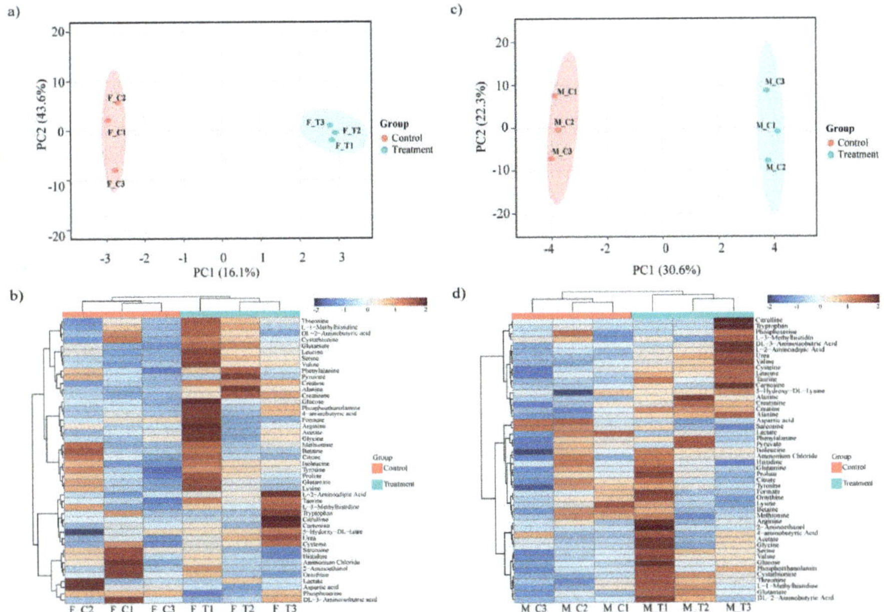


**Figure 5.** Metabolome analysis: (**a**) metabolic profiles of the female group visualized with principal component analysis; (**b**) heat map of all metabolites identified in the female group; (**c**) metabolic profiles of the male group visualized with principal component analysis; (**d**) heat map of all metabolites identified in the male group.

### *3.4. Metabolic Pathways Enriched in Response to Heat Stress*

Pathway analysis was performed with metabolites that significantly differed between the heat stress and control group using the pathway enrichment module in MetaboAnalyst 4.0 [36]. Twelve pathways were significantly ($p < 0.05$) enriched in the female group (Figure 6a), including "arginine and proline metabolism," "glycine, serine and threonine metabolism," "cysteine and methionine metabolism," "alanine, aspartate and glutamate metabolism" and "histidine metabolism". We used goseq [35] to identify metabolic pathways in KEGG database that were significantly altered in the transcriptome, in response to heat stress. Among the twelve pathways significantly enriched (Figure 6b), four pathways were found to be commonly enriched with DEGs and DEMs (differentially enriched metabolites) (Figure 6c); these were "arginine and proline metabolism," "glutathione metabolism," "selenoamino acid metabolism" and "histidine metabolism." Out of these, the first three pathways have been found to be enriched in response to oxidative stress induced by chronic HS in the pig [37–39].

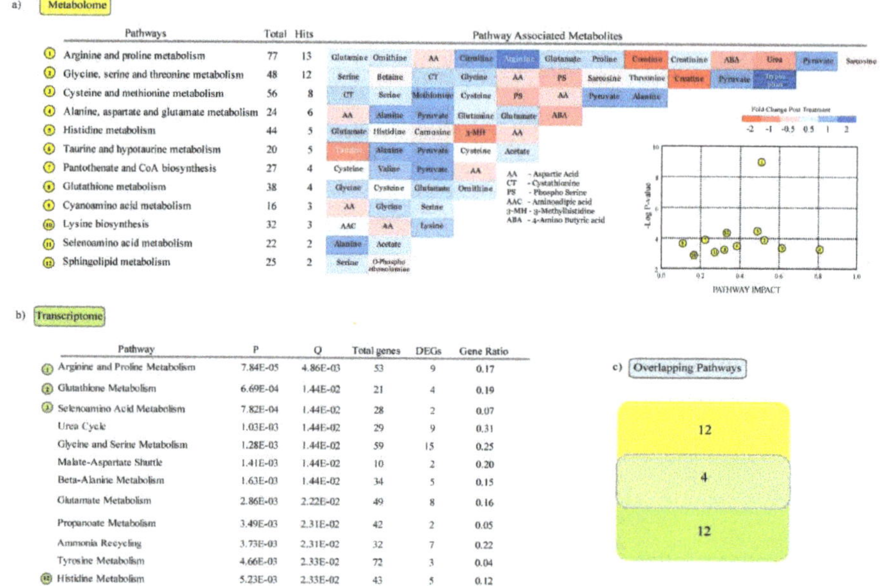


**Figure 6.** Metabolic pathways altered in the female group: (**a**) Metabolic pathway enrichment analyses of metabolites enriched in response to heat stress. (**b**) Metabolic pathway enrichment analyses of genes differentially expressed in response to heat stress. (**c**) Number of overlapping pathways enriched with DEGs and significantly differing metabolites in the heat stress group.

Among the male animals, 10 metabolic pathways were enriched with differently enriched metabolites, including "arginine and proline metabolism," "glycine, serine and threonine metabolism," "cysteine and methionine metabolism," "taurine and hypotaurine metabolism" and "histidine metabolism" (Figure 7a). The metabolic pathways enriched with DEGs (Figure 7b) included "inositol phosphate metabolism," "amino sugar and nucleotide sugar metabolism," "pantothenate and COA biosynthesis" and "riboflavin metabolism." Comparative analysis of the metabolic pathways enriched with DEMs and DEGs showed that three metabolic pathways were common (Figure 7c); those were "pantothenate and CoA biosynthesis," "sulfur metabolism" and "arginine proline metabolism." Pantothenate and CoA biosynthesis was previously found to be enriched in response to HS in the pig [5,20].

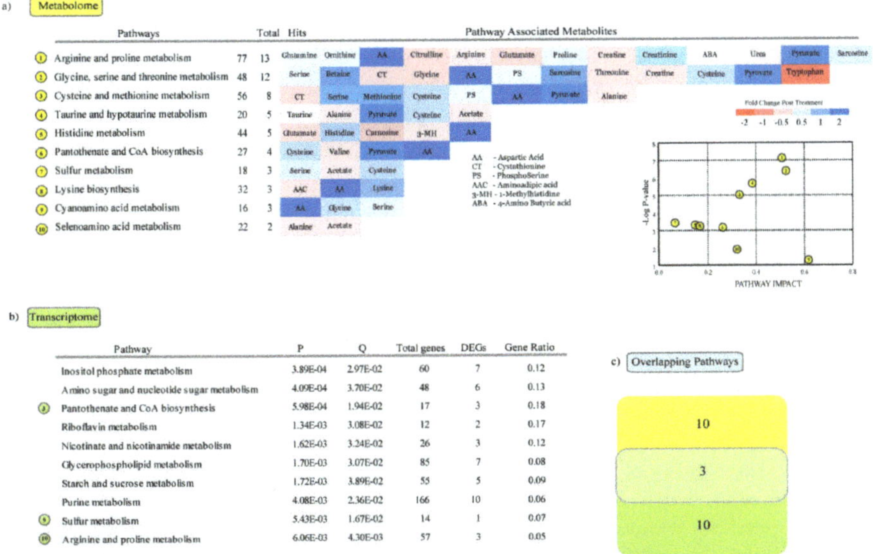


| Pathway | P | Q | Total genes | DEGs | Gene Ratio |
|---|---|---|---|---|---|
| Inositol phosphate metabolism | 3.89E-04 | 2.97E-02 | 60 | 7 | 0.12 |
| Amino sugar and nucleotide sugar metabolism | 4.09E-04 | 3.70E-02 | 48 | 6 | 0.13 |
| Pantothenate and CoA biosynthesis | 5.98E-04 | 1.94E-02 | 17 | 3 | 0.18 |
| Riboflavin metabolism | 1.34E-03 | 3.08E-02 | 12 | 2 | 0.17 |
| Nicotinate and nicotinamide metabolism | 1.62E-03 | 3.24E-02 | 26 | 3 | 0.12 |
| Glycerophospholipid metabolism | 1.70E-03 | 3.07E-02 | 85 | 7 | 0.08 |
| Starch and sucrose metabolism | 1.72E-03 | 3.89E-02 | 55 | 5 | 0.09 |
| Purine metabolism | 4.08E-03 | 2.36E-02 | 166 | 10 | 0.06 |
| Sulfur metabolism | 5.43E-03 | 1.67E-02 | 14 | 1 | 0.07 |
| Arginine and proline metabolism | 6.06E-03 | 4.30E-03 | 57 | 3 | 0.05 |

**Figure 7.** Metabolic pathways altered in the male group: (**a**) Metabolic pathway enrichment analyses of metabolites enriched in response to heat stress. (**b**) Metabolic pathway enrichment analyses of genes differentially expressed in response to heat stress. (**c**) Number of overlapping pathways enriched with DEGs and significantly differing metabolites in the heat stress group.

## 4. Discussion

Heat stress occurs when the rate of heat accumulation exceeds the rate of heat loss; i.e., when the animal is removed from the thermal comfort zone. HS causes the core body temperature of a pig to rise [40]. Temperature increase due to global warming will be detrimental to swine health and production. The global temperature has already been raising at an average of 0.13 °C over the past 50 years [41]. Pigs are particularly vulnerable due to their lack of functional sweat glands, and the thick layer of subcutaneous adipose tissue that they possess which impedes effective radiant heat loss [42]. Heat stress triggers an adaptive stress response by regulating a complex array of genes and metabolites [16,18,21]. In this study we found significant differences in water and feed intake post heat stress, and a significant difference in water intake post heat stress treatment between male and female pigs (Table S1).

Several studies have suggested that gender might have an effect on the heat stress response in pigs [40,43–45]. The sex-specific response to heat stress could be due to the difference in surface area to mass ratio or due to body compositional difference (lean meat to fat mass) which affects radiant heat loss [43]. Barrows are reported to generally have greater backfat thickness than gilts [46]. Though barrows consume more feed and gain body weight more rapidly, gilts deposit proportionally more muscle and less fat in their carcass [43,47]. This sex-specific difference in body composition could significantly alter the heat stress response, as it affects the ability of the animal to maintain the core body temperature through radiant heat loss. Moreover, several hormones, such as prostaglandins [48], endogenous opioids [48] and glucocorticoids have roles in the thermoregulatory mechanism [49]. Cortisol, a glucocorticoid, is the mostly widely used biomarker for detecting stress [50,51]. The concentration of cortisol is 15% higher in barrows than in gilts [52]. HS-induced increases in the concentrations of catecholamines and glucocorticoids have been observed in many species [53,54]; the effect of HS on immune cells is suggested to be dependent upon these hormones [55–57]. Similarly, Rudolph et al. [40], found that oxidative stress did not result in muscle injury in barrows, while previous studies in gilts reported that oxidative stress causes muscle injury [58–61]. This led them to suggest that muscle response to heat stress could be partially dependent upon sex of the pig, and that muscle tissues in males could be more resistant to heat stress than in females. The result of our study concurs with all these reports; our study shows that HS in pigs, as evidenced from transcriptome expression and metabolome concentration, is sexually dimorphic.

*4.1. Transcriptome Regulation in Response to Heat Stress*

Transcriptome analysis showed that heat stress caused a profound shift in gene expression in the male group: 879 genes were found to be significantly differentially regulated in the male group, while 552 genes were DEs (differentially expressed) in the female group. Among the DEGs were several genes involved in Heat shock response. Heat shock proteins (HSPs) are expressed upon encountering stress [62,63]. Heat is proteotoxic and causes proteins to denature [64], HSPs act as chaperones in assisting in protein folding, thereby helping in avoiding protein aggregation [65] and maintaining cellular homeostasis [66]. *HSPA5* (heat shock 70 kDa protein 5), *HSP90AA1* (heat shock protein 90 α family class a member 1) and *HSF4* (heat shock transcription factor 4) were up-regulated, in both male and female groups, while *FKBP5* (FKBP prolyl isomerase 5) was up-regulated only in the male group. *HSPA5* is a member of the HSP70 (heat shock protein 70) family of genes; it serves as an endoplasmic reticulum chaperone and as a sensor of protein misfolding [67]. *HSP90AA1* a member of the HSP90 family takes part in several cellular functions such as regulating protein activity, transport and also in activating different signaling pathways by forming complexes with steroid receptors [68]. *HSP90AA1* is essential for normal spermatogenesis in pigs [69], and regulation of *HSP90AA1* protects cells from heat shock [70]. *FKBP5* a member of immunophilin family, is a glucocorticoid receptor (GR)-regulating co-chaperone of heat shock protein 90, their expression are regulated by progestins and glucocorticoids [71], and are significantly correlated with plasma cortisol concentration in pig [72,73], suggesting that the male pigs were under considerable stress.

Several co-chaperones and genes involved in cellular response to heat stress were also DEs; these included *HMOX1* (heme oxygenase 1), *TRPM2* (transient receptor potential melastatin 2), which were up-regulated in the male group, *IL1A* (interleukin 1 α); up-regulated in both male and female groups, *SOD1* (superoxide dismutase 1) down-regulated in male group and TFEC (Transcription factor EC), DAXX (death domain associated protein), and TRPM2 which were down-regulated in the female group. *HMOX1* is an oxidative stress marker [74], and is up-regulated in response to oxidative stress [75]. *SOD1* catalyzes the removal of superoxide radicals that are generated due to biological oxidation [76]. *SOD* expression was found to be down-regulated in porcine skeletal muscle in response to long-term (3 days) heat stress [58]. In this study SOD expression was down-regulated in the male group following heat stress; this along with the increased expression of *HMOX1* suggests that the male group animals may have experienced significant oxidative stress following heat stress. Several genes involved in response to oxidative stress were up-regulated in the male group, including *TRPM2* [77], *LRRK2* [78], *MAPK14* [79] and *VNN1* [80]. In the female group *PRDX2* [81], *SLC7A11* [82] and *FOXO3* [83] which protects cells from ROX were up-regulated, while *TREX1* involved in cellular response to oxidative stress [84] and *SETX* (Senataxin) involved in defense against DNA damage due to oxidative stress [85] were down-regulated in the female group.

Immune response is closely linked with response to heat stress. Immune stimulation could occur either due to the hyper thermic effect on immune cells or due to indirect effects such as the activation of heat shock factors, which are potent immune modulators and can stimulate both innate and adaptive immune response [86]. Heat stress triggered a massive innate immune response in the male group, with several genes including *CD14*, *BMX*, *S100A8*, *TLR4*, *MYD88*, *SRC*, *TLR2*, *SLEC5A*, *S100A9*, *PTK2B* and *VNN1* being up-regulated; interestingly in the female group, 13 genes involved in innate immunity were down-regulated; those included, *TRIM5*, *ISG20*, *RSAD2*, *DDX58*, *PML*, *TRIM25*, *TEC*, *IRF3*, *TRIM26*, *SH2D1A*, *ANXA1*, *FADD* and *JAK3*. In the male group, among the DEGs several inflammatory response genes were up-regulated; these included *CD14*, *CCR1*, *TSPAN2*, *TLR4*, *THBS1*, *PTGER3*,

*MYD88*, *C5AR2*, *SELP*, *TLR2* and *CHI3L1*, while inflammatory response was not enriched amongst the DEGs in the female group. Anti-tumor immune response were also markedly enhanced due to heat stress, genes involved in proteoglycans in cancer, choline metabolism in cancer, pathways in cancer were up-regulated in both male and female groups, while apoptotic genes were only triggered in the female group; these included genes *PIK3CA*, *BCL2L1*, *AKT3* and *ATM*. Several studies have shown that heat stress triggers anti-tumor and apoptotic pathways possibly due to protein aggregation due to heat stress induced denaturation [16].

### 4.2. Metabolome Regulation in Response to Heat Stress

Metabolites are building blocks for growth and development; they are also key regulators and markers of animal health [87]. Blood metabolites are highly sensitive to environmental stress [88] and HS alters protein metabolism in a number of species [19,89–91]. Metabolome analysis showed that several plasma metabolites were enriched in response to HS in both male and female groups, this included, creatinine, histidine, lysine, methionine, ornithine, serine, proline and pyruvate, while 4-aminobutryic acid, Creatine, Taurine and Urea concentration was lower post HS. Plasma creatinine concentration has been found to be increased due to heat stress in several species including cattle, pigs and sheep [92–94]. Creatinine along with methylhistidine is a known indicator of muscle breakdown [95], the concentration of 1-Methylhistidine was also significantly increased in the female group. This indicates that the animals might be experiencing tissue break-down due to heat stress, the reason for such breakdown is not clear, however it could probably be due to increased protein catabolism, which is required for utilizing the carbon in amino acids for gluconeogenesis [96]. Methionine is a sulfur containing amino acid, and along with creatinine is involved in cellular antioxidant mechanism and is found abundantly on the surface of proteins exposed to very high oxidant fluxes [97,98]. The concentration of Pyruvate, an anti-oxidant [99] has been found to be increased under hyperthermia in pigs [100]. Pyruvate is also the precursor to alanine via alanine aminotransferase, and the entry of pyruvate into the TCA cycle through the pyruvate dehydrogenase complex is impaired due to heat stress, possibly due to the inactivation of PDH (Pyruvate dehydrogenase complex) complex due to the HS induced oxidative chain reactions generated by intracellular ROS (reactive oxygen species) [101]. Along with increased plasma concentration of Pyruvate due to the inactivation of PDH, the increased expression of *PDK4* (Pyruvate Dehydrogenase Kinase 4) is suggested to serve as a mechanism to reduce substrate oxidation and mitochondrial ROS production to protect HS induced cellular damage [6]. The expression of *PDK4* was significantly increased in the male group. Glycine stimulates protein synthesis, and has been found to inhibit oxidative stress in pig small intestine, the plasma concentration of glycine in the female group increased post HS while it reduced in the male group [102]. Similarly Alanine and Citrulline have protective effect against oxidative damage [103]. The difference in the plasma concentration of several of the above discussed metabolites together with increased expression of *PDK4*, *HMOX1* and down-regulation of *SOD1* indicates that the male group could have experience heat stress induced oxidative stress.

Several plasma metabolites were also oppositely enriched in the male and female groups. The plasma concentration of Tryptophan, 1-Methylhistidine, Acetate, Glycine, Aminoadipic acid, Alanine, Arginine, Citrulline, Cystathionine, Glutamate, Threonine, and Valine were all increased in the female group, while their concentration was reduced in the male group. Tryptophan metabolites are key neurotransmitters, that regulate immune response, increased plasma concentration of tryptophan could indicate intracellular protein degradation, and were found to increase the expression of apoptosis initiators in pig [104,105]. The plasma concentration of metabolites like glutamate, cystathionine and threonine are all associated with apoptosis and autophagy [106–108]. Several anti and pro apoptotic genes were also amongst the DEGs. This along with previously known effects of HS on protein denaturation and the proteotoxic effect of HS [16], and the increased concentration of amino acids that are by products of protein degradation suggest that HS induced cellular damage must have occurred in female group.

## 5. Conclusions

In conclusion, heat stress triggered a dynamic response in pigs; however, the response to heat stress was sex-specific. The reason for this sexual dimorphic response is not clear; however, evidence from other studies suggests that it could be due to the anatomical, physiological and hormonal differences. Future transcriptome and metabolome studies, along with blood parameters and hormonal analyses could provide more understanding about the effect of sex on the heat stress response in pigs. The results of this study along with increasing our understanding of porcine heat stress response will serve as a good reference for future studies. However, there are some limitations in this study, such as only six animals being used in this study for both transcriptome and metabolome analyses, and since only NMR was used for metabolite quantification, it resulted in limited detection of metabolites (only 48 metabolites) resulting in only a limited understanding of the effect of HS on metabolite accumulation, so complementing NMR analysis with mass spectrometric analysis might result in a more comprehensive understanding of the effect of HS on metabolite accumulation.

**Supplementary Materials:** The following are available online at http://www.mdpi.com/2073-4425/11/5/540/s1. File S1: List of primers used in the q-RT PCR analysis. File S2: Summary of the mapping and alignment statistics. File S3: List of DEGs identified in response to heat stress. File S4: List of metabolites found to be differentially enriched in response to heat stress. Figure S1: PCA using RNA-Seq and metabolome data from all samples. Table S1: Summary of the measurements of average daily feed and water intake.

**Author Contributions:** Conceptualization, K.S., S.D.L., G.-W.J. and J.-E.P.; methodology, K.S., S.Y.J., K.H.K., Y.K.L., M.K., Y.C.B. and S.D.L.; software, K.S.; validation, H.K. (Hana Kim); formal analysis, K.S., Y.C.B. and M.K.; investigation, K.S. and H.K. (Himansu Kumar); resources, J.-E.P. and S.D.L.; data curation, B.-H.C., G.-W.J. and S.Y.J.; writing—original draft preparation, K.S.; writing—review and editing, S.D.L. and J.-E.P.; visualization, K.S.; supervision, G.-W.J. and J.-E.P.; project administration, S.D.L.; funding acquisition, S.D.L. and J.-E.P. All authors have read and agreed to the published version of the manuscript.

**Funding:** This work was carried out with the support of the Cooperative Research Program for Agriculture Science and Technology Development (project number: PJ01277102), Rural Development Administration, Republic of Korea. K.S. and H.K.U. were supported by the 2020 RDA fellowship program of the National Institute of Animal Science, Rural Development Administration, Republic of Korea.

**Conflicts of Interest:** The authors declare no conflict of interest.

## References

1. Renaudeau, D.; Collin, A.; Yahav, S.; De Basilio, V.; Gourdine, J.; Collier, R. Adaptation to hot climate and strategies to alleviate heat stress in livestock production. *Animal* **2012**, *6*, 707–728. [CrossRef] [PubMed]
2. Sejian, V.; Bhatta, R.; Gaughan, J.; Dunshea, F.; Lacetera, N. Adaptation of animals to heat stress. *Animal* **2018**, *12*, s431–s444. [CrossRef] [PubMed]
3. Ross, J.W.; Hale, B.J.; Seibert, J.T.; Romoser, M.R.; Adur, M.K.; Keating, A.F.; Baumgard, L.H. Physiological mechanisms through which heat stress compromises reproduction in pigs. *Mol. Reprod. Dev.* **2017**, *84*, 934–945. [CrossRef]
4. Hao, Y.; Cui, Y.; Gu, X. Genome-wide DNA methylation profiles changes associated with constant heat stress in pigs as measured by bisulfite sequencing. *Sci. Rep.* **2016**, *6*, 27507. [CrossRef] [PubMed]
5. He, J.; Guo, H.; Zheng, W.; Xue, Y.; Zhao, R.; Yao, W. Heat stress affects fecal microbial and metabolic alterations of primiparous sows during late gestation. *J. Anim. Sci. Biotechnol.* **2019**, *10*, 1–12. [CrossRef] [PubMed]
6. Baumgard, L.H.; Rhoads Jr, R.P. Effects of heat stress on postabsorptive metabolism and energetics. *Annu. Rev. Anim. Biosci.* **2013**, *1*, 311–337. [CrossRef]
7. Brown-Brandl, T.; Eigenberg, R.; Nienaber, J.A.; Kachman, S.D. Thermoregulatory profile of a newer genetic line of pigs. *Livest. Prod. Sci.* **2001**, *71*, 253–260. [CrossRef]
8. Patience, J.; Umboh, J.; Chaplin, R.; Nyachoti, C. Nutritional and physiological responses of growing pigs exposed to a diurnal pattern of heat stress. *Livest. Prod. Sci.* **2005**, *96*, 205–214. [CrossRef]
9. Seibert, J.; Graves, K.; Hale, B.; Keating, A.; Baumgard, L.; Ross, J. Characterizing the acute heat stress response in gilts: I. Thermoregulatory and production variables. *J. Anim. Sci.* **2018**, *96*, 941–949. [CrossRef]

10. Pollmann, D. Seasonal effects on sow herds: industry experience and management strategies. *J. Anim. Sci* **2010**, *88*, 9.
11. St-Pierre, N.; Cobanov, B.; Schnitkey, G. Economic losses from heat stress by US livestock industries. *J. Dairy Sci.* **2003**, *86*, E52–E77. [CrossRef]
12. Quiniou, N.; Noblet, J. Influence of high ambient temperatures on performance of multiparous lactating sows. *J. Anim. Sci.* **1999**, *77*, 2124–2134. [CrossRef] [PubMed]
13. Dou, S.; Villa-Vialaneix, N.; Liaubet, L.; Billon, Y.; Giorgi, M.; Gilbert, H.; Gourdine, J.-L.; Riquet, J.; Renaudeau, D. 1HNMR-Based metabolomic profiling method to develop plasma biomarkers for sensitivity to chronic heat stress in growing pigs. *PLoS ONE* **2017**, *12*, e0188469. [CrossRef]
14. Mayorga, E.J.; Renaudeau, D.; Ramirez, B.C.; Ross, J.W.; Baumgard, L.H. Heat stress adaptations in pigs. *Anim. Front.* **2018**, *9*, 54–61. [CrossRef] [PubMed]
15. Lan, X.; Hsieh, J.C.; Schmidt, C.J.; Zhu, Q.; Lamont, S.J. Liver transcriptome response to hyperthermic stress in three distinct chicken lines. *Bmc Genom.* **2016**, *17*, 955. [CrossRef] [PubMed]
16. Srikanth, K.; Kwon, A.; Lee, E.; Chung, H. Characterization of genes and pathways that respond to heat stress in Holstein calves through transcriptome analysis. *Cell Stress Chaperones* **2017**, *22*, 29–42. [CrossRef]
17. Srikanth, K.; Lee, E.; Kwan, A.; Lim, Y.; Lee, J.; Jang, G.; Chung, H. Transcriptome analysis and identification of significantly differentially expressed genes in Holstein calves subjected to severe thermal stress. *Int. J. Biometeorol.* **2017**, *61*, 1993–2008. [CrossRef]
18. Hao, Y.; Feng, Y.; Yang, P.; Cui, Y.; Liu, J.; Yang, C.; Gu, X. Transcriptome analysis reveals that constant heat stress modifies the metabolism and structure of the porcine longissimus dorsi skeletal muscle. *Mol. Genet. Genom.* **2016**, *291*, 2101–2115. [CrossRef]
19. Qu, H.; Ajuwon, K.M. Metabolomics of heat stress response in pig adipose tissue reveals alteration of phospholipid and fatty acid composition during heat stress. *J. Anim. Sci.* **2018**, *96*, 3184–3195. [CrossRef]
20. He, J.; Zheng, W.; Lu, M.; Yang, X.; Xue, Y.; Yao, W. A controlled heat stress during late gestation affects thermoregulation, productive performance, and metabolite profiles of primiparous sow. *J. Therm. Boil.* **2019**, *81*, 33–40. [CrossRef]
21. Cui, Y.; Wang, C.; Hao, Y.; Gu, X.; Wang, H. Chronic Heat Stress Induces Acute Phase Responses and Serum Metabolome Changes in Finishing Pigs. *Animals* **2019**, *9*, 395. [CrossRef] [PubMed]
22. Close, W.; Le Dividich, J.; Duee, P. Influence of environmental temperature on glucose tolerance and insulin response in the new-born piglet. *Neonatology* **1985**, *47*, 84–91. [CrossRef] [PubMed]
23. Black, J.; Mullan, B.; Lorschy, M.; Giles, L. Lactation in the sow during heat stress. *Livest. Prod. Sci.* **1993**, *35*, 153–170. [CrossRef]
24. Jeong, J.Y.; Kim, M.S.; Jung, H.J.; Kim, M.J.; Lee, H.J.; Lee, S.D. NMR-based metabolomic profiling of the liver, serum, and urine of piglets treated with deoxynivalenol. *Korean J. Agric. Sci.* **2018**, *45*, 455–461.
25. Srikanth, K.; Kumar, H.; Park, W.; Byun, M.-J.; Lim, D.; Kemp, S.; Te Pas, M.; Kim, J.-M.; Park, J.-E. Cardiac and skeletal muscle transcriptome response to heat stress in Kenyan chicken ecotypes adapted to low and high altitudes reveal differences in thermal tolerance and stress response. *Front. Genet.* **2019**, *10*, 993. [CrossRef] [PubMed]
26. Andrews, S. FastQC: A Quality Control Tool for High Throughput Sequence Data. Available online: https://www.bioinformatics.babraham.ac.uk/projects/fastqc (accessed on 23 November 2019).
27. Bolger, A.M.; Lohse, M.; Usadel, B. Trimmomatic: a flexible trimmer for Illumina sequence data. *Bioinformatics* **2014**, *30*, 2114–2120. [CrossRef] [PubMed]
28. Kim, D.; Paggi, J.M.; Park, C.; Bennett, C.; Salzberg, S.L. Graph-based genome alignment and genotyping with HISAT2 and HISAT-genotype. *Nat. Biotechnol.* **2019**, *37*, 907–915. [CrossRef] [PubMed]
29. Liao, Y.; Smyth, G.K.; Shi, W. featureCounts: An efficient general purpose program for assigning sequence reads to genomic features. *Bioinformatics* **2013**, *30*, 923–930. [CrossRef]
30. Leek, J.T. Svaseq: Removing batch effects and other unwanted noise from sequencing data. *Nucleic Acids Res.* **2014**, *42*, e161. [CrossRef]
31. Love, M.I.; Huber, W.; Anders, S. Moderated estimation of fold change and dispersion for RNA-seq data with DESeq2. *Genome Biol.* **2014**, *15*, 550. [CrossRef]
32. Huang, D.W.; Sherman, B.T.; Lempicki, R.A. Systematic and integrative analysis of large gene lists using DAVID bioinformatics resources. *Nat. Protoc.* **2008**, *4*, 44. [CrossRef] [PubMed]

33. Bindea, G.; Mlecnik, B.; Hackl, H.; Charoentong, P.; Tosolini, M.; Kirilovsky, A.; Fridman, W.-H.; Pagès, F.; Trajanoski, Z.; Galon, J. ClueGO: A Cytoscape plug-in to decipher functionally grouped gene ontology and pathway annotation networks. *Bioinformatics* **2009**, *25*, 1091–1093. [CrossRef] [PubMed]
34. Shannon, P.; Markiel, A.; Ozier, O.; Baliga, N.S.; Wang, J.T.; Ramage, D.; Amin, N.; Schwikowski, B.; Ideker, T. Cytoscape: A software environment for integrated models of biomolecular interaction networks. *Genome Res.* **2003**, *13*, 2498–2504. [CrossRef] [PubMed]
35. Young, M.D.; Wakefield, M.J.; Smyth, G.K.; Oshlack, A. Gene ontology analysis for RNA-seq: accounting for selection bias. *Genome Biol.* **2010**, *11*, R14. [CrossRef] [PubMed]
36. Chong, J.; Soufan, O.; Li, C.; Caraus, I.; Li, S.; Bourque, G.; Wishart, D.S.; Xia, J. MetaboAnalyst 4.0: Towards more transparent and integrative metabolomics analysis. *Nucleic Acids Res.* **2018**, *46*, W486–W494. [CrossRef]
37. Ma, X.; Lin, Y.; Jiang, Z.; Zheng, C.; Zhou, G.; Yu, D.; Cao, T.; Wang, J.; Chen, F. Dietary arginine supplementation enhances antioxidative capacity and improves meat quality of finishing pigs. *Amino Acids* **2010**, *38*, 95–102. [CrossRef]
38. Cui, Y.; Hao, Y.; Li, J.; Bao, W.; Li, G.; Gao, Y.; Gu, X. Chronic heat stress induces immune response, oxidative stress response, and apoptosis of finishing pig liver: A proteomic approach. *Int. J. Mol. Sci.* **2016**, *17*, 393. [CrossRef]
39. Mahmoud, K.Z.; Edens, F. Influence of selenium sources on age-related and mild heat stress-related changes of blood and liver glutathione redox cycle in broiler chickens (Gallus domesticus). *Comp. Biochem. Physiol. Part B Biochem. Mol. Boil.* **2003**, *136*, 921–934. [CrossRef]
40. Rudolph, T.R.R.; Baumgard, L.; Selsby, J. aWhy we should sweat heat stress. *National Hog Farmer Daily*, 3 December 2019.
41. Min, L.; Zhao, S.; Tian, H.; Zhou, X.; Zhang, Y.; Li, S.; Yang, H.; Zheng, N.; Wang, J. Metabolic responses and "omics" technologies for elucidating the effects of heat stress in dairy cows. *Int. J. Biometeorol.* **2017**, *61*, 1149–1158. [CrossRef]
42. Ross, J.; Hale, B.; Gabler, N.; Rhoads, R.; Keating, A.; Baumgard, L. Physiological consequences of heat stress in pigs. *Anim. Prod. Sci.* **2015**, *55*, 1381–1390. [CrossRef]
43. Qu, H. *Mechanism of Adipose Tissue Specific Response to Heat Stress in Pigs*; Purdue University: West Lafayette, IN, USA, 2018.
44. Moeser, A. Gender and stress matter in pig gut health. *National Hog Farmer Daily*, 17 April 2018.
45. Renaudeau, D.; Giorgi, M.; Silou, F.; Weisbecker, J. Effect of breed (lean or fat pigs) and sex on performance and feeding behaviour of group housed growing pigs in a tropical climate. *Asian-Australas. J. Anim. Sci.* **2006**, *19*, 593–600. [CrossRef]
46. Jaturasitha, S.; Kamopas, S.; Suppadit, T.; Khiaosa-ard, R.; Kreuzer, M. The effect of gender of finishing pigs slaughtered at 110 kilograms on performance, and carcass and meat quality. *Sci. Asia* **2006**, *32*, 297–305. [CrossRef]
47. Ekstrom, K.; Miller, E.; Ullrey, D.; Lewis, A. Genetic and sex considerations in swine nutrition. In *Swine Nutrition*; Miller, E.R., Ullrey, D.E., Lewis, A.J., Eds.; Butterworth-Heinemann: Oxford, UK, 1991; pp. 415–424.
48. Singer, R.; Harker, C.T.; Vander, A.J.; Kluger, M.J. Hyperthermia induced by open-field stress is blocked by salicylate. *Physiol. Behav.* **1986**, *36*, 1179–1182. [CrossRef]
49. Dymond, K.E.; Fewell, J.E. Gender influences the core temperature response to a simulated open field in adult guinea pigs. *Physiol. Behav.* **1998**, *65*, 889–892. [CrossRef]
50. Martínez-Miró, S.; Tecles, F.; Ramón, M.; Escribano, D.; Hernández, F.; Madrid, J.; Orengo, J.; Martínez-Subiela, S.; Manteca, X.; Cerón, J.J. Causes, consequences and biomarkers of stress in swine: An update. *BMC Vet. Res.* **2016**, *12*, 171. [CrossRef]
51. Bottoms, G.; Roesel, O.; Rausch, F.; Akins, E. Circadian variation in plasma cortisol and corticosterone in pigs and mares. *Amer. J. Vet. Res.* **1972**, *33*, 785–790.
52. Ruis, M.A.; Te Brake, J.H.; Engel, B.; Ekkel, E.D.; Buist, W.G.; Blokhuis, H.J.; Koolhaas, J.M. The circadian rhythm of salivary cortisol in growing pigs: Effects of age, gender, and stress. *Physiol. Behav.* **1997**, *62*, 623–630. [CrossRef]
53. Burdick, N.; Randel, R.; Carroll, J.; Welsh, T. Interactions between temperament, stress, and immune function in cattle. *Int. J. Zool.* **2011**, *2011*, 1–9. [CrossRef]

54. Fagundes, A.C.A.; Negrão, J.A.; Silva, R.G.d.; Gomes, J.D.F.; Souza, L.W.d.O.; Fukushima, R.S. Environmental temperature and serum cortisol levels in growing-finishing pigs. *Braz. J. Vet. Res. Anim. Sci.* **2008**, *45*, 136–140. [CrossRef]
55. Devaraj, C.; Upadhyay, R. Effect of catecholamines and thermal exposure on lymphocyte proliferation, IL–1α & β in buffaloes. *Ital. J. Anim. Sci.* **2007**, *6*, 1336–1339.
56. DeKruyff, R.H.; Fang, Y.; Umetsu, D.T. Corticosteroids enhance the capacity of macrophages to induce Th2 cytokine synthesis in CD4+ lymphocytes by inhibiting IL-12 production. *J. Immunol.* **1998**, *160*, 2231–2237. [PubMed]
57. Salak, J.; McGlone, J.; Lyte, M. Effects of in vitro adrenocorticotrophic hormone, cortisol and human recombinant interleukin-2 on porcine neutrophil migration and luminol-dependent chemiluminescence. *Vet. Immunol. Immunopathol.* **1993**, *39*, 327–337. [CrossRef]
58. Montilla, S.I.R.; Johnson, T.P.; Pearce, S.C.; Gardan-Salmon, D.; Gabler, N.K.; Ross, J.W.; Rhoads, R.P.; Baumgard, L.H.; Lonergan, S.M.; Selsby, J.T. Heat stress causes oxidative stress but not inflammatory signaling in porcine skeletal muscle. *Temperature* **2014**, *1*, 42–50. [CrossRef]
59. Ganesan, S.; Summers, C.; Pearce, S.; Gabler, N.; Valentine, R.; Baumgard, L.; Rhoads, R.; Selsby, J. Short-term heat stress causes altered intracellular signaling in oxidative skeletal muscle. *J. Anim. Sci.* **2017**, *95*, 2438–2451. [CrossRef] [PubMed]
60. Ganesan, S.; Brownstein, A.J.; Pearce, S.C.; Hudson, M.B.; Gabler, N.K.; Baumgard, L.H.; Rhoads, R.P.; Selsby, J.T. Prolonged environment-induced hyperthermia alters autophagy in oxidative skeletal muscle in Sus scrofa. *J. Therm. Biol.* **2018**, *74*, 160–169. [CrossRef]
61. Brownstein, A.J.; Ganesan, S.; Summers, C.M.; Pearce, S.; Hale, B.J.; Ross, J.W.; Gabler, N.; Seibert, J.T.; Rhoads, R.P.; Baumgard, L.H. Heat stress causes dysfunctional autophagy in oxidative skeletal muscle. *Physiol. Rep.* **2017**, *5*, e13317. [CrossRef]
62. Pan, P.J.; Hsu, C.F.; Tsai, J.J.; Chiu, J.H. The role of oxidative stress response revealed in preconditioning heat stimulation in skeletal muscle of rats. *J. Surg. Res.* **2012**, *176*, 108–113. [CrossRef]
63. Nussbaum, E.L.; Locke, M. Heat shock protein expression in rat skeletal muscle after repeated applications of pulsed and continuous ultrasound. *Arch. Phys. Med. Rehabil.* **2007**, *88*, 785–790. [CrossRef]
64. Liu, S.; Wang, X.; Sun, F.; Zhang, J.; Feng, J.; Liu, H.; Rajendran, K.; Sun, L.; Zhang, Y.; Jiang, Y. RNA-Seq reveals expression signatures of genes involved in oxygen transport, protein synthesis, folding and degradation in response to heat stress in catfish. *Am. J. Physiol.-Heart Circ. Physiol.* **2013**, *45*, 462–476. [CrossRef]
65. Moseley, P.L. Heat shock proteins and heat adaptation of the whole organism. *J. Appl. Physiol.* **1997**, *83*, 1413–1417. [CrossRef]
66. Liu, Y.; Steinacker, J.M. Changes in skeletal muscle heat shock proteins: Pathological significance. *Front. Biosci.* **2001**, *6*, D12–D25. [PubMed]
67. Bertolotti, A.; Zhang, Y.; Hendershot, L.M.; Harding, H.P.; Ron, D. Dynamic interaction of BiP and ER stress transducers in the unfolded-protein response. *Nat. Cell Biol.* **2000**, *2*, 326. [CrossRef] [PubMed]
68. Richter, K.; Buchner, J. Hsp90: Chaperoning signal transduction. *J. Cell. Physiol.* **2001**, *188*, 281–290. [CrossRef] [PubMed]
69. Huang, S.-Y.; Tam, M.-F.; Hsu, Y.-T.; Lin, J.-H.; Chen, H.-H.; Chuang, C.-K.; Chen, M.-Y.; King, Y.-T.; Lee, W.-C. Developmental changes of heat-shock proteins in porcine testis by a proteomic analysis. *Theriogenology* **2005**, *64*, 1940–1955. [CrossRef]
70. Lei, L.; Yu, J.; Bao, E. Expression of heat shock protein 90 (Hsp90) and transcription of its corresponding mRNA in broilers exposed to high temperature. *Br. Poult. Sci.* **2009**, *50*, 504–511. [CrossRef]
71. Hubler, T.R.; Scammell, J.G. Intronic hormone response elements mediate regulation of FKBP5 by progestins and glucocorticoids. *Cell Stress Chaperones* **2004**, *9*, 243. [CrossRef]
72. Ponsuksili, S.; Du, Y.; Murani, E.; Schwerin, M.; Wimmers, K. Elucidating molecular networks that either affect or respond to plasma cortisol concentration in target tissues of liver and muscle. *Genetics* **2012**, *192*, 1109–1122. [CrossRef]
73. Sautron, V.; Terenina, E.; Gress, L.; Lippi, Y.; Billon, Y.; Larzul, C.; Liaubet, L.; Villa-Vialaneix, N.; Mormède, P. Time course of the response to ACTH in pig: Biological and transcriptomic study. *BMC Genom.* **2015**, *16*, 961. [CrossRef]

74. Weiß, M.; Kost, B.; Renner-Müller, I.; Wolf, E.; Mylonas, I.; Brüning, A. Efavirenz causes oxidative stress, endoplasmic reticulum stress, and autophagy in endothelial cells. *Cardiovasc. Toxicol.* **2016**, *16*, 90–99. [CrossRef]
75. Paul, C.; Teng, S.; Saunders, P.T. A single, mild, transient scrotal heat stress causes hypoxia and oxidative stress in mouse testes, which induces germ cell death. *Boil. Reprod.* **2009**, *80*, 913–919. [CrossRef]
76. Fridovich, I. Superoxide dismutases. *Adv. Enzymol. Relat. Areas. Mol. Biol.* **1986**, *58*, 61–97. [PubMed]
77. Nazıroğlu, M. New molecular mechanisms on the activation of TRPM2 channels by oxidative stress and ADP-ribose. *Neurochem. Res.* **2007**, *32*, 1990–2001. [CrossRef] [PubMed]
78. Nguyen, H.N.; Byers, B.; Cord, B.; Shcheglovitov, A.; Byrne, J.; Gujar, P.; Kee, K.; Schüle, B.; Dolmetsch, R.E.; Langston, W. LRRK2 mutant iPSC-derived DA neurons demonstrate increased susceptibility to oxidative stress. *Cell Stem Cell* **2011**, *8*, 267–280. [CrossRef] [PubMed]
79. Naito, Y.; Yoshikawa, T. Oxidative stress involvement and gene expression in indomethacin-induced gastropathy. *Redox Rep.* **2006**, *11*, 243–253. [CrossRef]
80. Berruyer, C.; Martin, F.; Castellano, R.; Macone, A.; Malergue, F.; Garrido-Urbani, S.; Millet, V.; Imbert, J.; Dupre, S.; Pitari, G. Vanin-1−/− mice exhibit a glutathione-mediated tissue resistance to oxidative stress. *Mol. Cell. Biol.* **2004**, *24*, 7214–7224. [CrossRef]
81. Zhou, S.; Han, Q.; Wang, R.; Li, X.; Wang, Q.; Wang, H.; Wang, J.; Ma, Y. PRDX2 protects hepatocellular carcinoma SMMC-7721 cells from oxidative stress. *Oncol. Lett.* **2016**, *12*, 2217–2221. [CrossRef]
82. Ali, D.; Mohammad, D.K.; Mujahed, H.; Jonson-Videsäter, K.; Nore, B.; Paul, C.; Lehmann, S. Anti-leukaemic effects induced by APR-246 are dependent on induction of oxidative stress and the NFE2L2/HMOX1 axis that can be targeted by PI3K and mTOR inhibitors in acute myeloid leukaemia cells. *Br. J. Haematol.* **2016**, *174*, 117–126. [CrossRef]
83. Webb, A.E.; Brunet, A. FOXO transcription factors: Key regulators of cellular quality control. *Trends Biochem. Sci.* **2014**, *39*, 159–169. [CrossRef]
84. Barizzone, N.; Monti, S.; Mellone, S.; Godi, M.; Marchini, M.; Scorza, R.; Danieli, M.G.; D'Alfonso, S. Rare variants in the TREX1 gene and susceptibility to autoimmune diseases. *BioMed Res. Int.* **2013**, *2013*, 471703. [CrossRef]
85. Suraweera, A.; Becherel, O.J.; Chen, P.; Rundle, N.; Woods, R.; Nakamura, J.; Gatei, M.; Criscuolo, C.; Filla, A.; Chessa, L. Senataxin, defective in ataxia oculomotor apraxia type 2, is involved in the defense against oxidative DNA damage. *J. Cell Boil.* **2007**, *177*, 969–979. [CrossRef]
86. Calderwood, S.K.; Theriault, J.R.; Gong, J. How is the immune response affected by hyperthermia and heat shock proteins? *Int. J. Hyperth.* **2005**, *21*, 713–716. [CrossRef] [PubMed]
87. Wang, L.; Urriola, P.E.; Luo, Z.h.; Rambo, Z.J.; Wilson, M.E.; Torrison, J.L.; Shurson, G.C.; Chen, C. Metabolomics revealed diurnal heat stress and zinc supplementation-induced changes in amino acid, lipid, and microbial metabolism. *Physiol. Rep.* **2016**, *4*, e12676. [CrossRef] [PubMed]
88. Sun, Q.; Zhao, W.; Wang, L.; Guo, F.; Song, D.; Zhang, Q.; Zhang, D.; Fan, Y.; Wang, J. Integration of metabolomic and transcriptomic profiles to identify biomarkers in serum of lung cancer. *J. Cell. Biochem.* **2019**, *120*, 11981–11989. [CrossRef] [PubMed]
89. Wheelock, J.; Rhoads, R.; VanBaale, M.; Sanders, S.; Baumgard, L. Effects of heat stress on energetic metabolism in lactating Holstein cows. *J. Dairy Sci.* **2010**, *93*, 644–655. [CrossRef]
90. Kamiya, M.; Kamiya, Y.; Tanaka, M.; Oki, T.; Nishiba, Y.; Shioya, S. Effects of high ambient temperature and restricted feed intake on urinary and plasma 3-methylhistidine in lactating Holstein cows. *Anim. Sci. J.* **2006**, *77*, 201–207. [CrossRef]
91. Azad, M.; Kikusato, M.; Maekawa, T.; Shirakawa, H.; Toyomizu, M. Metabolic characteristics and oxidative damage to skeletal muscle in broiler chickens exposed to chronic heat stress. *Comp. Biochem. Physiol. Part A Mol. Integr. Physiol.* **2010**, *155*, 401–406. [CrossRef]
92. Kellaway, R.; Colditz, P. The effect of heat stress on growth and nitrogen metabolism in Friesian and F1 Brahman× Friesian heifers. *Aust. J. Agric. Res.* **1975**, *26*, 615–622. [CrossRef]
93. Srikandakumar, A.; Johnson, E.; Mahgoub, O. Effect of heat stress on respiratory rate, rectal temperature and blood chemistry in Omani and Australian Merino sheep. *Small Rumin. Res.* **2003**, *49*, 193–198. [CrossRef]
94. Pearce, S.; Gabler, N.; Ross, J.; Escobar, J.; Patience, J.; Rhoads, R.; Baumgard, L. The effects of heat stress and plane of nutrition on metabolism in growing pigs. *J. Anim. Sci.* **2013**, *91*, 2108–2118. [CrossRef]

95. Pearce, S. The Effects of Heat Stress and Nutritional Status on Metabolism and Intestinal Integrity in Growing Pigs. Master's Thesis, Iowa State University, Ames, IA, USA, 2011.
96. Brockman, R. Pancreatic and adrenal hormonal regulation of metabolism. In *Control of Digestion and Metabolism in Ruminants*; Prentice-Hall: Upper Saddle River, NJ, USA, 1986.
97. Bender, A.; Hajieva, P.; Moosmann, B. Adaptive antioxidant methionine accumulation in respiratory chain complexes explains the use of a deviant genetic code in mitochondria. *Proc. Natl. Acad. Sci. USA* **2008**, *105*, 16496–16501. [CrossRef]
98. Lawler, J.M.; Barnes, W.S.; Wu, G.; Song, W.; Demaree, S. Direct antioxidant properties of creatine. *Biochem. Biophys. Res. Commun.* **2002**, *290*, 47–52. [CrossRef] [PubMed]
99. BV, S.K.; Ajeet, K.; Meena, K. Effect of heat stress in tropical livestock and different strategies for its amelioration. *J. Stress Physiol. Biochem.* **2011**, *7*, 45–54.
100. Hall, G.; Lucke, J.; Lovell, R.; Lister, D. Porcine malignant hyperthermia. VII: Hepatic metabolism. *Br. J. Anaesth.* **1980**, *52*, 11–17. [CrossRef] [PubMed]
101. Belhadj Slimen, I.; Najar, T.; Ghram, A.; Abdrrabba, M. Heat stress effects on livestock: molecular, cellular and metabolic aspects, a review. *J. Anim. Physiol. Anim. Nutr.* **2016**, *100*, 401–412. [CrossRef] [PubMed]
102. Wang, W.; Wu, Z.; Lin, G.; Hu, S.; Wang, B.; Dai, Z.; Wu, G. Glycine stimulates protein synthesis and inhibits oxidative stress in pig small intestinal epithelial cells. *J. Nutr.* **2014**, *144*, 1540–1548. [CrossRef] [PubMed]
103. Li, H.-T.; Feng, L.; Jiang, W.-D.; Liu, Y.; Jiang, J.; Li, S.-H.; Zhou, X.-Q. Oxidative stress parameters and anti-apoptotic response to hydroxyl radicals in fish erythrocytes: protective effects of glutamine, alanine, citrulline and proline. *Aquat. Toxicol.* **2013**, *126*, 169–179. [CrossRef] [PubMed]
104. Kim, C.J.; Kovacs-Nolan, J.A.; Yang, C.; Archbold, T.; Fan, M.Z.; Mine, Y. l-Tryptophan exhibits therapeutic function in a porcine model of dextran sodium sulfate (DSS)-induced colitis. *J. Nutr. Biochem.* **2010**, *21*, 468–475. [CrossRef]
105. Yao, K.; Fang, J.; Yin, Y.; Feng, Z.-M.; Tang, Z.-R.; Wu, G. Tryptophan metabolism in animals: important roles in nutrition and health. *Front. Biosci. (Schol Ed)* **2011**, *3*, 286–297.
106. Zhu, M.; Du, J.; Chen, S.; Liu, A.; Holmberg, L.; Chen, Y.; Zhang, C.; Tang, C.; Jin, H. L-cystathionine inhibits the mitochondria-mediated macrophage apoptosis induced by oxidized low density lipoprotein. *Int. J. Mol. Sci.* **2014**, *15*, 23059–23073. [CrossRef]
107. Liu, N.; Ma, X.; Luo, X.; Zhang, Y.; He, Y.; Dai, Z.; Yang, Y.; Wu, G.; Wu, Z. L-glutamine attenuates apoptosis in porcine enterocytes by regulating glutathione-related redox homeostasis. *J. Nutr.* **2018**, *148*, 526–534. [CrossRef]
108. Baird, C.H.; Niederlechner, S.; Beck, R.; Kallweit, A.R.; Wischmeyer, P.E. L-Threonine induces heat shock protein expression and decreases apoptosis in heat-stressed intestinal epithelial cells. *Nutrition* **2013**, *29*, 1404–1411. [CrossRef] [PubMed]

*Review*

# Survey of SNPs Associated with Total Number Born and Total Number Born Alive in Pig

**Siroj Bakoev [1], Lyubov Getmantseva [1,*], Faridun Bakoev [1,2], Maria Kolosova [1,3], Valeria Gabova [2], Anatoly Kolosov [1,3] and Olga Kostyunina [1]**

1 Federal Science Center for Animal Husbandry named after Academy Member L.K. Ernst, Dubrovitsy 142132, Russian; siroj1@yandex.ru (S.B.); bakoevfaridun@yandex.ru (F.B.); m.leonovaa@mail.ru (M.K.); kolosov777@gmail.com (A.K.); kostolan@yandex.ru (O.K.)

2 Department of Biology and Biotechnology, Southern Federal University, Rostov-on-Don 344006, Russia; gabova98@yandex.ru

3 Department of Biotechnology, Don State Agrarian University, Persianovski 346493, Russia

* Correspondence: ilonaluba@mail.ru; Tel.: +7-(4967)-65-11-01

Received: 12 April 2020; Accepted: 29 April 2020; Published: 30 April 2020

**Abstract:** Reproductive productivity depend on a complex set of characteristics. The number of piglets at birth (Total number born, Litter size, TNB) and the number of alive piglets at birth (Total number born alive, NBA) are the main indicators of the reproductive productivity of sows in pig breeding. Great hopes are pinned on GWAS (Genome-Wide Association Studies) to solve the problems associated with studying the genetic architecture of reproductive traits of pigs. This paper provides an overview of international studies on SNP (Single nucleotide polymorphism) associated with TNB and NBA in pigs presented in PigQTLdb as "Genome map association". Currently on the base of Genome map association results 306 SNPs associated with TNB (218 SNPs) and NBA (88 SNPs) have been identified and presented in the Pig QTLdb database. The results are based on research of pigs such as Large White, Yorkshire, Landrace, Berkshire, Duroc and Erhualian. The presented review shows that most SNPs found in chromosome areas where candidate genes or QTLs (Quantitative trait locus) have been identified. Further research in the given direction will allow to obtain new data that will become an impulse for creating breakthrough breeding technologies and increase the production efficiency in pig farming.

**Keywords:** pig; litter size; total number born alive; SNP (single nucleotide polymorphism); GWAS (Genome-Wide Association Studies)

---

## 1. Introduction

In the early 1990s the work on mapping the pig genome started and maps containing more than 1200 microsatellite markers appeared due to the development of the international project "PiGMap" and projects of the US Department of Agriculture and American agricultural institutes. These maps were used to identify the Quantitative Trait Loci (QTL) underlying the genetic architecture of pig productivity traits [1–3]. To date, an extended database has been created—Pig Quantitative Trait Locus Database (PigQTLdb), which presents 30170 QTLs for 688 pig trait of deferent classes [4].

The sows' reproductive potential is the basis for continuous and efficient production. In recent decades, the BLUP (Best linear unbiased prediction) method has made a significant contribution to improving reproduction rates [5,6]. However, low heritability coefficients of reproductive traits and their sex-limited phenotypic exhibition lead to developing new approaches revealing the biological nature of reproductive performance. Great hopes are pinned on GWAS (Genome-Wide Association Studies) to solve the problems associated with studying the genetic architecture of reproductive traits of pigs [7–10].

GWAS results can be represented by the information on detected associations with various genetic aberrations: chromosomal mutations (whole chromosomes or their fragments), large inserts or deletions (100–100,000 nucleotides), small inserts or deletions (1 to 100), single nucleotide polymorphisms (SNP) [11]. Each SNP is represented by at least two alleles: minor (rarer) and major. Genetic variations with minor allele frequency exceeding 0.01% are numbered and assigned to rs index [12].

Currently SNP BeadChip technology is more affordable for genome-wide research than sequencing is. SNP BeadChips have been developed to include high (HD), medium (MD), or low (LD) genome distribution of markers. Rigid structure is noted as a disadvantage, which allows us to analyze only what is already predetermined by the BeadChip design and in this connection the potentially important information can be omitted [13]. Markers do not have the same density across all chromosomes and not fully track structural genetic variations such as insertions and deletions [13]. However, despite these shortcomings, SNP BeadChip has recently gained great popularity in studies of the genetic architecture of quantitative traits of farm animals and pigs in particular [14–16].

The first SNP BeadChips with a resolution of about 60 thousand markers covering all autosomal and X chromosome genes PorcineSNP60 BeadChip v2 were presented by the American company Illumina (San Diego, CA, USA). In addition to SNP BeadChip with high density, LD SNP BeadChip with low density have been proposed to reduce genotyping costs. Commercial LD SNP BeadChip «GeneSeek/Neogen GPP-Porcine LD Illumina Bead Chip panel» were also developed by «GeneSeek/Neogen» (Lincoln, NE, USA). Besides this, the company introduced HD SNP BeadChips with a higher density (about 70 thousand markers). HD SNP BeadChips containing about 650 thousand markers and including all markers of the «Illumina PorcineSNP60 BeadChip v2» array, were produced by «Affymetrix» (Santa Clara, CA, USA). It should be noted that SNP BeadChip can be custom made including specific SNPs associated with given traits of productivity.

Reproductive productivity depend on a complex set of characteristics. The number of piglets at birth (Total number born, Litter size, TNB) and the number of alive piglets at birth (Total number born alive, NBA) are the main indicators of the reproductive productivity of sows in pig breeding [17–21]. These indicators reflect the level of all physiological processes associated with fertilization, intrauterine development of the fetus and a sow's labor, and are also quite easy to account for.

## 2. SNPs Associated with Total Number Born Alive

Currently on the base of Genome map association results 88 SNPs associated with NBA have been identified and presented in the Pig QTLdb database (Table 1). The results are based on a study of pigs such as Large White, Large White, Yorkshire, Landrace [7,14,22–26] Duroc [27] and Erhualian [28]. SNPs, associated with the NBA are represented in all Sus scrofa chromosomes (SSC) except SSCY.

**Table 1.** SNPs (Single nucleotide polymorphism) associated with NBA (Total number born alive) presented in the PigQTLdb (Pig Quantitative Trait Locus Database). Ensembl db, Sus Scrofa 11.1 (NA—This variant has not been mapped on Ensembl db, Sus Scrofa 11.1).

| | SNP | SSC | Location | Allele | Consequence | SYMBOL | Trait | Breed | Reference |
|---|---|---|---|---|---|---|---|---|---|
| 1 | rs339929690 | 1 | 59515751 | C | intergenic_variant | - | NBA | Landrace/Yorkshire | [14] |
| 2 | rs329624627 | 1 | 72605122 | C | intron_variant | CRYBG1 | NBA | Landrace/Yorkshire | [14] |
| 3 | rs338462676 | 1 | 74750265 | T | intron_variant | - | NBA | Landrace/Yorkshire | [14] |
| 4 | rs80982567 | 1 | 99668135 | G | intron_variant | - | NBA | Landrace/Yorkshire | [14] |
| 5 | rs329761313 | 1 | 116715803 | C | intron_variant | - | NBA | Landrace/Yorkshire | [14] |
| 6 | rs329145797 | 1 | 116791808 | T | intron_variant | - | NBA | Landrace/Yorkshire | [14] |
| 7 | rs80830052 | 1 | 139481542 | A | intron_variant | ALDH1A3 | NBA | Landace/Large White | [26] |
| 8 | rs80930659 | 1 | 139579572 | G | intron_variant | LRRK1 | NBA | Landace/Large White | [26] |
| 9 | rs80804265 | 1 | 139608452 | A | intron_variant | LRRK1 | NBA | Landace/Large White | [26] |
| 10 | rs80862569 | 1 | 139636710 | C | intergenic_variant | - | NBA | Landace/Large White | [26] |
| 11 | rs80846651 | 1 | 139655026 | A | intergenic_variant | - | NBA | Landace/Large White | [26] |
| 12 | rs81348779 | 1 | 141989297 | G | intron_variant | UBE3A | NBA | Large White | [22] |
| 13 | rs329931325 | 1 | 144835089 | T | intergenic_variant | - | NBA | Landrace/Yorkshire | [14] |
| 14 | rs326961952 | 1 | 145395812 | G | intergenic_variant | - | NBA | Landrace/Yorkshire | [14] |
| 15 | rs336474421 | 1 | 145452726 | T | intergenic_variant | - | NBA | Landrace/Yorkshire | [14] |
| 16 | rs322202112 | 1 | 146334522 | C | intron_variant | - | NBA | Landrace/Yorkshire | [14] |
| 17 | rs332924521 | 1 | 239558816 | C | intron_variant | XPA | NBA | Landrace/Yorkshire | [14] |
| 18 | rs330585697 | 1 | 239775818 | T | intron_variant | TRMO | NBA | Landrace/Yorkshire | [14] |
| 19 | rs334029855 | 1 | NA | | | | NBA | Landrace/Yorkshire | [14] |
| 20 | rs81291755 | 2 | 20563683 | G | intergenic_variant | - | NBA | Landrace/Yorkshire | [26] |
| 21 | rs81355894 | 2 | 20637563 | C | intergenic_variant | - | NBA | Landace/Large White | [26] |
| 22 | rs81355903 | 2 | 20665892 | T | intergenic_variant | - | NBA | Landace/Large White | [26] |
| 23 | rs81355915 | 2 | 20717076 | G | intergenic_variant | - | NBA | Landace/Large White | [26] |

**Table 1.** *Cont.*

| | SNP | SSC | Location | Allele | Consequence | SYMBOL | Trait | Breed | Reference |
|---|---|---|---|---|---|---|---|---|---|
| 24 | rs81356698 | 2 | 28380318 | G | intron_variant | EIF3M | NBA | Large White | [22] |
| 25 | rs81265647 | 2 | 125660328 | T | upstream_gene_var | - | NBA | Landace/Large White | [26] |
| 26 | rs328177895 | 2 | NA | | | | NBA | Large White | [23] |
| 27 | rs81379421 | 3 | 27307613 | G | intron_variant | XYLT1 | NBA | Landace/Large White | [7] |
| 28 | rs334867206 | 3 | 43312168 | T | intergenic_variant | - | NBA | Large White | [23] |
| 29 | rs319494663 | 3 | 43463318 | C | upstream_gene_var | ssc-let-7a-2 | NBA | Large White | [23] |
| 30 | rs80927364 | 4 | 130597207 | A | intron_variant | DDAH1 | NBA | Erhualian | [28] |
| 31 | rs80890206 | 5 | 31022891 | C | upstream_gene_var | GRIP1 | NBA | Erhualian | [28] |
| 32 | rs80867243 | 5 | 87337099 | C | intron_variant | ELK3 | NBA | Landace/Large White | [7] |
| 33 | rs81279319 | 5 | NA | | | | NBA | Landace/Large White | [7] |
| 34 | rs81383147 | 5 | NA | | | | NBA | Landace/Large White | [7] |
| 35 | rs81320475 | 6 | 70313133 | A | upstream_gene_var | RBP7 | NBA | Landace/Large White | [26] |
| 36 | rs81285644 | 6 | 70323076 | G | intron_variant | RBP7 | NBA | Landace/Large White | [26] |
| 37 | rs81275494 | 6 | 70408106 | G | intron_variant | UBE4B | NBA | Landace/Large White | [26] |
| 38 | rs81279050 | 6 | 70418172 | A | intron_variant | UBE4B | NBA | Landace/Large White | [26] |
| 39 | rs81270030 | 6 | 70428427 | T | intron_variant | UBE4B | NBA | Landace/Large White | [26] |
| 40 | rs81388947 | 6 | 80229040 | A | intergenic_variant | - | NBA | Duroc | [27] |
| 41 | rs704072370 | 6 | 80388745 | A | intergenic_variant | - | NBA | Duroc | [27] |
| 42 | rs81332505 | 6 | 83142163 | A | intron_variant | MAN1C1 | NBA | Duroc | [27] |
| 43 | rs81476037 | 6 | 83545020 | T | intron_variant | PDIK1L | NBA | Duroc | [27] |
| 44 | rs328276462 | 6 | 83863384 | G | downstream_gene_var | HMGN2 | NBA | Duroc | [27] |
| 45 | rs335265547 | 6 | 83871998 | C | upstream_gene_var | - | NBA | Duroc | [27] |
| 46 | rs81287462 | 6 | 83986459 | A | intergenic_variant | - | NBA | Duroc | [27] |

**Table 1.** *Cont.*

| | SNP | SSC | Location | Allele | Consequence | SYMBOL | Trait | Breed | Reference |
|---|---|---|---|---|---|---|---|---|---|
| 47 | rs81332455 | 6 | 84069079 | G | intron_variant | ARID1A | NBA | Duroc | [27] |
| 48 | rs336670754 | 6 | 84118086 | C | synonymous_variant | ARID1A | NBA | Duroc | [27] |
| 49 | rs342908929 | 6 | 159933806 | T | intron_variant | ZFYVE9 | NBA | Commercia | [25] |
| 50 | rs81345088 | 6 | 168897980 | A | intron_variant | ZMYND12 | NBA | Landace/Large White | [26] |
| 51 | rs81259198 | 6 | 168899114 | A | intron_variant | ZMYND12 | NBA | Landace/Large White | [26] |
| 52 | rs81245903 | 6 | 168916590 | T | intron_variant | RIMKLA | NBA | Landace/Large White | [26] |
| 53 | rs81273774 | 6 | 168931165 | C | intron_variant | RIMKLA | NBA | Landace/Large White | [26] |
| 54 | rs80882306 | 7 | 11003222 | G | upstream_gene_var | - | NBA | Erhualian | [28] |
| 55 | rs325729252 | 7 | 109101108 | C | intergenic_variant | - | NBA | Landace/Large White | [7] |
| 56 | rs81397142 | 7 | NA | | | | NBA | Large White | [22] |
| 57 | rs81397215 | 7 | NA | | | | NBA | Large White | [22] |
| 58 | rs318557169 | 8 | 95381096 | G | intergenic_variant | - | NBA | Landrace/Yorkshire | [14] |
| 59 | rs327655683 | 8 | 95402330 | A | intergenic_variant | - | NBA | Landrace/Yorkshire | [14] |
| 60 | rs81257618 | 9 | 13492371 | A | intergenic_variant | - | NBA | Landace/Large White | [7] |
| 61 | rs81417393 | 9 | 126541658 | C | intron_variant | - | NBA | Landace/Large White | [7] |
| 62 | rs81421148 | 10 | 16506621 | C | intron_variant | AKT3 | NBA | Landace/Large White | [7] |
| 63 | rs80978601 | 10 | 25256004 | G | intergenic_variant | - | NBA | Landace/Large White | [7] |
| 64 | rs81255997 | 10 | 28330355 | C | intergenic_variant | - | NBA | Landace/Large White | [7] |
| 65 | rs81274366 | 10 | 57593745 | G | intron_variant | PARD3 | NBA | Erhualian | [28] |
| 66 | rs81236069 | 10 | 58288430 | A | intergenic_variant | - | NBA | Landace/Large White | [7] |
| 67 | rs81302230 | 10 | 58290108 | G | intergenic_variant | - | NBA | Landace/Large White | [7] |
| 68 | rs81430147 | 11 | 836702 | C | downstream_gene_var | - | NBA | Landace/Large White | [7] |

**Table 1.** *Cont.*

| | SNP | SSC | Location | Allele | Consequence | SYMBOL | Trait | Breed | Reference |
|---|---|---|---|---|---|---|---|---|---|
| 69 | rs81430859 | 11 | 4296303 | C | intron_variant | WASF3 | NBA | Landace/Large White | [7] |
| 70 | rs81242222 | 11 | 67129570 | A | intergenic_variant | - | NBA | Landace/Large White | [7] |
| 71 | rs81434499 | 12 | 36207006 | G | intergenic_variant | - | NBA | Erhualian | [28] |
| 72 | rs80962240 | 13 | 52784022 | C | intron_variant | FOXP1 | NBA | Large White | [23] |
| 73 | rs81215583 | 13 | 71905692 | C | synonymous_variant | RPN1 | NBA | Erhualian | [28] |
| 74 | rs81447100 | 13 | 80866929 | A | intron_variant | CLSTN2 | NBA | Erhualian | [28] |
| 75 | rs81447231 | 13 | 82603578 | A | intron_variant | GRK7 | NBA | Erhualian | [28] |
| 76 | rs319258722 | 14 | 123766842 | T | intergenic_variant | - | NBA | Large White | [23] |
| 77 | rs45435330 | 15 | 118847390 | A | intron_variant | IGFBP2 | NBA | Berkshire | [24] |
| 78 | rs324003968 | 15 | NA | | | | NBA | Commercia | [25] |
| 79 | rs81459590 | 16 | 6025546 | A | intron_variant | MYO10 | NBA | Landace/Large White | [7] |
| 80 | rs80952566 | 17 | 27123793 | T | intron_variant | - | NBA | Erhualian | [28] |
| 81 | rs81469701 | 18 | 42920811 | T | upstream_gene_var | SCRN1 | NBA | Landace/Large White | [7] |
| 82 | rs81471172 | 18 | 51640938 | T | intron_variant | HECW1 | NBA | Large White | [23] |
| 83 | rs81473442 | X | 91880535 | C | intron_variant | TRPC5 | NBA | Landace/Large White | [26] |
| 84 | rs81323503 | X | 92070342 | A | intergenic_variant | - | NBA | Landace/Large White | [26] |
| 85 | rs81283192 | X | 92244402 | A | intron_variant | RTL4 | NBA | Landace/Large White | [26] |
| 86 | rs337547716 | X | 92330719 | A | intron_variant | RTL4 | NBA | Landace/Large White | [26] |
| 87 | rs80834138 | X | 92447181 | A | intron_variant | LHFPL1 | NBA | Landace/Large White | [26] |
| 88 | rs81339510 | X | 118854990 | T | intergenic_variant | - | NBA | Landace/Large White | [26] |

Search for SNPs associated with NBA of Chinese Erhualian pigs was carried out by Ma et al. [28]. Sows with high and low estimated breeding values (EBVs) were selected for genotyping. According to the research results, 9 SNPs associated with Pig QTLdb with NBA were presented. The greatest effect was found for SNP rs81447100 (SSC13), which was additionally tested on Erhualian pigs ($n = 313$), Sutai ($n = 173$) and Yorkshire ($n = 488$). In all groups under study, a significant association between allelic variants of SNP rs81447100 (SSC13) and NBA was determined. However, allele A was desirable for the Erhualian pigs and allele G for the Sutai and Yorkshire pigs.

According to research work of Wu et al. [14], conducted on Landrace and Yorkshire pigs the database contains 15 SNPs localized on SSC1 and SSC8 and associated with NBA. 11 of these SNPs are located in the QTL regions annotated earlier, and 4 SNPs are presented for the first time. All these 4 SNPs (rs329624627, rs339929690, rs322202112 and rs330585697) are located on the SSC1. The most significant effect was established for SNP rs332924521 (SSC1).

Coster et al. [23] conducted associative studies on Large White pigs from two commercial lines of «Hypor» and «Topigs», and revealed 4 SNPs associated with NBA located on SSC7 (rs81397142 and rs81397215), SSC1 (rs81348779) and SSC2 (rs81356698).

Bergfelder-Drüing et al. [7] conducted research using Large White and Landrace pigs. Preliminary calculations based on graphs of multidimensional scaling showed the genetic distance between the breeds. For analysis pigs were divided into two clusters taking into account the breed and intra-breed clusters taking into account the breeding economy. As a result, 17 SNPs associated with NBAs were identified, 5 of these SNPs had a minor allele frequency less than 1%. The study of Bergfelder-Drüing et al. [7] was the first to show an association with NBA for SNP rs81430147 (SSC11). All other SNPs were found in chromosome regions where candidate genes or QTLs affecting pig reproductive traits have already been identified. It should be noted that different SNPs were established for each breed cluster, and no associative communications were established simultaneously in two breeds. 13 SNPs were identified for Large White sows, and 4 SNPs for Landrace sows. In sows Landrace SNPs are localized on SSC7, SSC9, SSC11 and SSC16. Large White on SSC3, SSC5, SSC9, SSC10, SSC11 and SS18. The most significant effects on NBA of Large White sows are found for 4 SNPs: rs81379421 (SSC3), rs81417393 (SSC9), rs81242222 (SSC11) and rs81469701 (SSC18).

Research carried out by Wang et al. [22] on Large White pigs identified 6 SNPs associated with NBA located on SSC2, SSC3, SSC13, SSC14 and SSC18. The most significant effect was determined for SNP rs334867206 (SSC3), according to which pigs of AA genotype had more NBA, compared with analogues of GG genotype.

Suwannasing et al. [26] investigated Large White and Landrace pigs and established 25 SNPs for the NBA. Of these, 11 SNPs located in SSC1 and SSCX were defined for Large White pigs, 14 SNPs for Landrace pigs in SSC2 and SSC6. It is remarkable that SNPs on SSCX showed significance only at the NBA for Large White sows.

In the course of research on commercial pigs, Li et al. [25] established 2 SNPs (rs342908929 (SSC6) and rs324003968 (SSC15)) associated with NBA. According to the results of Chen et al. [27] obtained from the sows of Duroc in the database presents 9 SNPs and all of them are localized on SSC6.

An et al. [24] studied the *IGFBP2* (SSC15) and *IGFBP3* (SSC18) genes in Berkshire pigs. Their results showed significant SNPs in these genes. This work also analyzed the expression levels of *IGFBP2* and *IGFBP3* mRNA in the endometrium in pigs of various genotypes. Homozygous GG pigs expressed higher levels of *IGFBP3* mRNA in the endometrium than pigs of other genotypes, and a positive correlation was observed between litter size traits and *IGFBP3* but not *IGFBP2* expression level. These results suggest that SNPs in the *IGFBP2* and the *IGFBP3* gene are useful biomarkers for the little traits of pigs. According to the results of this work 2 SNPs are included in PigQTLdb rs45435330 on SSC15 as Genome map association for NBA and TNB.

## 3. SNPs Associated with Total Number Born

In general, in Pig QTLdb for TNB showed 218 SNPs (as Genome map association), of which 155 SNPs were detected by He et al. [8] and Ma et al. [28] in Erhualians, 52 SNPs are defined in Large White, Yorkshire, Landrace, 1 SNP in Berkshire and 10 SNPs in Duroc (Table 2) [14,18,22–25,29–33]. SNPs associated with TNB are represented in all Sus Scrofa Chromosome except for SSCY.

For an associative study He et al. [8] selected Erhualan sows with high and low EBV values. According to the results of their work, the most significant SNPs were detected on SSC2 chromosomes (rs81367039), SSC7 (rs80891106), SSC8 (rs81399474), SSC12 (rs81434499), SSC14 (rs80938898). Among them SNPs on chromosomes SSC2, SSC7, and SSC12 were annotated for the first time. To study the effect of significant SNPs additional studies were conducted on a livestock of 331 Erhualan sows. According to the results of additional testing, significant differences in TNB were found only for SNP rs81399474 (SSC8). In the studies of Ma et al. [28] 8 SNPs were identified as associated with TNB in Erhualian sows and located on SSC1, SSC4, SSC7, SSC8, SSC10, SSC12, SSC13and SSC16.

Studies of Large White pigs conducted by Sell-Kubiak et al. [30] allowed the identification of 10 SNPs associated with the number of piglets at birth and located on SSC1, SSC5, SSC8, SSC11, SSC13 and SSC18. SNPs (rs80989787 and rs81289355) located on SSC11 were annotated for the first time in this paper. The most significant effect has been determined for SNP rs80989787 (SSC11). According to the results of associative studies conducted by Uimari et al. [18] on Finnish Landrace pigs 10 SNPs were identified. All established SNPs are located on SSC9.

The most significant effect was established for SNP rs81300575 (SSC9), which amounted to about 1 piglet between two homozygous genotypes. In the studies of Uimari et al. [18] it was also noted that in the past 15 years the frequency of the desired SNP rs81300575 (SSC9) genotype in the studied population has increased from 0.14 to 0.22.

Zhang et al. [29] conducted research on Duroc pigs and identified 10 SNPs associated with TNB. The most significant SNPs were rs80979042 and rs80825112 located on SSC14. In addition, the remaining potential SNPs were located on SSC5, SSC6, SSC12 and SSC17.

In the studies of Coster et al. [23], Wu et al. [14], Wang et al. [22] and Li et al. [25] the effects of SNPs on TNBs were investigated along with the search of SNPs associated with NBAs. So according to the results of Coster et al. [23] the database contains 16 SNPs defined on SSC1, SSC2, SSC7, SSC14 and SSC18. Wu et al. [14] established 5 SNPs associated with TNB located on SSC8 and SSC14. Wang et al. [22] identified 11 SNPs for TNB on SSC1-SSC5, SSC13 and SSC18, and Li et al. [25] 2 SNPs (rs342908929 (SSC6) and rs324003968 (SSC15)). Besides this, Pig QTLdb presents SNPs associated with TNB according to the studies of Onteru et al. [31]—SNP rs81452018 SSC15, Wang et al. [32]—SNP rs345476947 SSC6 and Liu et al. [33]—rs rs55618224 SSC6.

**Table 2.** SNPs (Single nucleotide polymorphism) associated with TNB (Total number born alive) presented in the Pig QTLdb (Pig Quantitative Trait Locus Database). Ensembl db, Sus Scrofa 11.1 (NA—This variant has not been mapped on Ensembl db, Sus Scrofa 11.1).

| No | SNP | SSC | Location | Allele | Consequence | SYMBOL | Trait | Breed | Reference |
|---|---|---|---|---|---|---|---|---|---|
| 1 | rs80972494 | 1 | 7262020 | C | intergenic_variant | - | TNB | Large White | [22] |
| 2 | rs80787893 | 1 | 14869534 | C | intron_variant | ZBTB2 | TNB | Large White | [30] |
| 3 | rs80851003 | 1 | 20072414 | G | intergenic_variant | - | TNB | Erhualian | [8] |
| 4 | rs80970692 | 1 | 97934013 | A | intron_variant | ZBTB7C | TNB | Large White | [23] |
| 5 | rs80845110 | 1 | 139993309 | A | intron_variant | GABRG3 | TNB | Large White | [22] |
| 6 | rs80933698 | 1 | 140097794 | G | intron_variant | GABRG3 | TNB | Large White | [22] |
| 7 | rs80869858 | 1 | 140178409 | T | intron_variant | GABRG3 | TNB | Large White | [22] |
| 8 | rs80999701 | 1 | 140290677 | A | intron_variant | GABRG3 | TNB | Large White | [22] |
| 9 | rs81348717 | 1 | 140463533 | G | intron_variant | GABRG3 | TNB | Large White | [22] |
| 10 | rs81348724 | 1 | 140550299 | A | intron_variant | GABRA5 | TNB | Large White | [22] |
| 11 | rs81348751 | 1 | 140808020 | A | intron_variant | GABRB3 | TNB | Large White | [22] |
| 12 | rs80989931 | 1 | 140896691 | A | intron_variant | GABRB3 | TNB | Large White | [22] |
| 13 | rs81348779 | 1 | 141989297 | G | downstream_gene_var | UBE3A | TNB | Large White | [22] |
| 14 | rs80912860 | 1 | 164637575 | C | intergenic_variant | - | TNB | Large White | [30] |
| 15 | rs80956812 | 1 | 164674664 | A | intron_variant | SMAD6 | TNB | Large White | [30] |
| 16 | rs81267574 | 1 | 222278217 | G | intron_variant | PIP5K1B | TNB | Erhualian | [8] |
| 17 | rs80938435 | 1 | 245222778 | T | intergenic_variant | - | TNB | Erhualian | [8] |
| 18 | rs81296573 | 1 | 263022859 | A | intergenic_variant | - | TNB | Erhualian | [8] |
| 19 | rs80913204 | 1 | 270166629 | T | intron_variant | FNBP1 | TNB | Erhualian | [8] |
| 20 | rs81367039 | 1 | NA | | | | TNB | Erhualian | [8] |
| 21 | rs81330112 | 2 | 3058633 | A | upstream_gene_var | CTTN | TNB | Erhualian | [8] |
| 22 | rs81214065 | 2 | 7420349 | A | missense_variant | PYGM | TNB | Erhualian | [8] |
| 23 | rs81474834 | 2 | 9252507 | T | intron_variant | - | TNB | Erhualian | [8] |

Table 2. *Cont.*

| No | SNP | SSC | Location | Allele | Consequence | SYMBOL | Trait | Breed | Reference |
|---|---|---|---|---|---|---|---|---|---|
| 24 | rs81273273 | 2 | 24914867 | G | intron_variant | LDLRAD3 | TNB | Erhualian | [8] |
| 25 | rs81356698 | 2 | 28380318 | G | downstream_gene_var | | TNB | Large White | [22] |
| 26 | rs346316162 | 2 | 84269464 | T | intron_variant | ANKRD31 | TNB | Erhualian | [8] |
| 27 | rs81360234 | 2 | 85003252 | G | intron_variant | SV2C | TNB | Erhualian | [8] |
| 28 | rs81270902 | 2 | 136051929 | G | intron_variant | FSTL4 | TNB | Erhualian | [8] |
| 29 | rs81307772 | 2 | 142675693 | C | intron_variant | PCDHAC2 | TNB | Erhualian | [8] |
| 30 | rs81366836 | 2 | 145019988 | C | intergenic_variant | - | TNB | Erhualian | [8] |
| 31 | rs81367208 | 2 | 146880995 | T | intergenic_variant | - | TNB | Erhualian | [8] |
| 32 | rs81346993 | 2 | 148668538 | G | intron_variant | DPYSL3 | TNB | Erhualian | [8] |
| 33 | rs328177895 | 2 | NA | | | | TNB | Large White | [23] |
| 34 | rs81367039 | 2 | NA | | | | TNB | Erhualian | [8] |
| 35 | rs81367039 | 2 | NA | | | | TNB | Erhualian | [8] |
| 36 | rs334519198 | 3 | 5376605 | T | intron_variant | LMTK2 | TNB | Large White | [23] |
| 37 | rs81243084 | 3 | 14585248 | T | intron_variant | AUTS2 | TNB | Erhualian | [8] |
| 38 | rs81318451 | 3 | 20859502 | T | intron_variant | HS3ST4 | TNB | Erhualian | [8] |
| 39 | rs81319541 | 3 | 23688143 | G | downstream_gene_var | IGSF6 | TNB | Erhualian | [8] |
| 40 | rs81379942 | 3 | 30034842 | A | intron_variant | SHISA9 | TNB | Erhualian | [8] |
| 41 | rs81369361 | 3 | 43227069 | A | intergenic_variant | - | TNB | Large White | [23] |
| 42 | rs334867206 | 3 | 43312168 | T | intergenic_variant | - | TNB | Large White | [23] |
| 43 | rs319494663 | 3 | 43463318 | C | upstream_gene_variant | ssc-let-7a-2 | TNB | Large White | [23] |
| 44 | rs338135576 | 3 | 52266739 | T | intron_variant | IL1R1 | TNB | Erhualian | [8] |
| 45 | rs81370592 | 3 | 53699303 | C | intergenic_variant | - | TNB | Erhualian | [8] |
| 46 | rs81377897 | 3 | 124275280 | G | upstream_gene_variant | - | TNB | Erhualian | [8] |
| 47 | rs81272059 | 3 | 125892592 | G | intron_variant | NOL10 | TNB | Erhualian | [8] |

**Table 2.** *Cont.*

| No | SNP | SSC | Location | Allele | Consequence | SYMBOL | Trait | Breed | Reference |
|---|---|---|---|---|---|---|---|---|---|
| 48 | rs81319839 | 4 | 18194352 | A | intergenic_variant | - | TNB | Large White | [23] |
| 49 | rs81312912 | 4 | 18196598 | A | intergenic_variant | - | TNB | Large White | [23] |
| 50 | rs80986621 | 4 | 87990037 | C | missense_variant | UAP1 | TNB | Erhualian | [8] |
| 51 | rs80860510 | 4 | 89154886 | A | intron_variant | SDHC | TNB | Erhualian | [8] |
| 52 | rs80987610 | 4 | 89242132 | G | downstream_gene_var | APOA2 | TNB | Erhualian | [8] |
| 53 | rs80910021 | 4 | 96557774 | C | intergenic_variant | - | TNB | Erhualian | [8] |
| 54 | rs80999559 | 4 | 98260992 | C | downstream_gene_var | CERS2 | TNB | Erhualian | [8] |
| 55 | rs80927364 | 4 | 130597207 | A | intron_variant | DDAH1 | TNB | Erhualian | [8] |
| 56 | rs81385465 | 5 | 4131296 | G | intergenic_variant | - | TNB | Large White | [30] |
| 57 | rs336638152 | 5 | 11274122 | A | intron_variant | - | TNB | Duroc | [29] |
| 58 | rs327336155 | 5 | 31752997 | G | intergenic_variant | - | TNB | Large White | [23] |
| 59 | rs80890539 | 5 | 49084918 | T | intergenic_variant | - | TNB | Erhualian | [8] |
| 60 | rs80999110 | 5 | 65577418 | T | intergenic_variant | - | TNB | Duroc | [29] |
| 61 | rs328217833 | 5 | 78953555 | C | intergenic_variant | - | TNB | Erhualian | [8] |
| 62 | rs81303269 | 5 | 79020148 | A | intergenic_variant | - | TNB | Erhualian | [8] |
| 63 | rs81236331 | 6 | 4395948 | T | intron_variant | WFDC1 | TNB | Erhualian | [8] |
| 64 | rs81307446 | 6 | 4817058 | T | intron_variant | CDH13 | TNB | Erhualian | [8] |
| 65 | rs81393472 | 6 | 21728055 | A | intergenic_variant | - | TNB | Erhualian | [8] |
| 66 | rs345476947 | 6 | 54042595 | T | intron_variant | FUT2 | TNB | Large White | [32] |
| 67 | rs81322640 | 6 | 74250906 | T | intron_variant | KAZN | TNB | Erhualian | [8] |
| 68 | rs81283746 | 6 | 78467732 | C | downstream_gene_var | UBXN10 | TNB | Erhualian | [8] |
| 69 | rs55618224 | 6 | 80088638 | C | 3_prime_UTR_variant | - | TNB | Yorkshire | [33] |
| 70 | rs81318862 | 6 | 82375634 | A | intergenic_variant | - | TNB | Duroc | [29] |

**Table 2.** *Cont.*

| No | SNP | SSC | Location | Allele | Consequence | SYMBOL | Trait | Breed | Reference |
|---|---|---|---|---|---|---|---|---|---|
| 71 | rs329711941 | 6 | 84156205 | C | intron_variant | ZDHHC18 | TNB | Duroc | [29] |
| 72 | rs81391439 | 6 | 129931172 | A | intergenic_variant | - | TNB | Erhualian | [8] |
| 73 | rs342908929 | 6 | 159933806 | T | intron_variant | ZFYVE9 | TNB | Commercia | [25] |
| 74 | rs81319462 | 6 | 163598597 | G | intergenic_variant | - | TNB | Erhualian | [8] |
| 75 | rs81319428 | 6 | 163598619 | C | intergenic_variant | - | TNB | Erhualian | [8] |
| 76 | rs80882306 | 7 | 11003222 | G | upstream_gene_variant | - | TNB | Erhualian | [28] |
| 77 | rs80813007 | 7 | 21583296 | G | downstream_gene_var | - | TNB | Erhualian | [8] |
| 78 | rs80938431 | 7 | 22008244 | G | intron_variant | - | TNB | Erhualian | [8] |
| 79 | rs80933422 | 7 | 27905935 | T | intergenic_variant | - | TNB | Erhualian | [8] |
| 80 | rs80797074 | 7 | 29268803 | G | synonymous_variant | DST | TNB | Erhualian | [8] |
| 81 | rs340672537 | 7 | 38439290 | T | intron_variant | ABCC10 | TNB | Erhualian | [8] |
| 82 | rs81398070 | 7 | 58232894 | C | intron_variant | SIN3A | TNB | Erhualian | [8] |
| 83 | rs80824208 | 7 | 63161681 | G | intron_variant | SLC25A21 | TNB | Erhualian | [8] |
| 84 | rs326644823 | 7 | 63618208 | A | downstream_gene_var | - | TNB | Erhualian | [8] |
| 85 | rs80867596 | 7 | 70133128 | G | downstream_gene_var | - | TNB | Erhualian | [8] |
| 86 | rs80891106 | 7 | 73467314 | A | intergenic_variant | - | TNB | Erhualian | [8] |
| 87 | rs81295302 | 7 | 74333495 | T | intron_variant | STXBP6 | TNB | Erhualian | [8] |
| 88 | rs81398127 | 7 | 75705436 | G | 3_prime_UTR_variant | CMTM5 | TNB | Erhualian | [8] |
| 89 | rs336977324 | 7 | 81077680 | T | intron_variant | RYR3 | TNB | Erhualian | [8] |
| 90 | rs80927564 | 7 | 108030803 | C | intergenic_variant | - | TNB | Erhualian | [8] |
| 91 | rs80942529 | 7 | 110057727 | G | intergenic_variant | - | TNB | Erhualian | [8] |
| 92 | rs80969683 | 7 | NA | | | | TNB | Large White | [22] |
| 93 | rs81397215 | 7 | NA | | | | TNB | Large White | [22] |

**Table 2.** *Cont.*

| No | SNP | SSC | Location | Allele | Consequence | SYMBOL | Trait | Breed | Reference |
|---|---|---|---|---|---|---|---|---|---|
| 94 | rs81397142 | 7 | NA | | | | TNB | Large White | [22] |
| 95 | rs81403620 | 8 | 15474371 | T | intron_variant | KCNIP4 | TNB | Erhualian | [8] |
| 96 | rs81405013 | 8 | 16400498 | C | intergenic_variant | - | TNB | Erhualian | [8] |
| 97 | rs81342198 | 8 | 16560870 | C | intergenic_variant | - | TNB | Erhualian | [8] |
| 98 | rs81406385 | 8 | 17783561 | C | intron_variant | - | TNB | Erhualian | [8] |
| 99 | rs81343566 | 8 | 18863902 | C | intron_variant | CCDC149 | TNB | Erhualian | [8] |
| 100 | rs81399474 | 8 | 32370687 | C | downstream_gene_var | UCHL1 | TNB | Erhualian | [8] |
| 101 | rs81399527 | 8 | 32800253 | G | intergenic_variant | - | TNB | Erhualian | [8] |
| 102 | rs81399633 | 8 | 33508704 | A | splice_region_variant | ATP8A1 | TNB | Erhualian | [8] |
| 103 | rs81227962 | 8 | 34380373 | A | intergenic_variant | - | TNB | Erhualian | [8] |
| 104 | rs81476987 | 8 | 34885027 | C | intron_variant | KCTD8 | TNB | Erhualian | [8] |
| 105 | rs81399897 | 8 | 36628170 | C | intergenic_variant | - | TNB | Erhualian | [8] |
| 106 | rs81400131 | 8 | 38946698 | A | downstream_gene_var | CWH43 | TNB | Erhualian | [8] |
| 107 | rs81400868 | 8 | 63262973 | A | downstream_gene_var | U6 | TNB | Erhualian | [8] |
| 108 | rs81401375 | 8 | 73118551 | A | intergenic_variant | - | TNB | Erhualian | [8] |
| 109 | rs339466191 | 8 | 80345448 | A | intron_variant | NR3C2 | TNB | Landrace/ Yorkshire | [14] |
| 110 | rs333905163 | 8 | 80589152 | A | intron_variant | NR3C2 | TNB | Landrace/ Yorkshire | [14] |
| 111 | rs81294311 | 8 | 88225561 | T | intergenic_variant | - | TNB | Erhualian | [8] |
| 112 | rs319272490 | 8 | 103706053 | C | intergenic_variant | - | TNB | Landrace/ Yorkshire | [14] |
| 113 | rs81403286 | 8 | 107008700 | T | intergenic_variant | - | TNB | Erhualian | [8] |
| 114 | rs81403527 | 8 | 112530303 | A | downstream_gene_var | CASP6 | TNB | Erhualian | [8] |
| 115 | rs81403538 | 8 | 112566311 | T | intron_variant | MCUB | TNB | Erhualian | [8] |
| 116 | rs334180816 | 8 | 115856336 | C | intron_variant | NPNT | TNB | Landrace/ Yorkshire | [14] |
| 117 | rs80834695 | 8 | 120291881 | A | intergenic_variant | - | TNB | Erhualian | [8] |

**Table 2.** *Cont.*

| No | SNP | SSC | Location | Allele | Consequence | SYMBOL | Trait | Breed | Reference |
|---|---|---|---|---|---|---|---|---|---|
| 118 | rs81317149 | 8 | 127090631 | G | intergenic_variant | - | TNB | Large White | [30] |
| 119 | rs81413855 | 9 | 8393770 | G | 3_prime_UTR_variant | C2CD3 | TNB | Erhualian | [8] |
| 120 | rs81416386 | 9 | 11655571 | G | intron_variant | MYO7A | TNB | Erhualian | [8] |
| 121 | rs81407589 | 9 | 21695319 | C | intron_variant | - | TNB | Erhualian | [8] |
| 122 | rs81408950 | 9 | 34248222 | G | intergenic_variant | - | TNB | Erhualian | [8] |
| 123 | rs81409102 | 9 | 35438101 | C | upstream_gene_variant | GRIA4 | TNB | Erhualian | [8] |
| 124 | rs81413928 | 9 | 84593678 | A | intron_variant | AGMO | TNB | Landrace | [18] |
| 125 | rs81413949 | 9 | 85057918 | C | intergenic_variant | - | TNB | Landrace | [18] |
| 126 | rs81287478 | 9 | 85128043 | T | intergenic_variant | - | TNB | Landrace | [18] |
| 127 | rs81223525 | 9 | 85362483 | G | intergenic_variant | - | TNB | Landrace | [18] |
| 128 | rs81260290 | 9 | 87932749 | G | intron_variant | HDAC9 | TNB | Erhualian | [28] |
| 129 | rs81414623 | 9 | 96283483 | C | intron_variant | SEMA3A | TNB | Erhualian | [8] |
| 130 | rs81300575 | 9 | 102308556 | T | intron_variant | PHTF2 | TNB | Landrace | [18] |
| 131 | rs81331059 | 9 | 137095797 | G | intron_variant | - | TNB | Erhualian | [8] |
| 132 | rs81419264 | 9 | 137162279 | G | non_coding_transcript | - | TNB | Erhualian | [8] |
| 133 | rs81419315 | 9 | 137216947 | G | upstream_gene_variant | U6 | TNB | Erhualian | [8] |
| 134 | rs81315852 | 9 | 138107624 | T | intergenic_variant | - | TNB | Erhualian | [8] |
| 135 | rs81332239 | 9 | 138462123 | C | intergenic_variant | - | TNB | Erhualian | [8] |
| 136 | rs81417713 | 9 | NA | | | | TNB | Erhualian | [8] |
| 137 | rs81429231 | 10 | 10002753 | G | intron_variant | MARK1 | TNB | Erhualian | [8] |
| 138 | rs80895456 | 10 | 26509297 | T | intergenic_variant | - | TNB | Erhualian | [8] |
| 139 | rs341908955 | 10 | 48658434 | T | intergenic_variant | - | TNB | Erhualian | [8] |
| 140 | rs81329283 | 10 | 56421545 | A | intron_variant | NRP1 | TNB | Erhualian | [8] |
| 141 | rs81274366 | 10 | 57593745 | G | intron_variant | PARD3 | TNB | Erhualian | [28] |

**Table 2.** *Cont.*

| No | SNP | SSC | Location | Allele | Consequence | SYMBOL | Trait | Breed | Reference |
|---|---|---|---|---|---|---|---|---|---|
| 142 | rs81426281 | 10 | 57913635 | T | upstream_gene_variant | - | TNB | Erhualian | [8] |
| 143 | rs81314128 | 10 | 59665035 | T | intergenic_variant | - | TNB | Erhualian | [8] |
| 144 | rs80941850 | 11 | 8940153 | T | intron_variant | - | TNB | Erhualian | [8] |
| 145 | rs81477765 | 11 | 19542650 | G | intergenic_variant | - | TNB | Erhualian | [8] |
| 146 | rs81332839 | 11 | 19542846 | T | intergenic_variant | - | TNB | Erhualian | [8] |
| 147 | rs81285980 | 11 | 19544324 | A | intergenic_variant | - | TNB | Erhualian | [8] |
| 148 | rs81237348 | 11 | 22281193 | C | intron_variant | NUFIP1 | TNB | Erhualian | [8] |
| 149 | rs80989787 | 11 | 23362846 | T | intron_variant | ENOX1 | TNB | Large White | [30] |
| 150 | rs81289355 | 11 | 23410214 | C | intron_variant | ENOX1 | TNB | Large White | [30] |
| 151 | rs81293918 | 11 | 27751988 | A | intergenic_variant | - | TNB | Erhualian | [8] |
| 152 | rs80813604 | 11 | 48641897 | C | intergenic_variant | - | TNB | Erhualian | [8] |
| 153 | rs80950312 | 11 | 48749391 | T | intergenic_variant | - | TNB | Erhualian | [8] |
| 154 | rs80869156 | 11 | 64316157 | G | intergenic_variant | - | TNB | Erhualian | [8] |
| 155 | rs81329621 | 11 | 66354904 | G | intergenic_variant | - | TNB | Erhualian | [8] |
| 156 | rs81343376 | 12 | 6971812 | G | intergenic_variant | - | TNB | Erhualian | [8] |
| 157 | rs81332319 | 12 | 7414272 | C | intergenic_variant | - | TNB | Erhualian | [8] |
| 158 | rs81439394 | 12 | 9810809 | G | intergenic_variant | - | TNB | Duroc | [29] |
| 159 | rs81433045 | 12 | 26992374 | G | intron_variant | ANKRD40 | TNB | Erhualian | [8] |
| 160 | rs81433877 | 12 | 33577229 | G | intron_variant | MSI2 | TNB | Erhualian | [8] |
| 161 | rs81434044 | 12 | 33799100 | C | intron_variant | MSI2 | TNB | Erhualian | [8] |
| 162 | rs81434064 | 12 | 33903675 | A | intron_variant | MSI2 | TNB | Erhualian | [8] |
| 163 | rs81434489 | 12 | 35993482 | G | intron_variant | VMP1 | TNB | Erhualian | [8] |
| 164 | rs81434499 | 12 | 36207006 | G | intergenic_variant | - | TNB | Erhualian | [8] |

**Table 2.** *Cont.*

| No | SNP | SSC | Location | Allele | Consequence | SYMBOL | Trait | Breed | Reference |
|---|---|---|---|---|---|---|---|---|---|
| 165 | rs81309004 | 12 | 40519573 | T | intergenic_variant | - | TNB | Erhualian | [8] |
| 166 | rs81434931 | 12 | 40974978 | A | intron_variant | ASIC2 | TNB | Erhualian | [8] |
| 167 | rs81435036 | 12 | 41338150 | C | intron_variant | ASIC2 | TNB | Erhualian | [8] |
| 168 | rs81477883 | 12 | 43427846 | T | intron_variant | RAB11FIP4 | TNB | Erhualian | [8] |
| 169 | rs81447979 | 13 | 11362350 | A | upstream_gene_variant | THRB | TNB | Erhualian | [8] |
| 170 | rs80837222 | 13 | 13802573 | T | intron_variant | NEK10 | TNB | Erhualian | [8] |
| 171 | rs81445072 | 13 | 38880472 | C | intron_variant | ARHGEF3 | TNB | Erhualian | [8] |
| 172 | rs80964445 | 13 | 75637294 | G | intron_variant | CEP63 | TNB | Erhualian | [8] |
| 173 | rs81447100 | 13 | 80866929 | A | intron_variant | CLSTN2 | TNB | Erhualian | [8] |
| 174 | rs81441751 | 13 | 182451198 | C | intron_variant | CHODL | TNB | Large White | [30] |
| 175 | rs81442196 | 13 | 190614395 | G | intergenic_variant | - | TNB | Erhualian | [8] |
| 176 | rs80961068 | 13 | 200778735 | T | intron_variant | TTC3 | TNB | Large White | [23] |
| 177 | rs80841768 | 14 | 12118215 | T | intron_variant | PNOC | TNB | Erhualian | [8] |
| 178 | rs330621374 | 14 | 15126027 | A | intergenic_variant | - | TNB | Landrace/ Yorkshire | [14] |
| 179 | rs339777110 | 14 | 25374812 | G | intron_variant | TMEM132D | TNB | Duroc | [29] |
| 180 | rs334650508 | 14 | 26675179 | A | intergenic_variant | - | TNB | Erhualian | [8] |
| 181 | rs318344052 | 14 | 26907303 | T | intergenic_variant | - | TNB | Erhualian | [8] |
| 182 | rs80979042 | 14 | 62040147 | T | intron_variant | BICC1 | TNB | Duroc | [29] |
| 183 | rs80825112 | 14 | 62071531 | A | intron_variant | BICC1 | TNB | Duroc | [29] |
| 184 | rs80892145 | 14 | 73289391 | T | intron_variant | LRRC20 | TNB | Erhualian | [8] |
| 185 | rs80960182 | 14 | 84555165 | A | intergenic_variant | - | TNB | Erhualian | [8] |
| 186 | rs80873788 | 14 | 85481113 | A | upstream_gene_variant | RGR | TNB | Erhualian | [8] |
| 187 | rs80799654 | 14 | 98901034 | T | intergenic_variant | - | TNB | Erhualian | [8] |
| 188 | rs80800806 | 14 | 104041994 | T | intron_variant | IDE | TNB | Erhualian | [8] |

**Table 2.** *Cont.*

| No | SNP | SSC | Location | Allele | Consequence | SYMBOL | Trait | Breed | Reference |
|---|---|---|---|---|---|---|---|---|---|
| 189 | rs80968475 | 14 | 107031142 | T | intron_variant | SORBS1 | TNB | Erhualian | [8] |
| 190 | rs80959797 | 14 | 107336340 | G | intron_variant | ENTPD1 | TNB | Erhualian | [8] |
| 191 | rs80938898 | 14 | 107377294 | C | intron_variant | ENTPD1 | TNB | Erhualian | [8] |
| 192 | rs80971725 | 14 | 107584405 | C | intergenic_variant | - | TNB | Erhualian | [8] |
| 193 | rs80843720 | 14 | 107597908 | T | intron_variant | ZNF518A | TNB | Erhualian | [8] |
| 194 | rs80987149 | 14 | 107725686 | T | intergenic_variant | - | TNB | Erhualian | [8] |
| 195 | rs81449479 | 14 | 107794160 | G | 3_prime_UTR_variant | OPALIN | TNB | Erhualian | [8] |
| 196 | rs80867623 | 14 | 129792523 | G | intron_variant | SEC23IP | TNB | Large White | [22] |
| 197 | rs80983641 | 14 | 137193623 | G | intron_variant | PTPRE | TNB | Erhualian | [8] |
| 198 | rs80892643 | 14 | NA | | | | TNB | Erhualian | [8] |
| 199 | rs80782668 | 14 | NA | | | | TNB | Erhualian | [8] |
| 200 | rs80947288 | 14 | NA | | | | TNB | Duroc | [29] |
| 201 | rs80868389 | 15 | 9150979 | A | intergenic_variant | - | TNB | Erhualian | [8] |
| 202 | rs81452018 | 15 | 26337909 | G | intergenic_variant | - | TNB | Landace/Large White | [31] |
| 203 | rs80893476 | 15 | 82075550 | C | intron_variant | MTX2 | TNB | Erhualian | [8] |
| 204 | rs45435330 | 15 | 118847390 | A | intron_variant | IGFBP2 | TNB | Berkshire | [24] |
| 205 | rs324003968 | 15 | NA | | | | TNB | Commercia | [25] |
| 206 | rs81459026 | 16 | 41470716 | G | intron_variant | IPO11 | TNB | Erhualian | [8] |
| 207 | rs81289648 | 16 | 74903886 | A | intergenic_variant | - | TNB | Erhualian | [28] |
| 208 | rs81463092 | 16 | 74971494 | G | intergenic_variant | - | TNB | Erhualian | [8] |
| 209 | rs81463053 | 16 | NA | | | | TNB | Erhualian | [8] |
| 210 | rs331119584 | 17 | 1486143 | T | intron_variant | DLC1 | TNB | Erhualian | [8] |

**Table 2.** *Cont.*

| No | SNP | SSC | Location | Allele | Consequence | SYMBOL | Trait | Breed | Reference |
|---|---|---|---|---|---|---|---|---|---|
| 211 | rs328230332 | 17 | 34329730 | T | intergenic_variant | - | TNB | Duroc | [29] |
| 212 | rs81472303 | 18 | 18253338 | C | intron_variant | COPG2 | TNB | Large White | [22] |
| 213 | rs81471172 | 18 | 51640938 | T | intron_variant | HECW1 | TNB | Large White | [23] |
| 214 | rs81471381 | 18 | 53672799 | G | intron_variant | SUGCT | TNB | Large White | [30] |
| 215 | rs332595701 | 18 | 53697787 | G | intron_variant | SUGCT | TNB | Large White | [30] |
| 216 | rs81225238 | X | 15335129 | T | intron_variant | PHKA2 | TNB | Erhualian | [8] |
| 217 | rs80891661 | X | 101710962 | C | intergenic_variant | - | TNB | Erhualian | [8] |
| 218 | rs322114058 | X | 101975674 | A | intron_variant | TENM1 | TNB | Erhualian | [8] |

## 4. Potential Candidate Genes for Litter Traits of Pig

Accordingly, Pig QTLdb presents 306 SNPs associated with TNB (218 SNPs) and NBA (88 SNPs) in pigs of various breeds. We should note that 12 SNPs out of 306 SNPs provided in PigQTLdb as a Genome map association with TNB and NBA, are presented twice (add file 1). Perhaps this is due to the fact that the connection of SNPs with TNB and NBA was established in one study, for example, as for SNPs rs81348779 on SSC1 and rs81356698 on SSC2 [23]. These SNPs are localized in the intron of the *UBE3A* (SSC1) and *EIF3M* (SSC2) genes. The *UBE3A* gene encodes *ubiquitin protein ligase E3A*, which plays an essential role in the normal development and functioning of the nervous system, and helps regulate the balance between proteostasis synthesis and degradation in the joints between synapses. Human and mouse *UBE3A* is maternal imprinted [34]. However, there is no precise information regarding imprinting of pig *UBE3A*. According to the study performed by Wang et al. [35] *UBE3A* was not imprinted in the skeletal muscle of neonate pigs of Landrace boars and Laiwu sows cross. Further research will probably provide more information to clarify the effects of the *UBE3A* gene and its and its relation to pig fecundity. The *EIF3M* gene (the eukaryotic translation initiation factor 3 subunit M) is a complex translation initiation factor consisting of 13 subunits (*EIF3A-EIF3M*) and which is involved in mRNA modulation [36]. The *EIF3* complex is necessary for the key stages of protein synthesis initiation [37]. Previous studies have shown that *EIF3M* encodes a protein that is critical for mouse embryonic development [38].

In studies performed by Wang et al. [22] a relationship with TNB and NBA was defined for SNPs rs334867206 (intergenic_variant) and rs319494663 (upstream gene variant *ssc-let-7a-2*) on SSC3 and rs81471172 (intron variant *HECW1*) on SSC18. *Sus scrofa let-7a-2 stem-loop (ssc-let-7a-2)* belongs to miRNAs, a class of small non-coding RNAs (~21 nt) that regulate the mRNAs translation on the post-transcriptional level, mainly by binding their targets with the three prime untranslated region (3′-UTR) [39]. A variety of studies have shown that miRNAs can play a potential regulatory role in porcine ovary, testis and spermatogenesis [40–42].

The *HECW1* gene, also known as *NEDD4-like ubiquitin protein ligase 1* (*NEDL1*), is expressed in human neuronal tissues and enhances p53-mediated apoptotic cell death [43]. Supposedly, it regulates the bone morphogenetic protein signaling pathway during embryonic development and bone remodeling [44]. In the work of Li et al. [25] associations with TNB and NBA are defined for SNP rs342908929, which is localized on SSC6 in the intron of the *ZFYVE9* gene (*zinc finger FYVE-type containing 9 domain*). The protein encoded by *ZFYVE9* is involved in the signaling pathway of *transforming growth factor-beta* (*TGFB*) and directly interacts with *SMAD2* and *SMAD3*, needed for normal follicular development and ovulation [45].

In the studies performed by Ma et al. [28] a relation with TNB and NBA was defined for SNPs rs80882306 (intergenic variant) on SSC7 and rs319494663 (intron variant PARD3) on SSC10. *PARD3* (*PAR-3*) is a *scaffold-like PDZ-containing protein*. *PAR-3* forms a complex with *PAR-6* and *atypical protein kinase C* (*PAR-3-atypical protein kinase C-PAR-6 complex*) and is associated with the establishment of cell polarization [46–48]. McCole [49] argued that mutations in *PARD3* can also influence the recovery of wounds by weakening the response of the epithelial barrier to a damage or inflammation. Concerning its significant role in the regulation of various stages of ovarian development and the control of steroidogenesis in a ripening follicle An et al. [24] studied the *IGFBP-2* gene (SSC15) and defined SNP rs45435330 associated with TNB and NBA [50].

But more interesting are the variants repeated in two independent studies. Thus, rs80927364 (intron variant *DDAH1*) on SSC4, rs81434499 (intergenic variant) on SSC12, rs81447100 (intron variant *CLSTN2*) on SSC13 showed significant associations with TNB and NBA in the studies of He et al. [8] and Ma et al. [28] respectively.

The *DDAH1* gene (dimethylarginine dimethylaminohydrolase 1), along with other *DDAH* genes, is involved in the metabolic control of asymmetric dimethylarginine (ADMA), contributes to the maintenance of vascular homeostasis due to the expansion of blood vessels, suppression of inflammation and inhibiting vascular smooth muscle cells, adhesion and aggregation of platelets [51–53]. Data on the metabolic control of ADMA by *DDAH* genes and their effect on endothelial cells were obtained in animal studies. Transgenic mice with overexpression of *DDAH1* showed a twofold decrease of ADMA in plasma, associated with a twofold increase of NOS activity in tissue [54]. Conversely, the *DDAH1* Knock Out Mice exhibited increased pulmonary endothelial permeability as a result of increased ADMA, which was prevented by over-expression of *DDAH1* and *DDAH2* in endothelial cells [55].

*CLSTN2* (calsintenin 2) is associated with obesity in mammals, especially in the process of increasing adipocytes in visceral tissues and in subcutaneous fat [56]. Santana et al. [57] defined *CLSTN2* as a candidate gene associated with ultrasound-derived measurements of the rib-eye area, backfat thickness and rumpfat thickness in Nellore cattle. Adipose tissue is an active endocrine organ and proteins involved in forming adipose tissue is increasingly attracting attention as mammalian reproduction markers [58,59].

We can also note SNPs rs81463092 (intergenic variant) and rs81289648 (intergenic variant) defined in the works of He et al. [8] and Ma et al. [28], localized on SSC16 and lying in close proximity to each other (68 kb). The table shows the intervals between neighboring SNPs (add file 1). Hence, we can note a number of closely related SNPs identified in various studies. For example, rs55618224 (3 prime UTR variant) and rs81388947 (intergenic variant) on SSC6 are defined by Liu et al. [33] and Chen et al. [27] and the interval between them is 140 kb. SNP rs336670754 (synonymous variant *ARID1A*) and SNP rs329711941 (intron variant *ZDHHC18*) are located at a distance of about 39 kb from each other, defined on SSC6 by Chen et al., (2019) and Zhang et al. [29] respectively.

On the whole, we can note that SNPs presented in PigQTLdb as associations with TNB and NBA are more localized in genes (intron variant 60%). Wherein that, 15% are intergenic variants and this can be considered as an evidence of significance for intergenic variants in the genetic architecture of reproductive traits (add file 2).

In conclusion, we wanted to estimate the gene-based protein-protein interactions obtained with the studied SNPs being localized. The total number of genes was 127, but only 40 genes formed some pairs or chains (Figure 1). It is interesting that some chains are composed of genes (SNPs) identified in only one study (add file 3). For example, *RBP7, LRRK1, UBE4B, TRPC5, LHFPL1* genes were identified by Suwannasing et al. [26]; *DLC1, SEMA3A, DPYSL3, NRPI* by He et al. [8] similarly. However, the other chains includes genes identified in various studies such as *GRIA4* [8], *CLSTN2* [8], *GRIP1* [28], *GABRA5* [22], *UBE3A* [22], *COPG2* [22], *ANKRD40* [8], *MYO10* [7], *SHISA9* [8].

The presented review shows that most SNPs annotated from genome-wide studies are found in chromosome regions where candidate genes or QTLs have already been identified. Future research will be aimed at annotating sequences and analyzing these data, which can contribute to better understanding the mechanisms of reproductive traits formation.

However, the main problem in summarizing the results is the design of the experiment (consideration of traits, features of the studied populations, etc.). Some of the factors affecting the result are not controlled by the researcher (such as the trait inheritance level, the genetic architecture of the population, linkage disequilibrium, etc.). On the other hand, control over the reference data design (the choice of markers, the model for evaluating effects, etc.) that affects the accuracy of the results is controlled by the researcher. In the analyzed studies the majority of genotyping works were done by using Illumina PorcineSNP60 BeadChip or GeneSeek PorcineSNP80 BeadChip. There are numerous statistical approaches to conducting GWAS. However, the mixed model is more preferable, being implemented in various software packages (GEMMA, ASREML, GenABEL, etc.). There is still no consensus on the best method. Many researchers emphasize that the defined associations and their significance depend on the methodologies and details of data analysis, and we need to develop statistical approaches in order to improve the accuracy of the obtained information [16,19].

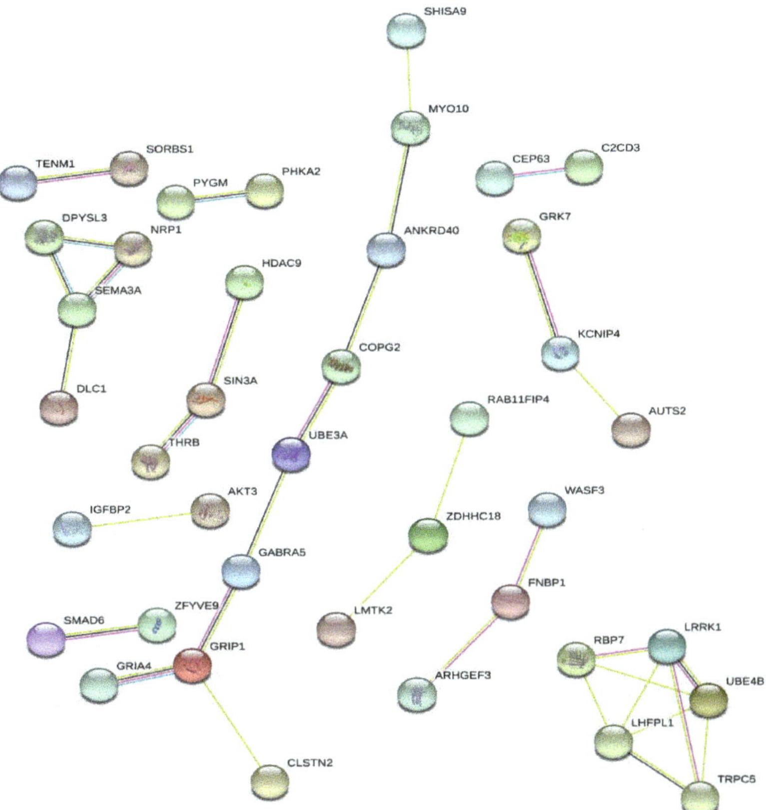


**Figure 1.** Protein-protein interactions obtained with the studied SNPs (STRING db).

## 5. Conclusions

The results of genome studies show the prospects of this approach to studying the genetic architecture of reproductive indicators in pigs. The material presented in this overview can be used in developing local test systems for a limited number of SNPs to estimate their effect on the own livestock. In addition, gene networks can be built on the basis of the presented SNPs to find potential candidate genes for reproductive signs. On the whole, these results help us to understand the genetic basis of pig reproductive traits and can be used in further studies. Further research in this direction will provide new data that will be a powerful impetus for creating breakthrough breeding technologies and improving the efficiency of breeding production in pig farming.

**Supplementary Materials:** The datasets produced and/or analyzed during the current study are available from the addition files. The following are available online at http://www.mdpi.com/2073-4425/11/5/491/s1, file 1: List of SNPs, file 2: Consequences of SNPs, file 3: List of genes in which SNP is defined.

**Author Contributions:** S.B., L.G., M.K.: Designed survey and wrote the paper, F.B., V.G., A.K., O.K.: Collected and analyzed works. All authors have read and agreed to the published version of the manuscript.

**Funding:** This research was supported by the Russian Scientific Foundation (RSF) within Project No. 19-76-10012.

**Conflicts of Interest:** The authors declare no conflict of interest.

**Ethics:** This article is original and contains unpublished materials. The corresponding author confirms that all of the other authors have read and approved the manuscript and no ethical issues involved.

## References


1. Choi, I.; Steibel, J.P.; Bates, R.O.; Raney, N.E.; Rumph, J.M.; Ernst, C.W. Identification of carcass and meat quality QTL in an F2 Duroc × Pietrain pig resource population using different least-squares analysis models. *Front. Genet.* **2011**, *2*, 18. [CrossRef] [PubMed]
2. Dragos-Wendrich, M.; Moser, G.; Bartenschlager, H.; Reiner, G.; Geldermann, H. Linkage and QTL mapping for Sus scrofa chromosome 10. *J. Anim. Breed. Genet.* **2003**, *120*, 82–88. [CrossRef]
3. Tuggle, C.K.; Wang, Y.F.; Couture, O. Advances in Swine Transcriptomics. *Int. J. Biol. Sci.* **2007**, *3*, 132–152. [CrossRef] [PubMed]
4. Pig QTLdb. Available online: https://www.animalgenome.org/cgi-bin/QTLdb/SS/index (accessed on 10 March 2020).
5. Rothschild, M.F.; Hu, Z.L.; Jiang, Z. Advances in QTL mapping in pigs. *Int J. Biol. Sci.* **2007**, *3*, 192–197. [CrossRef] [PubMed]
6. Samorè, A.B.; Fontanesi, L. Genomic selection in pigs: State of the art and perspectives. *Ital. J. Anim. Sci.* **2016**, *15*, 211–232. [CrossRef]
7. Bergfelder-Drüing, S.; Grosse-Brinkhaus, C.; Lind, B.; Erbe, M.; Schellander, K.; Simianer, H.; Tholen, E. A Genome-Wide Association Study in Large White and Landrace Pig Populations for Number Piglets Born Alive. *PLoS ONE* **2015**, *10*. [CrossRef] [PubMed]
8. He, L.C.; Li, P.H.; Ma, X.; Sui, S.P.; Gao, S.; Kim, S.W.; Gu, Y.Q.; Huang, Y.; Ding, N.S.; Huang, R.H. Identification of new single nucleotide polymorphisms affecting total number born and candidate genes related to ovulation rate in Chinese Erhualian pigs. *Anim. Genet.* **2016**, *48*, 48–54. [CrossRef]
9. Knol, E.F.; Nielsen, B.; Knap, P.W. Genomic selection in commercial pig breeding. *Anim. Front.* **2016**, *6*, 15–22. [CrossRef]
10. Lopes, M.S.; Bastiaansen, J.W.M.; Harlizius, B.; Knol, E.F.; Bovenhuis, H.A. Genome-Wide Association Study Reveals Dominance Effects on Number of Teats in Pigs. *PLoS ONE* **2014**, *9*. [CrossRef]
11. Stranger, B.; Stahl, E.; Raj, T. Progress and Promise of Genome-Wide Association Studies for Human Complex Trait Genetics. *Genetics* **2011**, *187*, 367–383. [CrossRef]
12. McEntyre, J.; Ostell, J. *The NCBI Handbook—Bethesda (MD)*; National Center for Biotechnology Information (US): Bethesda, MD, USA, 2015; Chapter 5.
13. Jiang, Z.; Wang, H.; Michal, J.J.; Zhou, X.; Liu, B.; Woods, L.C.S.; Fuchs, R.A. Genome Wide Sampling Sequencing for SNP Genotyping: Methods, Challenges and Future Development. *Int. J. Biol. Sci.* **2016**, *12*, 100–108. [CrossRef] [PubMed]
14. Wu, P.; Yang, Q.; Wang, K.; Zhou, J.; Ma, J.; Tang, Q.; Jin, L.; Xiao, W.; Jiang, A.; Jiang, Y.; et al. Single step genome-wide association studies based on genotyping by sequence data reveals novel loci for the litter traits of domestic pigs. *Genomics* **2018**, *110*, 171–179. [CrossRef] [PubMed]
15. Aliloo, H.; Pryce, J.E.; González-Recio, O.; Cocks, B.G.; Hayes, B.J. Accounting for dominance to improve genomic evaluations of dairy cows for fertility and milk production traits. *Genet. Sel. Evol.* **2016**, *48*, 186. [CrossRef] [PubMed]
16. Guo, Y.; Huang, Y.; Hou, L.; Ma, J.; Chen, C.; Ai, H.; Huang, L.; Ren, J. Genome-wide detection of genetic markers associated with growth and fatness in four pig populations using four approaches. *Genet. Sel. Evol. GSE* **2017**, *49*, 21. [CrossRef] [PubMed]
17. Häggman, J.; Uimari, P. Novel harmful recessive haplotypes for reproductive traits in pigs. *J. Anim. Breed. Genet.* **2017**, *134*, 129–135. [CrossRef]
18. Uimari, P.; Sironen, A.; Sevón-Aimonen, M.-L. Whole-genome SNP association analysis of reproduction traits in the Finnish Landrace pig breed. *Genet. Sel. Evol.* **2011**, *43*, 42. [CrossRef]
19. Wen, Y.J.; Zhang, H.; Ni, Y.L.; Huang, B.; Zhang, J.; Feng, J.Y.; Wang, S.B.; Dunwell, J.M.; Zhang, Y.M.; Wu, R. Methodological implementation of mixed linear models in multi-locus genome-wide association studies. *Brief. Bioinform.* **2017**, *19*, 7007–7012. [CrossRef]
20. Zhang, T.; Wang, L.-G.; Shi, H.-B.; Yan, H.; Zhang, L.-C.; Liu, X.; Lei, P.U.; Liang, J.; Zhang, Y.; Zhao, K.; et al. Hritabilities and genetic and phenotypic correlations of litter uniformity and litter size in Large White sows. *J. Integr. Agric.* **2016**, *15*, 848–854. [CrossRef]
21. Mucha, A.; Piórkowska, K.; Ropka-Molik, K.; Szyndler-Nędza, M. New polymorphic changes in the WNT7A gene and their effect on reproductive traits in pigs. *Ann. Anim. Sci.* **2018**, *18*, 375–385. [CrossRef]

22. Wang, Y.; Ding, X.; Tan, Z.; Xing, K.; Yang, T.; Wang, Y.; Sun, D.; Wang, C. Genome-wide association study for reproductive traits in a Large White pig population. *Anim. Genet.* **2018**, *49*, 127–131. [CrossRef]
23. Coster, A.; Madsen, O.; Heuven, H.C.M.; Dibbits, B.; Groenen, M.A.M.; van Arendonk, J.A.M.; Bovenhuis, H. The Imprinted Gene DIO3 Is a Candidate Gene for Litter Size in Pigs. *PLoS ONE* **2012**, *7*, e31825. [CrossRef]
24. An, S.M.; Hwang, J.H.; Kwon, S.; Yu, G.E.; Park, D.H.; Kang, D.G.; Kim, T.W.; Park, H.C.; Ha, J.; Kim, C.W. Effect of Single Nucleotide Polymorphisms in IGFBP2 and IGFBP3 Genes on Litter Size Traits in Berkshire Pigs. *Anim. Biotechnol.* **2017**, *29*, 301–308. [CrossRef] [PubMed]
25. Li, X.; Ye, J.; Han, X.; Qiao, R.; Li, X.; Lv, G.; Wang, K. Whole-genome sequencing identifies potential candidate genes for reproductive traits in pigs. *Genomics* **2020**, *112*, 199–206. [CrossRef] [PubMed]
26. Suwannasing, R.; Duangjinda, M.; Boonkum, W.; Taharnklaew, R.; Tuangsithtanon, K. The identification of novel regions for reproduction trait in Landrace and Large White pigs using a single step genome-wide association study. *Asian-Australas. J. Anim. Sci.* **2018**, *31*, 1852–1862. [CrossRef] [PubMed]
27. Chen, Z.; Ye, S.; Teng, J.; Diao, S.; Yuan, X.; Chen, Z.; Zhang, H.; Li, J.; Zhang, Z. Genome-wide association studies for the number of animals born alive and dead in duroc pigs. *Theriogenology* **2019**, *139*, 36–42. [CrossRef]
28. Ma, X.; Li, P.H.; Zhu, M.X.; He, L.C.; Sui, S.P.; Gao, S.; Su, G.S.; Ding, N.S.; Huang, Y.; Lu, Z.Q.; et al. Genome-wide association analysis reveals genomic regions on Chromosome 13 affecting litter size and candidate genes for uterine horn length in Erhualian pigs. *Animal* **2018**, *14*, 1–9. [CrossRef]
29. Zhang, Z.; Chen, Z.; Ye, S.; He, Y.; Huang, S.; Yuan, X.; Chen, Z.; Zhang, H.; Li, J. Genome-Wide Association Study for Reproductive Traits in a Duroc Pig Population. *Animals* **2019**, *26*, 732. [CrossRef]
30. Sell-Kubiak, E.; Duijvesteijn, N.; Lopes, M.S.; Janss, L.L.G.; Knol, E.F.; Bijmaand, P.; Mulder, H.A. Genome-wide association study reveals novel loci for litter size and its variability in a Large White pig population. *BMC Genomics* **2015**, *16*, 1049. [CrossRef]
31. Onteru, S.K.; Fan, B.; Du, Z.Q.; Garrick, D.J.; Stalder, K.J.; Rothschild, M.F. A whole-genome association study for pig reproductive traits. *Anim. Genet.* **2012**, *43*, 18–26. [CrossRef]
32. Wang, H.; Wu, S.; Wu, J.; Sun, S.; Wu, S.; Bao, W. Association analysis of the SNP (rs345476947) in the FUT2 gene with the production and reproductive traits in pigs. *Genes Genom.* **2018**, *40*, 199–206. [CrossRef]
33. Liu, R.; Deng, D.; Liu, X.; Xiao, Y.; Huang, J.; Wang, F.; Li, X.; Yu, M. A miR-18a binding-site polymorphism in CDC42 3'UTR affects CDC42 mRNA expression in placentas and is associated with litter size in pigs. *Mamm. Genom.* **2019**, *30*, 34–41. [CrossRef] [PubMed]
34. LaSalle, J.M.; Reiter, L.T.; Chamberlain, S.J. Epigenetic regulation of UBE3A and roles in human neurodevelopmental disorders. *Epigenomics* **2015**, *7*, 1213–1228. [CrossRef]
35. Wang, M.; Zhang, X.; Kang, L.; Jiang, C.; Jiang, Y. Molecular characterization of porcine NECD, SNRPN and UBE3A genes and imprinting status in the skeletal muscle of neonate pigs. *Mol. Biol. Rep.* **2012**, *39*, 9415–9422. [CrossRef] [PubMed]
36. Cate, J.H. Human eIF3: From 'blobology' to biological insight. *Philos. Trans. R. Soc. Lond. B Biol. Sci.* **2017**, *19*, 1716. [CrossRef] [PubMed]
37. Lee, A.S.; Kranzusch, P.J.; Doudna, J.A.; Cate, J.H. eIF3d is an mRNA cap-binding protein that is required for specialized translation initiation. *Nature* **2016**, *536*, 96–99. [CrossRef]
38. Zeng, L.; Wan, Y.; Li, D.; Wu, J.; Shao, M.; Chen, J.; Hui, L.; Ji, H.; Zhu, X. The m subunit of murine translation initiation factor EIF3Maintains the integrity of the eIF3 complex and is required for embryonic development, homeostasis, and organ size control. *J. Biol. Chem.* **2013**, *288*, 30087–30093. [CrossRef]
39. Ye, R.-S.; Xi, Q.-Y.; Qi, Q.; Cheng, X.; Chen, T.; Li, H.; Kallon, S.; Shu, G.; Wang, S.; Jiang, Q.; et al. Differentially Expressed miRNAs after GnRH Treatment and Their Potential Roles in FSH Regulation in Porcine Anterior Pituitary Cell. *PLoS ONE* **2013**, *8*, e57156. [CrossRef]
40. Curry, E.; Ellis, S.; Pratt, S. Detection of porcine sperm microRNAs using a heterologous microRNA microarray and reverse transcriptase polymerase chain reaction. *Mol. Reprod. Dev.* **2009**, *76*, 218–219. [CrossRef]
41. Curry, E.; Safranski, T.J.; Pratt, S.L. Differential expression of porcine sperm microRNAs and their association with sperm morphology and motility. *Theriogenology* **2011**, *76*, 1532–1539. [CrossRef]
42. Li, M.; Liu, Y.; Wang, T.; Guan, J.; Luo, Z.; Chen, H.; Wang, X.; Chen, L.; Ma, J.; Mu, Z.; et al. Repertoire of porcine microRNAs in adult ovary and testis by deep sequencing. *Int. J. Biol. Sci.* **2011**, *7*, 1045. [CrossRef]

43. Liu, Y.; Zheng, J.; Jia, J.; Li, H.; Hu, S.; Lin, Y.; Yan, T. Changes E3 ubiquitin protein ligase 1 gene mRNA expression correlated with IgA1 glycosylation in patients with IgA nephropathy. *Renal Fail.* **2019**, *41*, 370–376. [CrossRef] [PubMed]
44. Zhang, L.; Haraguchi, S.; Koda, T.; Hashimoto, K.; Nakagawara, A. Muscle atrophy and motor neuron degeneration in human nedl1 transgenic mice. *J. Biomed. Biotechnol.* **2011**, *2011*. [CrossRef] [PubMed]
45. Schneider, J.F.; Nonneman, D.J.; Wiedmann, R.T.; Vallet, J.L.; Rohrer, G.A. Genomewide association and identification of candidate genes for ovulation rate in swine. *J. Anim. Sci.* **2014**, *92*, 3792–3803. [CrossRef] [PubMed]
46. Mizuno, K.; Suzuki, A.; Hirose, T.; Kitamura, K.; Kutsuzawa, K.; Futaki, M.; Amano, Y.; Ohno, S. Self-association of PAR-3-mediated by the conserved N-terminal domain contributes to the development of epithelial tight junctions. *J. Biol. Chem.* **2003**, *278*, 31240–31250. [CrossRef] [PubMed]
47. Zen, K.; Yasui, K.; Gen, Y.; Dohi, O.; Wakabayashi, N.; Mitsufuji, S.; Itoh, Y.; Zen, Y.; Nakanuma, Y.; Taniwaki, M.; et al. Defective expression of polarity protein PAR-3 gene (PARD3) in esophageal squamous cell carcinoma. *Oncogene* **2009**, *28*, 2910–2918. [CrossRef] [PubMed]
48. Kim, S.K.; Lee, J.Y.; Park, H.J.; Kim, J.W.; Chung, J.H. Association study between polymorphisms of the PARD3 gene and schizophrenia. *Exp. Ther. Med.* **2012**, *3*, 881–885. [CrossRef]
49. McCole, D.F. IBD candidate genes and intestinal barrier regulation. *Inflamm. Bowel Dis.* **2014**, *20*, 1829–1849. [CrossRef]
50. Spitschak, M.; Hoeflich, A. Potential Functions of IGFBP-2 for Ovarian Folliculogenesis and Steroidogenesis. *Front. Endocrinol.* **2018**, *9*, 119. [CrossRef]
51. Dimmeler, S.; Hermann, C.; Galle, J.; Zeiher, A.M. Upregulation of superoxide dismutase and nitric oxide synthase mediates the apoptosis-suppressive effects of shear stress on endothelial cells. *Arterioscler. Thromb. Vasc. Biol.* **1999**, *19*, 656–664. [CrossRef]
52. Wang, B.Y.; Ho, H.K.; Lin, P.S.; Schwarzacher, S.P.; Pollman, M.J.; Gibbons, G.H.; Tsao, P.S.; Cooke, J.P. Regression of atherosclerosis: Role of nitric oxide and apoptosis. *Circulation* **1999**, *99*, 1236–1241. [CrossRef]
53. Abhary, S.; Burdon, K.P.; Kuot, A.; Javadiyan, S.; Whiting, M.J.; Kasmeridis, N.; Petrovsky, N.; Craig, J.E. Sequence Variation in DDAH1 and DDAH2 Genes Is Strongly and Additively Associated with Serum ADMA Concentrations in Individuals with Type 2 Diabetes. *PLoS ONE* **2010**, *5*, e9462. [CrossRef] [PubMed]
54. Dayoub, H.; Achan, V.; Adimoolam, S.; Jacobi, J.; Stuehlinger, M.C.; Wang, B.; Tsao, P.S.; Kimoto, M.; Vallance, P.; Patterson, A.J.; et al. Dimethylarginine dimethylaminohydrolase regulates nitric oxide synthesis: Genetic and physiological evidence. *Circulation* **2003**, *108*, 3042–3047. [CrossRef] [PubMed]
55. Wojciak-Stothard, B.; Torondel, B.; Zhao, L.; Renne, T.; Leiper, J.M. Modulation of Rac1 activity by ADMA/DDAH regulates pulmonary endothelial barrier function. *Mol. Biol. Cell* **2009**, *20*, 33–42. [CrossRef] [PubMed]
56. Ugi, S.; Maeda, S.; Kawamura, Y.; Kobayashi, M.-A.; Imam-ura, M.; Yoshizaki, T.; Morino, K.; Sekine, O.; Yamamoto, H.; Tani, T.; et al. CCDC3 is specifically upregulated in omental adi-pose tissue in subjects with abdominal obesity. *Obesity (Silver Spring)* **2014**, *22*, 1070–1077. [CrossRef]
57. Santana, M.H.; Ventura, R.V.; Utsunomiya, Y.T.; Neves, H.H.; Alexandre, P.A.; Oliveira Junior, G.A.; Gomes, R.C.; Bonin, M.N.; Coutinho, L.L.; Garcia, J.F.; et al. A genomewide association mapping study using ultrasound-scanned information identifies potential genomic regions and candidate genes affecting carcass traits in Nellore cattle. *J. Anim. Breed. Genet.* **2015**, *132*, 420–427. [CrossRef]
58. Getmantseva, L.; Kolosov, A.; Leonova, M.; Bakoev, S.; Klimenko, A.; Usatov, A. Polymorphism in obesity-related leptin gene and its assotiotion with reproductive traits of sows. *Bulg. J. Agric. Sci.* **2017**, *23*, 843–850.
59. Budak, E.; Fernández Sánchez, M.; Bellver, J.; Cerveró, A.; Simón, C.; Pellicer, A. Interactions of the hormones leptin, ghrelin, adiponectin, resistin, and PYY3–36 with the reproductive system. *Fertil. Steril.* **2006**, *85*, 1563–1581. [CrossRef]

*Article*

# AQP3 Facilitates Proliferation and Adipogenic Differentiation of Porcine Intramuscular Adipocytes

**Xiaoyu Wang, Jing Yang, Ying Yao, Xin'E Shi, Gongshe Yang and Xiao Li ***

Key Laboratory of Animal Genetics, Breeding and Reproduction of Shaanxi Province, College of Animal Science and Technology, Northwest A & F University, Yangling 712100, China; wangxy067@163.com (X.W.); yangjing@nwsuaf.edu.cn (J.Y.); yaoying@nwsuaf.edu.cn (Y.Y.); xineshi@163.com (X.S.); gsyang999@hotmail.com (G.Y.)

* Correspondence: nice.lixiao@gmail.com; Tel.: +86-29-870-81531

Received: 19 March 2020; Accepted: 13 April 2020; Published: 22 April 2020

**Abstract:** The meat quality of animal products is closely related to the intramuscular fat content. Aquaglyceroporin (AQP) defines a class of water/glycerol channels that primarily facilitate the passive transport of glycerol and water across biological membranes. In this study, the AQP3 protein of the AQP family was mainly studied in the adipogenic function of intramuscular adipocytes in pigs. Here, we found that AQP3 was increased at both mRNA and protein levels upon adipogenic stimuli in porcine intramuscular adipocytes in vitro. Western blot results showed knockdown of AQP3 by siRNA significantly suppressed the expression of adipogenic genes (PPARγ, aP2, etc.), repressed Akt phosphorylation, as well as reducing lipid accumulation. Furthermore, deletion of AQP3 by siRNA significantly downregulated expression of cell cycle genes (cyclin D, E), and decreased the number of EdU-positive cells as well as cell viability. Collectively, our data indicate that AQP3 is of great importance in both adipogenic differentiation and proliferation in intramuscular adipocytes, providing a potential target for modulating fat infiltration in skeletal muscles.

**Keywords:** AQP3; pig; intramuscular fat; adipogenesis; proliferation

---

## 1. Introduction

Ectopic fat deposition in skeletal muscle has attracted increasing attention in recent decades. In humans, excessive fat accumulation in skeletal muscle always represents muscle weakness, myopathy, and metabolic diseases such as obesity, diabetes, coronary heart disease, etc. [1]. Conversely, in livestock animals (cattle, pig, sheep, etc.), distribution of intramuscular fat (IMF), referred to as marbling by customers, is closely related to meat quality, and a moderate increase in IMF benefits the taste and flavor of meat products [2]. The high content of intramuscular fat can increase the tenderness and flavor of pork [3]. Marbling develops either by an increase in adipocyte number, or adipocyte volume, or both. Thus, the molecular mechanisms underlying the proliferation and differentiation of intramuscular adipocytes deserve further study.

Aquaglyceroporin (AQP) refers to a subgroup of aquaporins that conduct glycerol, water, and other small polar solutes in response to osmotic gradients. Glycerol is a necessary constituent of triglyceride (TG) backbones, and glycerol uptake together with release across the plasma membrane are two key steps in triglyceride synthesis (lipogenesis) and hydrolysis (lipolysis) in adipose, liver, and other metabolic organs, thus AQPs have emerged as key players in adipose biology and the development of obesity [4]. In this context, AQP7, the first identified AQP in adipose tissue, has been emerging as an important player in whole-body metabolism and the progress of obesity and diabetes [5]. In recent studies, expression of AQP3, AQP9, AQP10, and AQP11 has been detected in cultured adipocytes and adipose tissues [6–8]. Studies have shown that AQP3 can promote the transport of cellular glycerol and has affinity for glycerol [9]. AQP3 has been reported to regulate PPARα by adiponectin in hepatic stellate cells (HSC) [10]. The classical secretion factor, leptin, of adipocytes can improve systemic obesity and fatty liver in mice via AQP3 in ob/ob mice [11]. AQP3 mRNA expression in human adipose tissue was reported in another study [12]. These signs indicate that AQP3 may also have a regulatory role in IMF.

Our RNA-seq screening revealed that AQP3 [13], in addition to the well-known AQP7, was upregulated during adipogenic differentiation in porcine intramuscular preadipocytes (PIPAs), indicating a potential role of AQP3 in porcine IMF deposition. Work on the MDA-MB-231 breast cancer cell line has shown that knockdown of AQP3 modestly decreases water permeability (17%), but markedly decreases glycerol permeability (77%) [14], indicating AQP3 might be more permeable to glycerol. Currently, the most well-characterized role of AQP3 is the promotion of cancer metastasis, for AQP3 is abnormally escalated in various kinds of cancers [15] and knockdown of the AQP3 gene could significantly decrease cell proliferation, and increase cell death or apoptosis in cancer cells [14,16,17]. In addition, AQP3 deficiency can cause proliferation disorders and metabolic inhibition in gastric cancer cells [18]. More interestingly, another study showed that AQP3 in gastric cancer cells caused apoptosis in gastric cancer cells by downregulating cellular glycerol intake and inhibiting downstream adipogenesis [19]. These data encouraged us to explore the effects of AQP3 on adipogenic differentiation, lipid deposition, and proliferation, using porcine intramuscular preadipocytes as a model.

## 2. Materials and Methods

### 2.1. Animal Care

Piglets in our study were obtained from the experimental plot of Northwest A&F University (Yangling, China). Pigs were reared under standard light and temperature conditions and allowed food and water ad libitum. This project was approved by the Institutional Animal Care and Use Committee of Northwest A&F University.

### 2.2. Cell Culture

Porcine intramuscular preadipocytes were isolated from the longissimus dorsi muscle of 3-day-old piglets as previously described [20]. The specific method was as follows: LD (longissimus dorsi) muscles were quickly excised, rinsed twice in sterile pre-cooled phosphate-buffered saline (PBS), and then cut into 1 $mm^3$ sections. Muscle fragments were incubated in Dulbecco's Modified Eagle's Medium/F12 (DMEM/F12; Hyclone, Logan, UT, USA) containing 0.1% I type collagenase (270 U/mg; Gibco, Carlsbad, CA) for 1.5 h in a 37 °C water bath, with continuous shaking. The products were then sequentially passed through a 70 mesh (212 μm) and then a 200 mesh (75 μm) to obtain single cells. The cells were seeded in a dish with DMEM/F12 medium containing 10% fetal bovine serum (Gibco, Grand Island, NY, USA). After 2 h, we changed the medium to keep only adherent cells. In the proliferating stage, the primary preadipocytes were cultured in DMEM/F12 (Gibco, Grand Island, NY, USA) containing 10% FBS (Invitrogen, Carlsbad, CA, USA).

To induce adipogenic differentiation, when cells achieved 100% confluence, a mixture containing 10% FBS, 5 μg/mL (872 nM) insulin, 1 μM dexamethasone (DEX), and 0.5 mM isobutyl methylxanthine (IBMX; Sigma-Aldrich, St Louis, MO) was used to induce adipogenic differentiation. Two days later, a DMEM/F12 medium containing 10% fetal bovine serum (FBS) and 5 μg/mL (872 nM) insulin was changed to maintain differentiation.

### 2.3. Transfection with siRNA

Oligonucleotides of AQP3 siRNAs (forward: CCCUUAUCCUCGUGAUGUUTT, reverse: AACAUCACGAGGAUAAGGGTT) and NC (negative control, forward: UUCUCCGAACGUGUCACGUTT, reverse: ACGUGACACGUUCGGAGAATT) were obtained from GenePharma (Shanghai, China). Transfection was performed with Lipofectamine® RNAiMAX Reagent (ThermoFisher, Waltham, MA, USA) when cells reached proper confluence (40–50% for proliferation test, 70–80% for differentiation test). Negative control and siRNA were added to Opti-MEM (Gibco) and mixed with Lipofectamine® RNAiMAX Reagent. The mixture was allowed to stand for 20 min before being added to the cell culture plate. The culture medium was replaced after 24 h. The final concentration of siRNA or negative control was 50 nM. Cells were changed into fresh growth medium 24 h post-transfection.

### 2.4. RNA Isolation and RT-qPCR

Total RNA was purified using Trizol (TaKaRa Bio, Inc., Dalian, China) and was subjected to reverse transcription using the PrimeScriptTM RT reagent Kit (TaKaRa Bio, Inc., Dalian, China). A tissue sample (0.5 g) from a 180-day-old pig was weighed, and high-throughput grinding of the tissue sample in Trizol infiltration extracted total RNA. Each experimental group was subjected to the reverse transcription reaction with 500 ng of RNA. cDNA was subjected to the Multicolor Real-Time PCR detection system (iQ5, Bio-Rad Laboratories, Inc., Hercules, CA, USA) with SYBR Premix Ex TaqTM II kit (TaKaRa Bio, Inc., Dalian, China). The procedure of PCR reaction was pre-denaturation for 5 min, followed by denaturation for 10 s, annealing for 30 s, and extension of 30 s for 35 cycles. Primers targeting AQP3 [21], PPARγ [22] (peroxisome proliferator activated receptor γ), FABP4 [23] (adipocyte fatty-acid binding protein 4), mGPAT [24] (mitochondrial glycerol-3-phosphate acyltransferase), Perilipin 1 [25], cyclin B [26], and β-actin [27] were picked out in previous reports. Primers for SCD (stearoyl-CoA desaturase), CD36, C/EBPα (CCAAT/enhancer binding protein α), ELOVL6 (elongase of long-chain fatty acids family), FASN (fatty acid synthetase), ACACA (acetyl-CoA carboxylase), DGAT2 (diacylglycerol O-acyltransferase 2), cyclin E, and cyclin D were designed online (https://www.ncbi.nlm.nih.gov/tools/primer-blast/) and synthesized by Sangon Biotech (Shanghai, China). Relative expression of each gene was calculated using the 2−ΔΔ Ct method, using β-actin as the internal control. Sequences for all primers are shown in Table 1.

**Table 1.** Primer sequences for real-time qPCR.

| Gene | Accession Number | Primer Sequences | Production Length/bp |
|---|---|---|---|
| **AQP3** | NM_001110172.1 | F: CACCTCCATGGGCTTCAACT<br>R: TGCCCATTCGCATCTACTCC | 278 |
| **PPARγ** | NM_214379 | F: AGGACTACCAAAGTGCCATCAAA<br>R: GAGGCTTTATCCCCACAGACAC | 142 |
| **aP2** | NM_001002817.1 | F: GAGCACCATAACCTTAGATGGA<br>R: AAATTCTGGTAGCCGTGACA | 121 |
| **FASN** | NM_001099930.1 | F: GTCCTGCTGAAGCCTAACTC<br>R: TCCTTGGAACCGTCTGTG | 206 |
| **SCD** | NM_213781.1 | F: ACAAGAGGCCAAGACAAGTTCC<br>R: GCTGTAGGGAATGCTGGTTAGTTT | 142 |
| **ACACA** | NM_001114269.1 | F: TCCCAGTGCAAGCAGTATG<br>R: TGCCAATCCACACGAAGAC | 211 |
| **mGPAT** | XM_005671462.3 | F: ACTATCTCCTGCTCACTTTCA<br>R: CGTCTCATCTAGCCTCCGTC | 146 |
| **CD36** | NM_001044622.1 | F: ATCGTGCCTATCCTCTGG<br>R: CCAGGCCAAGGAGGTTAA | 103 |
| **C/EBPα** | XM_003127015.4 | F: AACAACTGAGCCGCGAACTG<br>R: GCTCCGGCAGTCTTGAGAT | 181 |
| **DGAT2** | NM_001160080.1 | F: GCAGGTGATCTTTGAGGAGG<br>R: GCTTGGAGTAGGGCATGAG | 140 |
| **ELOVL6** | XM_021100708.1 | F: ACCACATCACTGTGCTCCTC<br>R: CGAGTGCACGCCATAGTTCA | 95 |
| **Cyclin B** | NM_001170768.1 | F: AATCCCTTCTTGTGGTTA<br>R: CTTAGATGTGGCATACTTG | 104 |
| **Cyclin E** | XM_005653265.2 | F: AGAAGGAAAGGGATGCGAAGG<br>R: CCAAGGCTGATTGCCACACT | 173 |
| **Cyclin D** | XM_021082686.1 | F: TACACCGACAACTCCATCCG<br>R: GAGGGCGGGTTGGAAATGAA | 224 |
| **β-actin** | XM_021086047.1 | F: GGACTTCGAGCAGGAGATGG<br>R: AGGAAGGAGGGCTGGAAGAG | 138 |

### 2.5. Western Blot

Cells were scraped with RIPA (Radio Immunoprecipitation Assay) buffer (Beyotime, Shanghai, China) and lysates were subjected to SDS-PAGE and transferred to the PVDF (Polyvinylidene fluoride, Millipore, Burlington, MA, USA). Polyacrylamide gels were used to separate and mark proteins of different sizes. The proteins were then transferred to a PVDF membrane. Next, the membrane was soaked in 5% skim milk for 2 h and then incubated with primary antibodies overnight at 4 °C. After that, membranes were washed in Tris-buffered saline with Tween 20 and subsequently incubated with horseradish peroxidase-conjugated secondary antibodies. Finally, the stripes of target proteins were visualized by the enhanced chemiluminescent substrate (Millipore, MA) and observed using Gel Doc XR System (Bio-Rad). The densities of the brands were analyzed using Image Lab software (Bio-Rad). Target proteins were probed with primary antibodies (anti-AQP3, ab125219, Abcam, 1:1000; PPARγ, #2435, CST, 1:1000; FASN, sc-20140, Santa Cruz, 1:200; FABP4, sc-18661, Santa Cruz, 1:200; Akt, sc-8312, Santa Cruz, 1:200; p-Akt, sc-7985-R, Santa Cruz, 1:200; cyclin D, sc-753, Santa Cruz, 1:500; cyclin E, sc-247, Santa Cruz, 1:500; β-actin, sc-130656 Santa Cruz, 1:1000).

### 2.6. Oil Red O Staining

The well-differentiated cells were washed twice with PBS and fixed with 4% paraformaldehyde for 30 min, and then incubated with 1% filtered Oil Red O solution for 5 min. The stained lipid droplets in the cells were photographed (Nikon TE2000 microscope, Tokyo, Japan). For quantitative analysis, cellular Oil Red O was extracted by isopropanol and optical absorbance was detected at 510 nm.

### 2.7. EdU Staining

EdU assay was conducted with a Cell-Light™ EdU (5-ethynyl-2′-deoxyuridine) Apollo®567 In Vitro Imaging Kit (RiboBio, Guangzhou, China) as per the manufacturer's instructions. Porcine intramuscular preadipocytes in growth medium were incubated with 50 mM EdU for 2 h, and then fixed by paraformaldehyde. Then cells were labeled with Apollo reaction solution, and the nuclei were stained with Hoechst 33,342 (Thermo Fisher Scientific, Waltham, MA, USA). Cells were visualized using a Nikon TE2000 microscope (Nikon, Tokyo, Japan), and the images were processed with Image J software by the National Institutes of Health (NIH).

### 2.8. CCK-8 Assay

Porcine intramuscular preadipocytes were seeded on a 96-well plate at a density of $1 \times 10^3$ per well. Some 24 h later, 10% CCK-8 solution (Vazyme, Nanjing, China) was added, and after 4 h incubation, the absorbance was measured at 490 nm.

### 2.9. Statistical Analysis

All experiments were carried out in triplicate and the results were analyzed by one-way analysis of variance (ANOVA) using SPSS 18 software (SPSS Inc., Chicago, IL, USA). $P < 0.05$ was set as statistical significance. Data were presented as mean ± standard error (SE).

## 3. Results

### 3.1. AQP3 Is Upregulated during Adipogenesis

In order to explore the expression pattern of AQP3 gene in pig adipose tissue and PIPAs we selected 180-day-old pig tissue to test the expression of AQP3. RT-qPCR results showed that AQP3 is highly expressed in adipose tissue (Figure 1A). In subcutaneous adipose tissue of pigs of different ages, the expression of AQP3 reached the highest at 30 days of age, and then began to decline (Figure 1B).

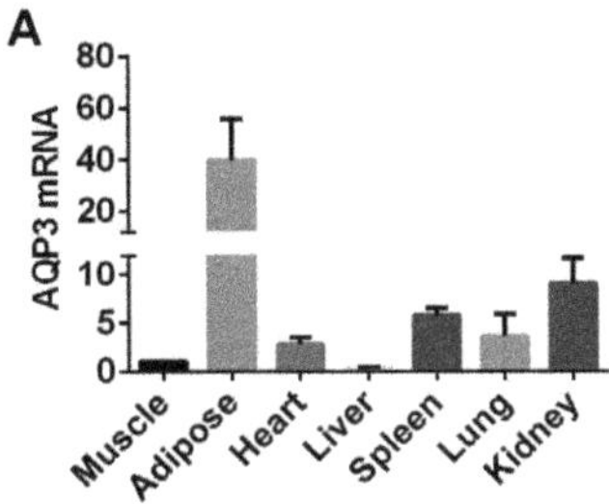


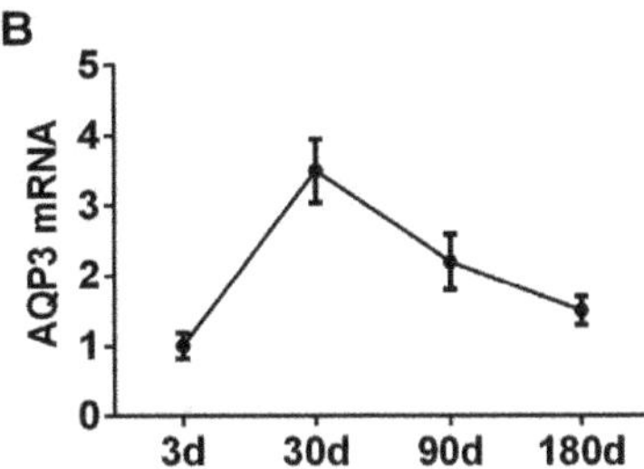


**Figure 1.** AQP3 expression pattern in pigs in vivo by RT-qPCR. (**A**) AQP3 pig tissue expression profile. (**B**) AQP3 expression during pig growth by RT-qPCR. β-actin was used as internal reference gene. $n = 3$.

In in vitro cell culture experiments, transcripts of AQP3 in PIPAs were rapidly increased upon adipogenic stimuli, reached a peak at 4 d post-differentiation, and then gradually decreased (Figure 2A). The expression of AQP3 proteins showed the same pattern (Figure 2B). For reference, the expression of PPARγ (Figure 2C), FASN (Figure 2D), and aP2 (Figure 2E) throughout the adipogenic process was profiled to represent the efficient differentiation of PIPAs in vitro. The data indicated a promising role of AQP3 in adipogenesis.

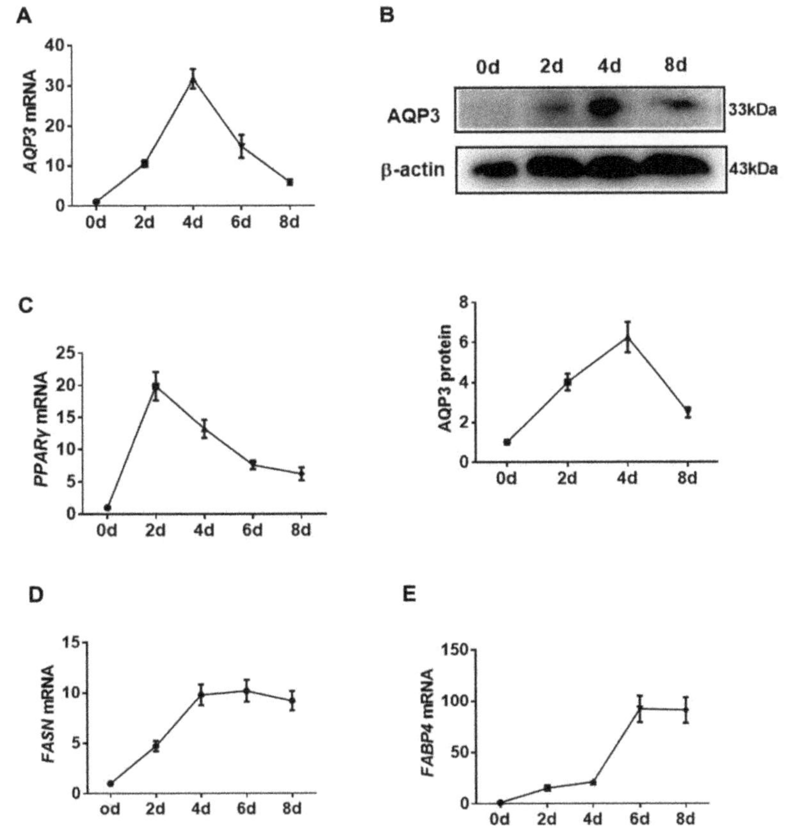


**Figure 2.** Expression pattern of porcine AQP3 in adipogenesis. The expression of AQP3 during adipogenesis was detected by RT-qPCR (**A**) and Western blot (**B**). The expression of PPARγ (**C**), FASN (**D**), and aP2 (**E**) was detected by RT-qPCR. β-actin was used as internal reference gene. $n = 3$.

### 3.2. Knockdown of AQP3 Blunts Adipogenesis

In view of the rising trend of AQP3 in adipocyte differentiation, siRNAs were employed to explore the role of AQP3 on adipogenic differentiation. Three siRNAs were designed, and only siRNA-1 showed >70% knockdown efficiency 24 h post-transfection (Figure 3A), and still significantly reduced AQP3 expression 48 h, 4 d, and 8 d post-differentiation (Figure 3B). Thus, siRNA-1 was used in the following study. RT-qPCR results showed that AQP3 siRNA significantly inhibited the expression of adipogenic markers, such as PPARγ, aP2, ACACA, SCD, DGAT2, mGPAT, ELOVL6, and FASN 4d post-differentiation (Figure 3C), and the genes (except PPARγ) detected above were still significantly downregulated 8 d post-differentiation (Figure 3D). Western blot results presented the expression of PPARγ, aP2, and FASN at protein levels and phosphorylated Akt (Figure 3E), and the gray level analysis showed that these proteins were significantly decreased (Figure 3F). Oil Red O staining showed that AQP3 siRNA significantly repressed triglyceride accumulation in intramuscular adipocytes (Figure 3G,H). These data indicated that AQP3 was essential for adipogenesis and lipid accumulation in PIPAs.

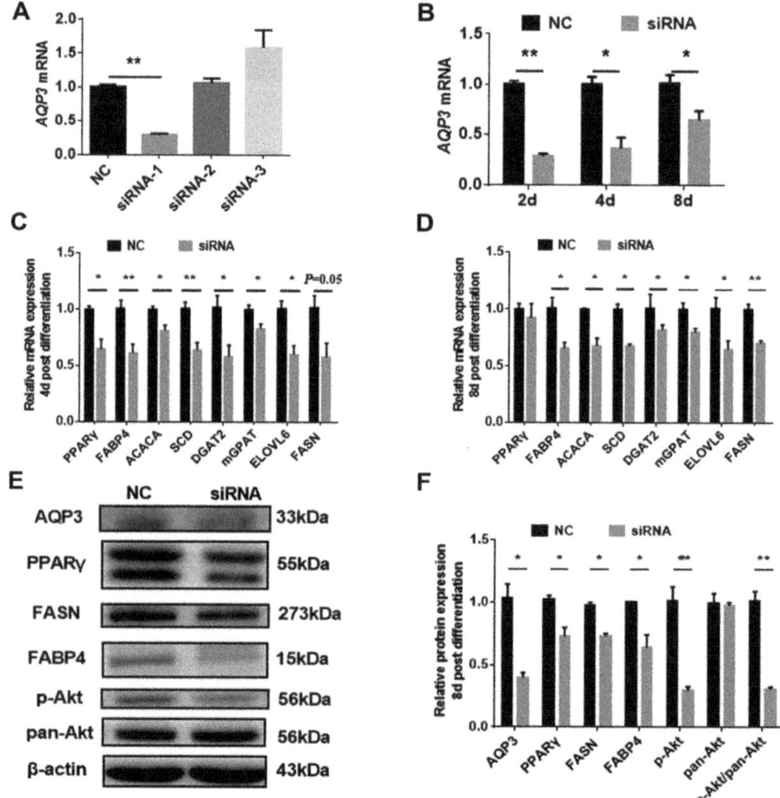


**Figure 3.** *Cont.*

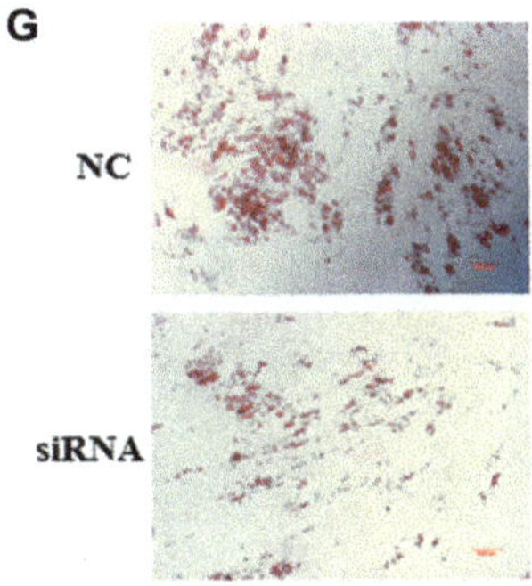


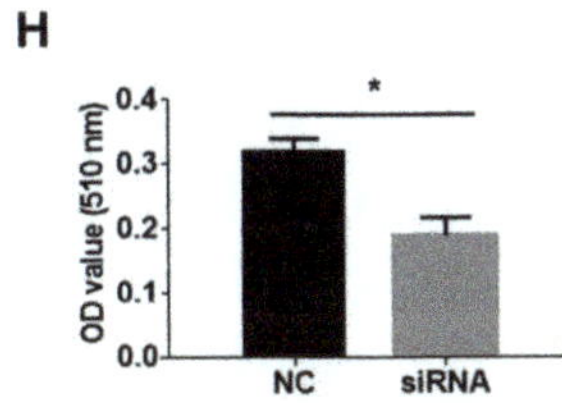


**Figure 3.** AQP3 silence repressed adipogenic differentiation in porcine intramuscular adipocytes. Cells were transfected with 3 candidate siRNAs targeting AQP3 when reaching 70–80% confluence, and only siRNA-1 could decrease AQP3 transcripts by 70% (**A**) 24 h post-transfection, and it also significantly repressed AQP3 expression 48 h, 4 d, and 8 d post-differentiation (**B**). Expression of adipogenic and lipogenic genes 4 d and 8 d post-differentiation was detected by RT-qPCR, using β-actin as reference gene (**C**). Expression of adipogenesis-related genes 8 d post-differentiation was detected by Western blot (**D**). Western blot images (**E**) and gray analysis statistics (**F**) of PPARγ, aP2, FASN, and Akt in PIPAs. Lipid accumulation was tested by Oil Red O staining (**G**) and quantified by isopropanol extraction (**H**). $n = 3$; * $p < 0.05$, ** $p < 0.01$.

### 3.3. AQP3 Deletion Inhibits Proliferation

AQP3 siRNA could significantly downregulate AQP3 mRNA expression in the proliferating porcine intramuscular preadipocytes 24 h post-transfection (Figure 4A), indicating that this siRNA can be used to inhibit the expression of AQP3 during the proliferation phase. The results of RT-qPCR showed that siRNA repressed the expression of cyclin B, cyclin D, cyclin E, and CDK4 mRNAs 48 h post-transfection (Figure 4B). Consistently, cyclin B, cyclin D, and proliferating cell nuclear antigen (PCNA) were significantly repressed by AQP3 siRNA at the protein level too (Figure 4C,D). Additionally, transfection of AQP3 siRNA significantly reduced the ratio of EdU-positive cells (Figure 4E,F) and cell viability (Figure 4G). In summary, the above results show that during the proliferation phase, AQP3 exerts this effect to promote cell proliferation.

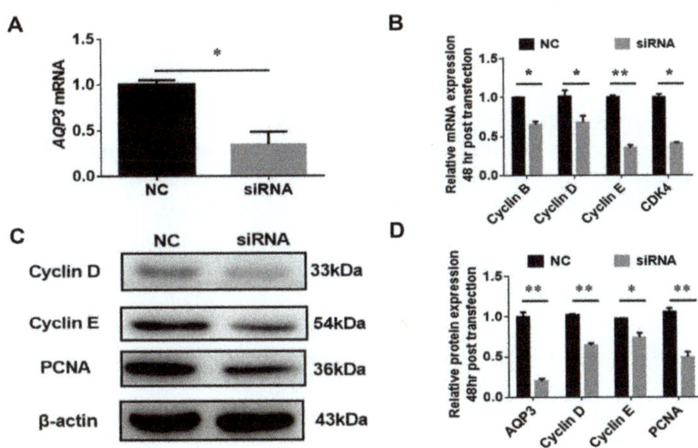


**Figure 4.** *Cont.*

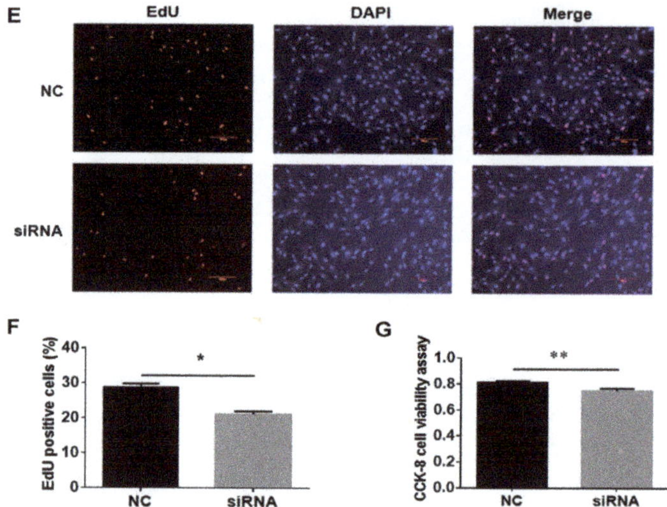


**Figure 4.** Knockdown of AQP3 inhibited the proliferation of porcine intramuscular adipocytes. Cells were transfected with AQP3 siRNA at 40–50% confluence, and the interference effect was over 70% 24 h post-transfection (**A**). Cell cycle genes were analyzed by RT-qPCR (**B**) and Western blot (**C**,**D**). EdU staining was captured (**E**) and EdU-positive cells were counted to monitor the proliferation of cells (**F**). CCK-8 was adopted to check cell viability (**G**). $n = 3$. CDK4, cyclin-dependent kinase 4; PCNA, proliferating cell nuclear antigen; DAPI, 4′,6-diamidino-2-phenylindole; * $p < 0.05$, ** $p < 0.01$.

## 4. Discussion

The relationship between AQP3 and adipose tissues or cells has been ignored for a long time, and its expression in adipose tissues or adipocytes remains ambiguous. In pioneering work, AQP3 mRNA was undetectable in adipose tissues of Meishan pigs by semi-quantitative RT-PCR [28]. Meanwhile, there were other studies that did not support the presence of AQP3 in mouse [29,30] or human adipose tissue [31]. However, mRNAs and proteins of AQP3 were later detected in human stroma vascular fraction of omental, subcutaneous adipocyte tissue, and also in freshly isolated adipocytes [6]. Additionally, the expression of AQP3 was confirmed in murine 3T3-L1 cell line [32,33]. The inconsistent reports might be due to the relatively lower expression levels of AQP3 in mature adipose tissues [6,12].

In the present study, RT-qPCR and Western blot uncovered the upregulation of AQP3 during the process of adipogenic differentiation, indicating that AQP3 may also play a regulatory role in intramuscular adipocytes. Furthermore, AQP3 knockdown by siRNA leads to reduced expression of adipogenic and lipogenic genes and defects of TG storage in porcine intramuscular preadipocytes. After AQP3 was silenced, the expression of the classic adipogenic factor PPARγ was significantly inhibited. PPARγ is an indispensable transcription factor for adipocyte differentiation [34], and a previous study has shown that AQP3 is the target of PPARγ in murine adipose cell line [35] and hepatic stellate cells [36]. The level of phosphorylation of AKT was also decreased, which indicates that the downstream signal of the classic insulin signaling pathway is weakened. At the same time, the expression levels of other adipogenic marker genes aP2, ACACA, etc. were also decreased, indicating that the silencing of AQP3 inhibits the adipogenic differentiation of porcine intramuscular preadipocytes from the overall gene expression level. A previous study has also shown that AQP3 is more permeable to glycerol compared with water [14], and AQP3 was increased in LPS-induced adipogenesis [32] and triglyceride sedimentation [37]. A combination of our and others' work supports a pivotal role of AQP3 in lipid accumulation in adipocytes.

Besides, AQP3 also has the potential to regulate cell proliferation. Accumulated studies have shown that AQPs are involved in tumor metastasis [15]. AQP3 and AQP5 can be used as new markers for breast cancer [38]. Another study showed that overexpression of AQP3 in mammalian cells can promote cell proliferation efficiency and cell cycle transition [39]. In our study, AQP3 siRNA weakened the proliferation ability of porcine intramuscular preadipocytes, which was reflected in the downregulation of genes such as cyclin B and cyclin D, and a decrease in the number of $EDU^+$ cells. This shows that AQP3 can not only regulate cell proliferation in cancer cells, but also in intramuscular preadipocytes. For IMF, the proliferation and differentiation of adipocytes are two very important fat deposition processes [40]. AQP3 shows regulatory capacity in both proliferation and differentiation, which further illustrates that AQP3 is a key gene for adipocytes.

In the AQPs family, AQP7 can promote the loss of glycerol from adipocytes and inhibit the accumulation of fat, while AQP9 is responsible for the uptake of glycerin [41]. Additionally, the weakening of AQP5 can inhibit the late adipogenesis of 3T3-L1 cells [42]. Compared with other AQP genes, the present study proves that AQP3 can regulate fat deposition in two ways, namely adipocyte proliferation and differentiation, revealing that AQP3 is likely to be a new key factor in regulating fat deposition, which supplements the regulation of adipocytes by AQPs.

## 5. Conclusions

Collectively, our work has identified AQP3 as a novel and essential modulator in proliferation, differentiation, and lipid accumulation in intramuscular adipocytes, providing a new theoretical basis for the regulation of skeletal muscle ectopic fat deposition. This conclusion may be due to the involvement of AQP3 in the transport of glycerol, but further research is needed.

**Author Contributions:** Wrote this paper and analyzed the data: X.W.; performed the cell culture and cell-based analysis: J.Y. and Y.Y.; gave critical suggestions about the experimental design and manuscript preparation: X.S. and G.Y.; designed the experiment and revised the manuscript: X.L. All authors have read and agreed to the published version of the manuscript.

**Funding:** This study was supported by Major Projects for Genetically Modified Organisms Breeding (2016ZX08006003) and National Natural Science Foundation (31501925).

**Conflicts of Interest:** The authors declare no conflict of interest.

## References

1. Hausman, G.J.; Basu, U.; Du, M.; Fernyhough-Culver, M.; Dodson, M.V. Intermuscular and intramuscular adipose tissues: Bad vs. good adipose tissues. *Adipocyte* **2014**, *3*, 242–255. [CrossRef] [PubMed]
2. Listrat, A.; Lebret, B.; Louveau, I.; Astruc, T.; Bonnet, M.; Lefaucheur, L.; Picard, B.; Bugeon, J. How Muscle Structure and Composition Influence Meat and Flesh Quality. *Scientific World J.* **2016**, *2016*, 3182746. [CrossRef]
3. Frank, D.; Joo, S.T.; Warner, R. Consumer Acceptability of Intramuscular Fat. *Korean J. Food Sci. Anim. Resour.* **2016**, *36*, 699–708. [CrossRef] [PubMed]
4. Da Silva, I.V.; Soveral, G. Aquaporins in Obesity. *Adv. Exp. Med. Biol.* **2017**, *969*, 227–238. [CrossRef] [PubMed]
5. Iena, F.M.; Lebeck, J. Implications of Aquaglyceroporin 7 in Energy Metabolism. *Int. J. Mol. Sci.* **2018**, *19*, 154. [CrossRef]
6. Rodriguez, A.; Catalan, V.; Gomez-Ambrosi, J.; Garcia-Navarro, S.; Rotellar, F.; Valenti, V.; Silva, C.; Gil, M.J.; Salvador, J.; Burrell, M.A.; et al. Insulin- and leptin-mediated control of aquaglyceroporins in human adipocytes and hepatocytes is mediated via the PI3K/Akt/mTOR signaling cascade. *J. Clin. Endocrinol. Metab.* **2011**, *96*, E586–E597. [CrossRef]
7. Laforenza, U.; Scaffino, M.F.; Gastaldi, G. Aquaporin-10 represents an alternative pathway for glycerol efflux from human adipocytes. *PLoS ONE* **2013**, *8*, e54474. [CrossRef]
8. Madeira, A.; de Almeida, A.; de Graaf, C.; Camps, M.; Zorzano, A.; Moura, T.F.; Casini, A.; Soveral, G. A gold coordination compound as a chemical probe to unravel aquaporin-7 function. *Chembiochem* **2014**, *15*, 1487–1494. [CrossRef]
9. Rodriguez, R.A.; Liang, H.; Chen, L.Y.; Plascencia-Villa, G.; Perry, G. Single-channel permeability and glycerol affinity of human aquaglyceroporin AQP3. *Biochim. Biophys. Acta Biomembr.* **2019**, *1861*, 768–775. [CrossRef]

10. Tardelli, M.; Claudel, T.; Bruschi, F.V.; Moreno-Viedma, V.; Trauner, M. Adiponectin regulates AQP3 via PPARalpha in human hepatic stellate cells. *Biochem. Biophys. Res. Commun.* **2017**, *490*, 51–54. [CrossRef]
11. Rodriguez, A.; Moreno, N.R.; Balaguer, I.; Mendez-Gimenez, L.; Becerril, S.; Catalan, V.; Gomez-Ambrosi, J.; Portincasa, P.; Calamita, G.; Soveral, G.; et al. Leptin administration restores the altered adipose and hepatic expression of aquaglyceroporins improving the non-alcoholic fatty liver of ob/ob mice. *Sci. Rep.* **2015**, *5*, 12067. [CrossRef] [PubMed]
12. Marzi, C.; Holdt, L.M.; Fiorito, G.; Tsai, P.C.; Kretschmer, A.; Wahl, S.; Guarrera, S.; Teupser, D.; Spector, T.D.; Iacoviello, L.; et al. Epigenetic Signatures at AQP3 and $SOCS_3$ Engage in Low-Grade Inflammation across Different Tissues. *PLoS ONE* **2016**, *11*, e0166015. [CrossRef] [PubMed]
13. Yang, J. Identification of Key Genes for Differential Deposition and Functional Analysis of AQP3 in Porcine Intramusclar Fat. Master's Thesis, NWAFU, Yangling, China, May 2018.
14. Arif, M.; Kitchen, P.; Conner, M.T.; Hill, E.J.; Nagel, D.; Bill, R.M.; Dunmore, S.J.; Armesilla, A.L.; Gross, S.; Carmichael, A.R.; et al. Downregulation of aquaporin 3 inhibits cellular proliferation, migration and invasion in the MDA-MB-231 breast cancer cell line. *Oncol. Lett.* **2018**, *16*, 713–720. [CrossRef] [PubMed]
15. Marlar, S.; Jensen, H.H.; Login, F.H.; Nejsum, L.N. Aquaporin-3 in Cancer. *Int. J. Mol. Sci.* **2017**, *18*, 2106. [CrossRef]
16. Wang, X.; Tao, C.; Yuan, C.; Ren, J.; Yang, M.; Ying, H. AQP3 small interfering RNA and PLD2 small interfering RNA inhibit the proliferation and promote the apoptosis of squamous cell carcinoma. *Mol. Med. Rep.* **2017**, *16*, 1964–1972. [CrossRef]
17. Xiong, G.; Chen, X.; Zhang, Q.; Fang, Y.; Chen, W.; Li, C.; Zhang, J. RNA interference influenced the proliferation and invasion of XWLC-05 lung cancer cells through inhibiting aquaporin 3. *Biochem. Biophys. Res. Commun.* **2017**, *485*, 627–634. [CrossRef] [PubMed]
18. Li, Z.; Li, B.; Zhang, L.; Chen, L.; Sun, G.; Zhang, Q.; Wang, J.; Zhi, X.; Wang, L.; Xu, Z.; et al. The proliferation impairment induced by AQP3 deficiency is the result of glycerol uptake and metabolism inhibition in gastric cancer cells. *Tumour Biol.* **2016**, *37*, 9169–9179. [CrossRef]
19. Chen, L.; Li, Z.; Zhang, Q.; Wei, S.; Li, B.; Zhang, X.; Zhang, L.; Li, Q.; Xu, H.; Xu, Z. Silencing of AQP3 induces apoptosis of gastric cancer cells via downregulation of glycerol intake and downstream inhibition of lipogenesis and autophagy. *OncoTargets Ther.* **2017**, *10*, 2791–2804. [CrossRef]
20. Li, X.; Huang, K.; Chen, F.; Li, W.; Sun, S.; Shi, X.E.; Yang, G. Verification of suitable and reliable reference genes for quantitative real-time PCR during adipogenic differentiation in porcine intramuscular stromal-vascular cells. *Animal* **2016**, *10*, 947–952. [CrossRef]
21. He, L.; Huang, N.; Li, H.; Tian, J.; Zhou, X.; Li, T.; Yao, K.; Wu, G.; Yin, Y. AMPK/α-Ketoglutarate Axis Regulates Intestinal Water and Ion Homeostasis in Young Pigs. *J. Agric. Food Chem.* **2017**, *65*, 2287–2298. [CrossRef]
22. Liu, J.B.; Chen, D.W.; Yu, B.; Mao, X.B. Effect of maternal folic acid supplementation on hepatic one-carbon unit associated gene expressions in newborn piglets. *Mol. Biol. Rep.* **2011**, *38*, 3849–3856. [CrossRef] [PubMed]
23. Cheng, J.; Song, Z.Y.; Pu, L.; Yang, H.; Zheng, J.M.; Zhang, Z.Y.; Shi, X.E.; Yang, G.S. Retinol binding protein 4 affects the adipogenesis of porcine preadipocytes through insulin signaling pathways. *Biochem. Cell Biol.* **2013**, *91*, 236–243. [CrossRef] [PubMed]
24. Lv, Y.; Zhang, S.; Guan, W.; Chen, F.; Zhang, Y.; Chen, J.; Liu, Y. Metabolic transition of milk triacylglycerol synthesis in response to varying levels of palmitate in porcine mammary epithelial cells. *Genes Nutr.* **2018**, *13*, 18. [CrossRef]
25. Yang, Y.; Ju, D.; Zhang, M.; Yang, G. Interleukin-6 stimulates lipolysis in porcine adipocytes. *Endocrine* **2008**, *33*, 261–269. [CrossRef] [PubMed]
26. Zhen, Y.H.; Wang, L.; Riaz, H.; Wu, J.B.; Yuan, Y.F.; Han, L.; Wang, Y.L.; Zhao, Y.; Dan, Y.; Huo, L.J. Knockdown of CEBPbeta by RNAi in porcine granulosa cells resulted in S phase cell cycle arrest and decreased progesterone and estradiol synthesis. *J. Steroid. Biochem. Mol. Biol.* **2014**, *143*, 90–98. [CrossRef] [PubMed]
27. Dozois, C.M.; Oswald, E.; Gautier, N.; Serthelon, J.P.; Fairbrother, J.M.; Oswald, I.P. A reverse transcription-polymerase chain reaction method to analyze porcine cytokine gene expression. *Vet. Immunol. Immunopathol.* **1997**, *58*, 287–300. [CrossRef]

28. Li, X.; Lei, T.; Xia, T.; Chen, X.; Feng, S.; Chen, H.; Chen, Z.; Peng, Y.; Yang, Z. Molecular characterization, chromosomal and expression patterns of three aquaglyceroporins (AQP3, 7, 9) from pig. *Comp. Biochem. Physiol. B Biochem. Mol. Biol.* **2008**, *149*, 468–476. [CrossRef]
29. Kishida, K.; Kuriyama, H.; Funahashi, T.; Shimomura, I.; Kihara, S.; Ouchi, N.; Nishida, M.; Nishizawa, H.; Matsuda, M.; Takahashi, M.; et al. Aquaporin adipose, a putative glycerol channel in adipocytes. *J. Biol. Chem.* **2000**, *275*, 20896–20902. [CrossRef]
30. Maeda, N.; Funahashi, T.; Hibuse, T.; Nagasawa, A.; Kishida, K.; Kuriyama, H.; Nakamura, T.; Kihara, S.; Shimomura, I.; Matsuzawa, Y. Adaptation to fasting by glycerol transport through aquaporin 7 in adipose tissue. *Proc. Natl. Acad. Sci. USA* **2004**, *101*, 17801–17806. [CrossRef]
31. Miranda, M.; Escote, X.; Ceperuelo-Mallafre, V.; Alcaide, M.J.; Simon, I.; Vilarrasa, N.; Wabitsch, M.; Vendrell, J. Paired subcutaneous and visceral adipose tissue aquaporin-7 expression in human obesity and type 2 diabetes: Differences and similarities between depots. *J. Clin. Endocrinol. Metab.* **2010**, *95*, 3470–3479. [CrossRef]
32. Chiadak, J.D.; Arsenijevic, T.; Gregoire, F.; Bolaky, N.; Delforge, V.; Perret, J.; Delporte, C. Involvement of JNK/NFkappaB Signaling Pathways in the Lipopolysaccharide-Induced Modulation of Aquaglyceroporin Expression in 3T3-L1 Cells Differentiated into Adipocytes. *Int. J. Mol. Sci.* **2016**, *17*, 1742. [CrossRef] [PubMed]
33. Choudhary, V.; Olala, L.O.; Kagha, K.; Pan, Z.Q.; Chen, X.; Yang, R.; Cline, A.; Helwa, I.; Marshall, L.; Kaddour-Djebbar, I.; et al. Regulation of the Glycerol Transporter, Aquaporin-3, by Histone Deacetylase-3 and p53 in Keratinocytes. *J. Investig. Dermatol.* **2017**, *137*, 1935–1944. [CrossRef] [PubMed]
34. Janani, C.; Ranjitha Kumari, B.D. PPAR $\gamma$ gene—A review. *Diabetes Metab. Syndr.* **2015**, *9*, 46–50. [CrossRef] [PubMed]
35. Kishida, K.; Shimomura, I.; Nishizawa, H.; Maeda, N.; Kuriyama, H.; Kondo, H.; Matsuda, M.; Nagaretani, H.; Ouchi, N.; Hotta, K.; et al. Enhancement of the aquaporin adipose gene expression by a peroxisome proliferator-activated receptor $\gamma$. *J. Biol. Chem.* **2001**, *276*, 48572–48579. [CrossRef] [PubMed]
36. Tardelli, M.; Bruschi, F.V.; Claudel, T.; Moreno-Viedma, V.; Halilbasic, E.; Marra, F.; Herac, M.; Stulnig, T.M.; Trauner, M. AQP3 is regulated by PPARgamma and JNK in hepatic stellate cells carrying PNPLA3 I148M. *Sci. Rep.* **2017**, *7*, 14661. [CrossRef] [PubMed]
37. Chiadak, J.D.; Gena, P.; Gregoire, F.; Bolaky, N.; Delforge, V.; Perret, J.; Calamita, G.; Delporte, C. Lipopolysaccharide Modifies Glycerol Permeability and Metabolism in 3T3-L1 Adipocytes. *Int. J. Mol. Sci.* **2017**, *18*, 2566. [CrossRef]
38. Zhu, Z.; Jiao, L.; Li, T.; Wang, H.; Wei, W.; Qian, H. Expression of AQP3 and AQP5 as a prognostic marker in triple-negative breast cancer. *Oncol. Lett.* **2018**, *16*, 2661–2667. [CrossRef]
39. Galan-Cobo, A.; Ramirez-Lorca, R.; Serna, A.; Echevarria, M. Overexpression of AQP3 Modifies the Cell Cycle and the Proliferation Rate of Mammalian Cells in Culture. *PLoS ONE* **2015**, *10*, e0137692. [CrossRef]
40. Sarantopoulos, C.N.; Banyard, D.A.; Ziegler, M.E.; Sun, B.; Shaterian, A.; Widgerow, A.D. Elucidating the Preadipocyte and Its Role in Adipocyte Formation: A Comprehensive Review. *Stem. Cell Rev. Rep.* **2018**, *14*, 27–42. [CrossRef]
41. Lebeck, J. Metabolic impact of the glycerol channels AQP7 and AQP9 in adipose tissue and liver. *J. Mol. Endocrinol.* **2014**, *52*, R165–R178. [CrossRef]
42. Madeira, A.; Mosca, A.F.; Moura, T.F.; Soveral, G. Aquaporin-5 is expressed in adipocytes with implications in adipose differentiation. *IUBMB Life* **2015**, *67*, 54–60. [CrossRef] [PubMed]

*Article*

# Transcriptional Regulation of *HMOX1* Gene in Hezuo Tibetan Pigs: Roles of WT1, Sp1, and C/EBPα

**Wei Wang [1], Qiaoli Yang [1], Kaihui Xie [1], Pengfei Wang [1], Ruirui Luo [1], Zunqiang Yan [1], Xiaoli Gao [1], Bo Zhang [1], Xiaoyu Huang [1] and Shuangbao Gun [1,2,*]**

1 College of Animal Science and Technology, Gansu Agricultural University, Lanzhou 730070, China; wangw@st.gsau.edu.cn (W.W.); yangql0112@163.com (Q.Y.); xkh34567@163.com (K.X.); wangpf815@163.com (P.W.); luoruirui628@163.com (R.L.); yanzunqiang@163.com (Z.Y.); gxl18892@163.com (X.G.); zhangb1662@163.com (B.Z.); huanghxy100@163.com (X.H.)

2 Gansu Research Center for Swine Production Engineering and Technology, Lanzhou 730070, China

* Correspondence: gunsbao056@126.com; Tel.: +86-931-763-1804

Received: 3 March 2020; Accepted: 24 March 2020; Published: 26 March 2020

**Abstract:** Heme oxygenase 1 (*HMOX1*) is a stress-inducing enzyme with multiple cardiovascular protective functions, especially in hypoxia stress. However, transcriptional regulation of swine *HMOX1* gene remains unclear. In the present study, we first detected tissue expression profiles of *HMOX1* gene in adult Hezuo Tibetan pig and analyzed the gene structure. We found that the expression level of *HMOX1* gene was highest in the spleen of the Hezuo Tibetan pig, followed by liver, lung, and kidney. A series of 5′ deletion promoter plasmids in pGL3-basic vector were used to identify the core promoter region and confirmed that the minimum core promoter region of swine *HMOX1* gene was located at −387 bp to −158 bp region. Then we used bioinformatics analysis to predict transcription factors in this region. Combined with site-directed mutagenesis and RNA interference assays, it was demonstrated that the three transcription factors WT1, Sp1 and C/EBPα were important transcription regulators of *HMOX1* gene. In summary, our study may lay the groundwork for further functional study of *HMOX1* gene.

**Keywords:** *HMOX1* gene; promoter; transcriptional regulation; Hezuo Tibetan pig

---

## 1. Introduction

Tibetan pigs are typical high-altitude pig breeds living on the Qinghai–Tibet Plateau, and are vital to the lives of Tibetan people [1]. The main producing areas of Hezuo Tibetan pigs are located in the southwestern of Gansu province in Gannan Tibetan Autonomous Prefecture, on the northeast edge of the Qinghai–Tibet Plateau, with an average elevation of about 3000 meters [2]. Hezuo Tibetan pigs belong to a group of Tibetan pigs, which can adapt to the harsh plateau environment and feeding conditions, mainly relying on grazing for their livelihood [3]. Genomic analyses found that there were different patterns of selection between domestic Duroc pigs and Titetan wild boars. Tibetan pigs have a higher proportion of lineage-specific genes, these specific genes were related to vascular smooth muscle contraction, disease resistance, and chemokine signaling pathway, which may reflect that Tibetan wild boar experienced natural selection to adapt to harsh environments [4]. Genome-wide SNP markers confirmed that the divergent evolution between Chinese and Western pigs, the interpopulation linkage disequilibrium, is much longer in Western pigs compared with Chinese pigs. In the genomic research to identify the candidate loci between Tibetan pigs and lowland pigs, several genes in Tibetan pigs are likely important for genetic adaptation to high altitude [5]. Previous research has shown that the adaptability of Tibetan pigs at high altitudes is significantly different from that of other domestic pigs. Tibetan pigs have evolved physiological characteristics to adapt to hypoxia in the plateau, such as a thicker alveolar septum, more developed capillaries [4], and larger and stronger hearts [5]. Therefore, an in-depth study of Tibetan pigs will help us understand the hypoxic adaptability of plateau species [6].

Heme oxygenase (HO) is a stress-inducing enzyme that catalyzes heme to produce free iron, carbon monoxide (CO), and biliverdin [7]. Heme oxygenase 1 (*HMOX1*), one of the main members of the HO family, has a variety of cardiovascular protective functions [8–10] and plays an important role in anti-inflammatory, anti-apoptotic, and antioxidant activities among others [11–13]. *HMOX1* exerts protective effects under stress conditions mainly through the active gas CO and antioxidant bilirubin produced by its decomposition [14,15]. In addition, as a protective gene, *HMOX1* has been reported to have protective effects in acute and chronic lung injury such as hyperoxia, hypoxia, ischemia, and hypertension [16–19]. Although there have been many reports on the structure and function of *HMOX1* gene, its transcriptional regulation mechanism has not been fully elucidated. Promoters play a major role in the regulation of gene transcription, and in-depth study of gene promoters is of great significance in explaining biological growth and development and disease defense [20].

Therefore, in the present study, with the Hezuo Tibetan pigs as the research object, we aim to explore the transcriptional regulation mechanism of the *HMOX1* gene, and understand the promoter structure and transcriptional activity of Tibetan *HMOX1* gene. First, the mRNA expression level of Hezuo Tibetan pig *HMOX1* gene in different tissues was quantified. Then, we cloned 5′ flanking promoter region and analyzed the sequence of *HMOX1* gene. In addition, a series of deletion recombinant plasmids were constructed for double-luciferase activity analysis. The transcription factor binding sites of the core promoter region was predicted by bioinformatics software. To further investigate the regulatory mechanism of this gene, site-directed mutation and RNA interference experiments were used to verify the critical transcription factors that regulate the *HMOX1* gene. This study will provide a reference for further studying the function of *HMOX1* gene and lay a foundation for research transcriptional regulation mechanism of pig *HMOX1* gene.

## 2. Materials and Methods

### 2.1. Ethics Statement

All animal experimental procedures used in this study were carried out following the guidelines of the China Council on Animal Care, and the protocols were subject to approval by the ethics committee of College of Animal Science and Technology, Gansu Agricultural University (approval number 2006-398).

### 2.2. Quantitative Real-Time PCR Analysis of Gene Expression Patterns

Fourteen tissues (heart, liver, spleen, lung, kidney, stomach, duodenum, jejunum, ileum, cecum, colon, rectum, psoas, dorsal longest muscle) were collected respectively from three adult Hezuo Tibetan pig in Hezuo, Tibetan Autonomous Prefecture of Gannan, China. Total RNA was extracted with *TransZol* Up reagent (TransGen Biotech, Beijing, China). For reverse transcriptional reaction, the first-strand cDNA was obtained according to the instructions of PrimeScript™ RT reagent kit with gDNA Eraser (TaKaRa, Dalian, China). The qRT-PCR was analyzed using TB Green Premix Ex Taq II (Tli RNaseH Plus) quantitative kit (TaKaRa, Dalian, China) and each reaction mixture was incubated in the Roche LightCycler 480 II instrument (Roche Applied Science, Penzberg, Germany). The primer information in this experiment is shown in Table 1. Relative expression levels of all genes were normalized with of β-actin (ACTB) expression, and the $2^{-\Delta\Delta Ct}$ method was used to calculate gene expressions [21].

### 2.3. HMOX1 Gene Promoter Region Cloning and Bioinformatics Analysis

The genomic DNA was extracted from Hezuo Tibetan pig blood using the TIANamp Blood DNA Kit (TIANGEN, Beijing, China), which as a template for PCR amplifications. An ~2 kb promoter region of the

Hezuo Tibetan pig *HMOX1* gene (NCBI accession no. NM_001004027.1, region from 1952829 to 1954953) was amplified using specific primers (HMOX1-P-F/R, Table 1). Transcription initiation site (TSS) was predicted using Neural Network Promoter Prediction (http://www.fruitfly.org/seq_tools/promoter.html) online software, AliBaba2.1 [22] (http://gene-regulation.com/pub/programs/alibaba2/index.html) and JASPAR [23] (http://jaspar.genereg.net/) were used to predict the potential transcription factor binding sites. The CpG islands were predicted using MethPrimer [24] (http://www.urogene.org/methprimer/). NCBI Conserved Domain Database (CDD) (https://www.ncbi.nlm.nih.gov/cdd) was used to analyze the conserved domain. Nucleotide homology analysis was performed using the core promoter sequence of *HMOX1* gene, mainly including six species: pig (*Sus scrofa*, NW_018084968.1), human (*Homo sapiens*, NC_000022.11), mouse (*Mus musculus*, NC_000074.6), cattle (*Bos taurus*, NC_037332.1), sheep (*Ovis aries*, NC_040254.1), and horse (*Equus caballus*, NC_009171.3). MEGA7.0 software was used for multi-sequence alignment and homologous tree construction.

**Table 1.** Primer information in this study

| Name | Primer Sequence(5′–3′) | Tm (°C) | Length | Region |
|---|---|---|---|---|
| HMOX1-P | F: GCCTCGTCCTTCTAAAGTCCC<br>R: CGCCAAAGCCCAAGTGACTG | 56 | 2125 bp | −1911/−1892<br>+214/+195 |
| HMOX1-RT | F: GACATGGCCTTCTGGTATGGG<br>R: CATGTAGCGGGTGTAGGCGT | 60 | 141 bp | 348–368<br>488–469 |
| WT1-RT | F: GGTGTCTTCAGGGGCATTCA<br>R: ACACATGAAGGGGCGTTTCT | 60 | 102 bp | 1127–1146<br>1228–1209 |
| Sp1-RT | F: TGTCTCTGGTGGGCAGTATG<br>R: TTGCCCATCAACCGTCTGG | 60 | 133 bp | 601–620<br>733–715 |
| C/EBPα-RT | F: GGCAAAGCCAAGAAGTCGGT<br>R: TCTGTTGAGTCTCCACGTTGC | 60 | 124 bp | 976–995<br>1099–1079 |
| β-actin-RT | F: ATATTGCTGCGCTCGTGGT<br>R: TAGGAGTCCTTCTGGCCCAT | 60 | 148 bp | 142–160<br>289–270 |
| HMOX1-P1 | F: CTA**GCTAGC**TATGACCGCTCCTCCTCCAC | 60 | 273 bp | −158/+115 |
| HMOX1-P2 | F: CTA**GCTAGC**CAGGGTTTGGGGTGCAGAAG | 60 | 502 bp | −387/+115 |
| HMOX1-P3 | F: CTA**GCTAGC**GCTCCCCTAGTAGTAACCTGC | 64 | 1034 bp | −919/+115 |
| HMOX1-P4 | F: CTA**GCTAGC**GGGCACTCGATACCTTGGTT | 60 | 1732 bp | −1617/+115 |
| HMOX1-P5 | F: CTA**GCTAGC**CAGCAACCCCCAAGTCTCT | 62 | 1993 bp | −1878/+115 |
| HMOX1-R | R: CCG**CTCGAG**TTGCCTGTTGGGCTGTGAG | | | |
| WT1-m | F: GAAGTCCCTGAGGTCGCC**GGTC**CCTCCCGCTCAGAGAAGC<br>R: GCTTCTCTGAGCCAGCGG**CCAG**GGAGGGCTCAGGGACTTC | | 229 bp | −387/−158 |
| Sp1-m | F: GGCACCATTCAGATCCGT**AAAA**GTGCTCAAGTCCCATCGC<br>R: GCGATGGGACTTAGGCA**TTTT**CACGAGCTGAATGGTGCC | | 229 bp | −387/−158 |
| C/EBPα-m | F: TCAGATTCCTAAAGTATC**AAAA**GCTTTGTTTTTAGTGTCC<br>R: GGACACTAAAAATCATAG**TTTT**CGAAACTTAGGAATCTGA | | 229 bp | −387/−158 |
| c-Ets-1-m | F: CTTTTGCTTTGTTTTTAG**ACAG**CTGTTTTAAACAGCTCTG<br>R: CAGAGCTGTTTAAAAATC**TGTC**GACAAAACAAAGVAAAAG | | 229 bp | −387/−158 |
| si-WT1 | F: GGGCUGCAAUAAGAGAUATT<br>R: AUAUCUCUUAUUGCAGCCCTT | | | |
| si-Sp1 | F: GCGGAUCUGCAGUCCAUUATT<br>R: UAAUGGACUGCAGAUCCGCTT | | | |
| si-C/EBPα | F: ACGAGACGUCCAUCGACAUTT<br>R: AUGUCGAUGGACGUCUCGUTT | | | |
| si-NC | F: UUCUCCGAACGUGUCACGUTT<br>R: UUAACUCAUCGCUUCUUGCTT | | | |

**GCTAGC:** The the bold and underlined represent restriction site. **GGTC:** the bold and underlined represent site-directed mutation site.

## 2.4. Cell Culture, Transfection, and Dual-Luciferase Reporter Assay

The 293T cells and Porcine alveolar macrophages (3D4/21) were purchased from BeNa Culture Collection (BNCC, Beijing, China). The cells were cultured in growth medium containing 90% DMEM (HyClone, New York, NY, USA), 10% fetal bovine serum (Invitrogen, Carlsbad, CA, USA), and 1% antibiotics (100 IU/ml penicillin and 100 μg/ml streptomycin) at 37 °C and 5% $CO_2$. When the cells reached 70–80% confluence, cells were incubated in 24-well plates. For transfection, cells were transfected using Lipofectamine™ 2000 reagent (Invitrogen, Carlsbad, CA, USA). In order to determine the core promoter of *HMOX1* gene, a series of promoter fragments (−1878/+115, −1617/+115, −919/+115, −387/+115, −158/+115) were amplified through 5′ unidirectional deletion specific primers containing *Nhe* I and *Xho* I restriction enzyme sites, respectively. The PCR products were cloned into pGL3-basic luciferase reporter vector (Progema, Madison, WI, USA) using T4 DNA Ligase (TaKaRa, Dalian, China). After enzyme digestion and sequencing identification, the recombinant plasmids were extracted using EndoFree Mini Plasmid Kit II (TIANGEN, Beijing, China), and named pGL3-1878/+115, pGL3-1617/+115, pGL3-919/+115, pGL3-387/+115, and pGL3-158/+115, respectively. In order to verify the promoter activity of different fragments, each recombinant plasmid (800 ng) was co-transfection with internal vector pRL-TK (20 ng) using Lipofectamine™ 2000 reagent according to the manufacturer's protocol. After 48 h post-transfection, the luciferase activity was detected using the Dual Luciferase Reporter Assay System (Promega, Madison, WI, USA), the pGL3-basic vector was considered as a negative control. Each group was performed three times.

### 2.5. Site-Directed Mutagenesis

The AliBaba2.1 and JASPAR online software were used to predict the transcription factor binding sites in core promoter region. In the present study, we mutated the potential transcription factor binding sites for WT1, Sp1, C/EBPα, and c-Ets-1 with the corresponding primers (Table 1) using Fast Site-Directed Mutagenesis Kit (TIANGEN, Beijing, China) according to the instruction manual.

### 2.6. RNA Interference

The siRNAs used in this experiment were designed and synthesized by GenePharma Company (Shanghai, China). The si-NC was regarded as a negative control. The interference sequences were shown in Table 1. Porcine alveolar macrophages (3D4/21) were cultured in 24-well plates and siRNAs (50 nM) co-transfected with pGL3-387/−158 plasmid (500 ng) according to the method mentioned before. After 48 h post-transfection, the relative mRNA expression and luciferase activity were measured as described above.

### 2.7. Statistical Analysis

The IBM SPSS Statistics (version 21.0) were used to analyze the relative mRNA expression levels of *HMOX1* gene in different tissues of Hezuo Tibetan pig, and the Duncan method was used to compare multiple groups. Independent sample *t*-test was used to analyze the relative luciferase activity among different promoter fragments. GraphPad Prism 8.0 software (Huntington, West Virginia, USA) was used for drawing. All values in this study were expressed as the mean ± standard deviation (SD). * indicates $p < 0.05$, ** indicates $p < 0.01$, n = 3.

## 3. Results

### 3.1. Tissue Expression Analysis of mRNA

Total RNA was extracted from fourteen tissues (heart, liver, spleen, lung, kidney, stomach, duodenum, jejunum, ileum, cecum, colon, rectum, psoas muscle, longissimus dorsi), qRT-PCR was used to analyzed the mRNA expression profiles of different tissues. Compared with the expression of *HMOX1* in the duodenum, the relative expression level of adults Hezuo Tibetan pig was shown in Figure 1. The *HMOX1* gene expression in the spleen was highest ($p < 0.01$) compared to other tissues. In addition, the liver, lung, kidney, and heart tissues also had the higher expression levels.

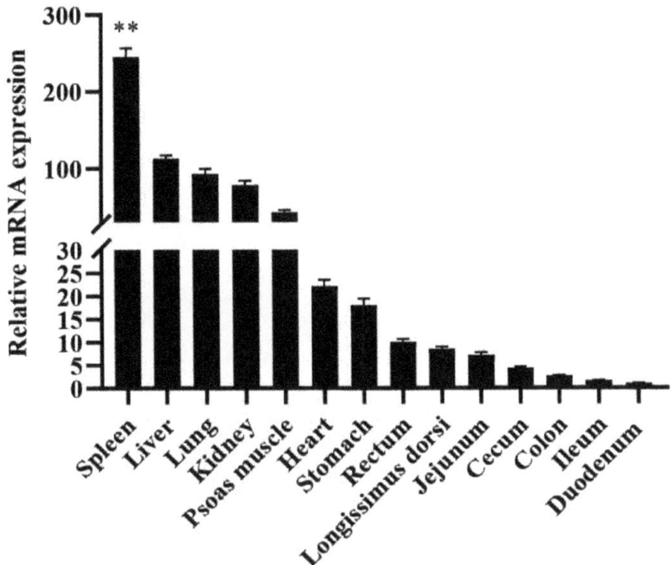


**Figure 1.** Relative expression patterns of Hezuo Tibetan pig *HMOX1* gene in different tissues. The result was normalized with *β-actin* gene and relative to gene expression in the duodenum group. ** indicates $p < 0.01$.

### 3.2. Promoter Region Cloning and Bioinformatics Analysis

Based on the pig *HMOX1* gene sequence published by GenBank database (NM_001004027.1), a 2125 bp 5′-flanking sequence spanning nucleotides from −1911 to +214 bp was amplified (Figure 2A). We used NCBI-BLAST program to perform a sequence alignment analysis between the cloned Hezuo Tibetan pig promoter region sequence and 2125 bp sequence upstream of the pig *HMOX1* gene in NCBI, and found that the sequence similarity was 96.63%. In addition, we summarized the gene structure of *HMOX1* and found that it contains five exons in genome, and the length is about 9 kb. The mRNA (NM_001004027.1) transcript length is 1552 bp. In addition, an open reading frame (ORF) of 867 bp, which encoded 288 amino acids (aa) with one conserved domain, the Heme_oxygenase domain (in aa 11 to 216) was identified (Figure 2B). The results are shown in. The transcription start site (TSS) was predicted through Neural Network Promoter Prediction and the adenine residue (A) proximal to 5′ untranslated regions was verified and designated as +1 (Figure 2D). The CpG island was predicted by MethPrimer online software and found that two CpG islands located in 197 bp (−1516~−1320) and 225 bp (−164~+61) at promoter region (Figure 2C).

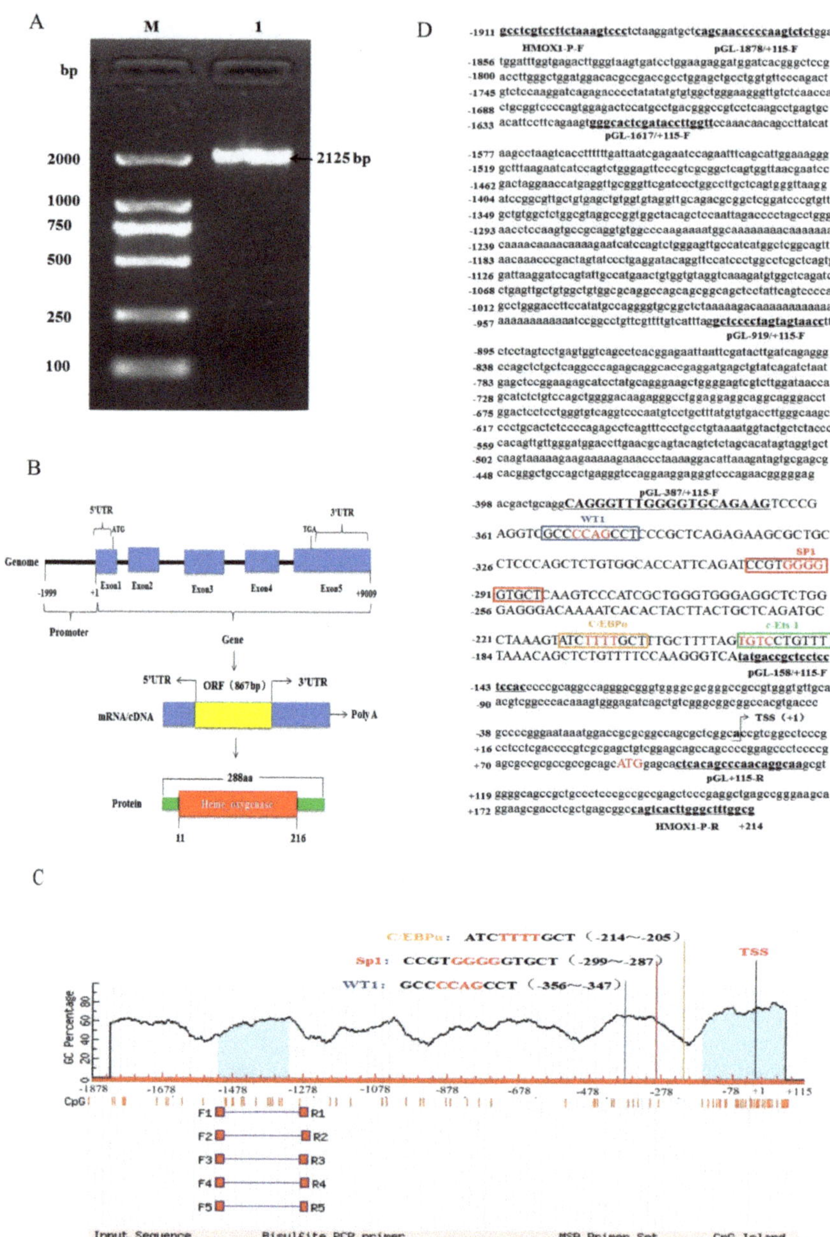

**Figure 2.** Promoter region cloning and bioinformatics analysis. (**A**) PCR amplification product of *HMOX1* gene promoter region. M, DL2000 DNA marker; 1, PCR product. (**B**) Genomic structure of *HMOX1* gene. (**C**) The predict results of CpG island in *HMOX1* gene. (**D**) The promoter sequence information of *HMOX1* gene. The box represents the transcription factor binding sites, and the 4 bp core binding sites are shown in red. The underline represents the amplification primer and the arrow represents the transcription start site (TSS).

### 3.3. Promoter Activity Analysis and Core Promoter Region Identification

Through enzyme digestion and sequencing, we successfully constructed five recombinant vectors as shown in Figure 3B. Then we measured luciferase activity to find the core promoter region of *HMOX1* gene. The results indicated that pGL3-387/+115 had the highest transcription activity compared to other regions and significantly increase compared to the pGL3-158/+115 recombinant plasmid ($p < 0.01$) (Figure 3A). In order to identify the minimum core promoter region, we constructed the pGL3-387/−158 recombinant vector by subcloning, and the luciferase activity was shown that the pGL3-387/−158 significantly increase compared to pGL-158/+115 ($p < 0.01$). This suggested that −387/−158 as the core promoter region of *HMOX1* gene (Figure 3C). We further analyzed transcriptional regulatory elements in this region using AliBaba2.1 and JASPAR. As a result, four essential transcription factors—WT1, Sp1, C/EBPα, and c-Ets-1—were predicted in the core promoter region (Figure 4A). In addition, multiple alignments of core promoter region sequences between six species (pig, human, mouse, cattle, sheep, and horse). The results showed that the WT1, Sp1, C/EBPα, and c-Ets-1 regulatory elements were conserved in livestock (Figure 4B).

A

-1878/+115 Luc
-1617/+115 Luc
-919/+115 Luc
-387/+115 Luc
-158/+115 Luc
pGL3-Basic Luc
**
**
0 5 10 15
Relative luciferase activity
293T
3D4/21

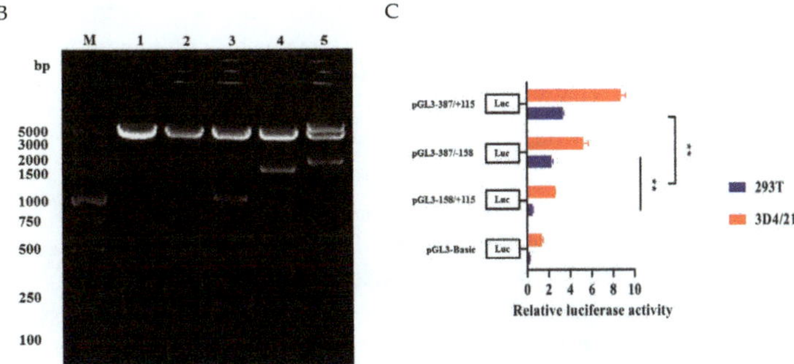


**Figure 3.** Construction of different deletion vectors of *HMOX1* gene promoter region and luciferase activity analysis. (**A**) Luciferase activity analysis of promoter with different deletion vectors (pGL3-1878/+115, pGL3-1617/+115, pGL3-919/+115, pGL3-387/+115, pGL3-158/+115), pGL3-Basic as a negative control. (**B**) Double enzyme digestion identification of different deletion vectors. M, DL5000 DNA marker, lanes 1–5 are different deletion fragments. (**C**) Analysis of luciferase activity in the core promoter region. All ** represents $p < 0.01$.

A

-387 →
CAGGGTTTGGGGTGCAGAAGTCCCTGAGGTCGCCCCAGCCTCCCGCTCAGAGAAGCGCTGCC
(WT1)
TCCCAGCTCTGTGGCACCATTCAGATCCGTGGGGGTGCTCAAGTCCCATCGCTGGGTGGGAGG
(SP1)
CTCTGGGAGGGACAAAATCACACTACTTACTGCTCAGATTCCTAAAGTATCTTTTGCTTTGTTTT
(C/EBPα)
TAGTGTCCTGTTTTAAACAGCTCTGTTTTCCAAGGGTCATATGACCGCTCCTCCTCCACCCCCGC
(c-Ets-1)
← -158
AGGCCAGGGGCGGGTGGGGCGCGGGCCGCCGTGGGTGTTGCAACGTCGGGCCACAAAGTGG
GAGATCGGCTGTCGGGCGGCGGCCACGTGACCCGCCCCGGGAATAAATGGACCGCGCGGCCA
GCGCTCGGCACCGTCGGCCTCCCGCCTCCTCGACCCCGTCGCGAGCTGTCGGAGCAGCCAGCC
TSS (+1)
CCGGAGCCCTCCCCGAGCGCCGCGCCGCCGCAGCATGGAGCACTCACAGCCCAACAGGCAA
+115

B

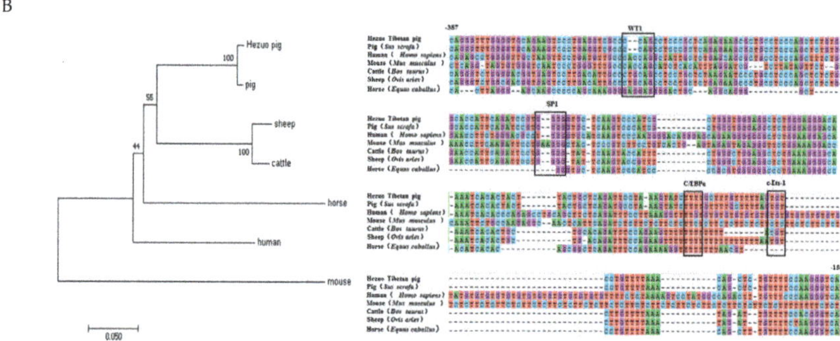


**Figure 4.** Transcription factor prediction and multi-species homology analysis in the core promoter region. (**A**) Transcription factor prediction of core promoter region. Four essential transcription factors WT1, Sp1, C/EBPα, and c-Ets-1 were predicted in −387/−158, the location of the 4 bp core binding is indicated in red. (**B**) Sequence alignment and phylogenetic construction of core promoter regions in six species. The box represents the 4 bp core binding site.

### *3.4. Roles of WT1, Sp1, C/EBPα, and c-Ets-1 in Transcriptional Regulation of HMOX1 Gene*

In order to explore the role of WT1, Sp1, C/EBPα, and c-Ets-1 in the regulation transcription of *HMOX1* gene, four recombinant plasmids with 4 bp point mutations in the WT1, Sp1, C/EBPα, and c-Ets-1 binding sites in pGL-387/-158 were constructed. Then, the four mutant vectors were transfected into 293T and 3D4/21 cells to detect the luciferase activity, respectively. The results showed that the luciferase activity of the mutated WT1, Sp1, and C/EBPα were significantly decreased ($p < 0.01$), while the luciferase activity of mutated c-Ets-1 had no significant changes (Figure 5A). To further validate the roles of WT1, Sp1, and C/EBPα binding sites in the core promoter region of the *HMOX1* gene, the selected transcription factors were silenced through siRNAs. First, the interference efficiency of the siRNAs (si-WT1, si-Sp1, and si-C/EBPα) were detected through negative control siRNA after transfection 24 h. The results showed that these siRNAs significantly reduced ($p < 0.01$) the mRNA expression levels of the WT1, Sp1, and C/EBPα as compared to the negative control (Figure 5B). In addition, the relative mRNA expression level of *HMOX1* gene had significantly decrease ($p < 0.05$ and $p < 0.01$) by WT1, Sp1, and C/EBPα inhibition, respectively (Figure 5C). Further co-transfection of si-WT1, si-Sp1, or si-C/EBPα with pGL-387/−158 also reduced luciferase activity (Figure 5D). In this context, we see that WT1, Sp1, and C/EBPα can promote the transcriptional activity of promoter and increase the expression of *HMOX1* gene. These three transcription factors may be bind to the core promoter region of *HMOX1* gene and play a role in activating *HMOX1* gene transcription.

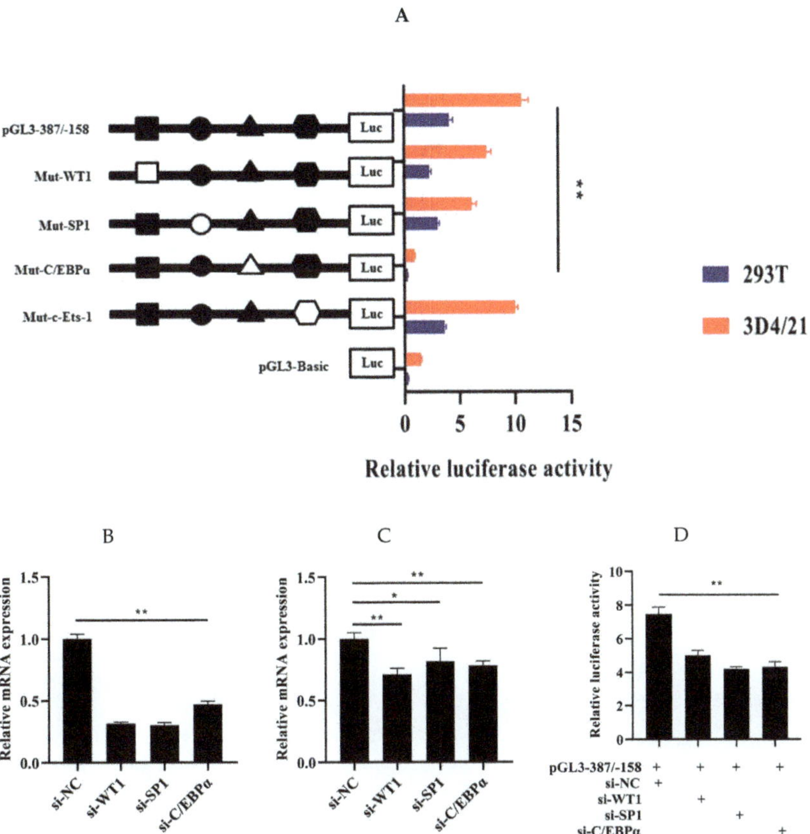


**Figure 5.** Role of WT1, Sp1, and C/EBPα in the transcription regulation of *HMOX1* gene. (**A**) Luciferase activity analysis with site-directed mutagenesis of WT1, Sp1, C/EBPα, and c-Ets-1 sites. Solid and hollow represent wild-type and mutant-type, respectively. (**B**) Interference efficiency of WT1, Sp1, and C/EBPα. (**C**) The mRNA expression level of *HMOX1* gene by inhibition with si-WT1, si-Sp1, and si-C/EBPα. (**D**) Luciferase activity after co- transfection of si-WT1, si-Sp1, and si-C/EBPα with pGL3-387/−158. (* indicates $p < 0.05$, ** indicates $p < 0.01$).

## 4. Discussion

Heme oxygenase (HO) family has three important members, including inducible *HMOX1* and constitutive *HMOX2* and *HMOX3* [25,26]. The *HMOX1* gene, also known as heat-shock protein 32 (HSP32), plays an important role in protecting the body from a variety of stimuli and pathological conditions [27–29]. Studies have reported that vascular endothelial growth factor (VEGF) can regulate the expression of *HMOX1* gene and play an important regulatory role in angiogenesis [30,31]. Wang et al. [32] found that *HMOX1* has a protective effect in fish hypoxic cells. In addition, Gou et al. [33] reported that increasing hemoglobin content is a key molecular mechanism of hypoxia adaptation in Tibetan chicken embryos. These studies suggest that the HO family genes may play an important protective role in the body's anti-inflammatory, antioxidant stress, and hypoxia injury activities.

In this study, the expression profiles of *HMOX1* gene in Hezuo Tibetan pigs were analyzed in 14 tissues and found that the expression level of *HMOX1* gene in spleen tissue was the highest, followed by liver, lung, and kidney. This results are consistent with the previous findings of Kovtunovych et al. [34] on mice and Park et al. [35] on transgenic pigs. At present, bioinformatics analysis and construction of luciferase reporter gene vectors are common methods to study promoter activity and core regulatory regions [36]. In this work, we successfully constructed recombinant vectors with different deletion lengths in the *HMOX1* gene promoter region and detected luciferase activity after transfection of 293T cells and 3D4/21 cells. It was found that the *HMOX1* gene promoter region -387/-158 had the highest activity, indicating that there were important positive regulatory transcription elements in this region. Subsequently, bioinformatics predicted that the promoter region −387 / −158 of *HMOX1* gene contains crucial transcription factor binding sites such as WT1, Sp1, C/EBPα, and c-Ets-1. Further combined with site-directed mutagenesis and RNA interference experiments, we have preliminarily confirmed that WT1, Sp1, and C/EBPα can activate the promoter activity of *HMOX1* gene. Transcription factors can bind to specific DNA sequences in the gene promoter region to control the transcription and expression of genes, thereby regulating various physiological activities of the body [36]. The siRNAs can specifically inhibit or degrade the expression of homologous mRNA, thereby mediating gene post-transcriptional silencing [37]. In this study, si-WT1, si-Sp1, and si-C/EBPα were designed and synthesized. After transfected into 3D4/21 cells, it was found that si-WT1, si-Sp1, and si-C/EBPα could effectively inhibit the expression of *HMOX1* mRNA level, and can reduce the *HMOX1* gene promoter activity. The previous study found that WT1 transcription factor (WT1) can play a role in heart and blood vessel formation [38]. It was also regarded as a key element in acute myeloid leukemia [39], and WT1 overexpression was an independent positive prognostic factor in adult B-cell acute lymphoblastic leukemia patients [40]. The transcription factor Sp1 (Sp1) can regulate the expression of multiple genes [41], which can play a role in the body's antioxidant stress [42]. Under hypoxic and ischemic conditions, regulating the expression of Sp1 can regulate the activation mechanism of coagulation in rats with hemorrhagic shock [43]. Sp1 can promote TGF-β1 expression and activate the SMAD2 pathway, the SP1/TGF-β1/SMAD2 pathway may enhance angiogenic processes in preosteoblasts [44]. Deshane et al. [45] showed that Sp1 also regulated the expression of *HMOX1* gene in human kidney cells. CCAAT/enhancer binding protein α (C/EBPα) can play a role in cell proliferation, apoptosis, inflammation, and other responses. According to previous study, the C/EBPα can play a role in the development of lipogenesis [46] and acute myeloid leukemia [47]. Liu et al. [48] found that C/EBPα was a key target gene of DNA methyltransferase 1 (Dnmt1) downstream, which can play a crucial role in maintain hematopoietic stem and progenitor cells in zebrafish. It has also been reported that C/EBPα can increase the expression of *HMOX1* gene under the induction of peptidoglycan [49]. Taken together, the transcription factors WT1, Sp1, and C/EBPα can regulate the expression of multiple genes and participate in a variety of biological processes related to hematopoietic function. Therefore, we speculate that transcription factors WT1, Sp1, and C/EBPα may be significant factors in regulating *HMOX1* gene, this research may provide a positive reference for further studying the transcriptional regulation mechanism of *HMOX1* gene.

## 5. Conclusions

In conclusion, in the present study, we analyzed *HMOX1* gene expression profiles and gene structure. We also cloned 5′ promoter region and predicted the transcription initiation site of Hezuo Tibetan pig *HMOX1* gene. In addition, we discovered that the core promoter region and three transcription factors WT1, Sp1, and C/EBPα are likely to play an important role in the expression of *HMOX1* gene in Hezuo Tibetan pig. Our results will provide a basic information for further research the transcriptional regulation mechanism of *HMOX1* gene.

**Author Contributions:** W.W. and Z.Y. conceived this study; W.W. and Q.Y. designed experiments; K.X. contributed to vectors construction; P.W., X.G., B.Z., and X.H. contributed to samples collection; W.W. and R.L. contributed to cell culture; W.W. wrote the paper; W.W., Q.Y., and S.G. checked and revised the paper. All authors have read and agreed to the published version of the manuscript.

**Funding:** This research was funded by special talents fund of Gansu Agricultural University (2017RCZX-15).

**Acknowledgments:** We acknowledge and thank the GENEWIZ (Suzhou, China) and GenePharma (Shanghai, China) biotechnology co. ltd. for its technical support in this experiment.

**Conflicts of Interest:** The authors declare no conflicts of interest.

## References

1. Gan, M.L.; Shen, L.Y.; Fan, Y.; Guo, Z.X.; Liu, B.; Chen, L.; Tang, G.Q.; Jiang, Y.Z.; Li, X.W.; Zhang, S.H.; et al. High altitude adaptability and meat quality in Tibetan pigs: A reference for local pork processing and genetic improvement. *Animals* **2019**, *9*, 1080. [CrossRef]
2. Zhang, B.; Qiangba, Y.Z.; Shang, P.; Wang, Z.X.; Ma, J.; Wang, L.Y.; Zhang, H. A comprehensive microRNA expression profile related to hypoxia adaptation in the Tibetan pig. *PLoS ONE* **2015**, *10*, e0143260. [CrossRef] [PubMed]
3. Ma, Y.F.; Han, X.M.; Huang, C.P.; Zhong, L.; Adeola, A.C.; Irwin, D.M.; Xie, H.B.; Zhang, Y.P. Population genomics analysis revealed origin and high-altitude adaptation of Tibetan pigs. *Sci. Rep.* **2019**, *9*, 11463. [CrossRef] [PubMed]
4. Li, M.Z.; Tian, S.L.; Jin, L.; Zhou, G.Y.; Li, Y.; Zhang, Y.; Wang, T.; Yeung, C.K.; Chen, L.; Ma, J.D.; et al. Genomic analyses identify distinct patterns of selection in domesticated pigs and Tibetan wild boars. *Nat. Genet.* **2013**, *45*, 1431–1438. [CrossRef] [PubMed]
5. Ai, H.S.; Yang, B.; Li, J.; Xie, X.H.; Chen, H.; Ren, J. Population history and genomic signatures for high-altitude adaptation in Tibetan pigs. *BMC Genom.* **2014**, *15*, 834. [CrossRef] [PubMed]
6. Kong, X.Y.; Dong, X.X.; Yang, S.L.; Qian, J.H.; Yang, J.F.; Jiang, Q.; Li, X.R.; Wang, B.; Yan, D.W.; Lu, S.X.; et al. Natural selection on TMPRSS6 associated with the blunted erythropoiesis and improved blood viscosity in Tibetan pigs. *Comp. Biochem. Physiol. B Biochem. Mol. Biol.* **2019**, *233*, 11–22. [CrossRef] [PubMed]
7. Otterbein, L.E.; Soares, M.P.; Yamashita, K.; Bach, F.H. Heme oxygenase-1: Unleashing the protective properties of heme. *Trends Immunol.* **2003**, *24*, 449–455. [CrossRef]
8. Kishimoto, Y.; Kondo, K.; Momiyama, Y. The Protective Role of Heme Oxygenase-1 in Atherosclerotic Diseases. *Int. J. Mol. Sci.* **2019**, *20*, 3628. [CrossRef]
9. Lin, H.H.; Chen, Y.H.; Chang, P.F.; Lee, Y.T.; Yet, S.F.; Chau, L.Y. Heme oxygenase-1 promotes neovascularization in ischemic heart by coinduction of VEGF and SDF-1. *J. Mol. Cell. Cardiol.* **2008**, *45*, 44–55. [CrossRef]
10. Singh, S.P.; Greenberg, M.; Glick, Y.; Bellner, L.; Favero, G.; Rezzani, R.; Rodella, L.F.; Agostinucci, K.; Shapiro, J.I.; Abraham, N.G. Adipocyte specific HO-1 gene therapy is effective in antioxidant treatment of insulin resistance and vascular function in an obese mice model. *Antioxidants* **2020**, *9*, 40. [CrossRef]
11. Inguaggiato, P.; Gonzalez-Michaca, L.; Croatt, A.J.; Haggard, J.J.; Alam, J.; Nath, K.A. Cellular overexpression of heme oxygenase-1 up-regulates p21 and confers resistance to apoptosis. *Kidney Int.* **2001**, *60*, 2181–2191. [CrossRef] [PubMed]
12. Takao, M.; Okinaga, T.; Ariyoshi, W.; Iwanaga, K.; Nakamichi, I.; Yoshioka, I.; Tominaga, K.; Nishihara, T. Role of heme oxygenase-1 in inflammatory response induced by mechanical stretch in synovial cells. *Inflamm. Res.* **2011**, *60*, 861–867. [CrossRef] [PubMed]
13. Yachie, A.; Niida, Y.; Wada, T.; Igarashi, N.; Kaneda, H.; Toma, T.; Ohta, K.; Kasahara, Y.; Koizumi, S. Oxidative stress causes enhanced endothelial cell injury in human heme oxygenase-1 deficiency. *J. Clin. Investig.* **1999**, *103*, 129–135. [CrossRef] [PubMed]
14. Brouard, S.; Otterbein, L.E.; Anrather, J.; Tobiasch, E.; Bach, F.H.; Choi, A.M.K.; Soares, M.P. Carbon monoxide generated by heme oxygenase 1 suppresses endothelial cell apoptosis. *J. Exp. Med.* **2000**, *192*, 1015–1026. [CrossRef]
15. Kawamura, K.; Ishikawa, K.; Wada, Y.; Kimura, S.; Matsumoto, H.; Kohro, T.; Itabe, H.; Kodama, T.; Maruyama, Y. Bilirubin from heme oxygenase-1 attenuates vascular endothelial activation and dysfunction. *Arter. Thromb Vasc Biol.* **2005**, *25*, 155–160. [CrossRef]
16. Clark, J.E.; Foresti, R.; Sarathchandra, P.; Kaur, H.; Green, C.J.; Motterlini, R. Heme oxygenase-1-derived bilirubin ameliorates postischemic myocardial dysfunction. *Am. J. Physiol. Heart Circ. Physiol.* **2000**, *278*, H643–H651. [CrossRef]

17. Fujiwara, A.; Hatayama, N.; Matsuura, N.; Yokota, N.; Fukushige, K.; Yakura, T.; Tarumi, S.; Go, T.; Hirai, S.; Naito, M.; et al. High-pressure carbon monoxide and oxygen mixture is effective for lung preservation. *Int. J. Mol. Sci.* **2019**, *20*, 2719. [CrossRef]
18. Rashid, I.; Baisvar, V.S.; Singh, M.; Kumar, R.; Srivastava, P.; Kushwaha, B.; Pathak, A.K. Isolation and characterization of hypoxia inducible heme oxygenase 1 (HMOX1) gene in Labeo rohita. *Genomics* **2020**, *112*, 2327–2333. [CrossRef]
19. Taylor, J.L.; Carraway, M.S.; Piantadosi, C.A. Lung-specific induction of heme oxygenase-1 and hyperoxic lung injury. *Am. J. Physiol.* **1998**, *274*, L582–L590. [CrossRef]
20. Mitchell, P.J.; Tjian, R. Transcriptional regulation in mammalian cells by sequence-specific DNA binding proteins. *Science* **1989**, *245*, 371–378. [CrossRef]
21. Livak, K.J.; Schmittgen, T.D. Analysis of relative gene expression data using real-time quantitative PCR and the $2^{-\Delta\Delta Ct}$ method. *Methods* **2000**, *25*, 402–408. [CrossRef] [PubMed]
22. Wingender, E.; Dietze, P.; Karas, H.; Knüppel, R. TRANSFAC: A database on transcription factors and their DNA binding sites. *Nucleic Acids Res.* **1996**, *24*, 238–241. [CrossRef] [PubMed]
23. Fornes, O.; Castro-Mondragon, J.A.; Khan, A.; van der Lee, R.; Zhang, X.; Richmond, P.A.; Modi, B.P.; Correard, S.; Gheorghe, M.; Baranašić, D.; et al. JASPAR 2020: Update of the open-access database of transcription factor binding profiles. *Nucleic Acids Res.* **2019**, *48*, D87–D92. [CrossRef] [PubMed]
24. Li, L.C.; Dahiya, R. MethPrimer: Designing primers for methylation PCRs. *Bioinformatics* **2002**, *11*, 1427–1431. [CrossRef]
25. Mccoubrey, W.K.; Ewing, J.F.; Maines, M.D. Human heme oxygenase-2: Characterization and expression of a full-length cDNA and evidence suggesting that the two HO-2 transcripts may differ by choice of polyadenylation signal. *Arch. Biochem. Biophys.* **1992**, *295*, 13–20. [CrossRef]
26. Mccoubrey, W.K.; Huang, T.J.; Maines, M.D. Isolation and Characterization of a cDNA from the Rat Brain that Encodes Hemoprotein Heme Oxygenase-3. *Eur. J. Biochem.* **1997**, *247*, 725–732. [CrossRef]
27. Li, L.; Li, C.M.; Wu, J.; Huang, S.; Wang, G.L. Heat shock protein 32/heme oxygenase-1 protects mouse Sertoli cells from hyperthermia-induced apoptosis by CO activation of sGC signalling pathways. *Cell Biol. Int.* **2014**, *38*, 64–71. [CrossRef]
28. Stuhlmeier, K.M. Activation and regulation of Hsp32 and Hsp70. *Eur. J. Biochem.* **2000**, *267*, 1161–1167. [CrossRef]
29. Taylor, L.; Hillman, A.R.; Midgley, A.W.; Peart, D.J.; Chrismas, B.; McNaughton, L.R. Hypoxia-mediated prior induction of monocyte-expressed HSP72 and HSP32 provides protection to the disturbances to redox balance associated with human sub-maximal aerobic exercise. *Amino Acids* **2012**, *43*, 1933–1944. [CrossRef]
30. Bussolati, B.; Mason, J.C. Dual role of VEGF-induced heme-oxygenase-1 in angiogenesis. *Antioxid. Redox Signal.* **2006**, *8*, 1153–1163. [CrossRef] [PubMed]
31. Dulak, J.; Loboda, A.; Zagórska, A.; Józkowicz, A. Complex role of heme oxygenase-1 in angiogenesis. *Antioxid. Redox Signal.* **2004**, *6*, 858–866. [CrossRef] [PubMed]
32. Wang, D.; Zhong, X.P.; Qiao, Z.X.; Gui, J.F. Inductive transcription and protective role of fish heme oxygenase-1 under hypoxic stress. *J. Exp. Biol.* **2008**, *211*, 2700–2706. [CrossRef]
33. Gou, X.; Li, N.; Lian, L.S.; Yan, D.W.; Zhang, H.; Wei, Z.H.; Wu, C.X. Hypoxic adaptations of hemoglobin in Tibetan chick embryo: High oxygen-affinity mutation and selective expression. *Comp. Biochem. Physiol. B Biochem. Mol. Biol.* **2007**, *147*, 147–155. [CrossRef] [PubMed]
34. Kovtunovych, G.; Eckhaus, M.A.; Ghosh, M.C.; Ollivierrewilson, H.; Rouault, T.A. Dysfunction of the heme recycling system in heme oxygenase 1–deficient mice: Effects on macrophage viability and tissue iron distribution. *Blood* **2010**, *116*, 6054–6062. [CrossRef] [PubMed]
35. Park, S.J.; Cho, B.; Koo, O.J.; Kim, H.; Kang, J.T.; Hurh, S.; Kim, S.J.; Yeom, H.J.; Moon, J.; Lee, E.M.; et al. Production and characterization of soluble human TNFRI-Fc and human HO-1 (HMOX1) transgenic pigs by using the F2A peptide. *Transgenic Res.* **2014**, *23*, 407–419. [CrossRef] [PubMed]
36. Alam, J. Functional analysis of the heme oxygenase-1 gene promoter. *Curr. Protoc. Toxicol.* **2000**, *6*. [CrossRef] [PubMed]
37. Elbashir, S.M.; Harborth, J.; Lendeckel, W.; Yalcin, A.; Weber, K.; Tuschl, T. Duplexes of 21-nucleotide RNAs mediate RNA interference in cultured mammalian cells. *Nature* **2001**, *411*, 494–498. [CrossRef]
38. Scholz, H.; Wagner, K.D.; Wagner, N. Role of the Wilms' tumour transcription factor, Wt1, in blood vessel formation. *Pflug. Arch.* **2009**, *458*, 315–323. [CrossRef]

39. Panuzzo, C.; Signorino, E.; Calabrese, C.; Ali, M.S.; Petiti, J.; Bracco, E.; Cilloni, D. Landscape of Tumor Suppressor Mutations in Acute Myeloid Leukemia. *J. Clin. Med.* **2020**, *9*, 802. [CrossRef]
40. Wang, S.J.; Wang, C.; Li, T.; Wang, W.Q.; Hao, Q.Q.; Xie, X.S.; Wang, D.M.; Jiang, Z.X.; Liu, Y.F. WT1 overexpression predicted good outcomes in adult B-cell acute lymphoblastic leukemia patients receiving chemotherapy. *Hematology* **2020**, *25*, 118–124. [CrossRef]
41. Bouwman, P.; Philipsen, S. Regulation of the activity of Sp1-related transcription factors. *Mol. Cell. Endocrinol.* **2002**, *195*, 27–38. [CrossRef]
42. Ryu, H.; Lee, J.; Zaman, K.; Kubilis, J.; Ferrante, R.J.; Ross, B.D.; Neve, R.; Ratan, R.R. Sp1 and Sp3 are oxidative stress-inducible, antideath transcription factors in cortical neurons. *J. Neurosci.* **2003**, *23*, 3597–3606. [CrossRef] [PubMed]
43. Xu, J.H.; Lu, S.J.; Wu, P.; Kong, L.C.; Ning, C.; Li, H.Y. Molecular mechanism whereby paraoxonase-2 regulates coagulation activation through endothelial tissue factor in rat haemorrhagic shock model. *Int. Wound J.* **2020**, 1–7. [CrossRef] [PubMed]
44. Ding, A.; Bian, Y.Y.; Zhang, Z.H. SP1/TGF-β1/SMAD2 pathway is involved in angiogenesis during osteogenesis. *Mol. Med. Rep.* **2020**, *21*, 1581–1589. [CrossRef] [PubMed]
45. Deshane, J.; Kim, J.; Bolisetty, S.; Hock, T.D.; Hill-Kapturczak, N.; Agarwal, A. Sp1 regulates chromatin looping between an intronic enhancer and distal promoter of the human heme oxygenase-1 gene in renal cells. *J. Biol. Chem.* **2010**, *285*, 16476–16486. [CrossRef] [PubMed]
46. Wang, X.Y.; Khan, R.; Raza, S.H.A.; Li, A.N.; Zhang, Y.; Liang, C.C.; Yang, W.C.; Wu, S.; Zan, L. Molecular characterization of ABHD5 gene promoter in intramuscular preadipocytes of Qinchuan cattle: Roles of Evi1 and C/EBPα. *Gene* **2019**, *690*, 38–47. [CrossRef]
47. Reckzeh, K.; Cammenga, J. Molecular mechanisms underlying deregulation of C/EBPα in acute myeloid leukemia. *Int. J. Hematol.* **2010**, *91*, 557–568. [CrossRef]
48. Liu, X.H.; Jia, X.E.; Yuan, H.; Ma, K.; Chen, Y.; Jin, Y.; Deng, M.; Pan, W.J.; Chen, S.J.; Chen, Z.; et al. DNA methyltransferase 1 functions through C/ebpa to maintain hematopoietic stem and progenitor cells in zebrafish. *J. Hematol. Oncol.* **2015**, *8*, 15. [CrossRef]
49. Hung, C.C.; Liu, X.L.; Kwon, M.Y.; Kang, Y.H.; Chung, S.W.; Perrella, M.A. Regulation of heme oxygenase-1 gene by peptidoglycan involves the interaction of Elk-1 and C/EBPα to increase expression. *Am. J. Physiol. Lung Cell Mol. Physiol.* **2010**, *298*, 870–879. [CrossRef]

MDPI
St. Alban-Anlage 66
4052 Basel
Switzerland
Tel. +41 61 683 77 34
Fax +41 61 302 89 18
www.mdpi.com

*Genes* Editorial Office
E-mail: genes@mdpi.com
www.mdpi.com/journal/genes

www.ingramcontent.com/pod-product-compliance
Lightning Source LLC
LaVergne TN
LVHW071533170726
843515LV00008B/2129
*9783036553580*